T0291648

CAMBRIDGE LIBRARY COLLECTION

Books of enduring scholarly value

Mathematical Sciences

From its pre-historic roots in simple counting to the algorithms powering modern desktop computers, from the genius of Archimedes to the genius of Einstein, advances in mathematical understanding and numerical techniques have been directly responsible for creating the modern world as we know it. This series will provide a library of the most influential publications and writers on mathematics in its broadest sense. As such, it will show not only the deep roots from which modern science and technology have grown, but also the astonishing breadth of application of mathematical techniques in the humanities and social sciences, and in everyday life.

Treatise on Natural Philosophy

'The term Natural Philosophy was used by Newton, and is still used in British Universities, to denote the investigation of laws in the material world, and the deduction of results not directly observed.' This definition, from the Preface to the second edition of 1879, defines the proposed scope of the work: the two volumes reissued here are the only completed part of a survey of the entirety of the physical sciences by Lord Kelvin and his fellow Scot, Peter Guthrie Tait, first published in 1867. Although the partnership ceased after eighteen years of collaboration, the published books, containing chapters on kinematics, dynamics and statics, had a great influence on the development of physics in the second half of the nineteenth century.

Treatise on Natural Philosophy

Volume 1

William Thomson, Baron Kelvin
Peter Guthrie Tait

CAMBRIDGE UNIVERSITY PRESS

Cambridge New York Melbourne Madrid Cape Town Singapore São Paolo Delhi

Published in the United States of America by Cambridge University Press, New York

www.cambridge.org
Information on this title: www.cambridge.org/9781108005357

This edition first published 1883
This digitally printed version 2009

ISBN 978-1-108-00535-7

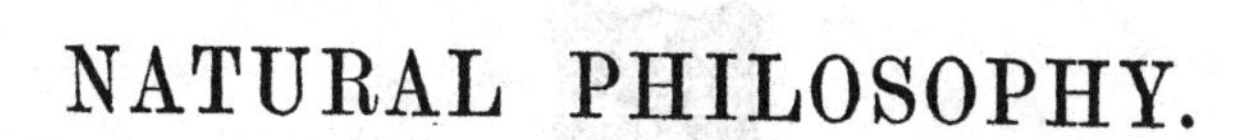

NATURAL PHILOSOPHY.

London:
CAMBRIDGE WAREHOUSE,
17, PATERNOSTER ROW,

Cambridge: DEIGHTON, BELL, AND CO.
Leipzig: F. A. BROCKHAUS.

TREATISE

ON

NATURAL PHILOSOPHY

BY

SIR WILLIAM THOMSON, LL.D., D.C.L., F.R.S.,

PROFESSOR OF NATURAL PHILOSOPHY IN THE UNIVERSITY OF GLASGOW,
FELLOW OF ST PETER'S COLLEGE, CAMBRIDGE,

AND

PETER GUTHRIE TAIT, M.A.,

PROFESSOR OF NATURAL PHILOSOPHY IN THE UNIVERSITY OF EDINBURGH,
FORMERLY FELLOW OF ST PETER'S COLLEGE, CAMBRIDGE.

VOL. I. PART I.

NEW EDITION.

Cambridge:
AT THE UNIVERSITY PRESS.
1879

Cambridge:
PRINTED BY C. J. CLAY, M.A.
AT THE UNIVERSITY PRESS.

PREFACE.

Les causes primordiales ne nous sont point connues; mais elles sont assujetties à des lois simples et constantes, que l'on peut découvrir par l'observation, et dont l'étude est l'objet de la philosophie naturelle.—FOURIER.

THE term Natural Philosophy was used by NEWTON, and is still used in British Universities, to denote the investigation of laws in the material world, and the deduction of results not directly observed. Observation, classification, and description of phenomena necessarily precede Natural Philosophy in every department of natural science. The earlier stage is, in some branches, commonly called Natural History; and it might with equal propriety be so called in all others.

Our object is twofold: to give a tolerably complete account of what is now known of Natural Philosophy, in language adapted to the non-mathematical reader; and to furnish, to those who have the privilege which high mathematical acquirements confer, a connected outline of the analytical processes by which the greater part of that knowledge has been extended into regions as yet unexplored by experiment.

We commence with a chapter on *Motion*, a subject totally independent of the existence of *Matter* and *Force*. In this we are naturally led to the consideration of the curvature and tortuosity of curves, the curvature of surfaces, distortions or strains, and various other purely geometrical subjects.

The *Laws of Motion*, the *Law of Gravitation and of Electric and Magnetic Attractions*, *Hooke's Law*, and other fundamental principles derived directly from experiment, lead by mathematical processes to interesting and useful results, for the full testing of which our most delicate experimental methods are as yet totally insufficient. A large part of the present volume is devoted to these deductions; which, though not immediately proved by experiment, are as certainly true as the elementary laws from which mathematical analysis has evolved them.

The analytical processes which we have employed are, as a rule, such as lead most directly to the results aimed at, and are therefore in great part unsuited to the general reader.

We adopt the suggestion of AMPÈRE, and use the term *Kinematics* for the purely geometrical science of motion in the abstract. Keeping in view the proprieties of language, and following the example of the most logical writers, we employ the term *Dynamics* in its true sense as the science which treats of the action of *force*, whether it maintains relative rest, or produces acceleration of relative motion. The two corresponding divisions of Dynamics are thus conveniently entitled *Statics* and *Kinetics*.

One object which we have constantly kept in view is the grand principle of the *Conservation of Energy*. According to modern experimental results, especially those of JOULE, Energy is as real and as indestructible as Matter. It is satisfactory to find that NEWTON anticipated, so far as the state of experimental science in his time permitted him, this magnificent modern generalization.

We desire it to be remarked that in much of our work, where we may appear to have rashly and needlessly interfered with methods and systems of proof in the present day generally accepted, we take the position of Restorers, and not of Innovators.

In our introductory chapter on Kinematics, the consideration of Harmonic Motion naturally leads us to *Fourier's Theorem*, one of the most important of all analytical results as regards usefulness in physical science. In the Appendices to that chapter we have introduced an extension of *Green's Theorem*, and a treatise on the remarkable functions known as *Laplace's Coefficients*. There can be but one opinion as to the beauty and utility of this analysis of Laplace; but the manner in which it has been hitherto presented has seemed repulsive to the ablest mathematicians, and difficult to ordinary mathematical students. In the simplified and symmetrical form in which we give it, it will be found quite within the reach of readers moderately familiar with modern mathematical methods.

In the second chapter we give NEWTON'S Laws of Motion in his own words, and with some of his own comments—every attempt that has yet been made to supersede them having ended in utter failure. Perhaps nothing so simple, and at the same time so comprehensive, has ever been given as the foundation of a system in any of the sciences. The dynamical use of the *Generalized Cöordinates* of LAGRANGE, and the *Varying Action* of HAMILTON, with kindred matter, complete the chapter.

The third chapter, "Experience," treats briefly of Observation and Experiment as the basis of Natural Philosophy.

The fourth chapter deals with the fundamental Units, and the chief Instruments used for the measurement of Time, Space, and Force.

Thus closes the First Division of the work, which is strictly preliminary, and to which we have limited the present issue.

This new edition has been thoroughly revised, and very considerably extended. The more important additions are to be found in the Appendices to the first chapter, especially that devoted to *Laplace's Coefficients;* also at the end of the second chapter, where a very full investigation of the "*cycloidal motion*" of systems is now given; and in Appendix B′, which describes a number of continuous calculating machines invented and constructed since the publication of our first edition. A great improvement has been made in the treatment of *Lagrange's Generalized Equations of Motion.*

We believe that the mathematical reader will especially profit by a perusal of the large type portion of this volume; as he will thus be forced to think out for himself what he has been too often accustomed to reach by a mere mechanical application of analysis. Nothing can be more fatal to progress than a too confident reliance on mathematical symbols; for the student is only too apt to take the easier course, and consider the *formula* and not the *fact* as the physical reality.

In issuing this new edition, of a work which has been for several years out of print, we recognise with legitimate satisfaction the very great improvement which has recently taken place in the more elementary works on Dynamics published in this country, and which we cannot but attribute, in great part, to our having effectually recalled to its deserved position Newton's system of elementary definitions, and Laws of Motion.

We are much indebted to Mr BURNSIDE and Prof. CHRYSTAL for the pains they have taken in reading proofs and verifying formulas; and we confidently hope that few erratums of serious consequence will now be found in the work.

W. THOMSON.
P. G. TAIT.

CONTENTS.

DIVISION I.—PRELIMINARY.

CHAPTER I.—KINEMATICS.

CHAPTER II.—DYNAMICAL LAWS AND PRINCIPLES.

DIVISION I.

PRELIMINARY.

CHAPTER I.—KINEMATICS.

1. THERE are many properties of motion, displacement, and deformation, which may be considered altogether independently of such physical ideas as force, mass, elasticity, temperature, magnetism, electricity. The preliminary consideration of such properties in the abstract is of very great use for Natural Philosophy, and we devote to it, accordingly, the whole of this our first chapter; which will form, as it were, the Geometry of our subject, embracing what can be observed or concluded with regard to actual motions, as long as the *cause* is not sought.

2. In this category we shall take up first the free motion of a point, then the motion of a point attached to an inextensible cord, then the motions and displacements of rigid systems—and finally, the deformations of surfaces and of solid or fluid bodies. Incidentally, we shall be led to introduce a good deal of elementary geometrical matter connected with the curvature of lines and surfaces.

Motion of a point.

3. When a point moves from one position to another it must evidently describe a *continuous* line, which may be curved or straight, or even made up of portions of curved and straight lines meeting each other at any angles. If the motion be that of a *material particle*, however, there cannot generally be any such abrupt changes of direction, since (as we shall afterwards see) this would imply the action of an *infinite* force, except in the case in which the velocity becomes zero at the angle. It is useful to consider at the outset various theorems connected

Motion of a point.

with the geometrical notion of the path described by a moving point, and these we shall now take up, deferring the consideration of Velocity to a future section, as being more closely connected with physical ideas.

4. The *direction* of motion of a moving point is at each instant the tangent drawn to its path, if the path be a curve, or the path itself if a straight line.

Curvature of a plane curve.

5. If the path be not straight the direction of motion changes from point to point, and the *rate* of this change, per unit of length of the curve $\left(\frac{d\theta}{ds}\text{ according to the notation below}\right)$, is called the *curvature*. To exemplify this, suppose two tangents drawn to a circle, and radii to the points of contact. The angle between the tangents is the change of direction required, and the rate of change is to be measured by the relation between this angle and the length of the circular arc. Let I be the angle, c the arc, and ρ the radius. We see at once that (as the angle between the radii is equal to the angle between the tangents)

$$\rho I = c,$$

and therefore $\frac{I}{c} = \frac{1}{\rho}$. Hence the curvature of a circle is inversely as its radius, and, measured in terms of the proper unit of curvature, is simply the reciprocal of the radius.

6. Any small portion of a curve may be approximately taken as a circular arc, the approximation being closer and closer to the truth, as the assumed arc is smaller. The curvature is then the reciprocal of the radius of this circle.

If $\delta\theta$ be the angle between two tangents at points of a curve distant by an arc δs, the definition of curvature gives us at once as its measure, the limit of $\frac{\delta\theta}{\delta s}$ when δs is diminished without limit; or, according to the notation of the differential calculus, $\frac{d\theta}{ds}$. But we have

$$\tan\theta = \frac{dy}{dx},$$

if, the curve being a plane curve, we refer it to two rectangular axes OX, OY, according to the Cartesian method, and if θ denote the inclination of its tangent, at any point x, y, to OX. Hence

Curvature of a plane curve.

$$\theta = \tan^{-1}\frac{dy}{dx};$$

and, by differentiation with reference to any independent variable t, we have

$$d\theta = \frac{d\left(\frac{dy}{dx}\right)}{1+\left(\frac{dy}{dx}\right)^2} = \frac{dx\,d^2y - dy\,d^2x}{dx^2+dy^2}.$$

Also, $$ds = (dx^2+dy^2)^{\frac{1}{2}}.$$

Hence, if ρ denote the radius of curvature, so that

$$\frac{1}{\rho} = \frac{d\theta}{ds} \quad\ldots\ldots\ldots\ldots\ldots\ldots\ldots\ldots\ldots (1),$$

we conclude $$\frac{1}{\rho} = \frac{dx\,d^2y - dy\,d^2x}{(dx^2+dy^2)^{\frac{3}{2}}} \quad\ldots\ldots\ldots\ldots\ldots\ldots (2).$$

Although it is generally convenient, in kinematical and kinetic formulæ, to regard time as the independent variable, and all the changing geometrical elements as functions of it, there are cases in which it is useful to regard the length of the arc or path described by a point as the independent variable. On this supposition we have

$$0 = d\,(ds^2) = d\,(dx^2+dy^2) = 2\,(dx\,d_s{}^2x + dy\,d_s{}^2y),$$

where we denote by the suffix to the letter d, the independent variable understood in the differentiation. Hence

$$\frac{dx}{d_s{}^2y} = -\frac{dy}{d_s{}^2x} = \frac{(dx^2+dy^2)^{\frac{1}{2}}}{\{(d_s{}^2y)^2+(d_s{}^2x)^2\}^{\frac{1}{2}}};$$

and using these, with $ds^2 = dx^2+dy^2$, to eliminate dx and dy from (2), we have

$$\frac{1}{\rho} = \frac{\{(d_s{}^2y)^2+(d_s{}^2x)^2\}^{\frac{1}{2}}}{ds^2};$$

or, according to the usual short, although not quite complete, notation,

$$\frac{1}{\rho} = \left\{\left(\frac{d^2y}{ds^2}\right)^2+\left(\frac{d^2x}{ds^2}\right)^2\right\}^{\frac{1}{2}}.$$

7. If all points of the curve lie in one plane, it is called a *plane curve*, and in the same way we speak of a *plane* polygon or broken line. If various points of the line do not lie in one plane, we have in one case what is called a *curve of double*

Tortuous curve.

1—2

Tortuous curve.

curvature, in the other a *gauche polygon*. The term 'curve of double curvature' is very bad, and, though in very general use, is, we hope, not ineradicable. The fact is, that there are not two curvatures, but only a curvature (as above defined), of which the plane is continuously changing, or twisting, round the tangent line; thus exhibiting a torsion. The course of such a curve is, in common language, well called 'tortuous;' and the measure of the corresponding property is conveniently called *Tortuosity*.

8. The nature of this will be best understood by considering the curve as a polygon whose sides are indefinitely small. Any two consecutive sides, of course, lie in a plane—and in that plane the curvature is measured as above, but in a curve which is not plane the third side of the polygon will not be in the same plane with the first two, and, therefore, the new plane in which the curvature is to be measured is different from the old one. The plane of the curvature on each side of any point of a tortuous curve is sometimes called the *Osculating Plane* of the curve at that point. As two successive positions of it contain the second side of the polygon above mentioned, it is evident that the osculating plane passes from one position to the next by revolving about the tangent to the curve.

Curvature and tortuosity.

9. Thus, as we proceed along such a curve, the curvature in general varies; and, at the same time, the plane in which the curvature lies is turning about the tangent to the curve. The tortuosity is therefore to be measured by the rate at which the osculating plane turns about the tangent, per unit length of the curve.

To express the radius of curvature, the direction cosines of the osculating plane, and the tortuosity, of a curve not in one plane, in terms of Cartesian triple co-ordinates, let, as before, $\delta\theta$ be the angle between the tangents at two points at a distance δs from one another along the curve, and let $\delta\phi$ be the angle between the osculating planes at these points. Thus, denoting by ρ the radius of curvature, and τ the tortuosity, we have

$$\frac{1}{\rho} = \frac{d\theta}{ds},$$

$$\tau = \frac{d\phi}{ds},$$

Curvature and tortuosity.

according to the regular notation for the limiting values of $\frac{\delta\theta}{\delta s}$, and $\frac{\delta\phi}{\delta s}$, when δs is diminished without limit. Let OL, OL' be lines drawn through any fixed point O parallel to any two successive positions of a moving line PT, each in the directions indicated by the order of the letters. Draw OS perpendicular to their plane in the direction from O, such that OL, OL', OS lie in the same relative order in space as the positive axes of co-ordinates, OX, OY, OZ. Let OQ bisect LOL', and let OR bisect the angle between OL' and LO produced through O.

Let the direction cosines of

OL	be	a, b, c;
OL'	,,	a', b', c';
OQ	,,	l, m, n;
OR	,,	α, β, γ;
OS	,,	λ, μ, ν:

and let $\delta\theta$ denote the angle LOL'. We have, by the elements of analytical geometry,

$$\cos\delta\theta = aa' + bb' + cc' \quad\ldots\ldots\ldots\ldots(3);$$

$$l = \frac{\frac{1}{2}(a+a')}{\cos\frac{1}{2}\delta\theta}, \quad m = \frac{\frac{1}{2}(b+b')}{\cos\frac{1}{2}\delta\theta}, \quad n = \frac{\frac{1}{2}(c+c')}{\cos\frac{1}{2}\delta\theta}\ldots\ldots(4);$$

$$\alpha = \frac{a'-a}{2\sin\frac{1}{2}\delta\theta}, \quad \beta = \frac{b'-b}{2\sin\frac{1}{2}\delta\theta}, \quad \gamma = \frac{c'-c}{2\sin\frac{1}{2}\delta\theta}\ldots\ldots(5);$$

$$\lambda = \frac{bc'-b'c}{\sin\delta\theta}, \quad \mu = \frac{ca'-c'a}{\sin\delta\theta}, \quad \nu = \frac{ab'-a'b}{\sin\delta\theta}\ldots\ldots(6).$$

Now let the two successive positions of PT be tangents to a curve at points separated by an arc of length δs. We have

$$\frac{1}{\rho} = \frac{\delta\theta}{\delta s} = \frac{2\sin\frac{1}{2}\delta\theta}{\delta s} = \frac{\sin\delta\theta}{\delta s}\ldots\ldots\ldots\ldots(7)$$

when δs is infinitely small; and in the same limit

$$l = \frac{dx}{ds}, \quad m = \frac{dy}{ds}, \quad n = \frac{dz}{ds};$$

$$a'-a = d\frac{dx}{ds}, \quad b'-b = d\frac{dy}{ds}, \quad c'-c = d\frac{dz}{ds}\ldots\ldots(8);$$

$$bc'-b'c = \frac{dy}{ds}d\frac{dz}{ds} - \frac{dz}{ds}d\frac{dy}{ds}, \text{ \&c.} \quad\ldots\ldots\ldots(9);$$

Curvature and tortuosity.

and α, β, γ become the direction cosines of the normal, PC, drawn towards the centre of curvature, C; and λ, μ, ν those of the perpendicular to the osculating plane drawn in the direction relatively to PT and PC, corresponding to that of OZ relatively to OX and OY. Then, using (8) and (9), with (7), in (5) and (6) respectively, we have

$$\alpha = \frac{d\frac{dx}{ds}}{\rho^{-1}ds}, \quad \beta = \frac{d\frac{dy}{ds}}{\rho^{-1}ds}, \quad \gamma = \frac{d\frac{dz}{ds}}{\rho^{-1}ds} \quad \ldots\ldots\ldots(10);$$

$$\lambda = \frac{\frac{dy}{ds}d\frac{dz}{ds} - \frac{dz}{ds}d\frac{dy}{ds}}{\rho^{-1}ds}, \quad \mu = \frac{\frac{dz}{ds}d\frac{dx}{ds} - \frac{dx}{ds}d\frac{dz}{ds}}{\rho^{-1}ds}, \quad \nu = \frac{\frac{dx}{ds}d\frac{dy}{ds} - \frac{dy}{ds}d\frac{dx}{ds}}{\rho^{-1}ds} \quad (11).$$

The simplest expression for the curvature, with choice of independent variable left arbitrary, is the following, taken from (10):

$$\frac{1}{\rho} = \frac{\sqrt{\left\{\left(d\frac{dx}{ds}\right)^2 + \left(d\frac{dy}{ds}\right)^2 + \left(d\frac{dz}{ds}\right)^2\right\}}}{ds} \quad \ldots\ldots\ldots(12).$$

This, modified by differentiation, and application of the formula

$$ds\,d^2s = dx\,d^2x + dy\,d^2y + dz\,d^2z \quad \ldots\ldots\ldots\ldots(13),$$

becomes

$$\frac{1}{\rho} = \frac{\sqrt{\{(d^2x)^2 + (d^2y)^2 + (d^2z)^2 - (d^2s)^2\}}}{ds^2} \quad \ldots\ldots\ldots\ldots(14).$$

Another formula for $\frac{1}{\rho}$ is obtained immediately from equations (11); but these equations may be put into the following simpler form, by differentiation, &c.,

$$\lambda = \frac{dy\,d^2z - dz\,d^2y}{\rho^{-1}ds^3}, \quad \mu = \frac{dz\,d^2x - dx\,d^2z}{\rho^{-1}ds^3}, \quad \nu = \frac{dx\,d^2y - dy\,d^2x}{\rho^{-1}ds^3} \quad (15);$$

from which we find

$$\rho^{-1} = \frac{\{(dy\,d^2z - dz\,d^2y)^2 + (dz\,d^2x - dx\,d^2z)^2 + (dx\,d^2y - dy\,d^2x)^2\}^{\frac{1}{2}}}{ds^3} \quad (16).$$

Each of these several expressions for the curvature, and for the directions of the relative lines, we shall find has its own special significance in the kinetics of a particle, and the statics of a flexible cord.

To find the tortuosity, $\frac{d\phi}{ds}$, we have only to apply the general equation above, with λ, μ, ν substituted for l, m, n, and $\frac{1}{\tau}\frac{d\lambda}{ds}$, $\frac{1}{\tau}\frac{d\mu}{ds}$, $\frac{1}{\tau}\frac{d\nu}{ds}$ for α, β, γ. Thus we have $\tau^2 = \left(\frac{d\lambda}{ds}\right)^2 + \left(\frac{d\mu}{ds}\right)^2 + \left(\frac{d\nu}{ds}\right)^2$,

Curvature and tortuosity.

$$\text{or}\quad \tau=\left\{\left(\mu\frac{d\nu}{ds}-\nu\frac{d\mu}{ds}\right)^2+\left(\nu\frac{d\lambda}{ds}-\lambda\frac{d\nu}{ds}\right)^2+\left(\lambda\frac{d\mu}{ds}-\mu\frac{d\lambda}{ds}\right)^2\right\}^{\frac{1}{2}},$$

where λ, μ, ν, denote the direction cosines of the osculating plane, given by the preceding formulæ.

Integral curvature of a curve (compare § 136).

10. The *integral curvature*, or *whole change of direction* of an arc of a plane curve, is the angle through which the tangent has turned as we pass from one extremity to the other. The *average curvature* of any portion is its whole curvature divided by its length. Suppose a line, drawn from a fixed point, to move so as always to be parallel to the direction of motion of a point describing the curve: the angle through which this turns during the motion of the point exhibits what we have thus defined as the integral curvature. In estimating this, we must of course take the enlarged modern meaning of an angle, including angles greater than two right angles, and also negative angles. Thus the integral curvature of any closed curve, whether everywhere concave to the interior or not, is four right angles, provided it does not cut itself. That of a Lemniscate, or figure of 8, is *zero*. That of the Epicycloid ◎ is eight right angles; and so on.

11. The definition in last section may evidently be extended to a plane polygon, and the integral change of direction, or the angle between the first and last sides, is then the sum of its exterior angles, all the sides being produced each in the direction in which the moving point describes it while passing round the figure. This is true whether the polygon be closed or not. If closed, then, as long as it is not crossed, this sum is four right angles,—an extension of the result in Euclid, where all *re-entrant* polygons are excluded. In the case of the star-shaped figure ☆, it is ten right angles, wanting the sum of the five acute angles of the figure; that is, eight right angles.

12. The *integral curvature* and the *average curvature* of a curve which is not plane, may be defined as follows:—Let successive lines be drawn from a fixed point, parallel to tangents at successive points of the curve. These lines will form a conical surface. Suppose this to be cut by a sphere of unit radius having its centre at the fixed point. The *length* of the

Integral curvature of a curve (compare § 186).

curve of intersection measures the *integral curvature* of the given curve. The *average curvature* is, as in the case of a plane curve, the integral curvature divided by the length of the curve. For a tortuous curve approximately plane, the integral curvature thus defined, approximates (not to the integral curvature according to the proper definition, § 10, for a plane curve, but) to the sum of the integral curvatures of all the parts of an approximately coincident plane curve, each taken as positive. Consider, for examples, varieties of James Bernouilli's plane elastic curve, § 611, and approximately coincident tortuous curves of fine steel piano-forte wire. Take particularly the plane lemniscate and an approximately coincident tortuous closed curve.

13. Two consecutive tangents lie in the osculating plane. This plane is therefore parallel to the tangent plane to the cone described in the preceding section. Thus the tortuosity may be measured by the help of the spherical curve which we have just used for defining integral curvature. We cannot as yet complete the explanation, as it depends on the theory of rolling, which will be treated afterwards (§§ 110—137). But it is enough at present to remark, that if a plane roll on the sphere, along the spherical curve, turning always round an instantaneous axis tangential to the sphere, the integral curvature of the curve of contact or trace of the rolling on the plane, is a proper measure of the *whole torsion*, or integral of tortuosity. From this and § 12 it follows that the curvature of this plane curve at any point, or, which is the same, the projection of the curvature of the spherical curve on a tangent plane of the spherical surface, is equal to the tortuosity divided by the curvature of the given curve.

Let $\frac{1}{\rho}$ be the curvature and τ the tortuosity of the given curve, and ds an element of its length. Then $\int \frac{ds}{\rho}$ and $\int \tau ds$, each integral extended over any stated length, l, of the curve, are respectively the integral curvature and the integral tortuosity. The mean curvature and the mean tortuosity are respectively

$$\frac{1}{l}\int \frac{ds}{\rho} \text{ and } \frac{1}{l}\int \tau ds.$$

Infinite tortuosity will be easily understood, by considering a helix, of inclination a, described on a right circular cylinder of radius r. The curvature in a circular section being $\frac{1}{r}$, that of the helix is, of course, $\frac{\cos^2 a}{r}$. The tortuosity is $\frac{\sin a \cos a}{r}$, or $\tan a \times$ curvature. Hence, if $a = \frac{\pi}{4}$ the curvature and tortuosity are equal.

Integral curvature of a curve (compare § 136).

Let the curvature be denoted by $\frac{1}{\rho}$, so that $\cos^2 a = \frac{r}{\rho}$. Let ρ remain finite, and let r diminish without limit. The *step* of the helix being $2\pi r \tan a = 2\pi \sqrt{\rho r} \left(1 - \frac{r}{\rho}\right)^{\frac{1}{2}}$, is, in the limit, $2\pi \sqrt{\rho r}$, which is infinitely small. Thus the motion of a point in the curve, though infinitely nearly in a straight line (the path being always at the infinitely small distance r from the fixed straight line, the axis of the cylinder), will have finite curvature $\frac{1}{\rho}$. The tortuosity, being $\frac{1}{\rho} \tan a$ or $\frac{1}{\sqrt{\rho r}} \left(1 - \frac{r}{\rho}\right)^{\frac{1}{2}}$, will in the limit be a mean proportional between the curvature of the circular section of the cylinder and the finite curvature of the curve.

The acceleration (or force) required to produce such a motion of a point (or material particle) will be afterwards investigated (§ 35 *d.*).

14. A chain, cord, or fine wire, or a fine fibre, filament, or hair, may suggest what is not to be found among natural or artificial productions, a perfectly *flexible and inextensible line.* The elementary kinematics of this subject require no investigation. The mathematical condition to be expressed in any case of it is simply that the distance measured along the line from any one point to any other, remains constant, however the line be bent.

Flexible line.

15. The use of a cord in mechanism presents us with many practical applications of this theory, which are in general extremely simple; although curious, and not always very easy, geometrical problems occur in connexion with it. We shall say nothing here about the theory of knots, knitting, weaving,

Flexible line.

plaiting, etc., but we intend to return to the subject, under vortex-motion in Hydrokinetics.

16. In the mechanical traoing of curves, a flexible and inextensible cord is often supposed. Thus, in drawing an ellipse, the focal property of the curve shows us that by fixing the ends of such a cord to the foci and keeping it stretched by a pencil, the pencil will trace the curve.

By a ruler moveable about one focus, and a string attached to a point in the ruler and to the other focus, the hyperbola may be described by the help of its analogous focal property; and so on.

Evolute.

17. But the consideration of evolutes is of some importance in Natural Philosophy, especially in certain dynamical and optical questions, and we shall therefore devote a section or two to this application of kinematics.

Def. If a flexible and inextensible string be fixed at one point of a plane curve, and stretched along the curve, and be then unwound in the plane of the curve, its extremity will describe an *Involute* of the curve. The original curve, considered with reference to the other, is called the *Evolute.*

18. It will be observed that we speak of *an* involute, and of *the* evolute, of a curve. In fact, as will be easily seen, a curve can have but one evolute, but it has an infinite number of involutes. For all that we have to do to vary an involute, is to change the point of the curve from which the tracing point starts, or consider the involutes described by different points of the string, and these will, in general, be different curves. The following section shows that there is but one evolute.

19. Let AB be any curve, PQ a portion of an involute, pP, qQ positions of the free part of the string. It will be seen at once that these must be tangents to the arc AB at p and q. Also (see § 90), the string at any stage, as pP, revolves about p. Hence pP is *normal* to the curve PQ. And thus the evolute of PQ is a definite curve, viz., the envelope of the normals drawn at every point of PQ,

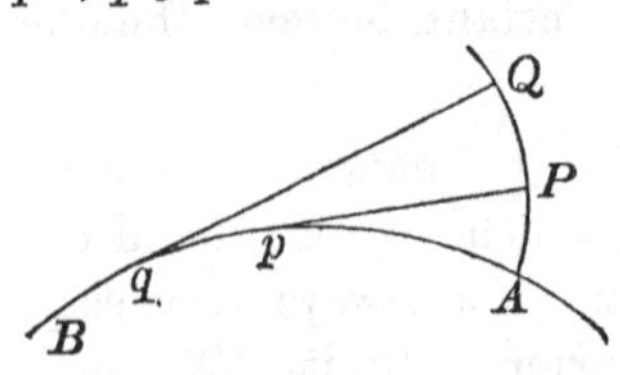

or, which is the same thing, the locus of the centres of curvature of the curve PQ. And we may merely mention, as an obvious result of the mode of tracing, that the arc pq is equal to the difference of qQ and pP, or that the arc pA is equal to pP. **Evolute.**

20. The rate of motion of a point, or its rate of change of position, is called its *Velocity*. It is greater or less as the space passed over in a given time is greater or less: and it may be *uniform*, *i.e.*, the same at every instant; or it may be *variable*. **Velocity.**

Uniform velocity is measured by the space passed over in unit of time, and is, in general, expressed in feet per second; if very great, as in the case of light, it is sometimes popularly reckoned in miles per second. It is to be observed, that time is here used in the abstract sense of a uniformly increasing quantity—what in the differential calculus is called an independent variable. Its physical definition is given in the next chapter.

21. Thus a point, which moves uniformly with velocity v, describes a space of v feet each second, and therefore vt feet in t seconds, t being any number whatever. Putting s for the space described in t seconds, we have

$$s = vt.$$

Thus with unit velocity a point describes unit of space in unit of time.

22. It is well to observe here, that since, by our formula, we have generally

$$v = \frac{s}{t};$$

and since nothing has been said as to the magnitudes of s and t, we may take these as small as we choose. Thus *we get the same result whether we derive v from the space described in a million seconds, or from that described in a millionth of a second.* This idea is very useful, as it makes our results intelligible when a variable velocity has to be measured, and we find ourselves obliged to approximate to its value by considering the space described in an interval so short, that during its lapse the velocity does not sensibly alter in value.

Velocity.

23. When the point does not move uniformly, the velocity is variable, or different at different successive instants; but we define the *average* velocity during any time as the space described in that time, divided by the time, and, the less the interval is, the more nearly does the average velocity coincide with the actual velocity at any instant of the interval. Or again, we define the exact velocity at any instant as the space which the point would have described in one second, if for one second its velocity remained unchanged. That there is at every instant a definite value of the velocity of any moving body, is evident to all, and is matter of everyday conversation. Thus, a railway train, after starting, gradually increases its speed, and every one understands what is meant by saying that at a particular instant it moves at the rate of ten or of fifty miles an hour,—although, in the course of an hour, it may not have moved a mile altogether. Indeed, we may imagine, at any instant during the motion, the steam to be so adjusted as to keep the train running for some time at a perfectly uniform velocity. This would be the velocity which the train had at the instant in question. Without supposing any such definite adjustment of the driving power to be made, we can evidently obtain an approximation to this instantaneous velocity by considering the motion for so short a time, that during it the actual variation of speed may be small enough to be neglected.

24. In fact, if v be the velocity at either beginning or end, or at any instant of the interval, and s the space actually described in time t, the equation $v=\frac{s}{t}$ is more and more nearly true, as the velocity is more nearly uniform during the interval t; so that if we take the interval small enough the equation may be made as nearly exact as we choose. Thus the set of values—

Space described in one second,
Ten times the space described in the first tenth of a second,
A hundred „ „ „ hundredth „

and so on, give nearer and nearer approximations to the velocity at the beginning of the first second. The whole foundation of the differential calculus is, in fact, contained in this simple question, "What is the rate at which the space described increases?" *i.e.*, What is the velocity of the moving point? Newton's notation for the velocity, *i.e.* the rate at which s increases, or the *fluxion* of s, is $\dot{s}$. This notation is very convenient, as it saves the introduction of a second letter.

Velocity.

Let a point which has described a space s in time t proceed to describe an additional space δs in time δt, and let v_1 be the greatest, and v_2 the least, velocity which it has during the interval δt. Then, evidently,

$$\delta s < v_1 \delta t, \quad \delta s > v_2 \delta t,$$

$$i.e., \ \frac{\delta s}{\delta t} < v_1, \quad \frac{\delta s}{\delta t} > v_2.$$

But as δt diminishes, the values of v_1 and v_2 become more and more nearly equal, and in the limit, each is equal to the velocity at time t. Hence

$$v = \frac{ds}{dt}.$$

25. The preceding definition of velocity is equally applicable whether the point move in a straight or curved line; but, since in the latter case the direction of motion continually changes, the mere amount of the velocity is not sufficient completely to describe the motion, and we must have in every such case additional data to remove the uncertainty.

Resolution of velocity.

In such cases as this the method commonly employed, whether we deal with velocities, or as we shall do farther on with accelerations and forces, consists mainly in studying, not the velocity, acceleration, or force, *directly*, but its components parallel to any three assumed directions at right angles to each other. Thus, for a train moving up an incline in a NE direction, we may have given the whole velocity and the steepness of the incline, or we may express the same ideas thus—the train is moving simultaneously northward, eastward, and upward—and the motion as to amount and direction will be completely known if we know separately the northward, eastward, and upward velocities—these being called the *components* of the whole velocity in the three mutually perpendicular directions N, E, and up.

Resolution of velocity.

In general the velocity of a point at x, y, z, is (as we have seen) $\frac{ds}{dt}$, or, which is the same, $\left\{\left(\frac{dx}{dt}\right)^2 + \left(\frac{dy}{dt}\right)^2 + \left(\frac{dz}{dt}\right)^2\right\}^{\frac{1}{2}}$.

Now denoting by u the rate at which x increases, or the velocity parallel to the axis of x, and so by v, w, for the other two; we have $u = \frac{dx}{dt}$, $v = \frac{dy}{dt}$, $w = \frac{dz}{dt}$. Hence, calling α, β, γ the angles which the direction of motion makes with the axes, and putting $q = \frac{ds}{dt}$, we have

$$\cos\alpha = \frac{dx}{ds} = \frac{\frac{dx}{dt}}{\frac{ds}{dt}} = \frac{u}{q}.$$

Hence $u = q\cos\alpha$, and therefore

26. A velocity in any direction may be resolved in, and perpendicular to, any other direction. The first component is found by multiplying the velocity by the cosine of the angle between the two directions—the second by using as factor the sine of the same angle. Or, it may be resolved into components in any three rectangular directions, each component being formed by multiplying the whole velocity by the cosine of the angle between its direction and that of the component.

It is useful to remark that if the axes of x, y, z are not rectangular, $\frac{dx}{dt}$, $\frac{dy}{dt}$, $\frac{dz}{dt}$ will still be the velocities parallel to the axes, but we shall no longer have

$$\left(\frac{ds}{dt}\right)^2 = \left(\frac{dx}{dt}\right)^2 + \left(\frac{dy}{dt}\right)^2 + \left(\frac{dz}{dt}\right)^2.$$

We leave as an exercise for the student the determination of the correct expression for the whole velocity in terms of its components.

If we resolve the velocity along a line whose inclinations to the axes are λ, μ, ν, and which makes an angle θ with the direction of motion, we find the two expressions below (which must of course be equal) according as we resolve q directly or by its components, u, v, w,

$$q\cos\theta = u\cos\lambda + v\cos\mu + w\cos\nu.$$

Substitute in this equation the values of u, v, w already given, § 25, and we have the well-known geometrical theorem for the angle between two straight lines which make given angles with the axes,

Resolution of velocity.

$$\cos\theta = \cos\alpha\cos\lambda + \cos\beta\cos\mu + \cos\gamma\cos\nu.$$

From the above expression we see at once that

27. The velocity resolved in any direction is the sum of the components (in that direction) of the three rectangular components of the whole velocity. And, if we consider motion in one plane, this is still true, only we have but *two* rectangular components. These propositions are virtually equivalent to the following obvious geometrical construction:—

Composition of velocities.

To compound any two velocities as OA, OB in the figure; from A draw AC parallel and equal to OB. Join OC:—then OC is the resultant velocity in magnitude and direction.

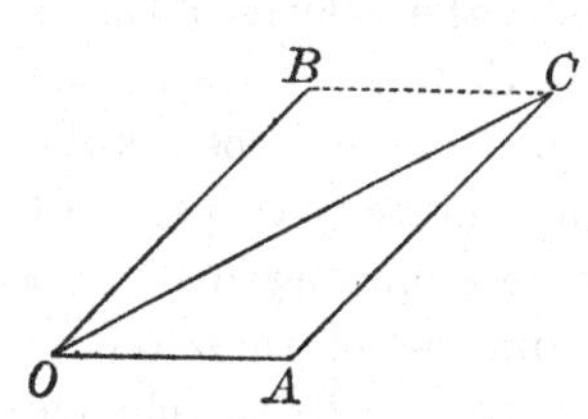

OC is evidently the diagonal of the parallelogram two of whose sides are OA, OB.

Hence the resultant of velocities represented by the sides of any closed polygon whatever, whether in one plane or not, taken all in the same order, is zero.

Hence also the resultant of velocities represented by all the sides of a polygon but one, taken in order, is represented by that one taken in the opposite direction.

When there are two velocities or three velocities in two or in three rectangular directions, the resultant is the square root of the sum of their squares—and the cosines of the inclination of its direction to the given directions are the ratios of the components to the resultant.

It is easy to see that as δs in the limit may be resolved into δr and $r\delta\theta$, where r and θ are polar co-ordinates of a plane curve, $\frac{dr}{dt}$ and $r\frac{d\theta}{dt}$ are the resolved parts of the velocity along, and perpendicular to, the radius vector. We may obtain the same result thus, $x = r\cos\theta$, $y = r\sin\theta$.

Composition of velocities.

Hence $\frac{dx}{dt} = \frac{dr}{dt}\cos\theta - r\sin\theta\frac{d\theta}{dt}$, $\frac{dy}{dt} = \frac{dr}{dt}\sin\theta + r\cos\theta\frac{d\theta}{dt}$.

But by § 26 the whole velocity along r is $\frac{dx}{dt}\cos\theta + \frac{dy}{dt}\sin\theta$,

i.e., by the above values, $\frac{dr}{dt}$. Similarly the transverse velocity is

$$\frac{dy}{dt}\cos\theta - \frac{dx}{dt}\sin\theta, \text{ or } r\frac{d\theta}{dt}.$$

Acceleration.

28. The velocity of a point is said to be accelerated or retarded according as it increases or diminishes, but the word *acceleration* is generally used in either sense, on the understanding that we may regard its quantity as either positive or negative. Acceleration of velocity may of course be either uniform or variable. It is said to be uniform when the velocity receives equal increments in equal times, and is then measured by the actual increase of velocity per unit of time. If we choose as the unit of acceleration that which adds a unit of velocity per unit of time to the velocity of a point, an acceleration measured by α will add α units of velocity in unit of time—and, therefore, αt units of velocity in t units of time. Hence if V be the change in the velocity during the interval t,

$$V = \alpha t, \text{ or } \alpha = \frac{V}{t}.$$

29. Acceleration is variable when the point's velocity does not receive equal increments in successive equal periods of time. It is then measured by the increment of velocity, which would have been generated in a unit of time had the acceleration remained throughout that interval the same as at its commencement. The *average* acceleration during any time is the whole velocity gained during that time, divided by the time. In Newton's notation $\dot{v}$ is used to express the acceleration in the direction of motion; and, if $v = \dot{s}$, as in § 24, we have

$$\alpha = \dot{v} = \ddot{s}.$$

Let v be the velocity at time t, δv its change in the interval δt, α_1 and α_2 the greatest and least values of the acceleration during the interval δt. Then, evidently,

$$\delta v < \alpha_1 \delta t, \quad \delta v > \alpha_2 \delta t,$$

Accelera-tion

$$\text{or} \quad \frac{\delta v}{\delta t} < \alpha_1, \quad \frac{\delta v}{\delta t} > \alpha_2.$$

As δt is taken smaller and smaller, the values of α_1 and α_2 approximate infinitely to each other, and to that of α the required acceleration at time t. Hence

$$\frac{dv}{dt} = \alpha.$$

It is useful to observe that we may also write (by changing the independent variable)

$$\alpha = \frac{dv}{ds}\frac{ds}{dt} = v\frac{dv}{ds}.$$

Since $v = \frac{ds}{dt}$, we have $\alpha = \frac{d^2s}{dt^2}$, and it is evident from similar reasoning that the component accelerations parallel to the axes are $\frac{d^2x}{dt^2}$, $\frac{d^2y}{dt^2}$, $\frac{d^2z}{dt^2}$. But it is to be carefully observed that $\frac{d^2s}{dt^2}$ is *not* generally the resultant of the three component accelerations, but is so only when either the curvature of the path, or the velocity is zero; for [§ 9 (14)] we have

$$\left(\frac{d^2s}{dt^2}\right)^2 = \left(\frac{d^2x}{dt^2}\right)^2 + \left(\frac{d^2y}{dt^2}\right)^2 + \left(\frac{d^2z}{dt^2}\right)^2 - \frac{1}{\rho}\frac{ds^2}{dt^2}.$$

The direction cosines of the tangent to the path at any point x, y, z are

$$\frac{1}{v}\frac{dx}{dt}, \quad \frac{1}{v}\frac{dy}{dt}, \quad \frac{1}{v}\frac{dz}{dt}.$$

Those of the line of resultant acceleration are

$$\frac{1}{f}\frac{d^2x}{dt^2}, \quad \frac{1}{f}\frac{d^2y}{dt^2}, \quad \frac{1}{f}\frac{d^2z}{dt^2},$$

where, for brevity, we denote by f the resultant acceleration. Hence the direction cosines of the plane of these two lines are

$$\frac{dyd^2z - dzd^2y}{\{(dyd^2z - dzd^2y)^2 + (dzd^2x - dxd^2z)^2 + (dxd^2y - dyd^2x)^2\}^{\frac{1}{2}}}, \text{ etc.}$$

These (§ 9) show that this plane is the osculating plane of the curve. Again, if θ denote the angle between the two lines, we have

$$\sin\theta = \frac{\{(dyd^2z - dzd^2y)^2 + (dzd^2x - dxd^2z)^2 + (dxd^2y - dyd^2x)^2\}^{\frac{1}{2}}}{vfdt^3},$$

Acceleration.

or, according to the expression for the curvature (§ 9),

$$\sin\theta = \frac{ds^3}{\rho v f dt^3} = \frac{v^2}{f\rho}.$$

Hence $$f\sin\theta = \frac{v^2}{\rho}.$$

Again, $$\cos\theta = \frac{1}{vf}\left(\frac{dx}{dt}\frac{d^2x}{dt^2} + \frac{dy}{dt}\frac{d^2y}{dt^2} + \frac{dz}{dt}\frac{d^2z}{dt^2}\right) = \frac{ds\,d^2s}{vfdt^3} = \frac{d^2s}{fdt^2}.$$

Hence $$f\cos\theta = \frac{d^2s}{dt^2},$$ and therefore

Resolution and composition of accelerations.

30. The whole acceleration in any direction is the sum of the components (in that direction) of the accelerations parallel to any three rectangular axes—each component acceleration being found by the same rule as component velocities, that is, by multiplying by the cosine of the angle between the direction of the acceleration and the line along which it is to be resolved.

31. When a point moves in a curve the whole acceleration may be resolved into two parts, one in the direction of the motion and equal to the acceleration of the velocity—the other towards the centre of curvature (perpendicular therefore to the direction of motion), whose magnitude is proportional to the square of the velocity and also to the curvature of the path. The former of these changes the velocity, the other affects only the form of the path, or the direction of motion. Hence if a moving point be subject to an acceleration, constant or not, whose direction is continually perpendicular to the direction of motion, the velocity will not be altered—and the only effect of the acceleration will be to make the point move in a curve whose curvature is proportional to the acceleration at each instant.

32. In other words, if a point move in a curve, whether with a uniform or a varying velocity, its change of direction is to be regarded as constituting an acceleration towards the centre of curvature, equal in amount to the square of the velocity divided by the radius of curvature. The whole acceleration will, in every case, be the resultant of the acceleration, thus measuring change of direction, and the acceleration of actual velocity along the curve.

Resolution and composition of accelerations.

We may take another mode of resolving acceleration for a plane curve, which is sometimes useful; along, and perpendicular to, the radius-vector. By a method similar to that employed in § 27, we easily find for the component along the radius-vector

$$\frac{d^2r}{dt^2} - r\left(\frac{d\theta}{dt}\right)^2,$$

and for that perpendicular to the radius-vector

$$\frac{1}{r}\frac{d}{dt}\left(r^2\frac{d\theta}{dt}\right).$$

33. If for any case of motion of a point we have given the whole velocity and its direction, or simply the components of the velocity in three rectangular directions, at any *time*, or, as is most commonly the case, for any *position*, the determination of the form of the path described, and of other circumstances of the motion, is a question of pure mathematics, and in all cases is capable, if not of an exact solution, at all events of a solution to any degree of approximation that may be desired.

Determination of the motion from given velocity or acceleration.

The same is true if the total acceleration and its direction at every instant, or simply its rectangular components, be given, provided the velocity and direction of motion, as well as the position, of the point at any one instant, be given.

For we have in the first case

$$\frac{dx}{dt} = u = q \cos a, \text{ etc.},$$

three simultaneous equations which can contain only x, y, z, and t, and which therefore suffice when integrated to determine x, y, and z in terms of t. By eliminating t among these equations, we obtain two equations among x, y, and z—each of which represents a surface on which lies the path described, and whose intersection therefore completely determines it.

In the second case we have

$$\frac{d^2x}{dt^2} = a, \quad \frac{d^2y}{dt^2} = \beta, \quad \frac{d^2z}{dt^2} = \gamma;$$

to which equations the same remarks apply, except that here each has to be twice integrated.

Determination of the motion from given velocity or acceleration.

The arbitrary constants introduced by integration are determined at once if we know the co-ordinates, and the components of the velocity, of the point at a given epoch.

Examples of velocity.

34. From the principles already laid down, a great many interesting results may be deduced, of which we enunciate a few of the most important.

a. If the velocity of a moving point be uniform, and if its direction revolve uniformly in a plane, the path described is a circle.

Let a be the velocity, and α the angle through which its direction turns in unit of time; then, by properly choosing the axes, we have

$$\frac{dx}{dt} = -a \sin \alpha t, \quad \frac{dy}{dt} = a \cos \alpha t,$$

whence
$$(x - A)^2 + (y - B)^2 = \frac{a^2}{\alpha^2}.$$

b. If a point moves in a plane, and if its component velocity parallel to each of two rectangular axes is proportional to its distance from that axis, the path is an ellipse or hyperbola whose principal diameters coincide with those axes; and the acceleration is directed to or from the origin at every instant.

$$\frac{dx}{dt} = \mu y, \quad \frac{dy}{dt} = \nu x.$$

Hence $\frac{d^2x}{dt^2} = \mu\nu x$, $\frac{d^2y}{dt^2} = \mu\nu y$, and the whole acceleration is towards or from O.

Also $\frac{dy}{dx} = \frac{\nu}{\mu}\frac{x}{y}$, from which $\mu y^2 - \nu x^2 = C$, an ellipse or hyperbola referred to its principal axes. (Compare § 65.)

c. When the velocity is uniform, but in direction revolving uniformly in a right circular cone, the motion of the point is in a circular helix whose axis is parallel to that of the cone.

Examples of acceleration.

35. *a.* When a point moves uniformly in a circle of radius R, with velocity V, the whole acceleration is directed towards the centre, and has the constant value $\frac{V^2}{R}$. See § 31.

b. With uniform acceleration in the direction of motion, a point describes spaces proportional to the squares of the times elapsed since the commencement of the motion.

Examples of acceleration.

In this case the space described in any interval is that which would be described in the same time by a point moving uniformly with a velocity equal to that at the middle of the interval. In other words, the average velocity (when the acceleration is uniform) is, during any interval, the arithmetical mean of the initial and final velocities. This is the case of a stone falling vertically.

For if the acceleration be parallel to x, we have

$$\frac{d^2x}{dt^2} = a, \text{ therefore } \frac{dx}{dt} = v = at, \text{ and } x = \tfrac{1}{2}at^2.$$

And we may write the equation (§ 29) $v\dfrac{dv}{dx} = a$, whence $\dfrac{v^2}{2} = ax.$

If at time $t = 0$ the velocity was V, these equations become at once

$$v = V + at, \quad x = Vt + \tfrac{1}{2}at^2, \text{ and } \frac{v^2}{2} = \frac{V^2}{2} + ax.$$

$$\begin{aligned} \text{And initial velocity} &= V, \\ \text{final} \quad \text{,,} \quad &= V + at; \\ \text{Arithmetical mean} &= V + \tfrac{1}{2}at, \\ &= \frac{x}{t}, \end{aligned}$$

whence the second part of the above statement.

c. When there is uniform acceleration in a constant direction, the path described is a parabola, whose axis is parallel to that direction. This is the case of a projectile moving in vacuum.

For if the axis of y be parallel to the acceleration a, and if the plane of xy be that of motion at any time,

$$\frac{d^2z}{dt^2} = 0, \quad \frac{dz}{dt} = 0, \quad z = 0,$$

and therefore the motion is wholly in the plane of xy.

Then $$\frac{d^2x}{dt^2} = 0, \quad \frac{d^2y}{dt^2} = a;$$

Examples of acceleration.

and by integration

$$x = Ut + a, \quad y = \tfrac{1}{2}at^2 + Vt + b,$$

where U, V, a, b are constants.

The elimination of t gives the equation of a parabola of which the axis is parallel to y, parameter $\frac{U^2}{2a}$, and vertex the point whose co-ordinates are

$$x = a - \frac{UV}{a}, \quad y = b - \frac{V^2}{2a}.$$

d. As an illustration of acceleration in a tortuous curve, we take the case of § 13, or of § 34, *c*.

Let a point move in a circle of radius r with uniform angular velocity ω (about the centre), and let this circle move perpendicular to its plane with velocity V. The point describes a helix on a cylinder of radius r, and the inclination α is given by

$$\tan \alpha = \frac{V}{r\omega}.$$

The curvature of the path is $\frac{1}{r}\frac{r^2\omega^2}{V^2 + r^2\omega^2}$ or $\frac{r\omega^2}{V^2 + r^2\omega^2}$, and the tortuosity $\frac{\omega}{V}\frac{V^2}{V^2 + r^2\omega^2} = \frac{V\omega}{V^2 + r^2\omega^2}$.

The acceleration is $r\omega^2$, directed perpendicularly towards the axis of the cylinder.—Call this A.

$$\text{Curvature} = \frac{A}{V^2 + Ar} = \frac{A}{V^2 + \frac{A^2}{\omega^2}}.$$

$$\text{Tortuosity} = \frac{V}{\sqrt{Ar}}\frac{A}{V^2 + Ar} = \frac{V\omega}{V^2 + \frac{A^2}{\omega^2}}.$$

Let A be finite, r indefinitely small, and therefore ω indefinitely great.

$$\text{Curvature (in the limit)} = \frac{A}{V^2}.$$

$$\text{Tortuosity (\quad ,, \quad)} = \frac{\omega}{V}.$$

Thus, if we have a material particle moving in the manner specified, and if we consider the force (see Chap. II.) required to produce the acceleration, we find that a finite force perpendicular to the line of motion, in a direction revolving with an infinitely great angular velocity, maintains constant infinitely small deflection (in a direction opposite to its own) from the line of undisturbed motion, *finite* curvature, and infinite tortuosity.

Examples of acceleration.

e. When the acceleration is perpendicular to a given plane and proportional to the distance from it, the path is a plane curve, which is the harmonic curve if the acceleration be *towards* the plane, and a more or less fore-shortened catenary (§ 580) if *from* the plane.

As in case *c*, $\frac{d^2z}{dt^2}=0$, $\frac{dz}{dt}=0$, and $z=0$, if the axis of z be perpendicular to the acceleration and to the direction of motion at any instant. Also, if we choose the origin *in* the plane,

$$\frac{d^2x}{dt^2}=0, \quad \frac{d^2y}{dt^2}=\mu y.$$

Hence $$\frac{dx}{dt}=\text{const.}=a \text{ (suppose)},$$

and $$\frac{d^2y}{dx^2}=\frac{\mu}{a^2}y=\mp\frac{y}{b^2}.$$

This gives, if μ is negative,

$$y=P\cos\left(\frac{x}{b}+Q\right), \text{ the harmonic curve, or curve of sines.}$$

If μ be positive, $$y=P\epsilon^{\frac{x}{b}}+Q\epsilon^{-\frac{x}{b}};$$

and by shifting the origin along the axis of x this can be put in the form

$$y=R\left(\epsilon^{\frac{x}{b}}+\epsilon^{-\frac{x}{b}}\right):$$

which is the catenary if $2R=b$; otherwise it is the catenary stretched or fore-shortened in the direction of y.

36. [Compare §§ 233—236 below.] *a.* When the acceleration is directed to a fixed point, the path is in a plane passing through that point; and in this plane the areas traced out by the radius-vector are proportional to the times employed. This includes the case of a satellite or planet revolving about its primary.

Acceleration directed to a fixed centre.

Evidently there is no acceleration perpendicular to the plane containing the fixed and moving points and the direction

Acceleration directed to a fixed centre.

of motion of the second at any instant; and, there being no velocity perpendicular to this plane at starting, there is therefore none throughout the motion; thus the point moves in the plane. And had there been no acceleration, the point would have described a straight line with uniform velocity, so that in this case the areas described by the radius-vector would have been proportional to the times. Also, the area actually described in any instant depends on the length of the radius-vector and the velocity perpendicular to it, and is shown below to be unaffected by an acceleration parallel to the radius-vector. Hence the second part of the proposition.

We have $$\frac{d^2x}{dt^2} = P\frac{x}{r}, \quad \frac{d^2y}{dt^2} = P\frac{y}{r}, \quad \frac{d^2z}{dt^2} = P\frac{z}{r},$$

the fixed point being the origin, and P being some function of x, y, z; in *nature* a function of r only.

Hence $$x\frac{d^2y}{dt^2} - y\frac{d^2x}{dt^2} = 0, \text{ etc.},$$

which give on integration

$$y\frac{dz}{dt} - z\frac{dy}{dt} = C_1, \quad z\frac{dx}{dt} - x\frac{dz}{dt} = C_2, \quad x\frac{dy}{dt} - y\frac{dx}{dt} = C_3.$$

Hence at once $C_1x + C_2y + C_3z = 0$, or the motion is in a plane through the origin. Take this as the plane of xy, then we have only the one equation

$$x\frac{dy}{dt} - y\frac{dx}{dt} = C_3 = h \text{ (suppose)}.$$

In polar co-ordinates this is

$$h = r^2\frac{d\theta}{dt} = 2\frac{dA}{dt}$$

if A be the area intercepted by the curve, a fixed radius-vector, and the radius-vector of the moving point. Hence the area increases uniformly with the time.

b. In the same case the velocity at any point is inversely as the perpendicular from the fixed point upon the tangent to the path, the momentary direction of motion.

For evidently the product of this perpendicular and the velocity gives double the area described in one second about the fixed point.

Acceleration directed to a fixed centre.

Or thus—if p be the perpendicular on the tangent,

$$p = x\frac{dy}{ds} - y\frac{dx}{ds},$$

and therefore
$$p\frac{ds}{dt} = x\frac{dy}{dt} - y\frac{dx}{dt} = h.$$

If we refer the motion to co-ordinates in its own plane, we have only the equations

$$\frac{d^2x}{dt^2} = \frac{Px}{r}, \quad \frac{d^2y}{dt^2} = \frac{Py}{r},$$

whence, as before,
$$r^2\frac{d\theta}{dt} = h.$$

If, by the help of this last equation, we eliminate t from $\frac{d^2x}{dt^2} = \frac{Px}{r}$, substituting polar for rectangular co-ordinates, we arrive at the polar differential equation of the path.

For variety, we may derive it from the formulæ of § 32.

They give
$$\frac{d^2r}{dt^2} - r\left(\frac{d\theta}{dt}\right)^2 = P, \quad r^2\frac{d\theta}{dt} = h.$$

Putting $\frac{1}{r} = u$, we have

$$\frac{d^2\left(\frac{1}{u}\right)}{dt^2} - \frac{1}{u}\left(\frac{d\theta}{dt}\right)^2 = P, \text{ and } \frac{d\theta}{dt} = hu^2.$$

But $\frac{d\left(\frac{1}{u}\right)}{dt} = hu^2\frac{d\left(\frac{1}{u}\right)}{d\theta} = -h\frac{du}{d\theta}$, therefore $\frac{d^2\left(\frac{1}{u}\right)}{dt^2} = -h^2u^2\frac{d^2u}{d\theta^2}$.

Also $\frac{1}{u}\left(\frac{d\theta}{dt}\right)^2 = h^2u^3$, the substitution of which values gives us

$$\frac{d^2u}{d\theta^2} + u = -\frac{P}{h^2u^2} \quad \text{.........................(1)},$$

the equation required. The integral of this equation involves *two* arbitrary constants besides h, and the remaining constant belonging to the two differential equations of the second order above is to be introduced on the farther integration of

$$\frac{d\theta}{dt} = hu^2 \quad \text{..............................(2)},$$

when the value of u in terms of θ is substituted from the equation of the path.

Other examples of these principles will be met with in the chapters on Kinetics.

Hodograph.

37. If from any fixed point, lines be drawn at every instant, representing in magnitude and direction the velocity of a point describing any path in any manner, the extremities of these lines form a curve which is called the *Hodograph.* The invention of this construction is due to Sir W. R. Hamilton. One of the most beautiful of the many remarkable theorems to which it led him is that of § 38.

Since the radius-vector of the hodograph represents the velocity at each instant, it is evident (§ 27) that an elementary arc represents the velocity which must be compounded with the velocity at the beginning of the corresponding interval of time, to find the velocity at its end. Hence the velocity in the hodograph is equal to the acceleration in the path; and the tangent to the hodograph is parallel to the direction of the acceleration in the path.

If x, y, z be the co-ordinates of the moving point, ξ, η, ζ those of the corresponding point of the hodograph, then evidently

$$\xi = \frac{dx}{dt}, \quad \eta = \frac{dy}{dt}, \quad \zeta = \frac{dz}{dt},$$

and therefore

$$\frac{d\xi}{\dfrac{d^2x}{dt^2}} = \frac{d\eta}{\dfrac{d^2y}{dt^2}} = \frac{d\zeta}{\dfrac{d^2z}{dt^2}},$$

or the tangent to the hodograph is parallel to the acceleration in the orbit. Also, if σ be the arc of the hodograph,

$$\frac{d\sigma}{dt} = \sqrt{\left(\frac{d\xi}{dt}\right)^2 + \left(\frac{d\eta}{dt}\right)^2 + \left(\frac{d\zeta}{dt}\right)^2}$$

$$= \sqrt{\left(\frac{d^2x}{dt^2}\right)^2 + \left(\frac{d^2y}{dt^2}\right)^2 + \left(\frac{d^2z}{dt^2}\right)^2},$$

or the velocity in the hodograph is equal to the rate of acceleration in the path.

Hodograph of planet or comet, deduced from Kepler's laws.

38. *The hodograph for the motion of a planet or comet is always a circle, whatever be the form and dimensions of the orbit.* In the motion of a planet or comet, the acceleration is directed towards the sun's centre. Hence (§ 36, *b*) the velocity is inversely as the perpendicular from that point upon the tangent to the orbit. The orbit we assume to be a conic section, whose focus is the sun's centre. But we know that the intersection of the perpendicular with the tangent lies in the circle whose diameter is the major axis, if the orbit be an ellipse or hyperbola; in the tangent at the vertex if a parabola. Measure off on the perpendicular a third proportional to its own length and any constant line; this portion will thus represent the velocity in magnitude and in a direction perpendicular to its own—so that the locus of the new points in each perpendicular will be the hodograph turned through a right angle. But we see by geometry* that the locus of these points is always a circle. Hence the proposition. The hodograph surrounds its origin if the orbit be an ellipse, passes through it if a parabola, and the origin is without the hodograph if the orbit is a hyperbola.

Hodograph of planet or comet, deduced from Kepler's laws.

For a projectile unresisted by the air, it will be shewn in Kinetics that we have the equations (assumed in § 35, c)

$$\frac{d^2x}{dt^2}=0,\quad \frac{d^2y}{dt^2}=-g,$$

if the axis of y be taken vertically upwards.

Hence for the hodograph

$$\frac{d\xi}{dt}=0,\quad \frac{d\eta}{dt}=-g,$$

or $\xi=C$, $\eta=C'-gt$, and the hodograph is a vertical straight line along which the describing point moves uniformly.

For the case of a planet or comet, instead of assuming as above that the orbit is a conic with the sun in one focus, assume (Newton's deduction from that and the law of areas) that the acceleration is in the direction of the radius-vector, and varies inversely as the square of the distance. We have obviously

Hodograph for planet or comet, deduced from Newton's law of force.

$$\frac{d^2x}{dt^2}=\frac{\mu x}{r^3},\quad \frac{d^2y}{dt^2}=\frac{\mu y}{r^3},$$

where $$r^2=x^2+y^2.$$

Hence, as in § 36, $$x\frac{dy}{dt}-y\frac{dx}{dt}=h\text{..........................}(1),$$

* See our smaller work, § 51.

Hodograph for planet or comet, deduced from Newton's law of force.

and therefore

$$\frac{d^2x}{dt^2} = \frac{\mu x}{h} \frac{x\dfrac{dy}{dt} - y\dfrac{dx}{dt}}{r^3},$$

$$= \frac{\mu}{h} \frac{(x^2+y^2)\dfrac{dy}{dt} - y\left(x\dfrac{dx}{dt} + y\dfrac{dy}{dt}\right)}{r^3}, \quad = \frac{\mu}{h} \frac{r^2\dfrac{dy}{dt} - yr\dfrac{dr}{dt}}{r^3}.$$

Hence $$\frac{dx}{dt} + A = \frac{\mu}{h}\frac{y}{r} \quad\text{.............................. (2).}$$

Similarly $$\frac{dy}{dt} + B = -\frac{\mu}{h}\frac{x}{r} \quad\text{.......................... (3).}$$

Hence for the hodograph

$$(\xi + A)^2 + (\eta + B)^2 = \frac{\mu^2}{h^2},$$

the circle as before stated.

We may merely mention that the equation of the orbit will be found at once by eliminating $\dfrac{dx}{dt}$ and $\dfrac{dy}{dt}$ among the three first integrals (1), (2), (3) above. We thus get

$$-h + Ay - Bx = \frac{\mu}{h} r,$$

a conic section of which the origin is a focus.

Applications of the hodograph.

39. The intensity of heat and light emanating from a point, or from an uniformly radiating spherical surface, diminishes with increasing distance according to the same law as gravitation. Hence the amount of heat and light, which a planet receives from the sun during any interval, is proportional to the time integral of the acceleration during that interval, *i.e.* (§ 37) to the corresponding arc of the hodograph. From this it is easy to see, for example, that if a comet move in a parabola, the amount of heat it receives from the sun in any interval is proportional to the angle through which its direction of motion turns during that interval. There is a corresponding theorem for a planet moving in an ellipse, but somewhat more complicated.

Curves of pursuit.

40. If two points move, each with a definite uniform velocity, one in a given curve, the other at every instant directing its course towards the first describes a path which is called a

Curves of pursuit.

Curve of Pursuit. The idea is said to have been suggested by the old rule of steering a privateer always directly for the vessel pursued. (Bouguer, *Mém. de l'Acad.* 1732.) It is the curve described by a dog running to its master.

The simplest cases are of course those in which the first point moves in a straight line, and of these there are three, for the velocity of the first point may be greater than, equal to, or less than, that of the second. The figures in the text below represent the curves in these cases, the velocities of the pursuer being $\frac{4}{3}$, 1, and $\frac{1}{2}$ of those of the pursued, respectively. In the second and third cases the second point can never overtake the first, and consequently the line of motion of the first is an asymptote. In the first case the second point overtakes the first, and the curve at that point touches the line of motion of the first. The remainder of the curve satisfies a modified form of statement of the original question, and is called the *Curve of Flight.*

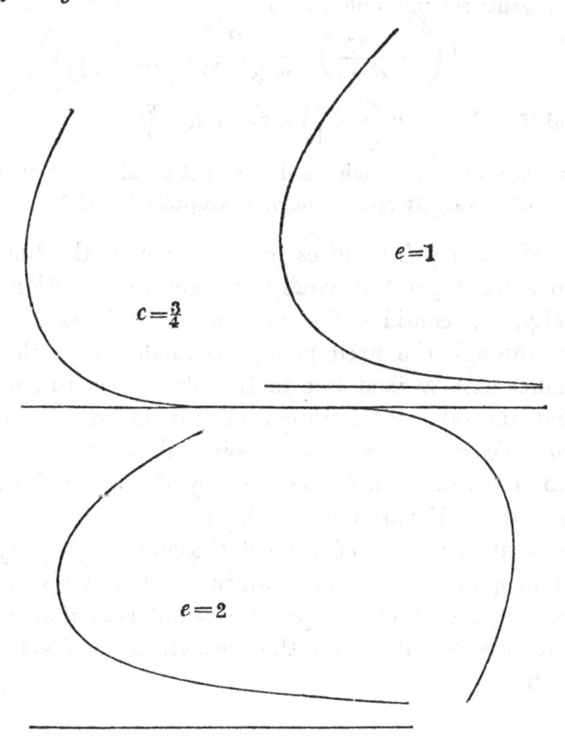

Curves of pursuit.

We will merely form the differential equation of the curve, and give its integrated form, leaving the work to the student.

Suppose Ox to be the line of motion of the first point, whose velocity is v, AP the curve of pursuit, in which the velocity is u, then the tangent at P always passes through Q, the instantaneous position of the first point. It will be evident, on a moment's consideration, that the curve AP must have a tangent perpendicular to Ox. Take this as the axis of y, and let $OA = a$. Then, if $OQ = \xi$, $AP = s$, and if x, y be the co-ordinates of P, we have

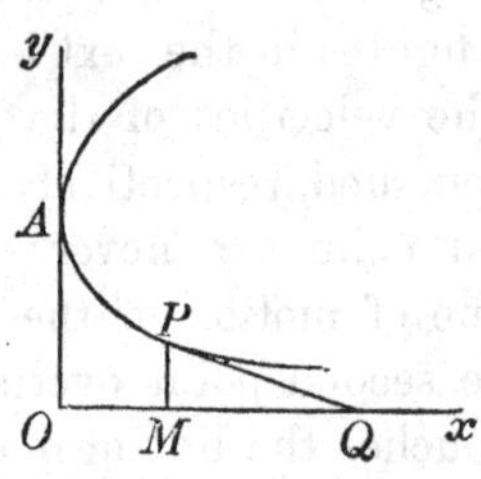

$$\frac{AP}{u} = \frac{OQ}{v},$$

because A, O and P, Q are pairs of simultaneous positions of the two points.

This gives $$\frac{v}{u}s = es = x - y\frac{dx}{dy}.$$

From this we find, unless $e = 1$,

$$2\left(x + \frac{ae}{e^2 - 1}\right) = \frac{y^{e+1}}{a^e(e+1)} + \frac{a^e}{y^{e-1}(e-1)};$$

and if $e = 1$, $$2\left(x + \frac{a}{4}\right) = \frac{y^2}{2a} - a\log_e\frac{y}{a},$$

the only case in which we do not get an algebraic curve. The axis of x is easily seen to be an asymptote if $e \not< 1$.

Angular velocity.

41. When a point moves in any manner, the line joining it with a fixed point generally changes its direction. If, for simplicity, we consider the motion as confined to a plane passing through the fixed point, the angle which the joining line makes with a fixed line in the plane is continually altering, and its rate of alteration at any instant is called the *Angular Velocity* of the first point about the second. If uniform, it is of course measured by the angle described in unit of time; if variable, by the angle which would have been described in unit of time if the angular velocity at the instant in question were maintained constant for so long. In this respect, the process is precisely similar to that which we have already explained for the measurement of velocity and acceleration.

Unit of angular velocity is that of a point which describes, or would describe, unit angle about a fixed point in unit of time. The usual unit angle is (as explained in treatises on plane trigonometry) that which subtends at the centre of a circle an arc whose length is equal to the radius; being an angle of

Angular velocity.

$$\frac{180^0}{\pi} = 57^0.29578\ldots = 57^0\ 17'\ 44''.8 \text{ nearly.}$$

For brevity we shall call this angle a radian.

42. The rate of increase or diminution of the angular velocity when variable is called the *angular acceleration*, and is measured in the same way and by the same unit.

Angular acceleration.

By methods precisely similar to those employed for linear velocity and acceleration we see that if θ be the angle-vector of a point moving in a plane—the

$$\text{Angular velocity is } \omega = \frac{d\theta}{dt}, \text{ and the}$$

$$\text{Angular acceleration is } \frac{d\omega}{dt} = \frac{d^2\theta}{dt^2} = \omega\frac{d\omega}{d\theta}.$$

Since (§ 27) $r\dfrac{d\theta}{dt}$ is the velocity perpendicular to the radius-vector, we see that

The angular velocity of a point in a plane is found by dividing the velocity perpendicular to the radius-vector by the length of the radius-vector.

43. When one point describes uniformly a circle about another, the time of describing a complete circumference being T, we have the angle 2π described uniformly in T; and, therefore, the angular velocity is $\dfrac{2\pi}{T}$. Even when the angular velocity is not uniform, as in a planet's motion, it is useful to introduce the quantity $\dfrac{2\pi}{T}$, which is then called the *mean* angular velocity.

Angular velocity.

When a point moves uniformly in a straight line its angular velocity evidently diminishes as it recedes from the point about which the angles are measured.

Angular velocity.

The polar equation of a straight line is

$$r = a \sec \theta.$$

But the length of the line between the limiting angles 0 and θ is $a \tan \theta$, and this increases with uniform velocity v. Hence

$$v = \frac{d}{dt}(a \tan \theta) = a \sec^2 \theta \frac{d\theta}{dt} = \frac{r^2}{a} \frac{d\theta}{dt}.$$

Hence $\frac{d\theta}{dt} = \frac{av}{r^2}$, and is therefore inversely as the square of the radius-vector.

Similarly for the angular acceleration, we have by a second differentiation,

$$\frac{d^2\theta}{dt^2} + 2 \tan \theta \left(\frac{d\theta}{dt}\right)^2 = 0,$$

i.e., $\frac{d^2\theta}{dt^2} = -\frac{2av^2}{r^3}\left(1 - \frac{a^2}{r^2}\right)^{\frac{1}{2}}$, and ultimately varies inversely as the third power of the radius-vector.

Angular velocity of a plane.

44. We may also talk of the angular velocity of a moving plane with respect to a fixed one, as the rate of increase of the angle contained by them—but unless their line of intersection remain fixed, or at all events parallel to itself, a somewhat more laboured statement is required to give definite information. This will be supplied in a subsequent section.

Relative motion.

45. All motion that we are, or can be, acquainted with, is *Relative* merely. We can calculate from astronomical data for any instant the direction in which, and the velocity with which we are moving on account of the earth's diurnal rotation. We may compound this with the similarly calculable velocity of the earth in its orbit. This resultant again we may compound with the (roughly known) velocity of the sun relatively to the so-called fixed stars; but, even if all these elements were accurately known, it could not be said that we had attained any idea of an *absolute* velocity; for it is only the sun's relative motion among the stars that we can observe; and, in all probability, sun and stars are moving on (possibly with very great rapidity) relatively to other bodies in space. We must therefore consider how, from the actual motions of a set of points, we may find their relative motions with regard to any one of them;

and how, having given the relative motions of all but one with regard to the latter, and the actual motion of the latter, we may find the actual motions of all. The question is very easily answered. Consider for a moment a number of passengers walking on the deck of a steamer. Their relative motions with regard to the deck are what we immediately observe, but if we compound with these the velocity of the steamer itself we get evidently their actual motion relatively to the earth. Again, in order to get the relative motion of all with regard to the deck, we *abstract our ideas from* the motion of the steamer altogether—that is, we alter the velocity of each by compounding it with the actual velocity of the vessel taken in a reversed direction.

Relative motion.

Hence to find the relative motions of any set of points with regard to one of their number, imagine, impressed upon each in composition with its own velocity, a velocity equal and opposite to the velocity of that one; it will be reduced to rest, and the motions of the others will be the same with regard to it as before.

Thus, to take a very simple example, two trains are running in opposite directions, say north and south, one with a velocity of fifty, the other of thirty, miles an hour. The relative velocity of the second with regard to the first is to be found by impressing on both a southward velocity of fifty miles an hour; the effect of this being to bring the first to rest, and to give the second a southward velocity of eighty miles an hour, which is the required relative motion.

Or, given one train moving north at the rate of thirty miles an hour, and another moving west at the rate of forty miles an hour. The motion of the second relatively to the first is at the rate of fifty miles an hour, in a south-westerly direction inclined to the due west direction at an angle of $\tan^{-1}\frac{3}{4}$. It is needless to multiply such examples, as they must occur to every one.

46. Exactly the same remarks apply to relative as compared with absolute acceleration, as indeed we may see at once, since accelerations are in all cases resolved and compounded by the same law as velocities.

Relative motion.

If x, y, z, and x', y', z', be the co-ordinates of two points referred to axes regarded as fixed; and ξ, η, ζ their relative co-ordinates—we have

$$\xi = x' - x, \quad \eta = y' - y, \quad \zeta = z' - z,$$

and, differentiating,

$$\frac{d\xi}{dt} = \frac{dx'}{dt} - \frac{dx}{dt}, \text{ etc.,}$$

which give the relative, in terms of the absolute, velocities; and

$$\frac{d^2\xi}{dt^2} = \frac{d^2x'}{dt^2} - \frac{d^2x}{dt^2}, \text{ etc.,}$$

proving our assertion about relative and absolute accelerations.

The corresponding expressions in polar co-ordinates in a plane are somewhat complicated, and by no means convenient. The student can easily write them down for himself.

47. The following proposition in relative motion is of considerable importance:—

Any two moving points describe similar paths relatively to each other, or relatively to any point which divides in a constant ratio the line joining them.

Let A and B be any simultaneous positions of the points. Take G or G' in AB such that the ratio $\frac{GA}{GB}$ or $\frac{G'A}{G'B}$ has a constant value. Then

G′ A G B

as the form of the relative path depends only upon the *length* and *direction* of the line joining the two points at any instant, it is obvious that these will be the same for A with regard to B, as for B with regard to A, saving only the inversion of the direction of the joining line. Hence B's path about A, is A's about B turned through two right angles. And with regard to G and G' it is evident that the directions remain the same, while the lengths are altered in a given ratio; but this is the definition of similar curves.

48. As a good example of relative motion, let us consider that of the two points involved in our definition of the curve of pursuit, § 40. Since, to find the relative position and motion of the pursuer with regard to the pursued, we must impress on both a velocity equal and opposite to that of the latter, we see

Relative motion.

at once that the problem becomes the same as the following. A boat crossing a stream is impelled by the oars with uniform velocity relatively to the water, and always towards a fixed point in the opposite bank; but it is also carried down stream at a uniform rate; determine the path described and the time of crossing. Here, as in the former problem, there are three cases, figured below. In the first, the boat, moving faster than the current, reaches the desired point; in the second, the velocities of boat and stream being equal, the boat gets across only after an infinite time—describing a parabola—but does not land at the desired point, which is indeed the focus of the parabola, while the landing point is the vertex. In the third case, its proper velocity being less than that of the water, it never reaches the other bank, and is carried indefinitely down stream. The comparison of the figures in § 40 with those in the present section cannot fail to be instructive. They are drawn to the same scale, and for the same relative velocities. The horizontal lines represent the farther bank of the river, and the vertical lines the path of the boat if there were no current.

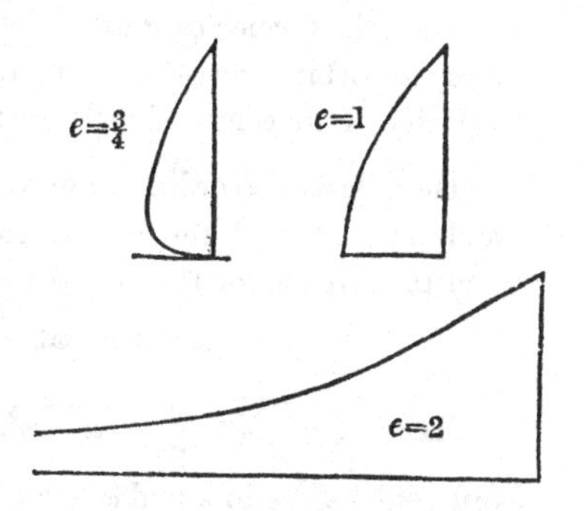

We leave the solution of this question as an exercise, merely noting that the equation of the curve is

$$\frac{y^{1+e}}{a^e} = \sqrt{x^2 + y^2} - x,$$

in one or other of the three cases, according as e is $>$, $=$, or <1.

When $e = 1$ this becomes

$$y^2 = a^2 - 2ax, \text{ the parabola.}$$

The time of crossing is

$$\frac{a}{u(1-e^2)},$$

which is finite only for $e<1$, because of course a negative value is inadmissible.

Relative motion.

49. Another excellent example of the transformation of relative into absolute motion is afforded by the family of cycloids. We shall in a future section consider their mechanical description, by the *rolling* of a circle on a fixed straight line or circle. In the mean time, we take a different form of enunciation, which, however, leads to precisely the same result.

Find the actual path of a point which revolves uniformly in a circle about another point—the latter moving uniformly in a straight line or circle in the same plane.

Take the former case first: let a be the radius of the relative circular orbit, and ω the angular velocity in it, v being the velocity of its centre along the straight line.

The relative co-ordinates of the point in the circle are $a \cos \omega t$ and $a \sin \omega t$, and the actual co-ordinates of the centre are vt and 0. Hence for the actual path

$$\xi = vt + a \cos \omega t, \quad \eta = a \sin \omega t.$$

Hence $\xi = \frac{v}{\omega} \sin^{-1} \frac{\eta}{a} + \sqrt{a^2 - \eta^2}$, an equation which, by giving different values to v and ω, may be made to represent the cycloid itself, or either form of trochoid. See § 92.

For the epicycloids, let b be the radius of the circle which B describes about A, ω_1 the angular velocity; a the radius of A's path, ω the angular velocity.

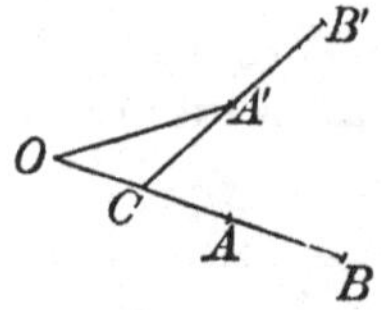

Also at time $t = 0$, let B be in the radius OA of A's path. Then at time t, if A', B' be the positions, we see at once that

$$\angle AOA' = \omega t, \quad \angle B'CA = \omega_1 t.$$

Hence, taking OA as axis of x,

$$x = a \cos \omega t + b \cos \omega_1 t, \quad y = a \sin \omega t + b \sin \omega_1 t,$$

which, by the elimination of t, give an algebraic equation between x and y whenever ω and ω_1 are commensurable.

Thus, for $\omega_1 = 2\omega$, suppose $\omega t = \theta$, and we have

$$x = a \cos \theta + b \cos 2\theta, \quad y = a \sin \theta + b \sin 2\theta,$$

or, by an easy reduction,

$$(x^2 + y^2 - b^2)^2 = a^2 \{(x + b)^2 + y^2\}.$$

Put $x-b$ for x, *i.e.*, change the origin to a distance AB to the left of O, the equation becomes

Relative motion.

$$a^2(x^2+y^2)=(x^2+y^2-2bx)^2,$$

or, in polar co-ordinates,

$$a^2=(r-2b\cos\theta)^2, \quad r=a+2b\cos\theta,$$

and when $2b=a$, $r=a(1+\cos\theta)$, the cardioid. (See § 94.)

50. As an additional illustration of this part of our subject, we may define as follows:—

Resultant motion.

If one point A executes any motion whatever with reference to a second point B; if B executes any other motion with reference to a third point C; and so on—the first is said to execute, with reference to the last, a movement which is the resultant of these several movements.

The relative position, velocity, and acceleration are in such a case the geometrical resultants of the various components combined according to preceding rules.

51. The following practical methods of effecting such a combination in the simple case of the movements of two points are useful in scientific illustrations and in certain mechanical arrangements. Let two moving points be joined by an elastic string; the middle point of this string will evidently execute a movement which is *half* the resultant of the motions of the two points. But for drawing, or engraving, or for other mechanical applications, the following method is preferable:—

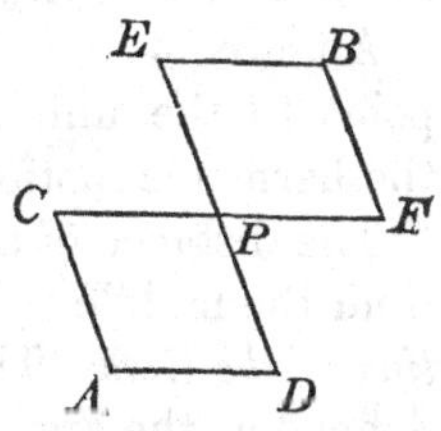

CF and ED are rods of equal length moving freely round a pivot at P, which passes through the middle point of each—CA, AD, EB, and BF, are rods of half the length of the two former, and so pivoted to them as to form a pair of equal rhombi CD, EF, whose angles can be altered at will. Whatever motions, whether in a plane, or in space of three dimensions, be given to A and B, P will evidently be subjected to half their resultant.

52. Amongst the most important classes of motions which we have to consider in Natural Philosophy, there is one, namely, *Harmonic Motion*, which is of such immense use, not only in

Harmonic motion.

Harmonic motion.

ordinary kinetics, but in the theories of sound, light, heat, etc., that we make no apology for entering here into considerable detail regarding it.

Simple harmonic motion.

53. *Def.* When a point Q moves uniformly in a circle, the perpendicular QP drawn from its position at any instant to a fixed diameter AA' of the circle, intersects the diameter in a point P, whose position changes by a *simple harmonic motion*.

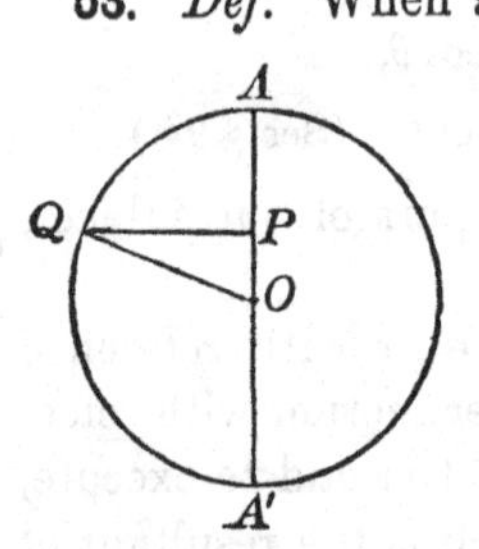

Thus, if a planet or satellite, or one of the constituents of a double star, supposed to move uniformly in a circular orbit about its primary, be viewed from a very distant position in the plane of its orbit, it will appear to move backwards and forwards in a straight line, with a simple harmonic motion. This is nearly the case with such bodies as the satellites of Jupiter when seen from the earth.

Physically, the interest of such motions consists in the fact of their being approximately those of the simplest vibrations of sounding bodies, such as a tuning-fork or pianoforte wire; whence their name; and of the various media in which waves of sound, light, heat, etc., are propagated.

54. The *Amplitude* of a simple harmonic motion is the range on one side or the other of the middle point of the course, *i.e.*, OA or OA' in the figure.

An arc of the circle referred to, measured from any fixed point to the uniformly moving point Q, is the *Argument* of the harmonic motion.

The distance of a point, performing a simple harmonic motion, from the middle of its course or range, is a *simple harmonic function of the time.* The *argument* of this function is what we have defined as the argument of the motion.

The *Epoch* in a simple harmonic motion is the interval of time which elapses from the era of reckoning till the moving point first comes to its greatest elongation in the direction reckoned as positive, from its mean position or the middle of its range. Epoch in angular measure is the angle described on the circle of reference in the period of time defined as the epoch.

The *Period* of a simple harmonic motion is the time which elapses from any instant until the moving point again moves in the same direction through the same position.

Simple harmonic motion.

The *Phase* of a simple harmonic motion at any instant is the fraction of the whole period which has elapsed since the moving point last passed through its middle position in the positive direction.

55. Those common kinds of mechanism, for producing rectilineal from circular motion, or *vice versa*, in which a crank moving in a circle works in a straight slot belonging to a body which can only move in a straight line, fulfil strictly the definition of a simple harmonic motion in the part of which the motion is rectilineal, if the motion of the rotating part is uniform.

Simple harmonic motion in mechanism.

The motion of the treadle in a spinning-wheel approximates to the same condition when the wheel moves uniformly; the approximation being the closer, the smaller is the angular motion of the treadle and of the connecting string. It is also approximated to more or less closely in the motion of the piston of a steam-engine connected, by any of the several methods in use, with the crank, provided always the rotatory motion of the crank be uniform.

56. The velocity of a point executing a simple harmonic motion is a simple harmonic function of the time, a quarter of a period earlier in phase than the displacement, and having its maximum value equal to the velocity in the circular motion by which the given function is defined.

Velocity in S. H. motion.

For, in the fig. of § 53, if V be the velocity in the circle, it may be represented by OQ in a direction perpendicular to its own, and therefore by OP and PQ in directions perpendicular to those lines. That is, the velocity of P in the simple harmonic motion is $\frac{V}{OQ} PQ$; which, when P is at O, becomes V.

57. The acceleration of a point executing a simple harmonic motion is at any time simply proportional to the displacement from the middle point, but in opposite direction, or always towards the middle point. Its maximum value is that with which a velocity equal to that of the circular motion would

Acceleration in S. H. motion.

Accelera-tion in S. H. motion.

be acquired in the time in which an arc equal to the radius is described.

For, in the fig. of § 53, the acceleration of Q (by § 35, a) is $\frac{V^2}{QO}$ along QO. Supposing, for a moment, QO to represent the magnitude of this acceleration, we may resolve it in QP, PO. The acceleration of P is therefore represented on the same scale by PO. Its magnitude is therefore $\frac{V^2}{QO} \cdot \frac{PO}{QO} = \frac{V^2}{QO^2} PO$, which is proportional to PO, and has at A its maximum value, $\frac{V^2}{QO}$, an acceleration under which the velocity V would be acquired in the time $\frac{QO}{V}$ as stated.

Let a be the amplitude, ϵ the epoch, and T the period, of a simple harmonic motion. Then if s be the displacement from middle position at time t, we have

$$s = a \cos\left(\frac{2\pi t}{T} - \epsilon\right).$$

Hence, for velocity, we have

$$v = \frac{ds}{dt} = -\frac{2\pi a}{T} \sin\left(\frac{2\pi t}{T} - \epsilon\right).$$

Hence V, the maximum value, is $\frac{2\pi a}{T}$, as above stated (§ 56).

Again, for acceleration,

$$\frac{dv}{dt} = -\frac{4\pi^2 a}{T^2} \cos\left(\frac{2\pi t}{T} - \epsilon\right) = -\frac{4\pi^2}{T^2} s. \quad \text{(See § 57.)}$$

Lastly, for the maximum value of the acceleration,

$$\frac{4\pi^2 a}{T^2} = \frac{V}{\frac{T}{2\pi}},$$

where, it may be remarked, $\frac{T}{2\pi}$ is the time of describing an arc equal to radius in the relative circular motion.

Composi-tion of S. H. M. in one line.

58. Any two simple harmonic motions in one line, and of one period, give, when compounded, a single simple harmonic motion; of the same period; of amplitude equal to the diagonal of a parallelogram described on lengths equal to their amplitudes measured on lines meeting at an angle equal to their difference of epochs; and of epoch differing from their epochs by angles equal to those which this diagonal makes with the two sides of the parallelogram. Let P and P' be two points executing simple harmonic motions of one period, and in one line $B'BCAA'$. Let Q and Q' be the uniformly moving points in the relative circles. On CQ and CQ' describe a parallelogram $SQCQ'$; and through S draw SR perpendicular to $B'A'$ produced. We have obviously $P'R=CP$ (being projections of the equal and parallel lines $Q'S, CQ$, on CR). Hence $CR=CP+CP'$; and therefore the point R executes the resultant of the motions P and P'. But CS, the diagonal of the parallelogram, is constant, and therefore the resultant motion is simple harmonic, of amplitude CS, and of epoch exceeding that of the motion of P, and falling short of that of the motion of P', by the angles QCS and SCQ' respectively.

Composition of S. H. M. in one line.

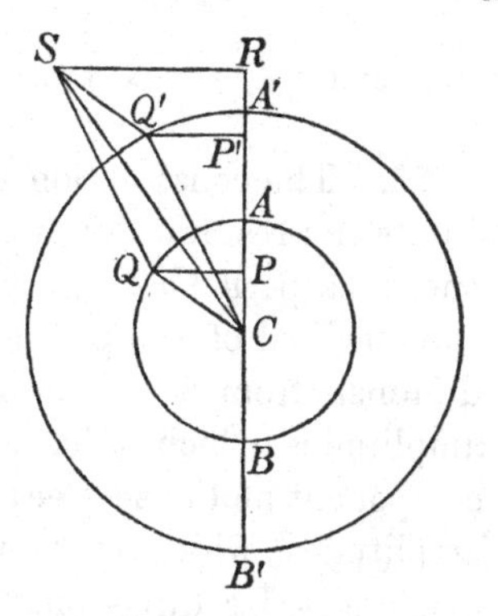

This geometrical construction has been usefully applied by the tidal committee of the British Association for a mechanical tide-indicator (compare § 60, below). An arm CQ' turning round C carries an arm $Q'S$ turning round Q'. Toothed wheels, one of them fixed with its axis through C, and the others pivoted on a framework carried by CQ', are so arranged that $Q'S$ turns very approximately at the rate of once round in 12 mean lunar hours, if CQ' be turned uniformly at the rate of once round in 12 mean solar hours. Days and half-days are marked by a counter geared to CQ'. The distance of S from a fixed line through C shows the deviation from mean sea-level due to the sum of mean solar and mean lunar tides for the time of day and year marked by CQ' and the counter.

An analytical proof of the same proposition is useful, being as follows:—

$$a\cos\left(\frac{2\pi t}{T}-\epsilon\right)+a'\cos\left(\frac{2\pi t}{T}-\epsilon'\right)$$

$$=(a\cos\epsilon+a'\cos\epsilon')\cos\frac{2\pi t}{T}+(a\sin\epsilon+a'\sin\epsilon')\sin\frac{2\pi t}{T}=r\cos\left(\frac{2\pi t}{T}-\theta\right),$$

Composition of S. H. M. in one line.

where $$r = \{(a\cos\epsilon + a'\cos\epsilon')^2 + (a\sin\epsilon + a'\sin\epsilon')^2\}^{\frac{1}{2}}$$
$$= \{a^2 + a'^2 + 2aa'\cos(\epsilon - \epsilon')\}^{\frac{1}{2}},$$

and $$\tan\theta = \frac{a\sin\epsilon + a'\sin\epsilon'}{a\cos\epsilon + a'\cos\epsilon'}.$$

59. The construction described in the preceding section exhibits the resultant of two simple harmonic motions, whether of the same period or not. Only, if they are not of the same period, the diagonal of the parallelogram will not be constant, but will diminish from a maximum value, the sum of the component amplitudes, which it has at the instant when the phases of the component motions agree; to a minimum, the difference of those amplitudes, which is its value when the phases differ by half a period. Its direction, which always must be nearer to the greater than to the less of the two radii constituting the sides of the parallelogram, will oscillate on each side of the greater radius to a maximum deviation amounting on either side to the angle whose sine is the less radius divided by the greater, and reached when the less radius deviates more than this by a quarter circumference from the greater. The full period of this oscillation is the time in which either radius gains a full turn on the other. The resultant motion is therefore not simple harmonic, but is, as it were, simple harmonic with periodically increasing and diminishing amplitude, and with periodical acceleration and retardation of phase. This view is particularly appropriate for the case in which the periods of the two component motions are nearly equal, but the amplitude of one of them much greater than that of the other.

To express the resultant motion, let s be the displacement at time t; and let a be the greater of the two component half-amplitudes.

$$s = a\cos(nt - \epsilon) + a'\cos(n't - \epsilon')$$
$$= a\cos(nt - \epsilon) + a'\cos(nt - \epsilon + \phi)$$
$$= (a + a'\cos\phi)\cos(nt - \epsilon) - a'\sin\phi\sin(nt - \epsilon),$$

if $$\phi = (n't - \epsilon') - (nt - \epsilon);$$

or, finally, $$s = r\cos(nt - \epsilon + \theta),$$

Composition of S. H. M. in one line.

if $$r = (a^2 + 2aa' \cos\phi + a'^2)^{\frac{1}{2}}$$

and $$\tan\theta = \frac{a' \sin\phi}{a + a'\cos\phi}.$$

The maximum value of $\tan\theta$ in the last of these equations is found by making $\phi = \frac{\pi}{2} + \sin^{-1}\frac{a'}{a}$, and is equal to $\frac{a'}{(a^2 - a'^2)^{\frac{1}{2}}}$, and hence the maximum value of θ itself is $\sin^{-1}\frac{a'}{a}$. The geometrical methods indicated above (§ 58) lead to this conclusion by the following very simple construction.

To find the time and the amount of the maximum acceleration or retardation of phase, let CA be the greater half-amplitude. From A as centre, with the less half-amplitude as radius, describe a circle. CB touching this circle is the generating radius of the most deviated resultant. Hence CBA is a right angle; and

$$\sin BCA = \frac{AB}{CA}.$$

Examples of composition of S. H. M. in one line.

60. A most interesting application of this case of the composition of harmonic motions is to the lunar and solar tides; which, except in tidal rivers, or long channels, or deep bays, follow each very nearly the simple harmonic law, and produce, as the actual result, a variation of level equal to the sum of variations that would be produced by the two causes separately.

The amount of the lunar equilibrium-tide (§ 812) is about 2·1 times that of the solar. Hence, if the actual tides conformed to the equilibrium theory, the spring tides would be 3·1, and the neap tides only 1·1, each reckoned in terms of the solar tide; and at spring and neap tides the hour of high water is that of the lunar tide alone. The greatest deviation of the actual tide from the phases (high, low, or mean water) of the lunar tide alone, would be about ·95 of a lunar hour, that is, ·98 of a solar hour (being the same part of 12 lunar hours that $28^0\ 26'$, or the angle whose sine is $\frac{1}{2 \cdot 1}$, is of 360^0). This maximum deviation would be in advance or in arrear according as the crown of the solar tide precedes or follows the crown of the lunar tide; and it would be exactly reached when the interval of phase between

Examples of composition of S. H. M. in one line.

the two component tides is 3·95 lunar hours. That is to say, there would be maximum advance of the time of high water $4\frac{1}{2}$ days after, and maximum retardation the same number of days before, spring tides (compare § 811).

61. We may consider next the case of equal amplitudes in the two given motions. If their periods are equal, their resultant is a simple harmonic motion, whose phase is at every instant the mean of their phases, and whose amplitude is equal to twice the amplitude of either multiplied by the cosine of half the difference of their phases. The resultant is of course nothing when their phases differ by half the period, and is a motion of double amplitude and of phase the same as theirs when they are of the same phase.

When their periods are very nearly, but not quite, equal (their amplitudes being still supposed equal), the motion passes very slowly from the former (zero, or no motion at all) to the latter, and back, in a time equal to that in which the faster has gone once oftener through its period than the slower has.

In practice we meet with many excellent examples of this case, which will, however, be more conveniently treated of when we come to apply kinetic principles to various subjects in acoustics, physical optics, and practical mechanics; such as the sympathy of pendulums or tuning-forks, the rolling of a turret ship at sea, the marching of troops over a suspension bridge, etc.

Mechanism for compounding S. H. motions in one line.

62. If any number of pulleys be so placed that a cord passing from a fixed point half round each of them has its free parts all in parallel lines, and if their centres be moved with simple harmonic motions of any ranges and any periods in lines parallel to those lines, the unattached end of the cord moves with a complex harmonic motion equal to twice the sum of the given simple harmonic motions. This is the principle of Sir W. Thomson's tide-predicting machine, constructed by the British Association, and ordered to be placed in South Kensington Museum, availably for general use in calculating beforehand for any port or other place on the sea for which the simple harmonic constituents of the tide have been determined by the "harmonic analysis" applied to previous observations*. We may exhibit, graphically, any case of single or compound simple harmonic motion in one line by curves in which the abscissæ represent intervals of time, and the

Graphical representation of harmonic motions in one line.

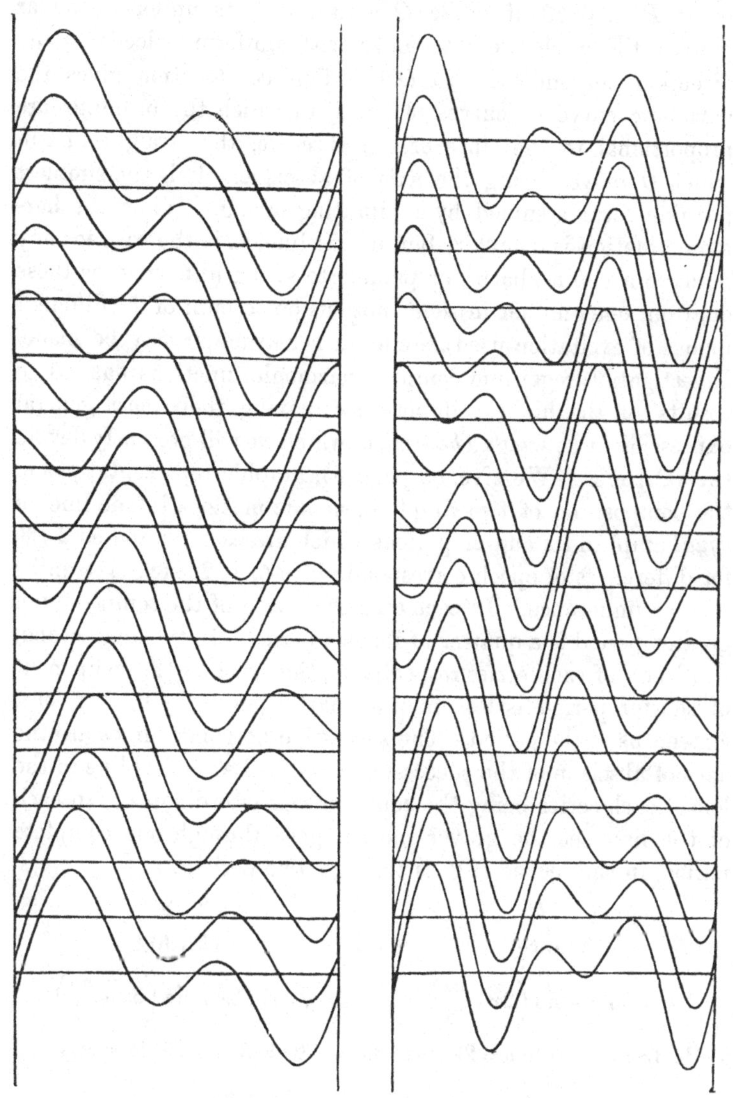

* See British Association Tidal Committee's Reports, 1868, 1872, 1875: or *Lecture on Tides*, by Sir W. Thomson (Collins, Glasgow, 1876).

Graphical representation of harmonic motions in one line.

ordinates the corresponding distances of the moving point from its mean position. In the case of a single simple harmonic motion, the corresponding curve would be that described by the point P in § 53, if, while Q maintained its uniform circular motion, the circle were to move with uniform velocity in any direction perpendicular to OA. This construction gives the harmonic curve, or curves of sines, in which the ordinates are proportional to the sines of the abscissæ, the straight line in which O moves being the axis of abscissæ. It is the simplest possible form assumed by a vibrating string. When the harmonic motion is complex, but in one line, as is the case for any point in a violin-, harp-, or pianoforte-string (differing, as these do, from one another in their motions on account of the different modes of excitation used), a similar construction may be made. Investigation regarding complex harmonic functions has led to results of the highest importance, having their most general expression in *Fourier's Theorem*, to which we will presently devote several pages. We give, on page 45, graphic representations of the composition of two simple harmonic motions in one line, of *equal* amplitudes and of periods which are as 1 : 2 and as 2 : 3, for differences of epoch corresponding to 0, 1, 2, etc., sixteenths of a circumference. In each case the epoch of the component of greater period is a quarter of its own period. In the first, second, third, etc., of each series respectively, the epoch of the component of shorter period is less than a quarter-period by 0, 1, 2, etc., sixteenths of the period. The successive horizontal lines are the axes of abscissæ of the successive curves; the vertical line to the left of each series being the common axis of ordinates. In each of the first set the graver motion goes through one complete period, in the second it goes through two periods.

1 : 2	2 : 3
(Octave)	(Fifth)
$y = \sin x + \sin\left(2x + \frac{n\pi}{8}\right)$.	$y = \sin 2x + \sin\left(3x + \frac{n\pi}{8}\right)$.

Both, from $x = 0$ to $x = 2\pi$; and for $n = 0, 1, 2 \ldots\ldots 15$, in succession.

These, and similar cases, when the periodic times are not commensurable, will be again treated of under Acoustics.

S. H. motions in different directions.

63. We have next to consider the composition of simple harmonic motions in different directions. In the first place, we see that any number of simple harmonic motions of one period, and of the same phase, superimposed, produce a single simple harmonic motion of the same phase. For, the displacement at any instant being, according to the principle of the composition of motions, the geometrical resultant (see above, § 50) of the displacements due to the component motions separately, these component displacements, in the case supposed, all vary in simple proportion to one another, and are in constant directions. Hence the resultant displacement will vary in simple proportion to each of them, and will be in a constant direction.

But if, while their periods are the same, the phases of the several component motions do not agree, the resultant motion will generally be elliptic, with equal areas described in equal times by the radius-vector from the centre; although in particular cases it may be uniform circular, or, on the other hand, rectilineal and simple harmonic.

64. To prove this, we may first consider the case in which we have two equal simple harmonic motions given, and these in perpendicular lines, and differing in phase by a quarter period. Their resultant is a uniform circular motion. For, let BA, $B'A'$ be their ranges; and from O, their common middle point, as centre, describe a circle through $AA'BB'$. The given motion of P in BA will be (§ 53) defined by the motion of a point Q round the circumference of this circle; and the same point, if moving in the direction indicated by the arrow, will give a simple harmonic motion of P', in $B'A'$, a quarter of a period behind that of the motion of P in BA. But, since $A'OA$, QPO, and $QP'O$ are right angles, the figure $QP'OP$ is a parallelogram, and therefore Q is in the position of the displacement compounded of OP and OP'. Hence two equal simple harmonic motions in perpendicular lines, of phases differing by a quarter period, are equivalent to a uniform circular motion of radius equal to the maximum displacement of either singly, and in the direction from the positive end of the range of

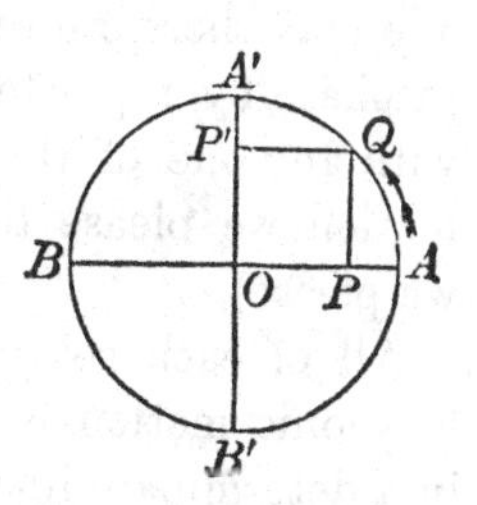

S. H. motions in different directions.

the component in advance of the other towards the positive end of the range of this latter.

65. Now, orthogonal projections of simple harmonic motions are clearly simple harmonic with unchanged phase. Hence, if we project the case of § 64 on any plane, we get motion in an ellipse, of which the projections of the two component ranges are conjugate diameters, and in which the radius-vector from the centre describes equal areas (being the projections of the areas described by the radius of the circle) in equal times. But the plane and position of the circle of which this projection is taken may clearly be found so as to fulfil the condition of having the projections of the ranges coincident with any two given mutually bisecting lines. Hence any two given simple harmonic motions, equal or unequal in range, and oblique or at right angles to one another in direction, provided only they differ by a quarter period in phase, produce elliptic motion, having their ranges for conjugate axes, and describing, by the radius-vector from the centre, equal areas in equal times (compare § 34, *b*).

66. Returning to the composition of any number of simple harmonic motions of one period, in lines in all directions and of all phases: each component simple harmonic motion may be determinately resolved into two in the same line, differing in phase by a quarter period, and one of them having any given epoch. We may therefore reduce the given motions to two sets, differing in phase by a quarter period, those of one set agreeing in phase with any one of the given, or with any other simple harmonic motion we please to choose (*i.e.*, having their epoch anything we please).

All of each set may (§ 58) be compounded into one simple harmonic motion of the same phase, of determinate amplitude, in a determinate line; and thus the whole system is reduced to two simple fully determined harmonic motions differing from one another in phase by a quarter period.

Now the resultant of two simple harmonic motions, one a quarter of a period in advance of the other, in different lines, has been proved (§ 65) to be motion in an ellipse of which the ranges of the component motions are conjugate axes, and in which equal areas are described by the radius-vector from the centre in equal times. Hence the general proposition of § 63.

S. H. motions in different directions.

Let
$$\left.\begin{aligned} x_1 &= l_1 a_1 \cos(\omega t - \epsilon_1), \\ y_1 &= m_1 a_1 \cos(\omega t - \epsilon_1), \\ z_1 &= n_1 a_1 \cos(\omega t - \epsilon_1), \end{aligned}\right\} \quad \text{.......................(1)}$$

be the Cartesian specification of the first of the given motions; and so with varied suffixes for the others;

l, m, n denoting the direction cosines,

a „ „ half amplitude,

ϵ „ „ epoch,

the proper suffix being attached to each letter to apply it to each case, and ω denoting the common relative angular velocity. The resultant motion, specified by x, y, z without suffixes, is

$$x = \Sigma l_1 a_1 \cos(\omega t - \epsilon_1) = \cos \omega t \Sigma l_1 a_1 \cos \epsilon_1 + \sin \omega t \Sigma l_1 a_1 \sin \epsilon_1,$$
$$y = \text{etc.}; \quad z = \text{etc.};$$

or, as we may write for brevity,

$$\left.\begin{aligned} x &= P \cos \omega t + P' \sin \omega t, \\ y &= Q \cos \omega t + Q' \sin \omega t, \\ z &= R \cos \omega t + R' \sin \omega t, \end{aligned}\right\} \quad \text{.......................(2)}$$

where
$$\left.\begin{aligned} P &= \Sigma\, l_1 a_1 \cos \epsilon_1, & P' &= \Sigma\, l_1 a_1 \sin \epsilon_1, \\ Q &= \Sigma m_1 a_1 \cos \epsilon_1, & Q' &= \Sigma m_1 a_1 \sin \epsilon_1, \\ R &= \Sigma\, n_1 a_1 \cos \epsilon_1, & R' &= \Sigma\, n_1 a_1 \sin \epsilon_1. \end{aligned}\right\} \quad \text{.............(3)}$$

The resultant motion thus specified, in terms of six component simple harmonic motions, may be reduced to two by compounding P, Q, R, and P', Q', R', in the elementary way. Thus if

$$\left.\begin{aligned} &\zeta = (P^2 + Q^2 + R^2)^{\frac{1}{2}}, \\ &\lambda = \frac{P}{\zeta}, \quad \mu = \frac{Q}{\zeta}, \quad \nu = \frac{R}{\zeta}, \\ &\zeta' = (P'^2 + Q'^2 + R'^2)^{\frac{1}{2}}, \\ &\lambda' = \frac{P'}{\zeta'}, \quad \mu' = \frac{Q'}{\zeta'}, \quad \nu' = \frac{R'}{\zeta'}, \end{aligned}\right\} \quad \text{..........................(4)}$$

the required motion will be the resultant of $\zeta \cos \omega t$ in the line (λ, μ, ν), and $\zeta' \sin \omega t$ in the line (λ', μ', ν'). It is therefore motion in an ellipse, of which 2ζ and $2\zeta'$ in those directions are

S. H. motions in different directions.

conjugate diameters; with radius-vector from centre tracing equal areas in equal times; and of period $\frac{2\pi}{\omega}$.

H. motions of different kinds and in different lines.

67. We must next take the case of the composition of simple harmonic motions of *different* periods and in different lines. In general, whether these lines be in one plane or not, the line of motion returns into itself if the periods are commensurable; and if not, not. This is evident without proof.

If a be the amplitude, ϵ the epoch, and n the angular velocity in the relative circular motion, for a component in a line whose direction cosines are λ, μ, ν—and if ξ, η, ζ be the co-ordinates in the resultant motion,

$$\xi = \Sigma . \lambda_1 a_1 \cos (n_1 t - \epsilon_1), \quad \eta = \Sigma . \mu_1 a_1 \cos (n_1 t - \epsilon_1), \quad \zeta = \Sigma . \nu_1 a_1 \cos (n_1 t - \epsilon_1).$$

Now it is evident that at time $t + T$ the values of ξ, η, ζ will recur as soon as $n_1 T$, $n_2 T$, etc., are multiples of 2π, that is, when T is the least common multiple of $\frac{2\pi}{n_1}$, $\frac{2\pi}{n_2}$, etc.

If there be such a common multiple, the trigonometrical functions may be eliminated, and the equations (or equation, if the motion is in one plane) to the path are algebraic. If not, they are transcendental.

68. From the above we see generally that the composition of any number of simple harmonic motions in any directions and of any periods, may be effected by compounding, according to previously explained methods, their resolved parts in each of any three rectangular directions, and then compounding the final resultants in these directions.

S. H. motions in two rectangular directions.

69. By far the most interesting case, and the simplest, is that of *two* simple harmonic motions of any periods, whose directions must of course be in one plane.

Mechanical methods of obtaining such combinations will be afterwards described, as well as cases of their occurrence in Optics and Acoustics.

We may suppose, for simplicity, the two component motions to take place in perpendicular directions. Also, as we can only have a re-entering curve when their periods are commensurable, it will be advisable to commence with such a case.

The following figures represent the paths produced by the combination of simple harmonic motions of *equal* amplitude in two rectangular directions, the periods of the components being as 1 : 2, and the epochs differing successively by 0, 1, 2, etc., sixteenths of a circumference.

S. H. motions in two rectangular directions.

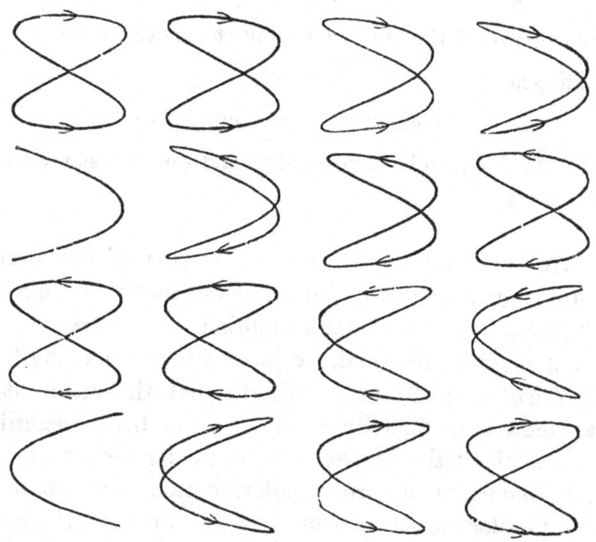

In the case of epochs equal, or differing by a multiple of π, the curve is a portion of a parabola, and is gone over twice in opposite directions by the moving point in each complete period.

For the case figured above,

$$x = a\cos(2nt - \epsilon), \quad y = a\cos nt.$$

Hence $$x = a\{\cos 2nt \cos\epsilon + \sin 2nt \sin\epsilon\}$$

$$= a\left\{\left(\frac{2y^2}{a^2} - 1\right)\cos\epsilon + 2\frac{y}{a}\sqrt{1 - \frac{y^2}{a^2}}\sin\epsilon\right\},$$

which for any given value of ϵ is the equation of the corresponding curve. Thus for $\epsilon = 0$,

$$\frac{x}{a} = \frac{2y^2}{a^2} - 1, \text{ or } y^2 = \frac{a}{2}(x + a), \text{ the parabola as above.}$$

S. H. motions in two rectangular directions.

For $\epsilon = \frac{\pi}{2}$ we have $\frac{x}{a} = 2\frac{y}{a}\sqrt{1-\frac{y^2}{a^2}}$, or $a^2x^2 = 4y^2(a^2 - y^2)$,

the equation of the 5th and 13th of the above curves.

In general

$$x = a\cos(nt+\epsilon), \quad y = a\cos(n_1t+\epsilon_1),$$

from which t is to be eliminated to find the Cartesian equation of the curve.

Composition of two uniform circular motions.

70. Another very important case is that of two groups of two simple harmonic motions in one plane, such that the resultant of each group is uniform circular motion.

If their periods are equal, we have a case belonging to those already treated (§ 63), and conclude that the resultant is, in general, motion in an ellipse, equal areas being described in equal times about the centre. As particular cases we may have simple harmonic, or uniform circular, motion. (Compare § 91.)

If the circular motions are in the *same* direction, the resultant is evidently circular motion in the same direction. This is the case of the motion of S in § 58, and requires no further comment, as its amplitude, epoch, etc., are seen at once from the figure.

71. If the periods of the two are very nearly equal, the resultant motion will be at any moment very nearly the circular motion given by the preceding construction. Or we may regard it as rigorously a motion in a circle with a varying radius decreasing from a maximum value, the sum of the radii of the two component motions, to a minimum, their difference, and increasing again, alternately; the direction of the resultant radius oscillating on each side of that of the greater component (as in corresponding case, § 59, above). Hence the angular velocity of the resultant motion is periodically variable. In the case of equal radii, next considered, it is constant.

72. When the radii of the two component motions are equal, we have the very interesting and important case figured below. Here the resultant radius bisects the angle between the component radii. The resultant angular velocity is the arithmetical mean of its components. We will explain in a future section

(§ 94) how this epitrochoid is traced by the rolling of one circle

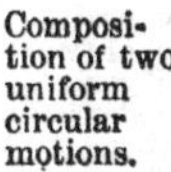
Composition of two uniform circular motions.

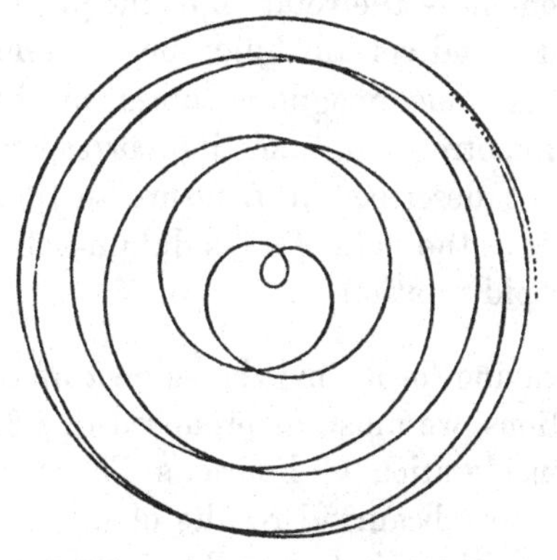

on another. (The particular case above delineated is that of a non-reëntrant curve.)

73. Let the uniform circular motions be in *opposite* directions; then, if the periods are equal, we may easily see, as before, § 66, that the resultant is in general elliptic motion, including the particular cases of uniform circular, and simple harmonic, motion.

If the periods are very nearly equal, the resultant will be easily found, as in the case of § 59.

74. If the radii of the component motions are equal, we have cases of *very* great importance in modern physics, one of which is figured below (like the preceding, a non-reëntrant curve).

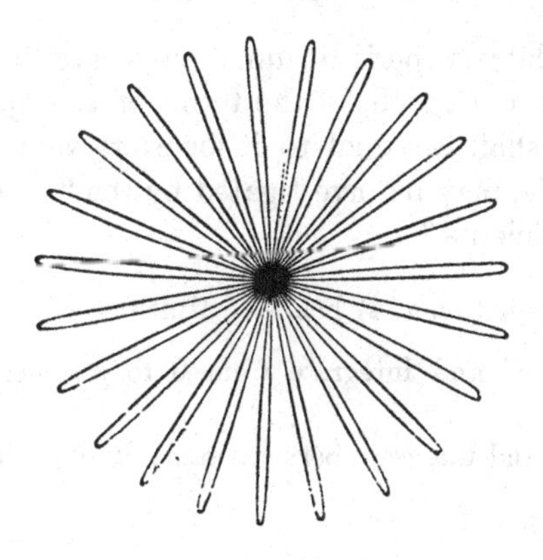

Composition of two uniform circular motions.

This is intimately connected with the explanation of two sets of important phenomena,—the rotation of the plane of polarization of light, by quartz and certain fluids on the one hand, and by transparent bodies under magnetic forces on the other. It is a case of the hypotrochoid, and its corresponding mode of description will be described in a future section. It will also appear in kinetics as the path of a pendulum-bob which contains a gyroscope in rapid rotation.

Fourier's Theorem.

75. Before leaving for a time the subject of the composition of harmonic motions, we must, as promised in § 62, devote some pages to the consideration of Fourier's Theorem, which is not only one of the most beautiful results of modern analysis, but may be said to furnish an indispensable instrument in the treatment of nearly every recondite question in modern physics. To mention only sonorous vibrations, the propagation of electric signals along a telegraph wire, and the conduction of heat by the earth's crust, as subjects in their generality intractable without it, is to give but a feeble idea of its importance. The following seems to be the most intelligible form in which it can be presented to the general reader:—

THEOREM.—*A complex harmonic function, with a constant term added, is the proper expression, in mathematical language, for any arbitrary periodic function; and consequently can express any function whatever between definite values of the variable.*

76. Any arbitrary periodic function whatever being given, the amplitudes and epochs of the terms of a complex harmonic function which shall be equal to it for every value of the independent variable, may be investigated by the "method of indeterminate coefficients."

Assume equation (14) below. Multiply both members first by $\cos \frac{2i\pi\xi}{p} d\xi$ and integrate from 0 to p: then multiply by $\sin \frac{2i\pi\xi}{p} d\xi$ and integrate between same limits. Thus instantly you find (13).

This investigation is sufficient as a solution of the problem, —to find a complex harmonic function expressing a given arbitrary periodic function,—when once we are assured that the problem is possible; and when we have this assurance, it proves that the resolution is determinate; that is to say, that no other complex harmonic function than the one we have found can satisfy the conditions.

Fourier's Theorem.

For description of an integrating machine by which the coefficients A_i, B_i in the Fourier expression (14) for any given arbitrary function may be obtained with exceedingly little labour, and with all the accuracy practically needed for the harmonic analysis of tidal and meteorological observations, see Proceedings of the Royal Society, Feb. 1876, or Chap. v. below.

77. The full theory of the expression investigated in § 76 will be made more intelligible by an investigation from a different point of view.

Let $F(x)$ be any periodic function, of period p. That is to say, let $F(x)$ be any function fulfilling the condition

$$F(x+ip) = F(x) \quad \text{..........................} \quad (1),$$

where i denotes any positive or negative integer. Consider the integral

$$\int_{c'}^{c} \frac{F(x)\,dx}{a^2+x^2},$$

where a, c, c' denote any three given quantities. Its value is less than $F(z)\int_{c'}^{c}\frac{dx}{a^2+x^2}$, and greater than $F(z')\int_{c'}^{c}\frac{dx}{a^2+x^2}$, if z and z' denote the values of x, either equal to or intermediate between the limits c and c', for which $F(x)$ is greatest and least respectively. But

$$\int_{c'}^{c}\frac{dx}{a^2+x^2} = \frac{1}{a}\left(\tan^{-1}\frac{c}{a} - \tan^{-1}\frac{c'}{a}\right) \quad \text{..................} \quad (2),$$

and therefore

$$\left.\begin{aligned} \int_{c'}^{c}\frac{F(x)\,a\,dx}{a^2+x^2} &< F(z)\left(\tan^{-1}\frac{c}{a} - \tan^{-1}\frac{c'}{a}\right), \\ \text{and} \qquad ,, \qquad &> F(z')\left(\tan^{-1}\frac{c}{a} - \tan^{-1}\frac{c'}{a}\right). \end{aligned}\right\} \quad \text{..........} \quad (3)$$

Fourier's Theorem.

Hence if A be the greatest of all the values of $F(x)$, and B the least,

$$\left.\begin{aligned}\int_c^\infty \frac{F(x)\,adx}{a^2+x^2} &< A\left(\frac{\pi}{2}-\tan^{-1}\frac{c}{a}\right),\\ \text{and}\qquad ,, \quad &> B\left(\frac{\pi}{2}-\tan^{-1}\frac{c}{a}\right).\end{aligned}\right\}\quad\text{..................(4)}$$

Also, similarly,

$$\left.\begin{aligned}\int_{-\infty}^{c'} \frac{F(x)\,adx}{a^2+x^2} &< A\left(\tan^{-1}\frac{c'}{a}+\frac{\pi}{2}\right),\\ \text{and}\qquad ,, \quad &> B\left(\tan^{-1}\frac{c'}{a}+\frac{\pi}{2}\right).\end{aligned}\right\}\quad\text{..................(5)}$$

Adding the first members of (3), (4), and (5), and comparing with the corresponding sums of the second members, we find

$$\left.\begin{aligned}\int_{-\infty}^\infty \frac{F(x)\,adx}{a^2+x^2} &< F(z)\left(\tan^{-1}\frac{c}{a}-\tan^{-1}\frac{c'}{a}\right)+A\left(\pi-\tan^{-1}\frac{c}{a}+\tan^{-1}\frac{c'}{a}\right),\\ \text{and}\quad ,, \quad &> F(z')\left(\tan^{-1}\frac{c}{a}-\tan^{-1}\frac{c'}{a}\right)+B\left(\pi-\tan^{-1}\frac{c}{a}+\tan^{-1}\frac{c'}{a}\right).\end{aligned}\right\}(6)$$

But, by (1),

$$\int_{-\infty}^\infty \frac{F(x)\,dx}{a^2+x^2} = \int_0^p F(x)\,dx\left\{\Sigma_{i=-\infty}^{i=\infty}\left(\frac{1}{a^2+(x+ip)^2}\right)\right\}\quad\text{..........(7).}$$

Now if we denote $\sqrt{-1}$ by v,

$$\frac{1}{a^2+(x+ip)^2} = \frac{1}{2av}\left(\frac{1}{x+ip-av}-\frac{1}{x+ip+av}\right),$$

and therefore, taking the terms corresponding to positive and equal negative values of i together, and the terms for $i=0$ separately, we have

$$\begin{aligned}\Sigma_{i=-\infty}^{i=\infty}\left(\frac{1}{a^2+(x+ip)^2}\right) &= \frac{1}{2av}\left\{\frac{1}{x-av}-2\Sigma_{i=1}^{i=\infty}\frac{x-av}{i^2p^2-(x-av)^2}\right.\\ &\qquad\left.-\frac{1}{x+av}+2\Sigma_{i=1}^{i=\infty}\frac{x+av}{i^2p^2-(x+av)^2}\right\}\\ &= \frac{\pi}{2apv}\left\{\cot\frac{\pi(x-av)}{p}-\cot\frac{\pi(x+av)}{p}\right\}\\ &= \frac{\dfrac{\pi}{2apv}\sin\dfrac{2\pi av}{p}}{\cos^2\dfrac{\pi av}{p}-\cos^2\dfrac{\pi x}{p}} = \frac{\dfrac{\pi}{apv}\sin\dfrac{2\pi av}{p}}{\cos\dfrac{2\pi av}{p}-\cos\dfrac{2\pi x}{p}}\\ &= \frac{\pi}{ap}\,\frac{\epsilon^{\frac{2\pi a}{p}}-\epsilon^{-\frac{2\pi a}{p}}}{\epsilon^{\frac{2\pi a}{p}}-2\cos\dfrac{2\pi x}{p}+\epsilon^{-\frac{2\pi a}{p}}}.\end{aligned}$$

Fourier's Theorem.

Hence,

$$\int_{-\infty}^{\infty}\frac{F(x)dx}{a^2+x^2}=\frac{\pi}{ap}\left(\epsilon^{\frac{2\pi a}{p}}-\epsilon^{-\frac{2\pi a}{p}}\right)\int_0^p\frac{F(x)dx}{\epsilon^{\frac{2\pi a}{p}}-2\cos\frac{2\pi x}{p}+\epsilon^{-\frac{2\pi a}{p}}}\ldots\ldots(8).$$

Next, denoting temporarily, for brevity, $\epsilon^{\frac{2\pi x v}{p}}$ by ζ, and putting

$$\epsilon^{-\frac{2\pi a}{p}}=e\ldots\ldots\ldots\ldots\ldots\ldots\ldots\ldots\ldots\ldots(9),$$

we have
$$\frac{1}{\epsilon^{\frac{2\pi a}{p}}-2\cos\frac{2\pi x}{p}+\epsilon^{-\frac{2\pi a}{p}}}=\frac{e}{1-e(\zeta+\zeta^{-1})+e^2}$$

$$=\frac{e}{1-e^2}\left(\frac{1}{1-e\zeta}+\frac{1}{1-e\zeta^{-1}}-1\right)$$

$$=\frac{e}{1-e^2}\{1+e(\zeta+\zeta^{-1})+e^2(\zeta^2+\zeta^{-2})+e^3(\zeta^3+\zeta^{-3})+\text{etc.}\}$$

$$=\frac{e}{1-e^2}\left(1+2e\cos\frac{2\pi x}{p}+2e^2\cos\frac{4\pi x}{p}+2e^3\cos\frac{6\pi x}{p}+\text{etc.}\right).$$

Hence, according to (8) and (9),

$$\int_{-\infty}^{\infty}\frac{F(x)dx}{a^2+x^2}=\frac{\pi}{ap}\int_0^p F(x)dx\left(1+2e\cos\frac{2\pi x}{p}+2e^2\cos\frac{4\pi x}{p}+\text{etc.}\right)\ldots(10).$$

Hence, by (6), we infer that

$$F(z)\left(\tan^{-1}\frac{c}{a}-\tan^{-1}\frac{c'}{a}\right)+A\left(\pi-\tan^{-1}\frac{c}{a}+\tan^{-1}\frac{c'}{a}\right)>$$

and
$$F(z')\left(\tan^{-1}\frac{c}{a}-\tan^{-1}\frac{c'}{a}\right)+B\left(\pi-\tan^{-1}\frac{c}{a}+\tan^{-1}\frac{c'}{a}\right)<$$

$$\frac{\pi}{p}\int_0^p F(x)dx\left(1+2e\cos\frac{2\pi x}{p}+\text{etc.}\right).$$

Now let $c'=-c$, and $x=\xi'-\xi$,

ξ' being a variable, and ξ constant, so far as the integration is concerned; and let

$$F(x)=\phi(x+\xi)=\phi(\xi'),$$

and therefore
$$F(z)=\phi(\xi+z),$$
$$F(z')=\phi(\xi+z').$$

Fourier's Theorem.

The preceding pair of inequalities becomes

$$\left.\begin{array}{l}\phi(\xi+z)\,.\,2\tan^{-1}\dfrac{c}{a}+A\left(\pi-2\tan^{-1}\dfrac{c}{a}\right)>\\ \text{and}\quad \phi(\xi+z')\,.\,2\tan^{-1}\dfrac{c}{a}+B\left(\pi-2\tan^{-1}\dfrac{c}{a}\right)<\\ \dfrac{\pi}{p}\left\{\displaystyle\int_0^p\phi(\xi')\,d\xi'+2\Sigma_{i=1}^{i=\infty}e^i\int_0^p\phi(\xi')\,d\xi'\cos\frac{2i\pi(\xi'-\xi)}{p}\right\},\end{array}\right\}\quad\ldots\ldots(11)$$

where ϕ denotes any periodic function whatever, of period p.

Now let c be a very small fraction of p. In the limit, where c is infinitely small, the greatest and least values of $\phi(\xi')$ for values of ξ' between $\xi+c$ and $\xi-c$ will be infinitely nearly equal to one another and to $\phi(\xi)$; that is to say,

$$\phi(\xi+z)=\phi(\xi+z')=\phi(\xi).$$

Next, let a be an infinitely small fraction of c. In the limit

$$\tan^{-1}\frac{c}{a}=\frac{\pi}{2},$$

and

$$e=\epsilon^{-\frac{2\pi a}{p}}=1.$$

Hence the comparison (11) becomes in the limit an equation which, if we divide both members by π, gives

$$\phi(\xi)=\frac{1}{p}\left\{\int_0^p\phi(\xi')\,d\xi'+2\Sigma_{i=1}^{i=\infty}\int_0^p\phi(\xi')\,d\xi'\cos\frac{2i\pi(\xi'-\xi)}{p}\right\}\ldots(12).$$

This is the celebrated theorem discovered by Fourier* for the development of an arbitrary periodic function in a series of simple harmonic terms. A formula included in it as a particular case had been given previously by Lagrange†.

If, for $\cos\dfrac{2i\pi(\xi'-\xi)}{p}$, we take its value

$$\cos\frac{2i\pi\xi'}{p}\cos\frac{2i\pi\xi}{p}+\sin\frac{2i\pi\xi'}{p}\sin\frac{2i\pi\xi}{p}$$

and introduce the following notation :—

$$\left.\begin{array}{l}A_0=\dfrac{1}{p}\displaystyle\int_0^p\phi(\xi)\,d\xi,\\ A_i=\dfrac{2}{p}\displaystyle\int_0^p\phi(\xi)\cos\frac{2i\pi\xi}{p}\,d\xi,\\ B_i=\dfrac{2}{p}\displaystyle\int_0^p\phi(\xi)\sin\frac{2i\pi\xi}{p}\,d\xi,\end{array}\right\}\quad\ldots\ldots\ldots\ldots\ldots\ldots\ldots(13)$$

* *Théorie analytique de la Chaleur.* Paris, 1822.

† *Anciens Mémoires de l'Académie de Turin.*

we reduce (12) to this form :—

Fourier's Theorem.

$$\phi(\xi) = A_0 + \Sigma_{i=1}^{i=\infty} A_i \cos\frac{2i\pi\xi}{p} + \Sigma_{i=1}^{i=\infty} B_i \sin\frac{2i\pi\xi}{p} \dots\dots\dots (14),$$

which is the general expression of an arbitrary function in terms of a series of cosines and of sines. Or if we take

$$P_i = (A_i^2 + B_i^2)^{\frac{1}{2}}, \quad \text{and } \tan\epsilon_i = \frac{B_i}{A_i} \dots\dots\dots\dots\dots\dots (15),$$

we have $\phi(\xi) = A_0 + \Sigma_{i=1}^{i=\infty} P_i \cos\left(\frac{2i\pi\xi}{p} - \epsilon_i\right) \dots\dots\dots\dots\dots\dots (16),$

which is the general expression in a series of single simple harmonic terms of the successive multiple periods.

Convergency of Fourier's series.

Each of the equations and comparisons (2), (7), (8), (10), and (11) is a true arithmetical expression, and may be verified by actual calculation of the numbers, for any particular case; provided only that $F(x)$ has no infinite value in its period. Hence, with this exception, (12) or either of its equivalents, (14), (16), is a true arithmetical expression; and the series which it involves is therefore convergent. Hence we may with perfect rigour conclude that even the extreme case in which the arbitrary function experiences an abrupt finite change in its value when the independent variable, increasing continuously, passes through some particular value or values, is included in the general theorem. In such a case, if any value be given to the independent variable differing however little from one which corresponds to an abrupt change in the value of the function, the series must, as we may infer from the preceding investigation, converge and give a definite value for the function. But if exactly the critical value is assigned to the independent variable, the series cannot converge to any definite value. The consideration of the limiting values shown in the comparison (11) does away with all difficulty in understanding how the series (12) gives definite values having a finite difference for two particular values of the independent variable on the two sides of a critical value, but differing infinitely little from one another.

If the differential coefficient $\frac{d\phi(\xi)}{d\xi}$ is finite for every value of ξ within the period, it too is arithmetically expressible by a series of harmonic terms, which cannot be other than the series obtained by differentiating the series for $\phi(\xi)$. Hence

Convergency of Fourier's series.

$$\frac{d\phi(\xi)}{d\xi} = -\frac{2\pi}{p}\Sigma_{i=1}^{i=\infty} iP_i \sin\left(\frac{2i\pi\xi}{p} - \epsilon_i\right)\ldots\ldots\ldots\ldots\ldots\ldots(17),$$

and this series is convergent; and we may therefore conclude that the series for $\phi(\xi)$ is more convergent than a harmonic series with

$$1,\ \tfrac{1}{2},\ \tfrac{1}{3},\ \tfrac{1}{4},\ \text{etc.},$$

for its coefficients. If $\frac{d^2\phi(\xi)}{d\xi^2}$ has no infinite values within the period, we may differentiate both members of (17) and still have an equation arithmetically true; and so on. We conclude that if the nth differential coefficient of $\phi(\xi)$ has no infinite values, the harmonic series for $\phi(\xi)$ must converge more rapidly than a harmonic series with

$$1,\ \frac{1}{2^n},\ \frac{1}{3^n},\ \frac{1}{4^n},\ \text{etc.},$$

for its coefficients.

Displacement of a rigid body.

78. We now pass to the consideration of the displacement of a rigid body or group of points whose relative positions are unalterable. The simplest case we can consider is that of the motion of a plane figure in its own plane, and this, as far as kinematics is concerned, is entirely summed up in the result of the next section.

Displacements of a plane figure in its plane.

79. If a plane figure be displaced in any way in its own plane, there is always (with an exception treated in § 81) one point of it common to any two positions; that is, it may be moved from any one position to any other by rotation in its own plane about one point held fixed.

To prove this, let A, B be any two points of the plane figure in its first position, A', B' the positions of the same two after a displacement. The lines AA', BB' will not be parallel, except in one case to be presently considered. Hence the line equidistant from A and A' will meet that equidistant from B and B' in some point O. Join OA, OB, OA', OB'. Then, evidently, because $OA' = OA$, $OB' = OB$ and $A'B' = AB$, the triangles $OA'B'$ and OAB are equal and similar. Hence O is similarly situated with regard to $A'B'$ and AB, and is therefore one and

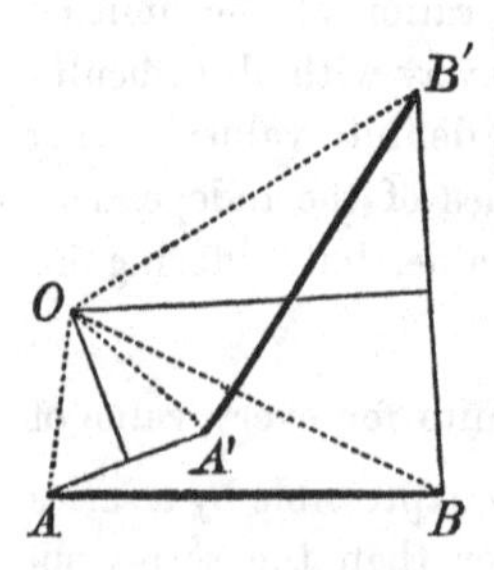

the same point of the plane figure in its two positions. If, for the sake of illustration, we actually trace the triangle OAB upon the plane, it becomes $OA'B'$ in the second position of the figure.

Displacements of a plane figure in its plane.

80. If from the equal angles $A'OB'$, AOB of these similar triangles we take the common part $A'OB$, we have the remaining angles AOA', BOB' equal, and each of them is clearly equal to the angle through which the figure must have turned round the point O to bring it from the first to the second position.

The preceding simple construction therefore enables us not only to demonstrate the general proposition, § 79, but also to determine from the two positions of one terminated line AB, $A'B'$ of the figure the common centre and the amount of the angle of rotation.

81. The lines equidistant from A and A', and from B and B', are parallel if AB is parallel to $A'B'$; and therefore the construction fails, the point O being infinitely distant, and the theorem becomes nugatory. In this case the motion is in fact a simple translation of the figure in its own plane without rotation—since, AB being parallel and equal to $A'B'$, we have AA' parallel and equal to BB'; and instead of there being one point of the figure common to both positions, the lines joining the two successive positions of all points in the figure are equal and parallel.

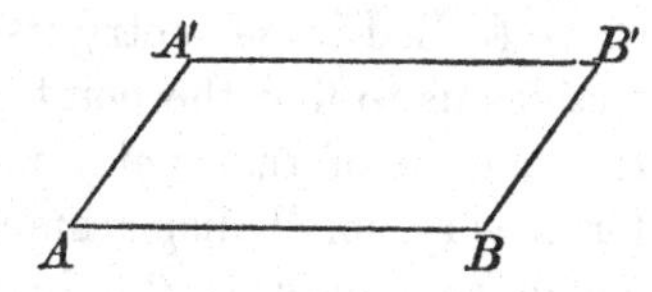

82. It is not necessary to suppose the figure to be a mere flat disc or plane—for the preceding statements apply to any one of a set of parallel planes in a rigid body, moving in any way subject to the condition that the points of any one plane in it remain always in a fixed plane in space.

83. There is yet a case in which the construction in § 79 is nugatory—that is when AA' is parallel to BB', but the lines of AB and $A'B'$ intersect. In this case, however, the point of intersection is the point O required, although the former method would not have enabled us to find it.

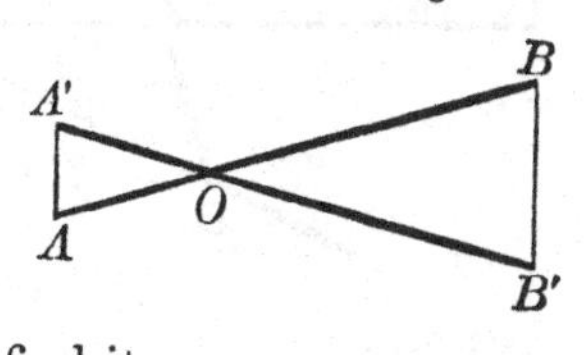

Examples of displacement in one plane.

84. Very many interesting applications of this principle may be made, of which, however, few belong strictly to our subject, and we shall therefore give only an example or two. Thus we know that if a line of given length AB move with its extremities always in two fixed lines OA, OB, any point in it as P describes an ellipse. It is required to find the direction of motion of P at any instant, *i.e.*, to draw a tangent to the ellipse. BA will pass to its next position by rotating about the point Q; found by the method of § 79 by drawing perpendiculars to OA and OB at A and B. Hence P for the instant revolves about Q, and thus its direction of motion, or the tangent to the ellipse, is perpendicular to QP. Also AB in its motion always touches a curve (called in geometry its envelop); and the same principle enables us to find the point of the envelop which lies in AB, for the motion of that point must evidently be ultimately (that is for a very small displacement) along AB, and the only point which so moves is the intersection of AB with the perpendicular to it from Q. Thus our construction would enable us to trace the envelop by points. (For more on this subject see § 91.)

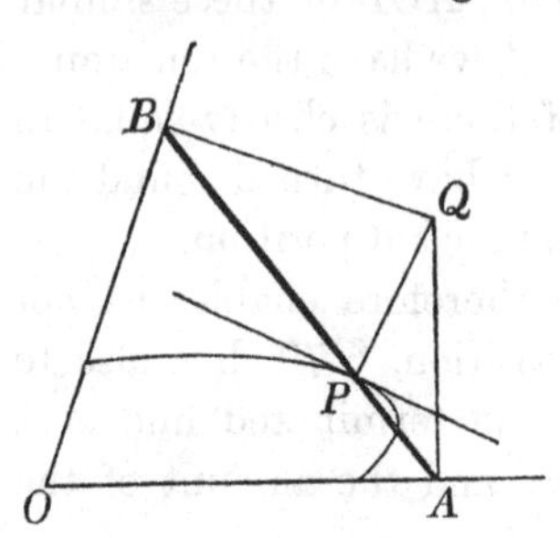

85. Again, suppose AB to be the beam of a stationary engine having a reciprocating motion about A, and by a link BD turning a crank CD about C. Determine the relation between the angular velocities of AB and CD in any position. Evidently the instantaneous direction of motion of B is transverse to AB, and of D transverse to CD—hence if AB, CD produced meet in O, the motion of BD is for an instant as if it turned about O. From this it may be easily seen that if the angular velocity of AB be ω, that of CD is $\frac{AB}{OB}\frac{OD}{CD}\omega$. A similar process is of course applicable to any combination of machinery, and we shall find it

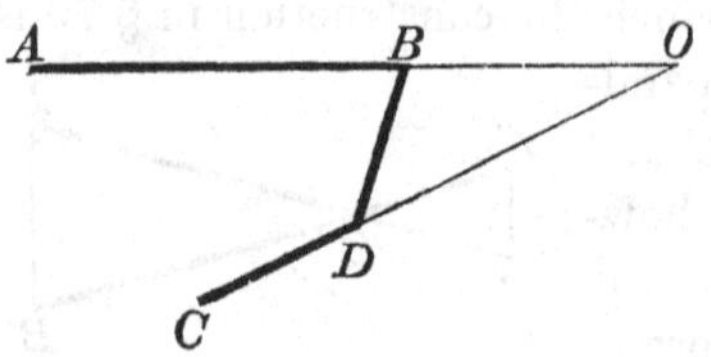

very convenient when we come to consider various dynamical problems connected with virtual velocities.

Examples of displacement in one plane.

Composition of rotations about parallel axes.

86. Since in general any movement of a plane figure in its plane may be considered as a rotation about one point, it is evident that two such rotations may in general be compounded into one; and therefore, of course, the same may be done with any number of rotations. Thus let A and B be the points of the figure about which in succession the rotations are to take place. By a rotation about A, B is brought say to B', and by a rotation about B', A is brought to A'. The construction of § 79 gives us at once the point O and the amount of rotation about it which singly gives the same effect as those about A and B in succession. But there is one case of exception, viz., when the rotations about A and B are of equal amount and in opposite directions. In this case $A'B'$ is evidently parallel to AB, and therefore the compound result is a *translation* only. That is, if a body revolve in succession through equal angles, but in opposite directions, about two parallel axes, it finally takes a position to which it could have been brought by a simple translation perpendicular to the lines of the body in its initial or final position, which were successively made axes of rotation; and inclined to their plane at an angle equal to half the supplement of the common angle of rotation.

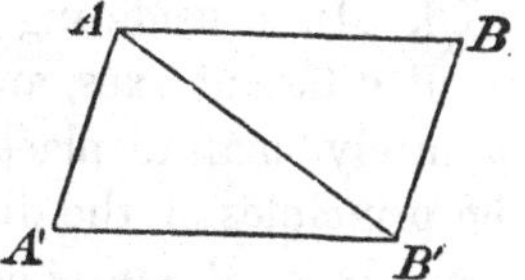

Composition of rotations and translations in one plane.

87. Hence to compound into an equivalent rotation a rotation and a translation, the latter being effected parallel to the plane of the former, we may decompose the translation into two rotations of equal amount and opposite direction, compound one of them with the given rotation by § 86, and then compound the other with the resultant rotation by the same process. Or we may adopt the following far simpler method. Let OA be the translation common to all points in the plane, and let BOC be the angle of rotation about O, BO being drawn so that OA bisects the exterior angle COB'. Take

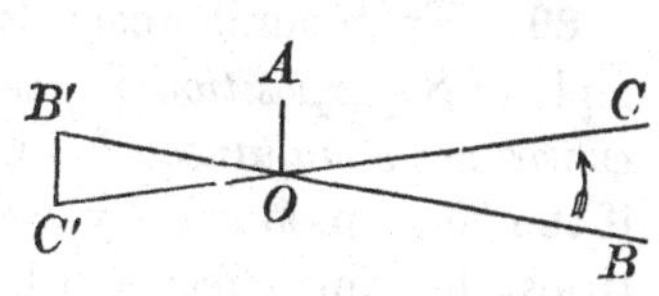

Composition of rotations and translations in one plane.

the point B' in BO produced, such that $B'C'$, the space through which the rotation carries it, is equal and opposite to OA. This point retains its former position after the performance of the compound operation; so that a rotation and a translation in one plane can be compounded into an equal rotation about a different axis.

In general, if the origin be taken as the point about which rotation takes place in the plane of xy, and if it be through an angle θ, a point whose co-ordinates were originally x, y will have them changed to

$$\xi = x\cos\theta - y\sin\theta, \quad \eta = x\sin\theta + y\cos\theta,$$

or, if the rotation be very small,

$$\xi = x - y\theta, \quad \eta = y + x\theta.$$

Omission of the second and higher orders of small quantities.

88. In considering the composition of angular velocities about different axes, and other similar cases, we may deal with infinitely small displacements only; and it results at once from the principles of the differential calculus, that if these displacements be of the *first* order of small quantities, any point whose displacement is of the *second* order of small quantities is to be considered as rigorously at rest. Hence, for instance, if a body revolve through an angle of the first order of small quantities about an axis (belonging to the body) which during the revolution is displaced through an angle or space, also of the first order, the displacement of any point of the body is rigorously what it would have been had the axis been fixed during the rotation about it, and its own displacement made either before or after this rotation. Hence in any case of motion of a rigid system the angular velocities about a system of axes moving *with* the system are the same at any instant as those about a system fixed in space, provided only that the latter coincide at the instant in question with the moveable ones.

Superposition of small motions.

89. From similar considerations follows also the general principle of *Superposition of small motions.* It asserts that if several causes act *simultaneously* on the same particle or rigid body, and if the effect produced by each is of the first order of small quantities, the joint effect will be obtained if we consider the causes to act *successively*, each taking the point or system in the position in which the preceding one left it. It is evident at once that this is an immediate deduction from the fact that the second order of infinitely small quantities may be with rigorous accuracy neglected. This principle is of very great use, as we shall find in the sequel; its applications are of constant occurrence.

Superposition of small motions.

A plane figure has given angular velocities about given axes perpendicular to its plane, find the resultant.

Let there be an angular velocity ω about an axis passing through the point a, b.

The consequent motion of the point x, y in the time δt is, as we have just seen (§ 87),

$$-(y-b)\omega\delta t \text{ parallel to } x, \quad \text{and } (x-a)\omega\delta t \text{ parallel to } y.$$

Hence, by the superposition of small motions, the whole motion parallel to x is

$$-(y\Sigma\omega - \Sigma b\omega)\delta t,$$

and that parallel to y $\quad (x\Sigma\omega - \Sigma a\omega)\delta t.$

Hence the point whose co-ordinates are

$$x' = \frac{\Sigma a\omega}{\Sigma\omega} \text{ and } y' = \frac{\Sigma b\omega}{\Sigma\omega}$$

is at rest, and the resultant axis passes through it. Any other point x, y moves through spaces

$$-(y\Sigma\omega - \Sigma b\omega)\delta t, \quad (x\Sigma\omega - \Sigma a\omega)\delta t.$$

But if the whole had turned about x', y' with velocity Ω, we should have had for the displacements of x, y,

$$-(y-y')\Omega\delta t, \quad (x-x')\Omega\delta t.$$

Comparing, we find $\Omega = \Sigma\omega$.

Hence if the sum of the angular velocities be zero, there is no rotation, and indeed the above formulæ show that there is then merely translation,

$$\Sigma(b\omega)\delta t \text{ parallel to } x, \quad \text{and } -\Sigma(a\omega)\delta t \text{ parallel to } y.$$

These formulæ suffice for the consideration of any problem on the subject.

Rolling of curve on curve.

90. Any motion whatever of a plane figure in its own plane might be produced by the rolling of a curve fixed to the figure upon a curve fixed in the plane.

For we may consider the whole motion as made up of successive elementary displacements, each of which corresponds, as we have seen, to an elementary rotation about some point in

Rolling of curve on curve.

the plane. Let o_1, o_2, o_3, etc., be the successive points of the moving figure about which the rotations take place, O_1, O_2, O_3, etc., the positions of these points when each is the instantaneous centre of rotation. Then the figure rotates about o_1 (or O_1, which coincides with it) till o_2 coincides with O_2, then about the latter till o_3 coincides with O_3, and so on. Hence, if we join o_1, o_2, o_3, etc., in the plane of the figure, and O_1, O_2, O_3, etc., in the fixed plane, the motion will be the same as if the polygon $o_1o_2o_3$, etc., rolled upon the fixed polygon $O_1O_2O_3$, etc. By supposing the successive displacements small enough the sides of these polygons gradually diminish, and the polygons finally become continuous curves. Hence the theorem.

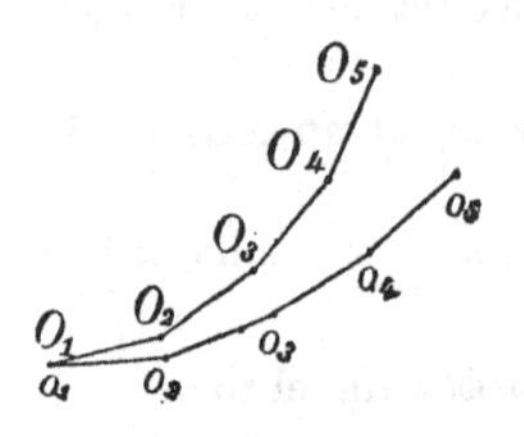

From this it immediately follows, that any displacement of a rigid solid, which is in directions wholly perpendicular to a fixed line, may be produced by the rolling of a cylinder fixed in the solid on another cylinder fixed in space, the axes of the cylinders being parallel to the fixed line.

91. As an interesting example of this theorem, let us recur to the case of § 84:—A circle may evidently be circumscribed about $OBQA$; and it must be of invariable magnitude, since in it a chord of given length AB subtends a given angle O at the circumference. Also OQ is a diameter of this circle, and is therefore constant. Hence, as Q is momentarily at rest, the motion of the circle circumscribing $OBQA$ is one of internal rolling on a circle of double its diameter. Hence if a circle roll internally on another of twice its diameter, any point in its circumference describes a diameter of the fixed circle, any other point in its plane an ellipse. This is precisely the same proposition as that of § 70, although the ways of arriving at it are very different. As it presents us with a particular case of the Hypocycloid, it warns us to return to the consideration of these and kindred curves, which give good instances of kinematical theorems, but which besides are of great use in physics generally.

Cycloids and Trochoids.

92. When a circle rolls upon a straight line, a point in its circumference describes a Cycloid; an internal point describes a

Prolate, an external one a Curtate, Cycloid. The two latter varieties are sometimes called Trochoids.

Cycloids and Trochoids.

The general form of these curves will be seen in the annexed figures; and in what follows we shall confine our remarks to the cycloid itself, as of immensely greater consequence than the others. The next section contains a simple investigation of those properties of the cycloid which are most useful in our subject.

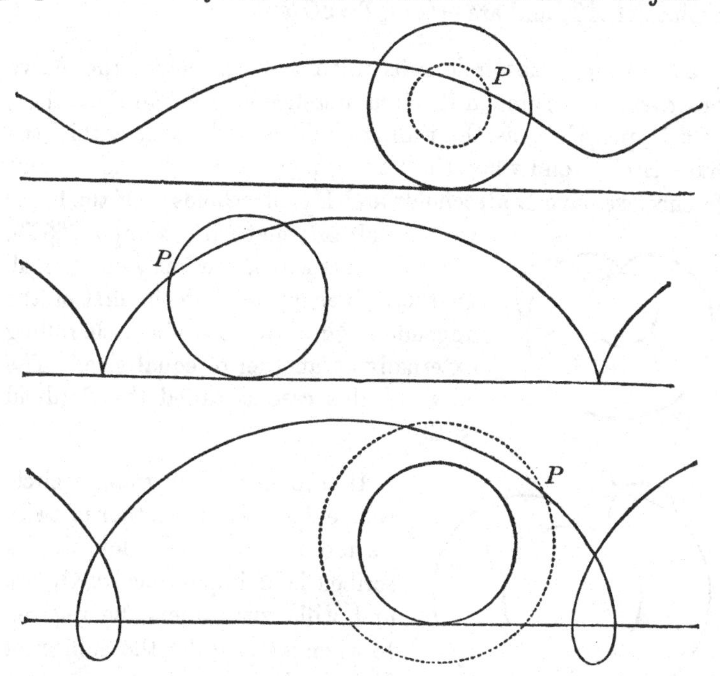

Properties of the cycloid.

93. Let AB be a diameter of the generating (or rolling) circle, BC the line on which it rolls. The points A and B describe similar and equal cycloids, of which AQC and BS are portions. If PQR be any subsequent position of the generating circle, Q and S the new positions of A and B, $\angle QPS$ is of course a right angle. If, therefore, QR be drawn parallel to PS, PR is a diameter

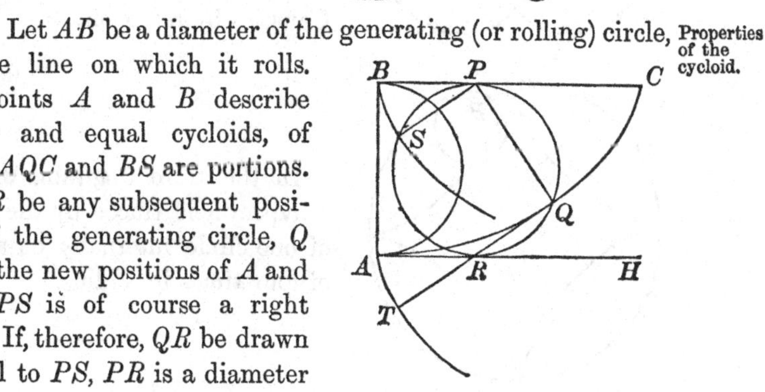

Properties of the cycloid.

of the rolling circle. Produce QR to T, making $RT=QR=PS$. Evidently the curve AT, which is the locus of T, is similar and equal to BS, and is therefore a cycloid similar and equal to AC. But QR is perpendicular to PQ, and is therefore the instantaneous direction of motion of Q, or is the tangent to the cycloid AQC. Similarly, PS is perpendicular to the cycloid BS at S, and so is therefore TQ to AT at T. Hence (§ 19) AQC is the evolute of AT, and arc $AQ=QT=2QR$.

Epicycloids, Hypocycloids, etc.

94. When the circle rolls upon another circle, the curve described by a point in its circumference is called an Epicycloid, or a Hypocycloid, as the rolling circle is without or within the fixed circle; and when the tracing point is not in the circumference, we have Epitrochoids and Hypotrochoids. Of the latter we have already met with examples, §§ 70, 91, and others will be presently mentioned. Of the former, we have in the first of the appended figures the case of a circle rolling externally on another of equal size. The curve in this case is called the Cardioid (§ 49).

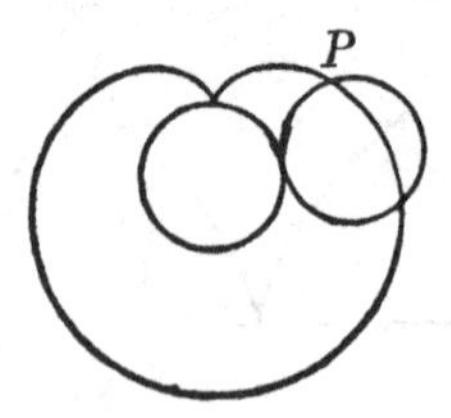

In the second diagram, a circle rolls externally on another of twice its radius. The epicycloid so described is of importance in Optics, and will, with others, be referred to when we consider the subject of Caustics by reflexion.

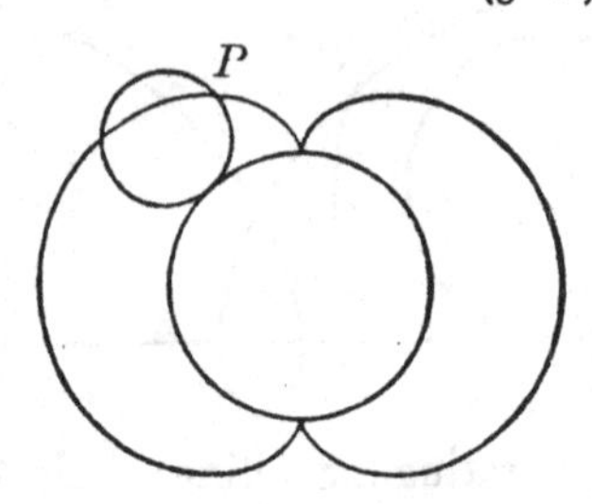

In the third diagram, we have a hypocycloid traced by the rolling of one circle internally on another of four times its radius.

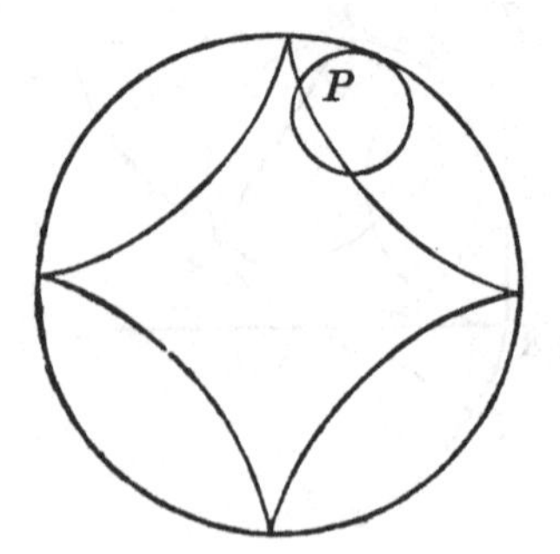

The curve figured in § 72 is an epitrochoid described by a point in the plane of a large circular disc which rolls upon a circular cylinder of small diameter, so that the point passes through the axis of the cylinder.

Epicycloids, Hypo-cycloids, etc.

That of § 74 is a hypotrochoid described by a point in the plane of a circle which rolls internally on another of rather more than twice its diameter, the tracing point passing through the centre of the fixed circle. Had the diameters of the circles been exactly as 1 : 2, § 72 or § 91 shows that this curve would have been reduced to a single straight line.

The general equations of this class of curves are

$$x = (a+b)\cos\theta - eb\cos\frac{a+b}{b}\theta,$$

$$y = (a+b)\sin\theta - eb\sin\frac{a+b}{b}\theta,$$

where a is the radius of the fixed, b of the rolling circle; and eb is the distance of the tracing point from the centre of the latter.

Motion about a fixed point.

95. If a rigid solid body move in any way whatever, subject only to the condition that one of its points remains fixed, there is always (without exception) one line of it through this point common to the body in any two positions. This most important theorem is due to Euler. To prove it, consider a spherical surface within the body, with its centre at the fixed point C. All points of this sphere attached to the body will move on a sphere fixed in space. Hence the construction of § 79 may be made, but with great circles instead of straight lines; and the same reasoning will apply to prove that the point O thus obtained is common to the body in its two positions. Hence every point of the body in the line OC, joining O with the fixed point, must be common to it in the two positions. Hence the body may pass from any one position to any other by rotating through a definite angle about a definite axis. Hence any position of the body may be specified by specifying the axis, and the angle, of rotation by which it may be brought to that position from a fixed position of reference, an idea due to Euler, and revived by Rodrigues.

Euler's theorem.

Rodrigues' co-ordinates.

Let OX, OY, OZ be any three fixed axes through the fixed point O round which the body turns. Let λ, μ, ν be the direction cosines, referred to these axes, of the axis OI round which the body must turn, and χ the angle through which it must turn round this axis, to bring it from some zero position to any other position. This other position, being specified by the four co-ordinates λ, μ, ν, χ (reducible, of course, to three by the relation $\lambda^2 + \mu^2 + \nu^2 = 1$), will be called for brevity $(\lambda, \mu, \nu, \chi)$. Let OA, OB, OC be three rectangular lines moving with the body, which in the "zero" position coincide respectively with OX, OY, OZ; and put

$$(XA), (YA), (ZA), (XB), (YB), (ZB), (XC), (YC), (ZC),$$

for the nine direction cosines of OA, OB, OC, each referred to OX, OY, OZ. These nine direction cosines are of course reducible to three independent co-ordinates by the well-known six relations. Let it be required now to express these nine direction cosines in terms of Rodrigues' co-ordinates λ, μ, ν, χ.

Let the lengths OX, ..., OA, ..., OI be equal, and call each unity: and describe from O as centre a spherical surface of unit radius; so that X, Y, Z, A, B, C, I shall be points on this surface. Let XA, YA, ... XB, denote arcs, and XAY, AXB, ... angles between arcs, in the spherical diagram thus obtained. We have $IA = IX = \cos^{-1}\lambda$, and $XIA = \chi$. Hence by the isosceles spherical triangle XIA,

$$\cos XA = \cos^2 IX + \sin^2 IX \cos\chi,$$

or

$$(XA) = \lambda^2 + (1 - \lambda^2)\cos\chi \quad\ldots\ldots\ldots\ldots\ldots\ldots\ldots\ldots (1).$$

And by the spherical triangle XIB,

$$\cos XB = \cos IX \cos IB + \sin IX \sin IB \cos XIB$$
$$= \lambda\mu + \sqrt{(1-\lambda^2)(1-\mu^2)}\cos XIB \quad\ldots\ldots\ldots (2).$$

Now $XIB = XIY + YIB = XIY + \chi$; and by the spherical triangle XIY we have

$$\cos XY = 0 = \cos IX \cos IY + \sin IX \sin IY \cos XIY$$
$$= \lambda\mu + \sqrt{(1-\lambda^2)(1-\mu^2)}\cos XIY.$$

Hence

$$\sqrt{(1-\lambda^2)(1-\mu^2)}\cos XIY = -\lambda\mu,$$

and

$$\sqrt{(1-\lambda^2)(1-\mu^2)}\sin XIY = \sqrt{(1-\lambda^2-\mu^2)} = \nu;$$

by which we have

$$\sqrt{(1-\lambda^2)(1-\mu^2)}\cos(XIY+\chi) = -\lambda\mu\cos\chi - \nu\sin\chi;$$

and using this in (2),

Rodrigues' co-ordinates.

$$\cos XB = \lambda\mu(1-\cos\chi) - \nu\sin\chi \qquad (3).$$

Similarly we find

$$\cos AY = \lambda\mu(1-\cos\chi) + \nu\sin\chi \qquad (4).$$

The other six formulæ may be written out by symmetry from (1), (3), and (4); and thus for the nine direction cosines we find

$$\left.\begin{array}{l}(XA)=\lambda^2+(1-\lambda^2)\cos\chi;\ (XB)=\lambda\mu(1-\cos\chi)-\nu\sin\chi;\ (YA)=\lambda\mu(1-\cos\chi)+\nu\sin\chi;\\ (YB)=\mu^2+(1-\mu^2)\cos\chi;\ (YC)=\mu\nu(1-\cos\chi)-\lambda\sin\chi;\ (ZB)=\mu\nu(1-\cos\chi)+\lambda\sin\chi;\\ (ZC)=\nu^2+(1-\nu^2)\cos\chi;\ (ZA)=\nu\lambda(1-\cos\chi)-\mu\sin\chi;\ (XC)=\nu\lambda(1-\cos\chi)+\mu\sin\chi.\end{array}\right\}(5).$$

Adding the three first equations of these three lines, and remembering that

$$\lambda^2+\mu^2+\nu^2=1 \qquad (6),$$

we deduce

$$\cos\chi = \tfrac{1}{2}[(XA)+(YB)+(ZC)-1] \qquad (7);$$

and then, by the three equations separately,

$$\left.\begin{array}{l}\lambda^2=\dfrac{1+(XA)-(YB)-(ZC)}{3-(XA)-(YB)-(ZC)},\\[2ex] \mu^2=\dfrac{1-(XA)+(YB)-(ZC)}{3-(XA)-(YB)-(ZC)},\\[2ex] \nu^2=\dfrac{1-(XA)-(YB)+(ZC)}{3-(XA)-(YB)-(ZC)}.\end{array}\right\} \qquad (8)$$

These formulæ, (8) and (7), express, in terms of (XA), (YB), (ZC), three out of the nine direction cosines (XA), ..., the direction cosines of the axis round which the body must turn, and the cosine of the angle through which it must turn round this axis, to bring it from the zero position to the position specified by those three direction cosines.

By aid of Euler's theorem above, successive or simultaneous rotations about any number of axes through the fixed point may be compounded into a rotation about one axis. Doing this for infinitely small rotations we find the law of composition of angular velocities.

Composition of rotations.

Let OA, OB be two axes about which a body revolves with angular velocities ϖ, ρ respectively.

Composition of angular velocities.

With radius unity describe the arc AB, and in it take any

Composition of angular velocities.

point I. Draw Ia, $I\beta$ perpendicular to OA, OB respectively. Let the rotations about the two axes be such that that about OB tends to *raise* I above the plane of the paper, and that about OA to depress it. In an infinitely short interval of time τ, the amounts of these displacements will be $\rho I\beta . \tau$ and $-\varpi Ia . \tau$. The point I, and therefore every point in the line OI, will be at rest during the interval τ if the sum of these displacements is zero, that is if $\rho . I\beta = \varpi . Ia$. Hence the line OI is instantaneously at rest, or the *two* rotations about OA and OB may be compounded into *one* about OI. Draw Ip, Iq, parallel to OB, OA respectively. Then, expressing in two ways the area of the parallelogram $IpOq$, we have

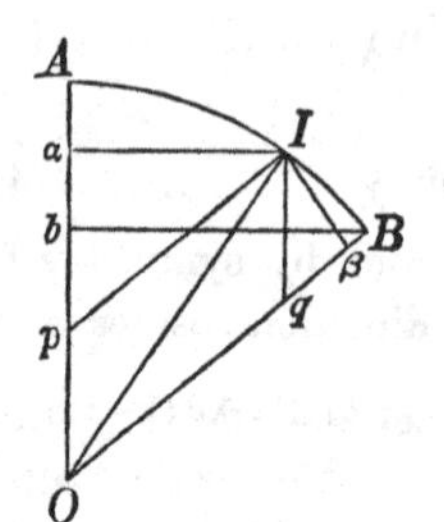

$$Oq . I\beta = Op . Ia,$$

$$Oq : Op :: \rho : \varpi.$$

Hence, if along the axes OA, OB, we measure off from O lines Op, Oq, proportional respectively to the angular velocities about these axes—the diagonal of the parallelogram of which these are contiguous sides is the resultant axis.

Again, if Bb be drawn perpendicular to OA, and if Ω be the angular velocity about OI, the whole displacement of B may evidently be represented either by $\varpi . Bb$ or $\Omega . I\beta$. Hence

$$\Omega : \varpi :: Bb : I\beta :: \sin BOA : \sin IOB :: \sin IpO : \sin pIO,$$
$$:: OI : Op.$$

Thus it is proved that,—

Parallelogram of angular velocities.

If lengths proportional to the respective angular velocities about them be measured off on the component and resultant axes, the lines so determined will be the sides and diagonal of a parallelogram.

Composition of angular velocities about axes meeting in a point.

96. Hence the single angular velocity equivalent to three co-existent angular velocities about three mutually perpendicular axes, is determined in magnitude, and the direction of its axis is found (§ 27), as follows :—The square of the resultant angular velocity is the sum of the squares of its components, and the ratios of the three components to the resultant are the direction cosines of the axis.

Composition of angular velocities about axes meeting in a point.

Hence simultaneous rotations about any number of axes meeting in a point may be compounded thus:—Let ω be the angular velocity about one of them whose direction cosines are l, m, n; Ω the angular velocity and λ, μ, ν the direction cosines of the resultant,

$$\lambda\Omega = \Sigma(l\omega), \quad \mu\Omega = \Sigma(m\omega), \quad \nu\Omega = \Sigma(n\omega),$$

whence
$$\Omega^2 = \Sigma^2(l\omega) + \Sigma^2(m\omega) + \Sigma^2(n\omega),$$

and
$$\lambda = \frac{\Sigma(l\omega)}{\Omega}, \quad \mu = \frac{\Sigma(m\omega)}{\Omega}, \quad \nu = \frac{\Sigma(n\omega)}{\Omega}.$$

Hence also, an angular velocity about any line may be resolved into three about any set of rectangular lines, the resolution in each case being (like that of simple velocities) effected by multiplying by the cosine of the angle between the directions.

Hence, just as in § 31 a uniform acceleration, perpendicular to the direction of motion of a point, produces a change in the *direction* of motion, but does not influence the *velocity;* so, if a body be rotating about an axis, and be subjected to an action tending to produce rotation about a perpendicular axis, the result will be a change of *direction* of the axis about which the body revolves, but no change in the *angular velocity.* On this kinematical principle is founded the dynamical explanation of the Precession of the Equinoxes (§ 107) and of some of the seemingly marvellous performances of gyroscopes and gyrostats.

The following method of treating the subject is useful in connexion with the ordinary methods of co-ordinate geometry. It contains also, as will be seen, an independent demonstration of the parallelogram of angular velocities:—

Angular velocities ϖ, ρ, σ about the axes of x, y, and z respectively, produce in time δt displacements of the point at x, y, z (§§ 87, 89),

$$(\rho z - \sigma y)\,\delta t \parallel x, \quad (\sigma x - \varpi z)\,\delta t \parallel y, \quad (\varpi y - \rho x)\,\delta t \parallel z.$$

Hence points for which

$$\frac{x}{\varpi} = \frac{y}{\rho} = \frac{z}{\sigma}$$

are not displaced. These are therefore the equations of the axis.

Composition of angular velocities about axes meeting in a point.

Now the perpendicular from any point x, y, z to this line is, by co-ordinate geometry,

$$\left[x^2+y^2+z^2-\frac{(\varpi x+\rho y+\sigma z)^2}{\varpi^2+\rho^2+\sigma^2}\right]^{\frac{1}{2}}$$

$$=\frac{1}{\sqrt{\varpi^2+\rho^2+\sigma^2}}\sqrt{(\rho z-\sigma y)^2+(\sigma x-\varpi z)^2+(\varpi y-\rho x)^2}$$

$$=\frac{\text{whole displacement of } x, y, z}{\sqrt{\varpi^2+\rho^2+\sigma^2}\,\delta t}.$$

The actual displacement of x, y, z is therefore the same as would have been produced in time δt by a single angular velocity, $\Omega=\sqrt{\varpi^2+\rho^2+\sigma^2}$, about the axis determined by the preceding equations.

Composition of successive finite rotations.

97. We give next a few useful theorems relating to the composition of successive *finite* rotations.

If a pyramid or cone of any form roll on a heterochirally similar* pyramid (the image in a plane mirror of the first position of the first) all round, it clearly comes back to its primitive position. This (as all rolling of cones) is conveniently exhibited by taking the intersection of each with a spherical surface. Thus we see that if a spherical polygon turns about its angular points in succession, always keeping on the spherical surface, and if the angle through which it turns about each point is twice the supplement of the angle of the polygon, or, which will come to the same thing, if it be in the other direction, but equal to twice the angle itself of the polygon, it will be brought to its original position.

The polar theorem (compare § 134, below) to this is, that a body, after successive rotations, represented by the doubles of the successive sides of a spherical polygon taken in order, is restored to its original position; which also is self-evident.

98. Another theorem is the following;—

If a pyramid rolls over all its sides on a plane, it leaves its track behind it as one plane angle, equal to the sum of the plane angles at its vertex.

* The similarity of a right-hand and a left-hand is called heterochiral: that of two right-hands, homochiral. Any object and its image in a plane mirror are heterochirally similar (Thomson, *Proc. R. S. Edinburgh*, 1873).

Otherwise:—in a spherical surface, a spherical polygon having rolled over all its sides along a great circle, is found in the same position as if the side first lying along that circle had been simply shifted along it through an arc equal to the polygon's periphery. The polar theorem is:—if a body be made to take successive rotations, represented by the sides of a spherical polygon taken in order, it will finally be as if it had revolved about the axis through the first angular point of the polygon through an angle equal to the spherical excess (§ 134) or area of the polygon.

Composition of successive finite rotations.

99. The investigation of § 90 also applies to this case; and it is thus easy to show that the most general motion of a spherical figure on a fixed spherical surface is obtained by the rolling of a curve fixed in the figure on a curve fixed on the sphere. Hence as at each instant the line joining C and O contains a set of points of the body which are momentarily at rest, the most general motion of a rigid body of which one point is fixed consists in the rolling of a cone fixed in the body upon a cone fixed in space—the vertices of both being at the fixed point.

Motion about a fixed point. Rolling cones.

100. Given at each instant the angular velocities of the body about three rectangular axes attached to it, determine its position in space at any time.

Position of the body due to given rotations.

From the given angular velocities about OA, OB, OC, we know, § 95, the position of the instantaneous axis OI with reference to the body at every instant. Hence we know the conical surface in the body which rolls on the cone fixed in space. The data are sufficient also for the determination of this other cone; and these cones being known, and the lines of them which are in contact at any given instant being determined, the position of the moving body is completely determined.

If λ, μ, ν be the direction cosines of OI referred to OA, OB, OC; ϖ, ρ, σ the angular velocities, and ω their resultant:

$$\frac{\lambda}{\varpi} = \frac{\mu}{\rho} = \frac{\nu}{\sigma} = \frac{1}{\omega},$$

by § 95. These equations, in which ϖ, ρ, σ, ω are given functions of t, express explicitly the position of OI relatively to OA, OB,

Position of the body due to given rotations.

OC, and therefore determine the cone fixed in the body. For the cone fixed in space: if r be the radius of curvature of its intersection with the unit sphere, r' the same for the rolling cone, we find from § 105 below, that if s be the length of the arc of either spherical curve from a common initial point,

$$\omega r' = \frac{1}{r}\frac{ds}{dt}\sin(\sin^{-1}r + \sin^{-1}r') = \frac{1}{r}\frac{ds}{dt}(r\sqrt{1-r'^2} + r'\sqrt{1-r^2}),$$

which, as s, r' and ω are known in terms of t, gives r in terms of t, or of s, as we please. Hence, by a single quadrature, the "intrinsic" equation of the fixed cone.

101. An unsymmetrical system of angular co-ordinates ψ, θ, ϕ, for specifying the position of a rigid body by aid of a line OB and a plane AOB moving with it, and a line OY and a plane YOX fixed in space, which is essentially proper for many physical problems, such as the Precession of the Equinoxes and the spinning of a top, the motion of a gyroscope and its gimbals, the motion of a compass-card and of its bowl and gimbals, is convenient for many others, and has been used by the greatest mathematicians often even when symmetrical methods would have been more convenient, must now be described.

ON being the intersection of the two planes, let $YON = \psi$, and $NOB = \phi$; and let θ be the angle from the fixed plane, produced through ON, to the portion NOB of the moveable plane. (Example, θ the "obliquity of the ecliptic," ψ the longitude of the autumnal equinox reckoned from OY, a fixed line in the plane of the earth's orbit supposed fixed; ϕ the hour-angle of the autumnal equinox; B being in the earth's equator and in the meridian of Greenwich: thus ψ, θ, ϕ are angular co-ordinates of the earth.) To show the relation of this to the symmetrical system, let OA be perpendicular to OB, and draw OC perpendicular to both; OX perpendicular to OY, and draw OZ perpendicular to OY and OX; so that OA, OB, OC are three rectangular axes fixed relatively to the body, and OX, OY, OZ fixed in space. The annexed diagram shows ψ, θ, ϕ in angles and arc, and in arcs and angles, on a spherical surface of unit radius with centre at O.

To illustrate the meaning of these angular co-ordinates, suppose A, B, C initially to coincide with X, Y, Z respectively.

Then, to bring the body into the position specified by θ, ϕ, ψ, rotate it round OZ through an angle equal to $\psi + \phi$, thus

Position of the body due to given rotations.

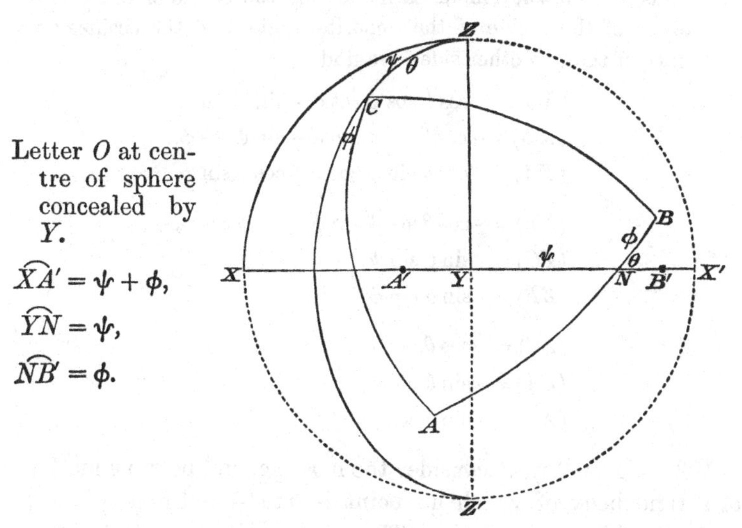

bringing A and B from X and Y to A' and B' respectively; and, (taking $\widehat{YN} = \psi$,) rotate the body round ON through an angle equal to θ, thus bringing A, B, and C from the positions A', B', and Z respectively, to the positions marked A, B, C in the diagram. Or rotate first round ON through θ, so bringing C from Z to the position marked C, and then rotate round OC through $\psi + \phi$. Or, while OC is turning from OZ to the position shown on the diagram, let the body turn round OC relatively to the plane $ZCZ'O$ through an angle equal to ϕ. It will be in the position specified by these three angles.

Let $\angle XZC = \psi$, $\angle ZCA = \pi - \phi$, and $ZC = \theta$, and ϖ, ρ, σ mean the same as in § 100. By considering in succession instantaneous motions of C along and perpendicular to ZC, and the motion of AB in its own plane, we have

$$\frac{d\theta}{dt} = \varpi \sin\phi + \rho\cos\phi, \qquad \sin\theta\,\frac{d\psi}{dt} = \rho\sin\phi - \varpi\cos\phi,$$

and
$$\frac{d\psi}{dt}\cos\theta + \frac{d\phi}{dt} = \sigma.$$

The nine direction cosines (XA), (YB), &c., according to the notation of § 95, are given at once by the spherical triangles

Position of the body due to given rotations.

XNA, YNB, &c.; each having N for one angular point, with θ, or its supplement or its complement, for the angle at this point. Thus, by the solution in each case for the cosine of one side in terms of the cosine of the opposite angle, and the cosines and sines of the two other sides, we find

$$(XA) = \cos\theta\cos\psi\cos\phi - \sin\psi\sin\phi,$$
$$(XB) = -\cos\theta\cos\psi\sin\phi - \sin\psi\cos\phi,$$
$$(YA) = \cos\theta\sin\psi\cos\phi + \cos\psi\sin\phi.$$

$$(YB) = -\cos\theta\sin\psi\sin\phi + \cos\psi\cos\phi,$$
$$(YC) = \sin\theta\sin\psi,$$
$$(ZB) = \sin\theta\sin\phi.$$

$$(ZC) = \cos\theta,$$
$$(ZA) = -\sin\theta\cos\phi,$$
$$(XC) = \sin\theta\cos\psi.$$

General motion of a rigid body.

102. We shall next consider the most general possible motion of a rigid body of which no point is fixed—and first we must prove the following theorem. There is one set of parallel planes in a rigid body which are parallel to each other in any two positions of the body. The parallel lines of the body perpendicular to these planes are of course parallel to each other in the two positions.

Let C and C' be any point of the body in its first and second positions. Move the body without rotation from its second position to a third in which the point at C' in the second position shall occupy its original position C. The preceding demonstration shows that there is a line CO common to the body in its first and third positions. Hence a line $C'O'$ of the body in its second position is parallel to the same line CO in the first position. This of course clearly applies to every line of the body parallel to CO, and the planes perpendicular to these lines also remain parallel.

Let S denote a plane of the body, the two positions of which are parallel. Move the body from its first position, without rotation, in a direction perpendicular to S, till S comes into the plane of its second position. Then to get the body into its actual position, such a motion as is treated in § 79 is farther required. But by § 79 this may be effected by rotation about a certain axis perpendicular to the plane S, unless the motion required belongs to the exceptional case of pure translation. Hence [this case excepted] the body may be brought from the first position to the second by translation through a determinate distance perpendicular to a given plane, and rotation through a determinate angle about a determinate axis perpendicular to that plane. This is precisely the motion of a screw in its nut.

General motion of a rigid body.

103. In the excepted case the whole motion consists of two translations, which can of course be compounded into a single one; and thus, in this case, there is no rotation at all, or every plane of it fulfils the specified condition for S of § 102.

104. Returning to the motion of a rigid body with one point fixed, let us consider the case in which the guiding cones, § 99, are both circular. The motion in this case may be called *Precessional Rotation.*

Precessional Rotation.

The plane through the instantaneous axis and the axis of the fixed cone passes through the axis of the rolling cone. This plane turns round the axis of the fixed cone with an angular velocity Ω (see § 105 below), which must clearly bear a constant ratio to the angular velocity ω of the rigid body about its instantaneous axis.

105. The motion of the plane containing these axes is called the *precession* in any such case. What we have denoted by Ω is the angular velocity of the precession, or, as it is sometimes called, the rate of precession.

The angular motions ω, Ω are to one another inversely as the distances of a point in the axis of the rolling cone from the instantaneous axis and from the axis of the fixed cone.

For, let OA be the axis of the fixed cone, OB that of the rolling cone, and OI the instantaneous axis. From any point P in OB draw PN perpendicular to OI, and PQ perpendicular to OA. Then we perceive that P moves always in the circle whose centre is Q, radius PQ, and plane perpendicular to OA. Hence

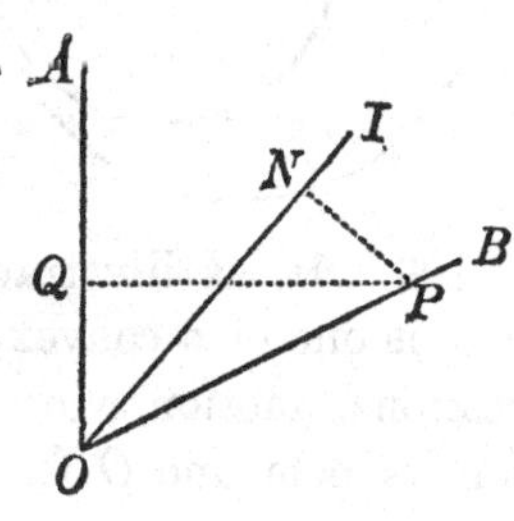

Precessional Rotation.

the actual velocity of the point P is ΩQP. But, by the principles explained above, § 99, the velocity of P is the same as that of a point moving in a circle whose centre is N, plane perpendicular to ON, and radius NP, which, as this radius revolves with angular velocity ω, is ωNP. Hence

$$\Omega \,.\, QP = \omega \,.\, NP, \text{ or } \omega : \Omega :: QP : NP.$$

Let α be the semivertical angle of the fixed, β of the rolling, cone. Each of these may be supposed for simplicity to be acute, and their sum or difference less than a right angle—though, of course, the formulæ so obtained are (like all trigonometrical results) applicable to every possible case. We have the following three cases:—

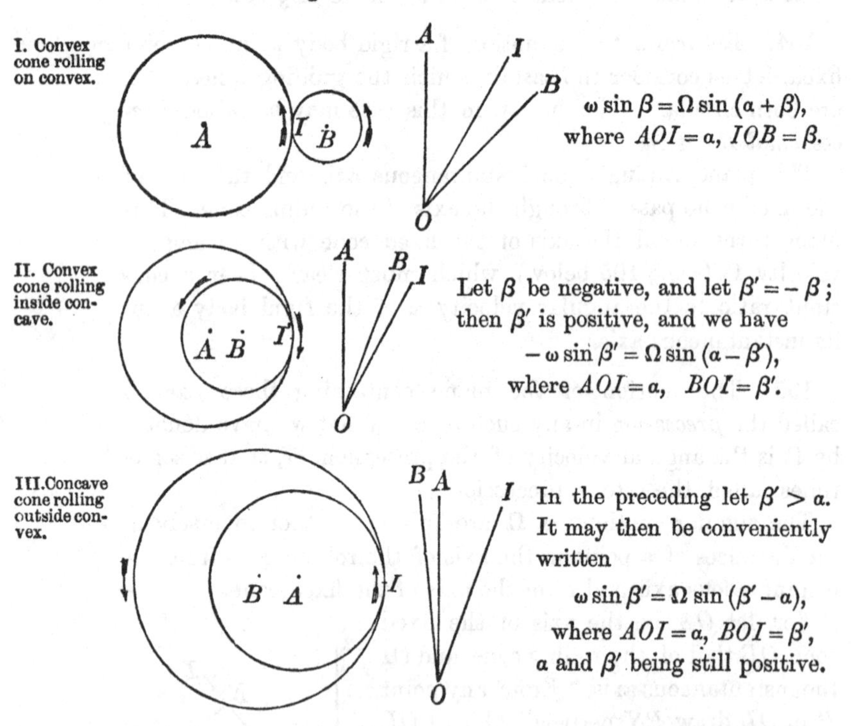

I. Convex cone rolling on convex.

$$\omega \sin \beta = \Omega \sin (\alpha + \beta),$$
where $AOI = \alpha$, $IOB = \beta$.

II. Convex cone rolling inside concave.

Let β be negative, and let $\beta' = -\beta$; then β' is positive, and we have
$$-\omega \sin \beta' = \Omega \sin (\alpha - \beta'),$$
where $AOI = \alpha$, $BOI = \beta'$.

III. Concave cone rolling outside convex.

In the preceding let $\beta' > \alpha$. It may then be conveniently written
$$\omega \sin \beta' = \Omega \sin (\beta' - \alpha),$$
where $AOI = \alpha$, $BOI = \beta'$, α and β' being still positive.

Cases of precessional rotation.

106. If, as illustrated by the first of these diagrams, the case is one of a convex cone rolling on a convex cone, the precessional motion, viewed on a hemispherical surface having A for its pole and O for its centre, is in a similar direction to that of the angular rotation about the instantaneous axis. This we shall call *positive* precessional rotation. It is the case of a common spinning-top (peery), spinning on a very fine point which remains at rest in a hollow or hole bored by itself; not sleeping upright, nor nodding, but sweeping its axis round in a circular cone whose axis is vertical. In Case III. also we have *positive* precession. A good example of this occurs in the case of a coin spinning on a table when its plane is nearly horizontal.

Cases of precessional rotation.

107. Case II., that of a convex cone rolling inside a concave one, gives an example of *negative* precession: for when viewed as before on the hemispherical surface the direction of angular rotation of the instantaneous axis is opposite to that of the rolling cone. This is the case of a symmetrical cup (or figure of revolution) supported on a point, and stable when balanced, *i.e.*, having its centre of gravity below the pivot; when inclined and set spinning non-nutationally. For instance, if a Troughton's top be placed on its pivot in any inclined position, and then spun off with very great angular velocity about its axis of figure, the nutation will be insensible; but there will be slow precession.

Model illustrating Precession of Equinoxes.

To this case also belongs the precessional motion of the earth's axis; for which the angle $\alpha = 23^\circ\ 27'\ 28''$, the period of the rotation ω the sidereal day; that of Ω is 25,868 years. If the second diagram represent a portion of the earth's surface round the pole, the arc $AI = 8{,}552{,}000$ feet, and therefore the circumference of the circle in which I moves $= 52{,}240{,}000$ feet. Imagine this circle to be the in-

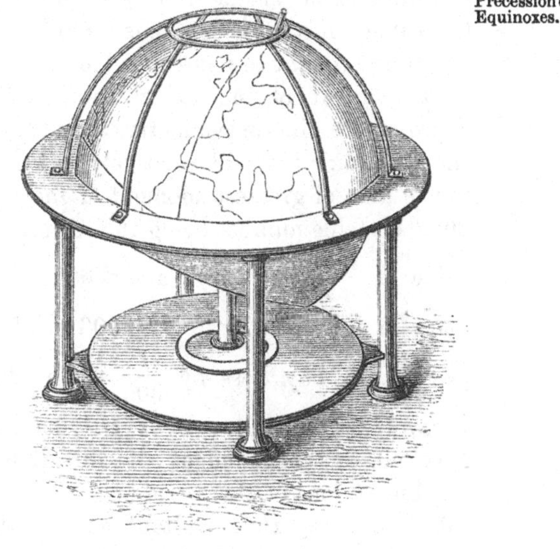

Precession of the equinoxes.

ner edge of a fixed ring in space (directionally fixed, that is to say, but having the same translational motion as the earth's centre), and imagine a circular post or pivot of radius BI to be fixed to the earth with its centre at B. This ideal pivot rolling on the inner edge of the fixed ring travels once round the 52,240,000 feet-circumference in 25,868 years, and therefore its own circumference must be 5·53 feet. Hence $BI = 0{\cdot}88$ feet; and angle BOI, or β, $= 0''{\cdot}00867$.

Free rotation of a body kinetically symmetrical about an axis.

108. Very interesting examples of Cases I. and III. are furnished by projectiles of different forms rotating about any axis. Thus the gyrations of an oval body or a rod or bar flung into the air belong to Class I. (the body having one axis of less moment of inertia than the other two, equal); and the seemingly irregular evolutions of an ill-thrown quoit belong to Class III. (the quoit having one axis of greater moment of inertia than the other two, which are equal). Case III. has therefore the following very interesting and important application.

If by a geological convulsion (or by the transference of a few million tons of matter from one part of the world to another) the earth's instantaneous axis OI (diagram III., § 105) were at any time brought to non-coincidence with its principal axis of least moment of inertia, which (§§ 825, 285) is an axis of approximate kinetic symmetry, the instantaneous axis will, and the fixed axis OA will, relatively to the solid, travel round the solid's axis of greatest moment of inertia in a period of about 306 days [this number being the reciprocal of the most probable value of $\frac{C-A}{C}$ (§ 828)]; and the motion is represented by the diagram of Case III. with $BI = 306 \times AI$. Thus in a very little less than a day (less by $\frac{1}{306}$ when BOI is a small angle) I revolves round A. It is OA, as has been remarked by Maxwell, that is found as the direction of the celestial pole by observations of the meridional zenith distances of stars, and this line being the resultant axis of the earth's moment of momentum (§ 267), would remain invariable in space did no external influence such as that of the moon and sun disturb the earth's rotation. When we neglect precession and nutation, the polar distances of the stars are constant notwithstanding the ideal motion of the fixed axis which we are now considering; and the effect of this motion will be to make a periodic variation of the latitude of every place on the earth's surface having for range on each side of its mean value the angle BOA, and for its period 306 days or thereabouts. Maxwell* examined a four years series of Greenwich observations of Polaris (1851-2-3-4), and concluded that there was during those years no variation exceeding half a second of angle on each side of mean range, but that the evidence did not disprove a variation of that amount, but on the contrary gave a very slight indication of a minimum latitude of Greenwich belonging to the set of months Mar. '51, Feb. '52, Dec. '52, Nov. '53, Sept. '54.

Free rotation of a body kinetically symmetrical about an axis.

"This result, however, is to be regarded as very doubtful......
"and more observations would be required to establish the
"existence of so small a variation at all.

"I therefore conclude that the earth has been for a long time
"revolving about an axis very near to the axis of figure, if not
"coinciding with it. The cause of this near coincidence is
"either the original softness of the earth, or the present fluidity
"of its interior [or the existence of water on its surface].
"The axes of the earth are so nearly equal that a con-
"siderable elevation of a tract of country might produce a
"deviation of the principal axis within the limits of observa-
"tion, and the only cause which would restore the uniform
"motion, would be the action of a fluid which would gradually
"diminish the oscillations of latitude. The permanence of
"latitude essentially depends on the inequality of the earth's
"axes, for if they had all been equal, any alteration in the
"crust of the earth would have produced new principal axes,
"and the axis of rotation would travel about those axes, alter-

* On a Dynamical Top, *Trans. R. S. E.*, 1857, p. 559.

Free rotation of a body kinetically symmetrical about an axis.

"ing the latitudes of all places, and yet not in the least altering "the position of the axis of rotation among the stars."

Perhaps by a more extensive "search and analysis of the "observations of different observatories, the nature of the "periodic variation of latitude, if it exist, may be determined. "I am not aware* of any calculations having been made to prove "its non-existence, although, on dynamical grounds, we have "every reason to look for some very small variation having the "periodic time of 325·6 days nearly" [more nearly 306 days], "a period which is clearly distinguished from any other astro- "nomical cycle, and therefore easily recognised†."

The periodic variation of the earth's instantaneous axis thus anticipated by Maxwell must, if it exists, give rise to a tide of 306 days period (§ 801). The amount of this tide at the equator would be a rise and fall amounting only to $5\frac{1}{2}$ centimetres above and below mean for a deviation of the instantaneous axis amounting to 1″ from its mean position OB, or for a deviation BI on the earth's surface amounting to 31 metres. This, although discoverable by elaborate analysis of long-continued and accurate tidal observations, would be less easily discovered than the periodic change of latitude by astronomical observations according to Maxwell's method‡.

* (Written twenty years ago).

† Maxwell; *Transactions of the Royal Society of Edinburgh*, 20th April, 1857.

‡ Prof. Maxwell now refers us to Peters (*Recherches sur la parallaxe des étoiles fixes*, St Petersburgh Observatory Papers, Vol. I., 1853), who seems to have been the first to raise this interesting and important question. He found from the Pulkova observations of Polaris from March 11, 1842 till April 30, 1843 an angular radius of 0″·079 (probable error 0″·017), for the circle round its mean position described by the instantaneous axis, and for the time, within that interval, when the latitude of Pulkova was a maximum, Nov. 16, 1842. The period (calculated from the dynamical theory) which Peters assumed was 304 mean solar days: the rate therefore 1·201 turns per annum, or, nearly enough, 12 turns per ten years. Thus if Peters' result were genuine, and remained constant for ten years, the latitude of Pulkova would be a maximum about the 16th of Nov. again in 1852, and Pulkova being in 30° East longitude from Greenwich, the latitude of Greenwich would be a maximum $\frac{1}{12}$ of the period, or about 25 days earlier, that is to say about Oct. 22, 1852. But Maxwell's examination of observations seemed to indicate more nearly the minimum latitude of Greenwich about the same time. This discrepance is altogether in accordance with a continuation of Peters' investigation by Dr Nysen of the Pulkova Ob-

109. In various illustrations and arrangements of apparatus useful in Natural Philosophy, as well as in Mechanics, it is required to connect two bodies, so that when either turns about a certain axis, the other shall turn with an equal angular velocity about another axis in the same plane with the former, but inclined to it at any angle. This is accomplished in mechanism by means of equal and similar bevelled wheels, or rolling cones; when the mutual inclination of two axes is not to be varied. It is approximately accomplished by means of Hooke's joint, when the two axes are nearly in the same line, but are required to be free to vary in their mutual inclination. A chain of an infinitely great number of Hooke's joints may be imagined as constituting a perfectly flexible, untwistable cord, which, if its end-links are rigidly attached to the two bodies, connects them so as to fulfil the condition rigorously without the restriction that the two axes remain in one plane. If we imagine an infinitely short length of such a chain (still, however, having an infinitely great number of links) to have its ends attached to two bodies, it will fulfil rigorously the condition stated, and at the same time keep a definite point of one body infinitely near a definite point of the other; that is to say, it will accomplish precisely for every angle of inclination what Hooke's joint does approximately for small inclinations.

Communication of angular velocity equally between inclined axes.

Hooke's joint.

Flexible but untwistable cord.

Universal flexure joint.

The same is dynamically accomplished with perfect accuracy for every angle, by a short, naturally straight, elastic wire of

Elastic universal flexure joint.

servatory, in which, by a careful scrutiny of several series of Pulkova observations between the years 1842...1872, he concluded that there is no constancy of magnitude or phase in the deviation sought for. A similar negative conclusion was arrived at by Professor Newcomb of the United States Naval Observatory, Washington, who at our request kindly undertook an investigation of the ten-month period of latitude from the Washington Prime Vertical Observations from 1862 to 1867. His results, as did those of Peters and Nysen and Maxwell, seemed to indicate real variations of the earth's instantaneous axis amounting to possibly as much as $\frac{1}{2}''$ or $\frac{1}{4}''$ from its mean position, but altogether irregular both in amount and direction; in fact, just such as might be expected from irregular heapings up of the oceans by winds in different localities of the earth.

We intend to return to this subject and to consider cognate questions regarding irregularities of the earth as a timekeeper, and variations of its figure and of the distribution of matter within it, of the ocean on its surface, and of the atmosphere surrounding it, in §§ 267, 276, 405, 406, 830, 832, 845, 846.

Elastic universal flexurejoint.

truly circular section, provided the forces giving rise to any resistance to equality of angular velocity between the two bodies are infinitely small. In many practical cases this mode of connexion is useful, and permits very little deviation from the conditions of a true universal flexure joint. It is used, for instance, in the suspension of the gyroscopic pendulum (§ 74) with perfect success. The dentist's tooth-mill is an interesting illustration of the elastic universal flexure joint. In it a long spiral spring of steel wire takes the place of the naturally straight wire suggested above.

Moving body attached to a fixed object by a universal flexure joint.

Of two bodies connected by a universal flexure joint, let one be held fixed. The motion of the other, as long as the angle of inclination of the axes remains constant, will be exactly that figured in Case I., § 105, above, with the angles α and β made equal. Let O be the joint; AO the axis of the fixed body; OB the axis of the moveable body. The supplement of the angle AOB is the mutual inclination of the axes; and the angle AOB itself is bisected by the instantaneous axis of the moving body. The diagram shows a case of this motion, in which the mutual inclination, θ, of the axes is acute. According to the formulæ of Case I., § 105, we have

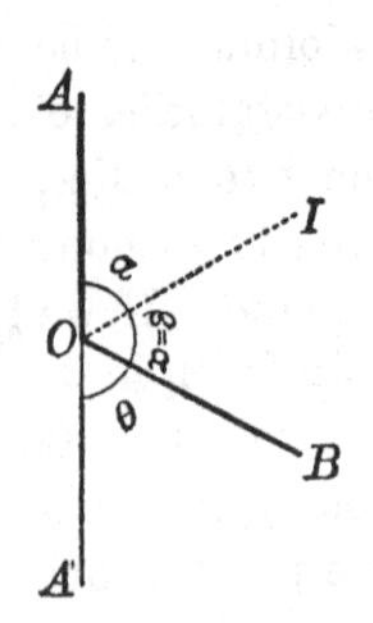

$$\omega \sin \alpha = \Omega \sin 2\alpha,$$

or

$$\omega = 2\Omega \cos \alpha = 2\Omega \sin \frac{\theta}{2},$$

where ω is the angular velocity of the moving body about its instantaneous axis, OI, and Ω is the angular velocity of its precession; that is to say, the angular velocity of the plane through the fixed axis AA', and the moving axis OB of the moving body.

Two degrees of freedom to move enjoyed by a body thus suspended.

Besides this motion, the moving body may clearly have any angular velocity whatever about an axis through O perpendicular to the plane AOB, which, compounded with ω round OI, gives the resultant angular velocity and instantaneous axis.

Two co-ordinates, $\theta = A'OB$, and ϕ measured in a plane perpendicular to AO, from a fixed plane of reference to the plane

AOB, fully specify the position of the moveable body in this case.

110. Suppose a rigid body bounded by any curved surface to be touched at any point by another such body. Any motion of one on the other must be of one or more of the forms *sliding*, *rolling*, or *spinning*. The consideration of the first is so simple as to require no comment.

General motion of one rigid body touching another.

Any motion in which there is no slipping at the point of contact must be rolling or spinning separately, or combined.

Let one of the bodies rotate about successive instantaneous axes, all lying in the common tangent plane at the point of instantaneous contact, and each passing through this point—the other body being fixed. This motion is what we call rolling, or simple rolling, of the moveable body on the fixed.

On the other hand, let the instantaneous axis of the moving body be the common normal at the point of contact. This is pure spinning, and does not change the point of contact.

Let the moving body move, so that its instantaneous axis, still passing through the point of contact, is neither in, nor perpendicular to, the tangent plane. This motion is combined rolling and spinning.

111. When a body rolls and spins on another body, the *trace of either on the other* is the curved or straight line along which it is successively touched. If the instantaneous axis is in the normal plane perpendicular to the traces, the rolling is called *direct*. If not direct, the rolling may be resolved into a direct rolling, and a rotation or twisting round the tangent line to the traces.

Traces of rolling.

Direct rolling.

When there is *no spinning* the projections of the two traces on the common tangent plane at the point of contact of the two surfaces have equal and same-way directed curvature: or they have "contact of the second order." When there *is spinning*, the two projections still touch one another, but with contact of the first order only: their curvatures differ by a quantity equal to the angular velocity of spinning divided by the velocity of the point of contact. This last we see by noticing that the rate of change of direction along the pro-

Direct rolling.

jection of the fixed trace must be equal to the rate of change of direction along the projection of the moving trace if held fixed plus the angular velocity of the spinning.

At any instant let $2z = Ax^2 + 2Cxy + By^2$ ....(1)

and $2z' = A'x^2 + 2C'xy + B'y^2$ ....(2)

be the equations of the fixed and moveable surfaces S and S' infinitely near the point of contact O, referred to axes OX, OY in their common tangent plane, and OZ perpendicular to it: let ϖ, ρ, σ be the three components of the instantaneous angular velocity of S'; and let x, y, be co-ordinates of P, the point of contact at an infinitely small time t, later: the third co-ordinate, z, is given by (1).

Let P' be the point of S' which at this later time coincides with P. The co-ordinates of P' at the first instant are $x + \sigma yt$, $y - \sigma xt$; and the corresponding value of z' is given by (2). This point is infinitely near to (x, y, z'), and therefore at the first instant the direction cosines of the normal to S' through it differ but infinitely little from

$$-(A'x + C'y), \quad -(C'x + B'y), \quad 1.$$

But at time t the normal to S' at P' coincides with the normal to S at P, and therefore its direction cosines change from the preceding values, to

$$-(Ax + Cy), \quad -(Cx + By), \quad 1:$$

that is to say, it rotates through angles

$$(C' - C)\,x + (B' - B)\,y \text{ round } OX,$$

and $$-\{(A' - A)x + (C' - C)\,y\} \quad ,, \quad OY.$$

Hence
$$\left.\begin{aligned} \varpi t &= (C' - C)\,x + (B' - B)\,y \\ \rho t &= -\{(A' - A)\,x + (C' - C)\,y\} \end{aligned}\right\} \text{..........(3)},$$

or
$$\left.\begin{aligned} \varpi &= (C' - C)\,\dot{x} + (B' - B)\,\dot{y} \\ \rho &= -\{(A' - A)\,\dot{x} + (C' - C)\,\dot{y}\} \end{aligned}\right\} \text{..................(4)},$$

if $\dot{x}$, $\dot{y}$ denote the component velocities of the point of contact.

Put $$q = \sqrt{(\dot{x}^2 + \dot{y}^2)} \text{..........................(5)},$$

and take components of ϖ and ρ round the tangent to the traces and the perpendicular to it in the common tangent plane of the two surfaces, thus:

(twisting component)......$\dfrac{\dot{x}}{q}\varpi + \dfrac{\dot{y}}{q}\rho$

$$= (C' - C)\frac{\dot{x}^2 - \dot{y}^2}{q} + [(B' - B) - (A' - A)]\frac{\dot{x}\dot{y}}{q} \text{ ......(6)},$$

and

Direct rolling.

(direct-rolling component)...... $\frac{\dot{y}}{q}\varpi - \frac{\dot{x}}{q}\rho$

$$= \frac{1}{q}[(A'-A)\dot{x}^2 + 2(C'-C)\dot{x}\dot{y} + (B'-B)\dot{y}^2] \ldots\ldots (7).$$

Choose OX, OY so that $C - C' = 0$, and put $A' - A = \alpha$, $B' - B = \beta$ (6) and (7) become

(twisting component) .... $\frac{\dot{x}}{q}\varpi + \frac{\dot{y}}{q}\rho = (\beta - \alpha)\frac{\dot{x}\dot{y}}{q}$ ....(8),

(direct-rolling component)...... $\frac{\dot{y}}{q}\varpi - \frac{\dot{x}}{q}\rho = \frac{1}{q}(\alpha\dot{x}^2 + \beta\dot{y}^2)$......(9).

[Compare below, § 124 (2) and (1).]

And for σ, the angular velocity of spinning, the obvious proposition stated in the preceding large print gives

$$\sigma = q\left(\frac{1}{\gamma} - \frac{1}{\gamma'}\right) \ldots\ldots\ldots\ldots\ldots\ldots (10),$$

if $\frac{1}{\gamma}$ and $\frac{1}{\gamma'}$, be the curvatures of the projections on the tangent plane of the fixed and moveable traces. [Compare below, § 124 (3).]

From (1) and (2) it follows that

When one of the surfaces is a plane, and the trace on the other is a line of curvature (§ 130), the rolling is direct.

When the trace on each body is a line of curvature, the rolling is direct. *Generally*, the rolling is direct when the twists of infinitely narrow bands (§ 120) of the two surfaces, along the traces, are equal and in the same direction.

112. Imagine the traces constructed of rigid matter, and all the rest of each body removed. We may repeat the motion with these curves alone. The difference of the circumstances now supposed will only be experienced if we vary the direction of the instantaneous axis. In the former case, we can only do this by introducing more or less of spinning, and if we do so we *alter the trace* on each body. In the latter, we have always the same moveable curve rolling on the same fixed curve; and therefore a determinate line perpendicular to their common tangent for one component of the rotation; but along with this we may give arbitrarily any velocity of twisting round the common tangent. The consideration of this case is very in-

Curve rolling on curve.

structive. It may be roughly imitated in practice by two stiff wires bent into the forms of the given curves, and prevented from crossing each other by a short piece of elastic tube clasping them together.

First, let them be both plane curves, and kept in one plane. We have then *rolling*, as of one cylinder on another.

Let ρ' be the radius of curvature of the rolling, ρ of the fixed, cylinder; ω the angular velocity of the former, V the linear velocity of the point of contact. We have

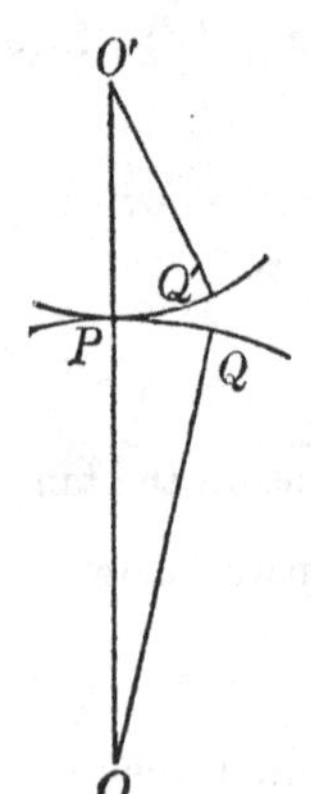

$$\omega = \left(\frac{1}{\rho} + \frac{1}{\rho'}\right) V.$$

For, in the figure, suppose P to be at any time the point of contact, and Q and Q' the points which are to be in contact after an infinitely small interval t; O, O' the centres of curvature; $POQ = \theta$; $PO'Q' = \theta'$.

Then $PQ = PQ' =$ space described by point of contact. In symbols $\rho\theta = \rho'\theta' = Vt$.

Also, before $O'Q'$ and OQ can coincide in direction, the former must evidently turn through an angle $\theta + \theta'$.

Therefore $\omega t = \theta + \theta'$; and by eliminating θ and θ', and dividing by t, we get the above result.

It is to be understood, that as the radii of curvature have been considered positive here when both surfaces are convex, the negative sign must be introduced for either radius when the corresponding curve is concave.

Angular velocity of rolling in a plane.

Hence the angular velocity of the rolling curve is in this case equal to the product of the linear velocity of the point of contact by the sum or difference of the curvatures, according as the curves are both convex, or one concave and the other convex.

Plane curves not in same plane.

113. When the curves are both plane, but in different planes, the plane in which the rolling takes place divides the angle between the plane of one of the curves, and that of the other produced through the common tangent line, into parts whose sines are inversely as the curvatures in them respectively; and the angular velocity is equal to the linear velocity of the point of contact multiplied by the difference of the projections of the two curvatures on this plane. The projections of the circles of the two curvatures on the plane of the common tangent and of the instantaneous axis coincide.

Plane curves not in same plane.

For, let PQ, Pp be equal arcs of the two curves as before, and let PR be taken in the common tangent (*i.e.*, the intersection of the planes of the curves) equal to each. Then QR, pR are ultimately perpendicular to PR.

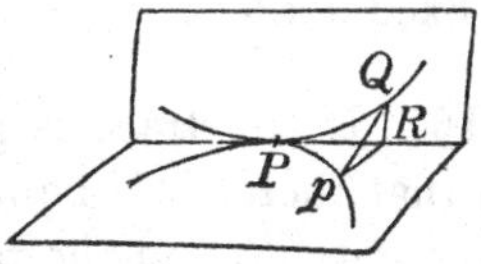

Hence $$pR = \frac{PR^2}{2\sigma},$$

$$QR = \frac{PR^2}{2\rho}.$$

Q R P p

Also, $\angle\, QRp = \alpha$, the angle between the planes of the curves.

We have $$Qp^2 = \frac{PR^4}{4}\left(\frac{1}{\sigma^2} + \frac{1}{\rho^2} - \frac{2}{\sigma\rho}\cos\alpha\right).$$

Therefore if ω be the velocity of rotation as before,

$$\omega = V\sqrt{\frac{1}{\sigma^2} + \frac{1}{\rho^2} - \frac{2\cos\alpha}{\sigma\rho}}.$$

Also the instantaneous axis is evidently perpendicular, and therefore the plane of rotation parallel, to Qp. Whence the above. In the case of $\alpha = \pi$, this agrees with the result of § 112.

A good example of this is the case of a coin spinning on a table (mixed rolling and spinning motion), as its plane becomes gradually horizontal. In this case the curvàtures become more and more nearly equal, and the angle between the planes of the curves smaller and smaller. Thus the resultant angular velocity becomes exceedingly small, and the motion of the point of contact very great compared with it.

114. The preceding results are, of course, applicable to tortuous as well as to plane curves; it is merely requisite to substitute the osculating plane of the former for the plane of the latter.

Curve rolling on curve: two degrees of freedom.

115. We come next to the case of a curve rolling, with or without spinning, on a surface.

Curve rolling on surface: three degrees of freedom.

It may, of course, roll on any curve traced on the surface. When this curve is given, the moving curve may, while rolling along it, revolve arbitrarily round the tangent. But the com-

Curve rolling on surface: three degrees of freedom.

ponent instantaneous axis perpendicular to the common tangent, that is, the axis of the direct rolling of one curve on the other, is determinate, § 113. If this axis does not lie in the surface, there is spinning. Hence, when the trace on the surface is given, there are two independent variables in the motion; the space traversed by the point of contact, and the inclination of the moving curve's osculating plane to the tangent plane of the fixed surface.

Trace prescribed and no spinning permitted.

116. If the trace is given, and it be prescribed as a condition that there shall be no spinning, the angular position of the rolling curve round the tangent at the point of contact is determinate. For in this case the instantaneous axis must be in the tangent plane to the surface. Hence, if we resolve the rotation into components round the tangent line, and round an axis perpendicular to it, the latter must be in the tangent plane. Thus the rolling, as of curve on curve, must be in a normal plane to the surface; and therefore (§§ 114, 113) the rolling curve must be always so situated relatively to its trace on the surface that the projections of the two curves on the tangent plane may be of coincident curvature.

Two degrees of freedom.

The curve, as it rolls on, must continually revolve about the tangent line to it at the point of contact with the surface, so as in every position to fulfil this condition.

Let a denote the inclination of the plane of curvature of the trace, to the normal to the surface at any point, a' the same for the plane of the rolling curve; $\frac{1}{\rho}$, $\frac{1}{\rho'}$ their curvatures. We reckon a as obtuse, and a' acute, when the two curves lie on opposite sides of the tangent plane. Then

$$\frac{1}{\rho'}\sin a' = \frac{1}{\rho}\sin a,$$

which fixes a' or the position of the rolling curve when the point of contact is given.

Angular velocity of direct rolling.

Let ω be the angular velocity of rolling about an axis perpendicular to the tangent, ϖ that of twisting about the tangent, and let V be the linear velocity of the point of contact. Then, since $\frac{1}{\rho'}\cos a'$

and $-\frac{1}{\rho}\cos\alpha$ (each positive when the curves lie on opposite sides of the tangent plane) are the projections of the two curvatures on a plane through the normal to the surface containing their common tangent, we have, by § 112,

Angular velocity of direct rolling.

$$\omega = V\left(\frac{1}{\rho'}\cos\alpha' - \frac{1}{\rho}\cos\alpha\right),$$

α' being determined by the preceding equation. Let τ and τ' denote the tortuosities of the trace, and of the rolling curve, respectively. Then, first, if the curves were both plane, we see that one rolling on the other about an axis always perpendicular to their common tangent could never change the inclination of their planes. Hence, secondly, if they are both tortuous, such rolling will alter the inclination of their osculating planes by an indefinitely small amount $(\tau - \tau')\,ds$ during rolling which shifts the point of contact over an arc ds. Now α is a known function of s if the trace is given, and therefore so also is α'. But $\alpha - \alpha'$ is the inclination of the osculating planes, hence

Angular velocity round tangent.

$$V\left\{\frac{d(\alpha-\alpha')}{ds} - (\tau - \tau')\right\} = \varpi.$$

117. Next, for one surface rolling and spinning on another. First, if the trace on each is given, we have the case of § 113 or § 115, one curve rolling on another, with this farther condition, that the former must *revolve* round the tangent to the two curves so as to keep the tangent planes of the two surfaces coincident.

Surface on surface.

It is well to observe that when the points in contact, and the two traces, are given, the position of the moveable surface is quite determinate, being found thus:—Place it in contact with the fixed surface, the given points together, and *spin* it about the common normal till the tangent lines to the traces coincide.

Both traces prescribed: one degree of freedom.

Hence when both the traces are given the condition of no spinning cannot be imposed. During the rolling there must in general be spinning, such as to keep the tangents to the two traces coincident. The rolling along the trace is due to rotation round the line perpendicular to it in the tangent plane. The whole rolling is the resultant of this rotation and a rotation about the tangent line required to keep the two tangent planes coincident.

Surface on surface, both traces prescribed; one degree of freedom.

In this case, then, there is but one independent variable—the space passed over by the point of contact: and when the velocity of the point of contact is given, the resultant angular velocity, and the direction of the instantaneous axis of the rolling body are determinate. We have thus a sufficiently clear view of the general character of the motion in question, but it is right that we consider it more closely, as it introduces us very naturally to an important question, the measurement of the *twist* of a rod, wire, or narrow plate, a quantity wholly distinct from the *tortuosity* of its axis (§ 7).

118. Suppose all of each surface cut away except an infinitely narrow strip, including the trace of the rolling. Then we have the rolling of one of these strips upon the other, each having at every point a definite curvature, tortuosity, and twist.

Twist.

119. Suppose a flat bar of small section to have been bent (the requisite amount of stretching and contraction of its edges being admissible) so that its axis assumes the form of any plane or tortuous curve. If it be unbent without twisting, *i.e.*, if the curvature of each element of the bar be removed by bending it through the requisite angle in the osculating plane, and it be found untwisted when thus rendered straight, it had no *twist* in its original form. This case is, of course, included in the general theory of *twist*, which is the subject of the following sections.

Axis and transverse.

120. A bent or straight rod of circular or any other form of section being given, a line through the centres, or any other chosen points of its sections, may be called its *axis*. Mark a line on its side all along its length, such that it shall be a straight line parallel to the axis when the rod is unbent and untwisted. A line drawn from any point of the axis perpendicular to this side line of reference, is called the *transverse* of the rod at this point.

The whole twist of any length of a straight rod is the angle between the transverses of its ends. The average twist is the integral twist divided by the length. The twist at any point is the average twist in an infinitely short length through this point; in other words, it is the rate of rotation of its transverse per unit of length along it.

The twist of a curved, plane or tortuous, rod at any point is the rate of component rotation of its transverse round its tangent line, per unit of length along it. **Twist.**

If t be the twist at any point, $\int tds$ over any length is the integral twist in this length.

121. Integral twist in a curved rod, although readily defined, as above, in the language of the integral calculus, cannot be exhibited as the angle between any two lines readily constructible. The following considerations show how it is to be reckoned, and lead to a geometrical construction exhibiting it in a spherical diagram, for a rod bent and twisted in any manner:—

122. If the axis of the rod forms a plane curve lying in one plane, the integral twist is clearly the difference between the inclinations of the transverse at its ends to its plane. For if it be simply unbent, without altering the twist in any part, the inclination of each transverse to the plane in which its curvature lay will remain unchanged; and as the axis of the rod now has become a straight line in this plane, the mutual inclination of the transverses at any two points of it has become equal to the difference of their inclinations to the plane. **Estimation of integral twist: In a plane curve;**

123. No simple application of this rule can be made to a tortuous curve, in consequence of the change of the plane of curvature from point to point along it; but, instead, we may proceed thus:—

First, Let us suppose the plane of curvature of the axis of the wire to remain constant through finite portions of the curve, and to change abruptly by finite angles from one such portion to the next (a supposition which involves no angular points, that is to say, no infinite curvature, in the curve). Let planes parallel to the planes of curvature of three successive portions, PQ, QR, RS (not shown in the diagram), cut a spherical surface in the great circles GAG', ACA', CE. The radii of the sphere parallel to the tangents at the points Q and R of the curve where its curvature changes will cut its surface in A and C, the intersections of these circles. **In a curve consisting of plane portions in different planes.**

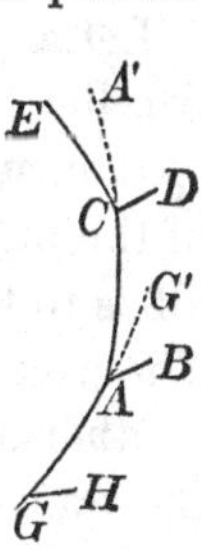

Let G be the point in which the radius of the sphere parallel to the tangent at P cuts the surface; and let GH, AB, CD (lines necessarily in tangent planes to the spherical surface), be parallels to the transverses of the bar drawn from the points P, Q, R of its axis. Then (§ 122) the twist from P to Q is equal to the difference of the angles HGA and BAG'; and the twist from Q to R is equal to the difference between BAC and DCA'. Hence the whole twist from P to R is equal to

Estimation of integral twist: in a curve consisting of plane portions in different planes.

$$HGA - BAG' + BAC - DCA',$$

or, which is the same thing,

$$A'CE + G'AC - (DCE - HGA).$$

Continuing thus through any length of rod, made up of portions curved in different planes, we infer that the integral twist between any two points of it is equal to the sum of the exterior angles in the spherical diagram, wanting the excess of the inclination of the transverse at the second point to the plane of curvature at the second point above the inclination at the first point to the plane of curvature at the first point. The sum of those exterior angles is what is defined below as the "change of direction in the spherical surface" from the first to the last side of the polygon of great circles. When the polygon is closed, and the sum includes all its exterior angles, it is (§ 134) equal to 2π wanting the area enclosed if the radius of the spherical surface be unity. The construction we have made obviously holds in the limiting case, when the lengths of the plane portions are infinitely small, and is therefore applicable to a wire forming a tortuous curve with continuously varying plane of curvature, for which it gives the following conclusion:—

In a continuously tortuous curve.

Let a point move uniformly along the axis of the bar: and, parallel to the tangent at every instant, draw a radius of a sphere cutting the spherical surface in a curve, the hodograph of the moving point. From points of this hodograph draw parallels to the transverses of the corresponding points of the bar. The excess of the change of direction (§ 135) from any point to another of the hodograph, above the increase of its inclination to the transverse, is equal to the twist in the corresponding part of the bar.

Estimation of integral twist: in a continuously tortuous curve.

The annexed diagram, showing the hodograph and the parallels to the transverses, illustrates this rule. Thus, for instance, the excess of the change of direction in the spherical surface along the hodograph from A to C, above $DCS—BAT$, is equal to the twist in the bar between the points of it to which A and C correspond. Or, again, if we consider a portion of the bar from any point of it, to another point at which the tangent to its axis is parallel to the tangent at its first point, we shall have a closed curve as the spherical hodograph; and if A be the point of the hodograph corresponding to them, and AB and AB' the parallels to the transverses, the whole twist in the included part of the bar will be equal to the change of direction all round the hodograph, wanting the excess of the exterior angle $B'AT$ above the angle BAT; that is to say, the whole twist will be equal to the excess of the angle BAB' above the area enclosed by the hodograph.

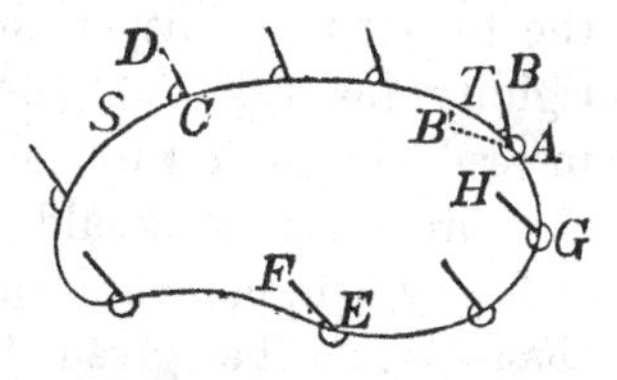

The principles of twist thus developed are of vital importance in the theory of rope-making, especially the construction and the dynamics of wire ropes and submarine cables, elastic bars, and spiral springs.

Dynamics of twist in kinks.

For example: take a piece of steel pianoforte-wire carefully straightened, so that when free from stress it is straight: bend it into a circle and join the ends securely so that there can be no turning of one relatively to the other. Do this first without torsion: then twist the ring into a figure of 8, and tie the two parts together at the crossing. The area of the spherical hodograph is zero, and therefore there is one full turn (2π) of twist; which (§ 600 below) is uniformly distributed throughout the length of the wire. The form of the wire, (which is not in a plane,) will be investigated in § 610. Meantime we can see that the "torsional couples" in the normal sections farthest from the crossing give rise to forces by which the tie at the crossing is pulled in opposite directions perpendicular to the plane of the crossing. Thus if the tie is cut the wire springs back into the circular form. Now do the same thing again,

Dynamics of twist in kinks.

beginning with a straight wire, but giving it one full turn (2π) of twist before bending it into the circle. The wire will stay in the 8 form without any pull on the tie. Whether the circular or the 8 form is stable or unstable depends on the relations between torsional and flexural rigidity. If the torsional rigidity is small in comparison with the flexural rigidity [as (§§ 703, 704, 705, 709) would be the case if, instead of round wire, a rod of + shaped section were used], the circular form would be stable, the 8 unstable.

Lastly, suppose any degree of twist, either more or less than 2π, to be given before bending into the circle. The circular form, which is always a figure of free equilibrium, may be stable or unstable, according as the ratio of torsional to flexural rigidity is more or less than a certain value depending on the actual degree of twist. The tortuous 8 form is not (except in the case of whole twist $= 2\pi$, when it becomes the plane elastic lemniscate of Fig. 4, § 610,) a continuous figure of free equilibrium, but involves a positive pressure of the two crossing parts on one another when the twist $> 2\pi$, and a negative pressure (or a pull on the tie) between them when twist $< 2\pi$: and with this force it is a figure of stable equilibrium.

Surface rolling on surface; both traces given.

124. Returning to the motion of one surface rolling and spinning on another, the trace on each being given, we may consider that, of each, the curvature (§ 6), the tortuosity (§ 7), and the twist reckoned according to transverses in the tangent plane of the surface, are known; and the subject is fully specified in § 117 above.

Let $\frac{1}{\rho'}$ and $\frac{1}{\rho}$ be the curvatures of the traces on the rolling and fixed surfaces respectively; α' and α the inclinations of their planes of curvature to the normal to the tangent plane, reckoned as in § 116; τ' and τ their tortuosities; t' and t their twists; and q the velocity of the point of contact. All these being known, it is required to find:—

ω the angular velocity of rotation about the transverse of the traces; that is to say, the line in the tangent plane perpendicular to their tangent line,

ϖ the angular velocity of rotation about the tangent line, and

σ ,, ,, of spinning.

We have

Surface rolling on surface; both traces given.

$$\omega = q\left(\frac{1}{\rho'}\cos\alpha' - \frac{1}{\rho}\cos\alpha\right) \qquad (1),$$

$$\varpi = q(t - t') = q\left\{\frac{d(\alpha - \alpha')}{ds} - (\tau - \tau')\right\} \qquad (2),$$

and

$$\sigma = q\left(\frac{1}{\rho'}\sin\alpha' - \frac{1}{\rho}\sin\alpha\right) \qquad (3).$$

These three formulas are respectively equivalent to (9), (8), and (10) of § 111.

125. In the same case, suppose the trace on *one* only of the surfaces to be given. We may evidently impose the condition of no spinning, and then the trace on the other is determinate. This case of motion is thoroughly examined in § 137, below.

Surface rolling on surface without spinning.

The condition is that the projections of the curvatures of the two traces on the common tangent plane must coincide.

If $\frac{1}{r'}$ and $\frac{1}{r}$ be the curvatures of the rolling and stationary surfaces in a normal section of each through the tangent line to the trace, and if α, α', ρ, ρ' have their meanings of § 124,

$\rho' = r'\cos\alpha'$, $\rho = r\cos\alpha$ (Meunier's Theorem, § 129, below).

But $\frac{1}{\rho'}\sin\alpha' = \frac{1}{\rho}\sin\alpha$, hence $\tan\alpha' = \frac{r'}{r}\tan\alpha$, the condition required.

126. If a straight rod with a straight line marked on one side of it be bent along any curve on a spherical surface, so that the marked line is laid in contact with the spherical surface, it acquires no twist in the operation. For if it is laid so along any finite arc of a small circle there will clearly be no twist. And no twist is produced in continuing from any point along another small circle having a common tangent with the first at this point.

Examples of tortuosity and twist.

If a rod be bent round a cylinder so that a line marked along one side of it may lie in contact with the cylinder, or if, what presents somewhat more readily the view now de-

Examples of tortuosity and twist.

sired, we wind a straight ribbon spirally on a cylinder, the axis of bending is parallel to that of the cylinder, and therefore oblique to the axis of the rod or ribbon. We may therefore resolve the instantaneous rotation which constitutes the bending at any instant into two components, one round a line perpendicular to the axis of the rod, which is pure bending, and the other round the axis of the rod, which is pure twist.

The twist at any point in a rod or ribbon, so wound on a circular cylinder, and constituting a uniform helix, is

$$\frac{\cos\alpha\sin\alpha}{r},$$

if r be the radius of the cylinder and α the inclination of the spiral. For if V be the velocity at which the bend proceeds along the previously straight wire or ribbon, $\frac{V\cos\alpha}{r}$ will be the angular velocity of the instantaneous rotation round the line of bending (parallel to the axis), and therefore

$$\frac{V\cos\alpha}{r}\sin\alpha \text{ and } \frac{V\cos\alpha}{r}\cos\alpha$$

are the angular velocities of twisting and of pure bending respectively.

From the latter component we may infer that the curvature of the helix is

$$\frac{\cos^2\alpha}{r},$$

a known result, which agrees with the expression used above (§ 13).

127. The hodograph in this case is a small circle of the sphere. If the specified condition as to the mode of laying on of the rod on the cylinder is fulfilled, the transverses of the spiral rod will be parallel at points along it separated by one or more whole turns. Hence the integral twist in a single turn is equal to the excess of four right angles above the spherical area enclosed by the hodograph. If α be the inclination of the spiral, $\frac{1}{2}\pi - \alpha$ will be the arc-radius of the hodograph, and therefore its area is $2\pi(1-\sin\alpha)$. Hence the integral twist in a turn of the spiral is $2\pi\sin\alpha$, which agrees with the result previously obtained (§ 126).

128. As a preliminary to the further consideration of the rolling of one surface on another, and as useful in various parts of our subject, we may now take up a few points connected with the curvature of surfaces.

Curvature of surface.

The tangent plane at any point of a surface may or may not cut it at that point. In the former case, the surface bends away from the tangent plane partly towards one side of it, and partly towards the other, and has thus, in some of its normal sections, curvatures oppositely directed to those in others. In the latter case, the surface on every side of the point bends away from the same side of its tangent plane, and the curvatures of all normal sections are similarly directed. Thus we may divide curved surfaces into *Anticlastic* and *Synclastic*. A saddle gives a good example of the former class; a ball of the latter. Curvatures in opposite directions, with reference to the tangent plane, have of course different signs. The outer portion of an anchor-ring is synclastic, the inner anticlastic.

Synclastic, and anticlastic surfaces.

129. *Meunier's Theorem.*—The curvature of an oblique section of a surface is equal to that of the normal section through the same tangent line multiplied by the secant of the inclination of the planes of the sections. This is evident from the most elementary considerations regarding projections.

Curvature of oblique sections.

130. *Euler's Theorem.*—There are at every point of a synclastic surface two normal sections, in one of which the curvature is a maximum, in the other a minimum; and these are at right angles to each other.

Principal curvatures.

In an anticlastic surface there is maximum curvature (but in opposite directions) in the two normal sections whose planes bisect the angles between the lines in which the surface cuts its tangent plane. On account of the difference of sign, these may be considered as a maximum and a minimum.

Generally the sum of the curvatures at a point, in any two normal planes at right angles to each other, is independent of the position of these planes.

Sum of curvatures in normal sections at right angles to each other.

Taking the tangent plane as that of x, y, and the origin at the point of contact, and putting

$$\left(\frac{d^2z}{dx^2}\right)_0 = A, \left(\frac{d^2z}{dxdy}\right)_0 = B, \left(\frac{d^2z}{dy^2}\right)_0 = C;$$

we have $$z = \frac{1}{2}(Ax^2 + 2Bxy + Cy^2) + \text{etc.} \qquad (1)$$

The curvature of the normal section which passes through the point x, y, z is (in the limit)

$$\frac{1}{r} = \frac{2z}{x^2+y^2} = \frac{Ax^2 + 2Bxy + Cy^2}{x^2+y^2}.$$

If the section be inclined at an angle θ to the plane of XZ, this becomes

$$\frac{1}{r} = A\cos^2\theta + 2B\sin\theta\cos\theta + C\sin^2\theta. \qquad (2)$$

Hence, if $\frac{1}{r}$ and $\frac{1}{s}$ be curvatures in normal sections at right angles to each other,

$$\frac{1}{r} + \frac{1}{s} = A + C = \text{constant}.$$

(2) may be written

$$\frac{1}{r} = \frac{1}{2}\{A(1+\cos 2\theta) + 2B\sin 2\theta + C(1-\cos 2\theta)\}$$

$$= \frac{1}{2}\{\overline{A+C} + \overline{A-C}\cos 2\theta + 2B\sin 2\theta\},$$

or if $$\frac{1}{2}(A-C) = R\cos 2\alpha, \quad B = R\sin 2\alpha,$$

that is $$R = \sqrt{\left\{\frac{1}{4}(A-C)^2 + B^2\right\}}, \text{ and } \tan 2\alpha = \frac{2B}{A-C},$$

we have $$\frac{1}{r} = \frac{1}{2}(A+C) + \sqrt{\left\{\frac{1}{4}(A-C)^2 + B^2\right\}}\cos 2(\theta - \alpha).$$

Principal normal sections.

The maximum and minimum curvatures are therefore those in normal places at right angles to each other for which $\theta = \alpha$ and $\theta = \alpha + \frac{\pi}{2}$, and are respectively

$$\frac{1}{2}(A+C) \pm \sqrt{\left\{\frac{1}{4}(A-C)^2 + B^2\right\}}.$$

Hence their product is $AC - B^2$.

If this be positive we have a synclastic, if negative an anticlastic, surface. If it be zero we have one curvature only, and the surface is *cylindrical* at the point considered. It is demonstrated

(§ 152, below) that if this condition is fulfilled at every point, the surface is "developable" (§ 139, below).

Principal normal sections.

By (1) a plane parallel to the tangent plane and very near it cuts the surface in an ellipse, hyperbola, or two parallel straight lines, in the three cases respectively. This section, whose nature informs us as to whether the curvature be synclastic, anticlastic, or cylindrical, at any point, was called by Dupin the *Indicatrix*.

Definition of Line of Curvature.

A line of curvature of a surface is a line which at every point is cotangential with normal section of maximum or minimum curvature.

Shortest line between two points on a surface.

131. Let P, p be two points of a surface infinitely near to each other, and let r be the radius of curvature of a normal section passing through them. Then the radius of curvature of an oblique section through the same points, inclined to the former at an angle α, is (§ 129) $r \cos \alpha$. Also the length along the normal section, from P to p, is less than that along the oblique section—since a given chord cuts off an arc from a circle, longer the less the radius of that circle.

If a be the length of the chord Pp, we have

$$\text{Distance } Pp \text{ along normal section} = 2r \sin^{-1} \frac{a}{2r} = a\left(1 + \frac{a^2}{24r^2}\right),$$

$$\text{,, \quad ,, \quad oblique section} = a\left(1 + \frac{a^2}{24r^2 \cos^2 a}\right).$$

132. Hence, if the shortest possible line be drawn from one point of a surface to another, its plane of curvature is everywhere perpendicular to the surface.

Geodetic Lines.

Such a curve is called a *Geodetic* line. And it is easy to see that it is the line in which a flexible and inextensible string would touch the surface if stretched between those points, the surface being supposed smooth.

133. If an infinitely narrow ribbon be laid on a surface along a geodetic line, its twist is equal to the tortuosity of its axis at each point. We have seen (§ 125) that when one body rolls on another without spinning, the projections of the traces on the common tangent plane agree in curvature at the point

Shortest line between two points on a surface.

of contact. Hence, if one of the surfaces be a plane, and the trace on the other be a geodetic line, the trace on the plane is a straight line. Conversely, if the trace on the plane be a straight line, that on the surface is a geodetic line.

And, quite generally, if the given trace be a geodetic line, the other trace is also a geodetic line.

Spherical excess.

134. The area of a spherical triangle (on a sphere of unit radius) is known to be equal to the "spherical excess," *i.e.*, the excess of the sum of its angles over two right angles, or the excess of four right angles over the sum of its exterior angles.

Area of spherical polygon.

The area of a spherical polygon whose n sides are portions of great circles—*i.e.*, geodetic lines—is to that of the hemisphere as the excess of four right angles over the sum of its exterior angles is to four right angles. (We may call this the "spherical excess" of the polygon.)

For the area of a spherical triangle is known to be equal to

$$A + B + C - \pi.$$

Divide the polygon into n such triangles, with a common vertex, the angles about which, of course, amount to 2π.

$$\begin{aligned} \text{Area} &= \text{sum of interior angles of triangles} - n\pi \\ &= 2\pi + \text{sum of interior angles of polygon} - n\pi \\ &= 2\pi - \text{sum of exterior angle of polygon.} \end{aligned}$$

Reciprocal polars on a sphere.

Given an open or closed spherical polygon, or line on the surface of a sphere composed of consecutive arcs of great circles. Take either pole of the first of these arcs, and the corresponding poles of all the others (all the poles to be on the right hand, or all on the left, of a traveller advancing along the given great circle arcs in order). Draw great circle arcs from the first of these poles to the second, the second to the third, and so on in order. Another closed or open polygon, constituting what is called the polar diagram to the given polygon, is thus obtained. The sides of the second polygon are evidently equal to the exterior angles in the first; and the exterior angles of the second are equal to the sides of the first. Hence the relation between the two diagrams is reciprocal, or each is polar to the other. The polar figure to any continuous curve on a spherical surface is the locus of the ultimate intersections of great circles equatorial to points taken infinitely near each other along it.

Reciprocal polars on a sphere.

The area of a closed spherical figure is, consequently, according to what we have just seen, equal to the excess of 2π above the periphery of its polar, if the radius of the sphere be unity.

Integral change of direction in a surface.

135. If a point move on a surface along a figure whose sides are geodetic lines, the sum of the exterior angles of this polygon is defined to be the *integral change of the direction in the surface.*

In great circle sailing, unless a vessel sail on the equator, or on a meridian, her course, as indicated by points of the compass (true, not magnetic, for the latter change even on a meridian), perpetually changes. Yet just as we say her direction does not change if she sail in a meridian, or in the equator, so we ought to say her direction does not change if she moves in *any* great circle. Now, the great circle is the geodetic line on the sphere, and by extending these remarks to other curved surfaces, we see the connexion of the above definition with that in the case of a plane polygon (§ 10).

Change of direction in a surface, of any arc traced on it.

Note.—We cannot define integral change of direction here by any angle directly constructible from the first and last tangents to the path, as was done (§ 10) in the case of a plane curve or polygon; but from §§ 125 and 133 we have the following statement:—The whole change of direction in a curved surface, from one end to another of any arc of a curve traced on it, is equal to the change of direction from end to end of the trace of this arc on a plane by pure rolling.

Integral curvature.

136. *Def.* The excess of four right angles above the integral change of direction from one side to the same side next time in going round a closed polygon of geodetic lines on a curved surface, is the *integral curvature* of the enclosed portion of surface. This excess is zero in the case of a polygon traced on a plane. We shall presently see that this corresponds exactly to what Gauss has called the *curvatura integra.*

Curvatura integra.

Def. (Gauss.) The *curvatura integra* of any given portion of a curved surface, is the area enclosed on a spherical surface of unit radius by a straight line drawn from its centre, parallel to a normal to the surface, the normal being carried round the boundary of the given portion.

Horograph.

The curve thus traced on the sphere is called the *Horograph* of the given portion of curved surface.

The *average curvature* of any portion of a curved surface is the integral curvature divided by the area. The *specific curvature* of a curved surface at any point is the average curvature of an infinitely small area of it round that point.

Change of direction round the boundary in the surface, together with area of the horograph, equals four right angles: or "Integral Curvature" equals "Curvatura Integra."

137. The excess of 2π above the change of direction, in a surface, of a point moving round any closed curve on it, is equal to the area of the horograph of the enclosed portion of surface.

Let a tangent plane roll without spinning on the surface over every point of the bounding line. (Its instantaneous axis will always lie in it, and pass through the point of contact, but will not, as we have seen, be at right angles to the given bounding curve, except when the twist of a narrow ribbon of the surface along this curve is nothing.) Considering the auxiliary sphere of unit radius, used in Gauss's definition, and the moving line through its centre, we perceive that the motion of this line is, at each instant, in a plane perpendicular to the instantaneous axis of the tangent plane to the given surface. The direction of motion of the point which cuts out the area on the spherical surface is therefore perpendicular to this instantaneous axis. Hence, if we roll a tangent plane on the spherical surface also, making it keep time with the other, the trace on this tangent plane will be a curve always perpendicular to the instantaneous axis of each tangent plane. The change of direction, in the spherical surface, of the point moving round and cutting out the area, being equal to the change of direction in its own trace on its own tangent plane (§ 135), is therefore equal to the change of direction of the instantaneous axis in the tangent plane to the given surface reckoned from a line fixed relatively to this plane. But having rolled all round, and being in position to roll round again, the instantaneous axis of the fresh start must be inclined to the trace at the same angle as in the beginning. Hence the change of direction of the instantaneous axis in either tangent plane is equal to the change of direction, in the given surface, of

Curvatura integra, and horograph.

a point going all round the boundary of the given portion of it (§ 135); to which, therefore, the change of direction, in the spherical surface, of the point going all round the spherical area is equal. But, by the well-known theorem (§ 134) of the "spherical excess," this change of direction subtracted from 2π leaves the spherical area. Hence the spherical area, called by Gauss the *curvatura integra*, is equal to 2π wanting the change of direction in going round the boundary.

Curvatura integra, and horograph.

It will be perceived that when the two rollings we have considered are each complete, each tangent plane will have come back to be parallel to its original position, but any fixed line in it will have changed direction through an angle equal to the equal changes of direction just considered.

Note.—The two rolling tangent planes are at each instant parallel to one another, and a fixed line relatively to one drawn at any time parallel to a fixed line relatively to the other, remains parallel to the last-mentioned line.

If, instead of the closed curve, we have a closed polygon of geodetic lines on the given surface, the trace of the rolling of its tangent plane will be an unclosed rectilineal polygon. If each geodetic were a plane curve (which could only be if the given surface were spherical), the instantaneous axis would be always perpendicular to the particular side of this polygon which is rolled on at the instant; and, of course, the spherical area on the auxiliary sphere would be a similar polygon to the given one. But the given surface being other than spherical, there must (except in the particular case of some of the geodetics being lines of curvature) be tortuosity in every geodetic of the closed polygon; or, which is the same thing, twist in the corresponding ribbons of the surface. Hence the portion of the whole trace on the second rolling tangent plane which corresponds to any one side of the given geodetic polygon, must in general be a curve; and as there will generally be finite angles in the second rolling corresponding to (but not equal to) those in the first, the trace of the second on its tangent plane will be an unclosed polygon of curves. The trace of the same rolling on the spherical surface in which it takes place will generally be a spherical polygon, not of great circle arcs, but of other curves. The sum of the exterior angles of this polygon, and of the changes of direction from one end to the other of each of its sides, is the whole change of direction considered, and is, by the proper

Curvatura integra, and horograph.

application of the theorem of § 134, equal to 2π wanting the spherical area enclosed.

Or again, if, instead of a geodetic polygon as the given curve, we have a polygon of curves, each fulfilling the condition that the normal to the surface through any point of it is parallel to a fixed plane; one plane for the first curve, another for the second, and so on; then the figure on the auxiliary spherical surface will be a polygon of arcs of great circles; its trace on its tangent plane will be an unclosed rectilineal polygon; and the trace of the given curve on the tangent plane of the first rolling will be an unclosed polygon of curves. The sum of changes of direction in these curves, and of exterior angles in passing from one to another of them, is of course equal to the change of direction in the given surface, in going round the given polygon of curves on it. The change of direction in the other will be simply the sum of the exterior angles of the spherical polygon, or of its rectilineal trace. Remark that in this case the instantaneous axis of the first rolling, being always perpendicular to that plane to which the normals are all parallel, remains parallel to one line, fixed with reference to the tangent plane, during rolling along each curved side, and also remains parallel to a fixed line in space.

Lastly, remark that although the whole change of direction of the trace in one tangent plane is equal to that in the trace on the other, when the rolling is completed round the given circuit; the changes of direction in the two are generally unequal in any part of the circuit. They may be equal for particular parts of the circuit, viz., between those points, if any, at which the instantaneous axis is equally inclined to the direction of the trace on the first tangent plane.

Any difficulty which may have been felt in reading this Section will be removed if the following exercises on the subject be performed.

(1) Find the horograph of an infinitely small circular area of any continuous curved surface. It is an ellipse or a hyperbola according as the surface is synclastic or anticlastic (§ 128). Find the axes of the ellipse or hyperbola in either case.

(2) Find the horograph of the area cut off a synclastic surface by a plane parallel to the tangent plane at any given point of it, and infinitely near this point. Find and interpret the corresponding result for the case in which the surface is anticlastic in the neighbourhood of the given point.

(3) Let a tangent plane roll without spinning over the boundary of a given closed curve or geodetic polygon on any curved surface. Show that the points of the trace in the tangent plane which successively touch the same point of the given surface are at equal distances successively on the circumference of a circle, the angular values of the intermediate arcs being each $2\pi - K$ if taken in the direction in which the trace is actually described, and K if taken in the contrary direction, K being the "integral curvature" of the portion of the curved surface enclosed by the given curve or geodetic polygon. Hence if K be commensurable with π the trace on the tangent plane, however complicatedly autotomic it may be, is a finite closed curve or polygon.

Curvatura integra, and horograph.

(4) The trace by a tangent plane rolling successively over three principal quadrants bounding an eighth part of the circumference of an ellipsoid is represented in the accompanying diagram, the whole of which is traced when the tangent plane is

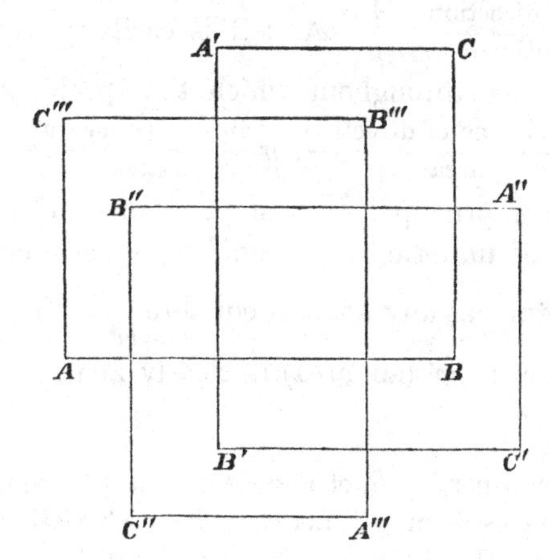

rolled four times over the stated boundary. A, B, C; A', B', C', &c. represent the points of the tangent plane touched in order by ends of the mean principal axis (A), the greatest principal axis (B), and least principal axis (C), and AB, BC, CA' are the lengths of the three principal quadrants.

138. It appears from what precedes, that the same equality or identity subsists between "whole curvature" in a plane arc and the excess of π above the angle between the terminal

Analogy between lines and surfaces as regards curvature.

Analogy between lines and surfaces as regards curvature.

tangents, as between "whole curvature" and excess of 2π above change of direction along the bounding line in the surface for any portion of a curved surface.

Or, according to Gauss, whereas the whole curvature in a plane arc is the angle between two lines parallel to the terminal normals, the whole curvature of a portion of curve surface is the solid angle of a cone formed by drawing lines from a point parallel to all normals through its boundary.

Again, average curvature in a plane curve is $\frac{\text{change of direction}}{\text{length}}$; and specific curvature, or, as it is commonly called, curvature, at any point of it $= \frac{\text{change of direction in infinitely small length}}{\text{length}}$.

Thus average curvature and specific curvature are for surfaces analogous to the corresponding terms for a plane curve.

Lastly, in a plane arc of uniform curvature, *i.e.*, in a circular arc, $\frac{\text{change of direction}}{\text{length}} = \frac{1}{\rho}$. And it is easily proved (as below) that, in a surface throughout which the specific curvature is uniform, $\frac{2\pi - \text{change of direction}}{\text{area}}$, or $\frac{\text{integral curvature}}{\text{area}}$, $= \frac{1}{\rho\rho'}$, where ρ and ρ' are the principal radii of curvature. Hence in a surface, whether of uniform or non-uniform specific curvature, the specific curvature at any point is equal to $\frac{1}{\rho\rho'}$. In geometry of three dimensions, $\rho\rho'$ (an area) is clearly analogous to ρ in a curve and plane.

Horograph.

Consider a portion S, of a surface of any curvature, bounded by a given closed curve. Let there be a spherical surface, radius r, and centre C. Draw a radius CQ, parallel to the normal at any point P of S. If this be done for every point of the boundary, the line so obtained encloses the spherical area used in Gauss's definition. Now let there be an infinitely small rectangle on S, at P, having for its sides arcs of angles ζ and ζ', on the normal sections of greatest and least curvature, and let their radii of curvature be denoted by ρ and ρ'. The lengths of these sides will be $\rho\zeta$ and $\rho'\zeta$ respectively. Its area will therefore be $\rho\rho'\zeta\zeta'$. The corresponding figure at Q on the spherical surface will be bounded by arcs of angles equal to those, and therefore of lengths $r\zeta$ and $r\zeta'$ respectively, and its area will be $r^2\zeta\zeta'$. Hence if $d\sigma$ denote this area, the area of the infinitely small portion of the given surface will be $\frac{\rho\rho' d\sigma}{r^2}$. In a surface for which $\rho\rho'$ is constant, the area is therefore $= \frac{\rho\rho'}{r^2}\iint d\sigma = \rho\rho' \times$ integral curvature.

Area of the horograph.

139. A perfectly flexible but inextensible surface is suggested, although not realized, by paper, thin sheet metal, or cloth, when the surface is plane; and by sheaths of pods, seed vessels, or the like, when it is not capable of being stretched flat without tearing. The process of changing the form of a surface by bending is called "*developing*." But the term "*Developable Surface*" is commonly restricted to such inextensible surfaces as can be developed into a plane, or, in common language, "smoothed flat."

Flexible and inextensible surface.

140. The geometry or kinematics of this subject is a great contrast to that of the flexible line (§ 14), and, in its merest elements, presents ideas not very easily apprehended, and subjects of investigation that have exercised, and perhaps even overtasked, the powers of some of the greatest mathematicians.

141. Some care is required to form a correct conception of what is a perfectly flexible inextensible surface. First let us consider a plane sheet of paper. It is very flexible, and we can easily form the conception from it of a sheet of ideal matter perfectly flexible. It is very inextensible; that is to say, it yields very little to any application of force tending to pull or stretch it in any direction, up to the strongest it can bear without tearing. It does, of course, stretch a little. It is easy to test that it stretches when under the influence of force, and that it contracts again when the force is removed, although not always to its original dimensions, as it may and generally does remain to some sensible extent permanently stretched. Also, flexure stretches one side and condenses the other temporarily; and, to a less extent, permanently. Under elasticity (§§ 717, 718, 719) we shall return to this. In the meantime, in considering illustrations of our kinematical propositions, it is necessary to anticipate such physical circumstances.

Surface inextensible in two directions.

142. Cloth woven in the simple common way, very fine muslin for instance, illustrates a surface perfectly inextensible in two directions (those of the warp and the woof), but susceptible of any amount of extension from 1 up to $\sqrt{2}$ along one diagonal, with contraction from 1 to 0 (each degree of extension along one diagonal having a corresponding determinate degree of contraction along the other, the relation being $e^2 + e'^2 = 2$, where $1 : e$ and $1 : e'$ are the ratios of elongation, which will be contraction in the case in which e or e' is < 1) in the other.

"Elastic finish" of muslin goods.

143. The flexure of a surface fulfilling any case of the geometrical condition just stated, presents an interesting subject for investigation, which we are reluctantly obliged to forego. The moist paper drapery that Albert Dürer used on his little lay figures must hang very differently from cloth. Perhaps the stiffness of the drapery in his pictures may be to some extent owing to the fact that he used the moist paper in preference to cloth on account of its superior flexibility, while unaware of the great distinction between them as regards extensibility. Fine muslin, prepared with starch or gum, is, during the process of drying, kept moving by a machine, which, by producing a to-and-fro relative angular motion of warp and woof, stretches and contracts the diagonals of its structure alternately, and thus prevents the parallelograms from becoming stiffened into rectangles.

Flexure of inextensible developable.

144. The flexure of an inextensible surface which can be plane, is a subject which has been well worked by geometrical investigators and writers, and, in its elements at least, presents little difficulty. The first elementary conception to be formed is, that such a surface (if perfectly flexible), taken plane in the first place, may be bent about any straight line ruled on it, so that the two plane parts may make any angle with one another.

Such a line is called a "generating line" of the surface to be formed.

Next, we may bend one of these plane parts about any other line which does not (within the limits of the sheet) intersect the former; and so on. If these lines are infinite in number, and the angles of bending infinitely small, but such that their sum may be finite, we have our plane surface bent into a curved surface, which is of course "developable" (§ 139).

Flexure of inextensible developable.

145. Lift a square of paper, free from folds, creases, or ragged edges, gently by one corner, or otherwise, without crushing or forcing it, or very gently by two points. It will hang in a form which is very rigorously a developable surface; for although it is not absolutely inextensible, yet the forces which tend to stretch or tear it, when it is treated as above described, are small enough to produce no sensible stretching. Indeed the greatest stretching it can experience without tearing, in any direction, is not such as can affect the form of the surface much when sharp flexures, singular points, etc., are kept clear of.

146. Prisms and cylinders (when the lines of bending, § 144, are parallel, and finite in number with finite angles, or infinite in number with infinitely small angles), and pyramids and cones (the lines of bending meeting in a point if produced), are clearly included.

147. If the generating lines, or line-edges of the angles of bending, are not parallel, they must meet, since they are in a plane when the surface is plane. If they do not meet all in one point, they must meet in several points: in general, each one meets its predecessor and its successor in different points.

148. There is still no difficulty in understanding the form of, say a square, or circle, of the plane surface when bent as explained above, provided it does not include any of these points of intersection. When the number is infinite, and the surface finitely curved, the developable lines will in general be tangents to a curve (the locus of the points of intersection when the number is infinite). This curve is called the *edge of regression*. The surface must clearly, when *complete* (according to mathematical ideas), consist of two sheets meeting in this edge

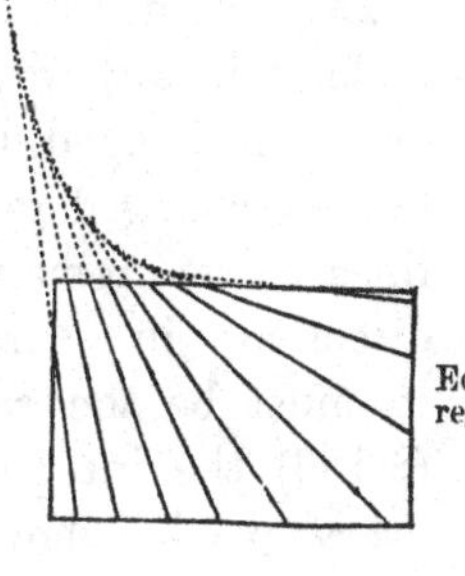

Edge of regression.

Edge of regression.

of regression (just as a cone consists of two sheets meeting in the vertex), because each tangent may be produced beyond the point of contact, instead of stopping at it, as in the annexed diagram.

Practical construction of a developable from its edge.

149. To construct a complete developable surface in two sheets from its edge of regression—

Lay one piece of perfectly flat, unwrinkled, smooth-cut paper on the top of another. Trace any curve on the upper,

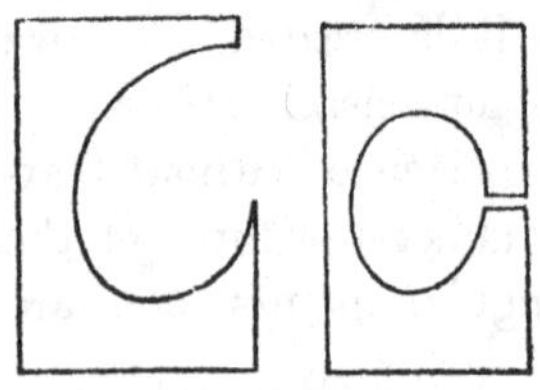

and let it have no point of inflection, but everywhere finite curvature. Cut the two papers along the curve and remove the convex portions. If the curve traced is closed, it must be cut open (see second diagram).

Attach the two sheets together by very slight paper or muslin clamps gummed to them along the common curved

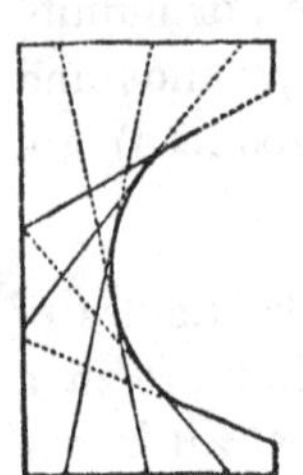

edge. These must be so slight as not to interfere sensibly with the flexure of the two sheets. Take hold of one corner of one sheet and lift the whole. The two will open out into the two sheets of a developable surface, of which the curve, bending into a curve of double curvature, is the edge of regression. The tangent to the curve drawn in one direction from the point of contact, will always lie in one of the sheets, and its continuation on the other side in the other sheet. Of course a double-sheeted developable polyhedron can be constructed by this process, by starting from a polygon instead of a curve.

General property of inextensible surface.

150. A flexible but perfectly inextensible surface, altered in form in any way possible for it, must keep any line traced on it unchanged in length; and hence any two intersecting lines unchanged in mutual inclination. Hence, also, geodetic lines must remain geodetic lines. Hence "the change of direction" in a surface, of a point going round any portion of it, must be the same, however this portion is bent. Hence (§ 136) the integral curvature remains the same in any and every portion however the surface is bent. Hence (§ 138,

Gauss's Theorem) the product of the principal radii of curvature at each point remains unchanged.

General property of inextensible surface.

151. The general statement of a converse proposition, expressing the condition that two given areas of curved surfaces may be bent one to fit the other, involves essentially some mode of specifying corresponding points on the two. A full investigation of the circumstances would be out of place here.

Surface of constant specific curvature.

152. In one case, however, a statement in the simplest possible terms is applicable. Any two surfaces, in each of which the specific curvature is the same at all points, and equal to that of the other, may be bent one to fit the other. Thus any surface of uniform positive specific curvature (*i.e.*, wholly convex one side, and concave the other) may be bent to fit a sphere whose radius is a mean proportional between its principal radii of curvature at any point. A surface of uniform negative, or anticlastic, curvature would fit an imaginary sphere, but the interpretation of this is not understood in the present condition of science. But practically, of any two surfaces of uniform anticlastic curvature, either may be bent to fit the other.

Geodetic triangles on such a surface.

153. It is to be remarked, that geodetic trigonometry on any surface of uniform positive, or synclastic, curvature, is identical with spherical trigonometry.

If $a=\frac{s}{\sqrt{\rho\rho'}}$, $b=\frac{t}{\sqrt{\rho\rho'}}$, $c=\frac{u}{\sqrt{\rho\rho'}}$, where s, t, u are the lengths of three geodetic lines joining three points on the surface, and if A, B, C denote the angles between the tangents to the geodetic lines at these points; we have six quantities which agree perfectly with the three sides and the three angles of a certain spherical triangle. A corresponding anticlastic trigonometry exists, although we are not aware that it has hitherto been noticed, for any surface of uniform anticlastic curvature. In a geodetic triangle on an anticlastic surface, the sum of the three angles is of course less than three right angles, and the difference, or "anticlastic defect" (like the "spherical excess"), is equal to the area divided by $\rho \times -\rho'$, where ρ and $-\rho'$ are positive.

Strain.

154. We have now to consider the very important kinematical conditions presented by the changes of volume or figure

Strain.

experienced by a solid or liquid mass, or by a group of points whose positions with regard to each other are subject to known conditions. Any such definite alteration of form or dimensions is called a *Strain*.

Thus a rod which becomes longer or shorter is strained. Water, when compressed, is strained. A stone, beam, or mass of metal, in a building or in a piece of framework, if condensed or dilated in any direction, or bent, twisted, or distorted in any way, is said to experience a strain. A ship is said to "strain" if, in launching, or when working in a heavy sea, the different parts of it experience relative motions.

Definition of homogeneous strain.

155. If, when the matter occupying any space is strained in any way, all pairs of points of its substance which are initially at equal distances from one another in parallel lines remain equidistant, it may be at an altered distance; and in parallel lines, altered, it may be, from their initial direction; the strain is said to be homogeneous.

Properties of homogeneous strain.

156. Hence if any straight line be drawn through the body in its initial state, the portion of the body cut by it will continue to be a straight line when the body is homogeneously strained. For, if ABC be any such line, AB and BC, being parallel to one line in the initial, remain parallel to one line in the altered, state; and therefore remain in the same straight line with one another. Thus it follows that a plane remains a plane, a parallelogram a parallelogram, and a parallelepiped a parallelepiped.

157. Hence, also, similar figures, whether constituted by actual portions of the substance, or mere geometrical surfaces, or straight or curved lines passing through or joining certain portions or points of the substance, similarly situated (*i.e.*, having corresponding parameters parallel) when altered according to the altered condition of the body, remain similar and similarly situated among one another.

158. The lengths of parallel lines of the body remain in the same proportion to one another, and hence all are altered in the same proportion. Hence, and from § 156, we infer that any plane figure becomes altered to another plane figure which is a diminished or magnified orthographic projection of the first on some plane. For example, if an ellipse be altered into a circle, its principal axes become radii at right angles to one another.

Properties of homogeneous strain.

The elongation of the body along any line is the proportion which the addition to the distance between any two points in that line bears to their primitive distance.

159. Every orthogonal projection of an ellipse is an ellipse (the case of a circle being included). Hence, and from § 158, we see that an ellipse remains an ellipse; and an ellipsoid remains a surface of which every plane section is an ellipse; that is, remains an ellipsoid.

A plane curve remains (§ 156) a plane curve. A system of two or of three straight lines of reference (Cartesian) remains a rectilineal system of lines of reference; but, in general, a rectangular system becomes oblique.

Let
$$\frac{x^2}{a^2}+\frac{y^2}{b^2}=1$$

be the equation of an ellipse referred to any rectilineal conjugate axes, in the substance, of the body in its initial state. Let α and β be the proportions in which lines respectively parallel to OX and OY are altered. Thus, if we call ξ and η the altered values of x and y, we have

$$\xi=\alpha x,\quad \eta=\beta y.$$

Hence
$$\frac{\xi^2}{(\alpha a)^2}+\frac{\eta^2}{(\beta b)^2}=1,$$

which also is the equation of an ellipse, referred to oblique axes at, it may be, a different angle to one another from that of the given axes, in the initial condition of the body.

Or again, let
$$\frac{x^2}{a^2}+\frac{y^2}{b^2}+\frac{z^2}{c^2}=1$$

be the equation of an ellipsoid referred to three conjugate diametral planes, as oblique or rectangular planes of reference, in the initial condition of the body. Let α, β, γ be the proportion in which lines parallel to OX, OY, OZ are altered; so that if ξ, η, ζ be the altered values of x, y, z, we have

$$\xi=\alpha x,\quad \eta=\beta y,\quad \zeta=\gamma z.$$

Thus
$$\frac{\xi^2}{(\alpha a)^2}+\frac{\eta^2}{(\beta b)^2}+\frac{\zeta^2}{(\gamma c)^2}=1,$$

Properties of homogeneous strain.

which is the equation of an ellipsoid, referred to conjugate diametral planes, altered it may be in mutual inclination from those of the given planes of reference in the initial condition of the body.

Strain ellipsoid.

160. The ellipsoid which any surface of the body initially spherical becomes in the altered condition, may, to avoid circumlocutions, be called the strain ellipsoid.

161. In any absolutely unrestricted homogeneous strain there are three directions (the three principal axes of the strain ellipsoid), at right angles to one another, which remain at right angles to one another in the altered condition of the body (§ 158). Along one of these the elongation is greater, and along another less, than along any other direction in the body. Along the remaining one, the elongation is less than in any other line in the plane of itself and the first mentioned, and greater than along any other line in the plane of itself and the second.

Note.—Contraction is to be reckoned as a negative elongation: the maximum elongation of the preceding enunciation may be a minimum contraction: the minimum elongation may be a maximum contraction.

162. The ellipsoid into which a sphere becomes altered may be an ellipsoid of revolution, or, as it is called, a spheroid, prolate, or oblate. There is thus a maximum or minimum elongation along the axis, and equal minimum or maximum elongation along all lines perpendicular to the axis.

Or it may be a sphere; in which case the elongations are equal in all directions. The effect is, in this case, merely an alteration of dimensions without change of figure of any part.

Change of volume.

The original volume (sphere) is to the new (ellipsoid) evidently as $1 : \alpha\beta\gamma$.

Axes of a Strain.

163. The principal axes of a strain are the principal axes of the ellipsoid into which it converts a sphere. The principal elongations of a strain are the elongations in the direction of its principal axes.

164. When the position of the principal axes, and the magnitudes of the principal elongations of a strain are given, the elongation of any line of the body, and the alteration of angle between any two lines, may be obviously determined by a simple geometrical construction,

Elongation and change of direction of any line of the body.

Analytically thus:—let $\alpha - 1$, $\beta - 1$, $\gamma - 1$ denote the principal elongations, so that α, β, γ may be now the ratios of alteration along the three principal axes, as we used them formerly for the ratios for any three oblique or rectangular lines. Let l, m, n be the direction cosines of any line, with reference to the three principal axes. Thus,

$$lr,\ mr,\ nr$$

being the three initial co-ordinates of a point P, at a distance $OP = r$, from the origin in the direction l, m, n; the co-ordinates of the same point of the body, with reference to the same rectangular axes, become, in the altered state,

$$\alpha lr,\ \beta mr,\ \gamma nr.$$

Hence the altered length of OP is

$$(\alpha^2 l^2 + \beta^2 m^2 + \gamma^2 n^2)^{\frac{1}{2}} r,$$

and therefore the "elongation" of the body in that direction is

$$(\alpha^2 l^2 + \beta^2 m^2 + \gamma^2 n^2)^{\frac{1}{2}} - 1.$$

For brevity, let this be denoted by $\zeta - 1$, *i.e.*

let $$\zeta = (\alpha^2 l^2 + \beta^2 m^2 + \gamma^2 n^2)^{\frac{1}{2}}.$$

The direction cosines of OP in its altered position are

$$\frac{\alpha l}{\zeta},\ \frac{\beta m}{\zeta},\ \frac{\gamma n}{\zeta};$$

and therefore the angles XOP, YOP, ZOP are altered to having their cosines of these values respectively, from having them of the values l, m, n.

The cosine of the angle between any two lines OP and OP', specified in the initial condition of the body by the direction cosines l', m', n', is

$$ll' + mm' + nn',$$

in the initial condition of the body, and becomes

$$\frac{\alpha^2 ll' + \beta^2 mm' + \gamma^2 nn'}{(\alpha^2 l^2 + \beta^2 m^2 + \gamma^2 n^2)^{\frac{1}{2}}\ (\alpha^2 l'^2 + \beta^2 m'^2 + \gamma^2 n'^2)^{\frac{1}{2}}}$$

in the altered condition.

Change of plane in the body.

165. With the same data the alteration of angle between any two planes of the body may also be easily determined, either geometrically or analytically.

Let l, m, n be the cosines of the angles which a plane makes with the planes YOZ, ZOX, XOY, respectively, in the initial condition of the body. The effects of the change being the same on all parallel planes, we may suppose the plane in question to pass through O; and therefore its equation will be

$$lx + my + nz = 0.$$

In the altered condition of the body we shall have, as before,

$$\xi = \alpha x, \quad \eta = \beta y, \quad \zeta = \gamma z,$$

for the altered co-ordinates of any point initially x, y, z. Hence the equation of the altered plane is

$$\frac{l\xi}{\alpha} + \frac{m\eta}{\beta} + \frac{n\zeta}{\gamma} = 0.$$

But the planes of reference are still rectangular, according to our present supposition. Hence the cosines of the inclinations of the plane in question, to YOZ, ZOX, XOY, in the altered condition of the body, are altered from l, m, n to

$$\frac{l}{\alpha\vartheta}, \quad \frac{m}{\beta\vartheta}, \quad \frac{n}{\gamma\vartheta},$$

respectively, where for brevity

$$\vartheta = \left(\frac{l^2}{\alpha^2} + \frac{m^2}{\beta^2} + \frac{n^2}{\gamma^2}\right)^{\frac{1}{2}}.$$

If we have a second plane similarly specified by l', m', n', in the initial condition of the body, the cosine of the angle between the two planes, which is

$$ll' + mm' + nn'$$

in the initial condition, becomes altered to

$$\frac{\dfrac{ll'}{\alpha^2} + \dfrac{mm'}{\beta^2} + \dfrac{nn'}{\gamma^2}}{\left(\dfrac{l^2}{\alpha^2} + \dfrac{m^2}{\beta^2} + \dfrac{n^2}{\gamma^2}\right)^{\frac{1}{2}} \left(\dfrac{l'^2}{\alpha^2} + \dfrac{m'^2}{\beta^2} + \dfrac{n'^2}{\gamma^2}\right)^{\frac{1}{2}}}.$$

Conical surface of equal elongation.

166. Returning to elongations, and considering that these are generally different in different directions, we perceive that all lines through any point, in which the elongations have any one value intermediate between the greatest and least, must lie on a determinate conical surface. This is easily proved to be in general a cone of the second degree.

Conical surface of equal elongation.

For, in a direction denoted by direction cosines l, m, n, we have

$$\alpha^2 l^2 + \beta^2 m^2 + \gamma^2 n^2 = \zeta^2,$$

where ζ denotes the ratio of elongation, intermediate between α the greatest and γ the least. This is the equation of a cone of the second degree, l, m, n being the direction cosines of a generating line.

167. In one particular case this cone becomes two planes, the planes of the circular sections of the strain ellipsoid.

Two planes of no distortion,

Let $\zeta = \beta$. The preceding equation becomes

$$\alpha^2 l^2 + \gamma^2 n^2 - \beta^2 (1 - m^2) = 0,$$

or, since

$$1 - m^2 = l^2 + n^2,$$

$$(\alpha^2 - \beta^2) l^2 - (\beta^2 - \gamma^2) n^2 = 0.$$

The first member being the product of two factors, the equation is satisfied by putting either $= 0$, and therefore the equation represents the two planes whose equations are

$$l (\alpha^2 - \beta^2)^{\frac{1}{2}} + n (\beta^2 - \gamma^2)^{\frac{1}{2}} = 0,$$

and

$$l (\alpha^2 - \beta^2)^{\frac{1}{2}} - n (\beta^2 - \gamma^2)^{\frac{1}{2}} = 0,$$

respectively.

This is the case in which the given elongation is equal to that along the mean principal axis of the strain ellipsoid. The two planes are planes through the mean principal axis of the ellipsoid, equally inclined on the two sides of either of the other axes. The lines along which the elongation is equal to the mean principal elongation, all lie in, or parallel to, either of these two planes. This is easily proved as follows, without any analytical investigation.

being the circular sections of the strain ellipsoid.

168. Let the ellipse of the annexed diagram represent the section of the strain ellipsoid through the greatest and least principal axes. Let $S'OS$, $T'OT$ be the two diameters of this ellipse, which are equal to the mean principal axis of the ellipsoid. Every plane through O, perpendicular to the plane of the diagram, cuts the ellipsoid in an ellipse of which

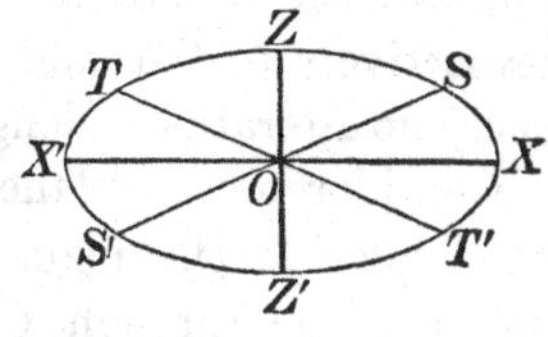

Two planes of no distortion, being the circular sections of the strain ellipsoid.

one principal axis is the diameter in which it cuts the ellipse of the diagram, and the other, the mean principal diameter of the ellipsoid. Hence a plane through either SS', or TT', perpendicular to the plane of the diagram, cuts the ellipsoid in an ellipse of which the two principal axes are equal, that is to say, in a circle. Hence the elongations along all lines in either of these planes are equal to the elongation along the mean principal axis of the strain ellipsoid.

Distortion in parallel planes without change of volume.

169. The consideration of the circular sections of the strain ellipsoid is highly instructive, and leads to important views with reference to the analysis of the most general character of a strain. First, let us suppose there to be no alteration of volume on the whole, and neither elongation nor contraction along the mean principal axis. That is to say, let $\beta = 1$, and $\gamma = \frac{1}{\alpha}$ (§ 162).

Let OX and OZ be the directions of elongation $\alpha - 1$ and contraction $1 - \frac{1}{\alpha}$ respectively. Let A be any point of the

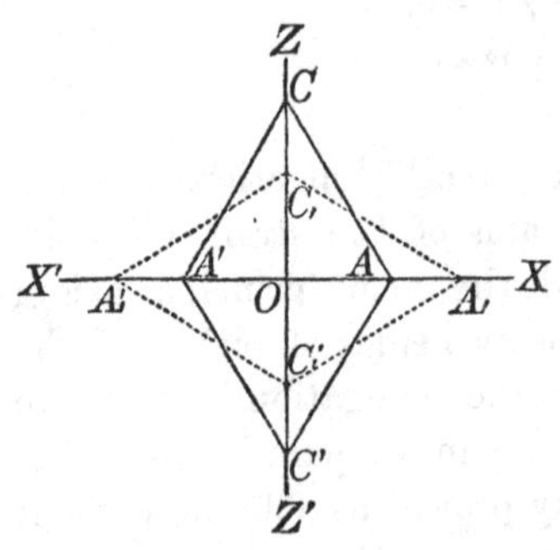

body in its primitive condition, and $A_{,}$ the same point of the altered body, so that $OA_{,} = \alpha OA$. Now, if we take $OC = OA_{,}$, and if $C_{,}$ be the position of that point of the body which was in the position C initially, we shall have $OC_{,} = \frac{1}{\alpha} OC$, and therefore $OC_{,} = OA$. Hence the two triangles COA and $C_{,}OA_{,}$ are equal and similar.

Initial and altered position of lines of no elongation.

Hence CA experiences no alteration of length, but takes the altered position $C_{,}A_{,}$ in the altered position of the body. Similarly, if we measure on XO produced, OA' and $OA_{,}'$ equal respectively to OA and $OA_{,}$, we find that the line CA' experiences no alteration in length, but takes the altered position $C_{,}A_{,}'$.

Consider now a plane of the body initially through CA perpendicular to the plane of the diagram, which will be altered into a plane through $C_{,}A_{,}$, also perpendicular to the plane of the diagram. All lines initially perpendicular to the plane of the diagram remain so, and remain unaltered in length. AC has just been proved to remain unaltered in length. Hence (§ 158) all lines in the plane we have just drawn remain unaltered in length and in mutual inclination. Similarly we see that all lines in a plane through CA', perpendicular to the plane of the diagram, altering to a plane through $C_{,}A_{,}'$, perpendicular to the plane of the diagram, remain unaltered in length and in mutual inclination.

Initial and altered position of lines of no elongation.

170. The precise character of the strain we have now under consideration will be elucidated by the following:—Produce CO, and take OC' and $OC_{,}'$ respectively equal to OC and $OC_{,}$. Join $C'A$, $C'A'$, $C_{,}'A_{,}$, and $C_{,}'A_{,}'$, by plain and dotted lines as in the diagram. Then we see that the rhombus $CAC'A'$ (plain lines) of the body in its initial state becomes the rhombus $C_{,}A_{,}C_{,}'A_{,}'$ (dotted) in the altered condition. Now imagine the body thus strained to be moved as a rigid body (*i. e.*, with its state of strain kept unchanged) till $A_{,}$ coincides with A, and $C_{,}'$ with C', keeping all the lines of the diagram still in the same plane. $A_{,}'C_{,}$ will take a position in CA' produced, as shown in the new diagram, and the original and the altered parallelogram will be on the same base AC', and between the same parallels AC' and $CA_{,}'$, and their other sides will be equally inclined on the two sides of a perpendicular to these parallels. Hence, irrespectively of any rotation, or other absolute motion of the body not involving change of form or dimensions, the strain under consideration may be produced by holding fast and unaltered the plane of the body through AC' perpendicular to the plane of the diagram, and making every plane parallel to it slide, keeping the same distance, through a space proportional to this distance (*i. e.*, different planes parallel to the fixed plane slide through spaces proportional to their distances).

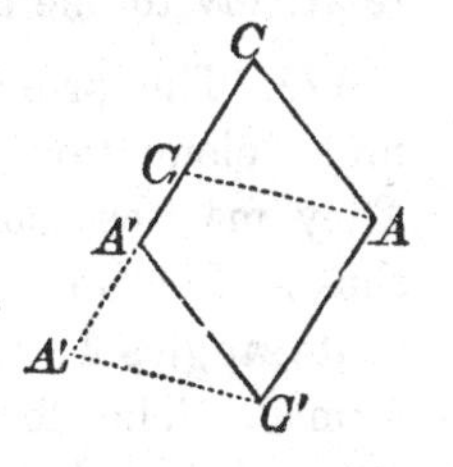

171. This kind of strain is called a *simple shear*. The plane of a shear is a plane perpendicular to the undistorted planes, and parallel to the lines of their relative motion. It

Simple shear.

Simple shear.

has (1) the property that one set of parallel planes remain each unaltered in itself; (2) that another set of parallel planes remain each unaltered in itself. This other set is found when the first set and the degree or amount of shear are given, thus:—Let $CC_{,}$ be the motion of one point of one plane, relative to a plane KL held fixed—the diagram being in a plane of the shear. Bisect $CC_{,}$ in N. Draw NA perpendicular to it. A plane perpendicular to the plane of the diagram, initially through AC, and finally through $AC_{,}$, remains unaltered in its dimensions.

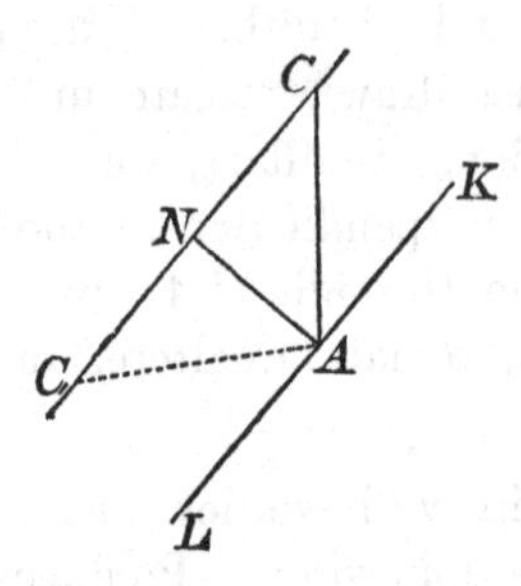

172. One set of parallel undistorted planes, and the amount of their relative parallel shifting having been given, we have just seen how to find the other set. The shear may be otherwise viewed, and considered as a shifting of this second set of parallel planes, relative to any one of them. The amount of this relative shifting is of course equal to that of the first set, relatively to one of them.

Axes of a shear.

173. The principal axes of a shear are the lines of maximum elongation and of maximum contraction respectively. They may be found from the preceding construction (§ 171), thus:—In the plane of the shear bisect the obtuse and acute angles between the planes destined not to become deformed. The former bisecting line is the principal axis of elongation, and the latter is the principal axis of contraction, in their initial positions. The former angle (obtuse) becomes equal to the latter, its supplement (acute), in the altered condition of the body, and the lines bisecting the altered angles are the principal axes of the strain in the altered body.

Otherwise, taking a plane of shear for the plane of the diagram, let AB be a line in which it is cut by one of either set of parallel planes of no distortion. On any portion AB of this as diameter, describe a semicircle. Through C, its middle point, draw, by the preceding construction, CD the initial, and CE

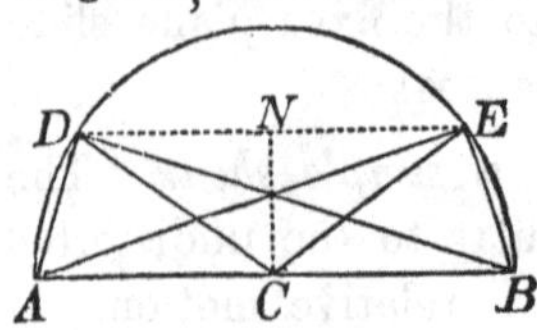

the final, position of an unstretched line. Join DA, DB, EA, EB. DA, DB are the initial, and EA, EB the final, positions of the principal axes.

Axes of a shear.

174. The ratio of a shear is the ratio of elongation or contraction of its principal axes. Thus if one principal axis is elongated in the ratio $1 : \alpha$, and the other therefore (§ 169) contracted in the ratio $\alpha : 1$, α is called the ratio of the shear. It will be convenient generally to reckon this as the ratio of elongation; that is to say, to make its numerical measure greater than unity.

Measure of a shear.

In the diagram of § 173, the ratio of DB to EB, or of EA to DA, is the ratio of the shear.

175. The amount of a shear is the amount of relative motion per unit distance between planes of no distortion.

It is easily proved that this is equal to the excess of the ratio of the shear above its reciprocal.

Since $DCA = 2DBA$, and $\tan DBA = \frac{1}{a}$ we have $\tan DCA = \frac{2a}{a^2-1}$.

But $DE = 2CN \tan DCN = 2CN \cot DCA$.

Hence $$\frac{DE}{CN} = 2\frac{a^2-1}{2a} = a - \frac{1}{a}.$$

176. The planes of no distortion in a simple shear are clearly the circular sections of the strain ellipsoid. In the ellipsoid of this case, be it remembered, the mean axis remains unaltered, and is a mean proportional between the greatest and the least axis.

Ellipsoidal specification of a shear.

177. If we now suppose all lines perpendicular to the plane of the shear to be elongated or contracted in any proportion, without altering lengths or angles in the plane of the shear, and if, lastly, we suppose every line in the body to be elongated or contracted in some other fixed ratio, we have clearly (§ 161) the most general possible kind of strain. Thus if s be the ratio of the simple shear, for which case s, 1, $\frac{1}{s}$ are the three principal ratios, and if we elongate lines perpendicular to its plane in the

Shear, simple elongation, and expansion combined.

Shear, simple elongation, and expansion, combined.

ratio $1 : m$, without any other change, we have a strain of which the principal ratios are

$$s, m, \frac{1}{s}.$$

If, lastly, we elongate all lines in the ratio $1 : n$, we have a strain in which the principal ratios are

$$ns, nm, \frac{n}{s},$$

where it is clear that ns, nm, and $\frac{n}{s}$ may have any values whatever. It is of course not necessary that nm be the mean principal ratio. Whatever they are, if we call them α, β, γ respectively, we have

$$s = \sqrt{\frac{\alpha}{\gamma}}\,;\ n = \sqrt{\alpha\gamma}\,;\ \text{and}\ m = \frac{\beta}{\sqrt{\alpha\gamma}}.$$

Analysis of a strain.

178. Hence any strain (α, β, γ) whatever may be viewed as compounded of a uniform dilatation in all directions, of linear ratio $\sqrt{\alpha\gamma}$, superimposed on a simple elongation $\frac{\beta}{\sqrt{\alpha\gamma}}$ in the direction of the principal axis to which β refers, superimposed on a simple shear, of ratio $\sqrt{\frac{\alpha}{\gamma}}\left(\text{or of amount } \sqrt{\frac{\alpha}{\gamma}} - \sqrt{\frac{\gamma}{\alpha}}\right)$ in the plane of the two other principal axes.

179. It is clear that these three elementary component strains may be applied in any other order as well as that stated. Thus, if the simple elongation is made first, the body thus altered must get just the same shear in planes perpendicular to the line of elongation, as the originally unaltered body gets when the order first stated is followed. Or the dilatation may be first, then the elongation, and finally the shear, and so on.

Displacement of a body, rigid or not, one point of which is held fixed.

180. In the preceding sections on strains, we have considered the alterations of lengths of lines of the body, and of angles between lines and planes of it; and we have, in particular cases, founded on particular suppositions (the principal axes of the strain remaining fixed in direction, § 169, or one of either set of undistorted planes in a simple shear remaining fixed, § 170), considered the actual displacements of parts of the body from their original positions. But to complete the kinematics of a non-rigid solid, it is necessary to take a more general view of the relation between displacements and strains. It will be sufficient for us to suppose one point of the body to remain fixed, as it is easy to see the effect of superimposing upon any motion with one point fixed, a motion of translation without strain or rotation.

Displacement of a body, rigid or not, one point of which is held fixed.

181. Let us therefore suppose one point of a body to be held fixed, and any displacement whatever given to any point or points of it, subject to the condition that the whole substance if strained at all is homogeneously strained.

Let OX, OY, OZ be any three rectangular axes, fixed with reference to the initial position and condition of the body. Let x, y, z be the initial co-ordinates of any point of the body, and x_1, y_1, z_1 be the co-ordinates of the same point of the altered body, with reference to those axes unchanged. The condition that the strain is homogeneous throughout is expressed by the following equations:—

$$\left.\begin{aligned} x_1 &= [Xx]\,x + [Xy]\,y + [Xz]\,z, \\ y_1 &= [Yx]\,x + [Yy]\,y + [Yz]\,z, \\ z_1 &= [Zx]\,x + [Zy]\,y + [Zz]\,z, \end{aligned}\right\} \qquad (1)$$

where $[Xx]$, $[Xy]$, etc., are nine quantities, of absolutely arbitrary values, the same for all values of x, y, z.

$[Xx]$, $[Yx]$, $[Zx]$ denote the three final co-ordinates of a point originally at unit distance along OX, from O. They are, of course, proportional to the direction-cosines of the altered position of the line primitively coinciding with OX. Similarly for $[Xy]$, $[Yy]$, $[Zy]$, etc.

Let it be required to find, if possible, a line of the body which remains unaltered in direction, during the change specified by $[Xx]$, etc. Let x, y, z, and x_1, y_1, z_1, be the co-ordinates of the primitive and altered position of a point in such a line. We must have $\frac{x_1}{x} = \frac{y_1}{y} = \frac{z_1}{z} = 1 + \epsilon$, where ϵ is the elongation of the line in question.

Displacement of a body, rigid or not, one point of which is held fixed.

Thus we have $x_1 = (1+\epsilon)x$, etc., and therefore if $\eta = 1 + \epsilon$

$$\left.\begin{array}{l}\{[Xx]-\eta\}x + [Xy]y + [Xz]z = 0,\\ [Yx]x + \{[Yy]-\eta\}y + [Yz]z = 0,\\ [Zx]x + [Zy]y + \{[Zz]-\eta\}z = 0.\end{array}\right\} \quad (2)$$

From these equations, by eliminating the ratios $x:y:z$ according to the well-known algebraic process, we find

$$\begin{array}{l}([Xx]-\eta)([Yy]-\eta)([Zz]-\eta)\\ -[Yz][Zy]([Xx]-\eta)-[Zx][Xz]([Yy]-\eta)-[Xy][Yx]([Zz]-\eta)\\ \qquad +[Xz][Yx][Zy]+[Xy][Yz][Zx]=0.\end{array}$$

This cubic equation is necessarily satisfied by at least one real value of η, and the two others are either both real or both imaginary. Each real value of η gives a real solution of the problem, since any two of the preceding three equations with it, in place of η, determine real values of the ratios $x:y:z$. If the body is rigid (*i.e.*, if the displacements are subject to the condition of producing no strain), we know (*ante*, § 95) that there is just one line common to the body in its two positions, the axis round which it must turn to pass from one to the other, except in the peculiar cases of *no* rotation, and of rotation through *two* right angles, which are treated below. Hence, in this case, the cubic equation has only one real root, and therefore it has two imaginary roots. The equations just formed solve the problem of finding the axis of rotation when the data are the actual displacements of the points primitively lying in three given fixed axes of reference, OX, OY, OZ; and it is worthy of remark, that the practical solution of this problem is founded on the one real root of a cubic which has two imaginary roots.

Again, on the other hand, let the given displacements be made so as to produce a strain of the body with no angular displacement of the principal axes of the strain. Thus three lines of the body remain unchanged. Hence there must be three real roots of the equation in η, one for each such axis; and the three lines determined by them are necessarily at right angles to one another.

But if neither of these conditions holds, we may have three real solutions and three oblique lines of directional identity; or we may have only one real root and only one line of directional identity.

An analytical proof of these conclusions may easily be given; thus we may write the cubic in the form—

Displacement of a body, rigid or not, one point of which is held fixed.

$$\begin{vmatrix}[Xx], & [Xy], & [Xz]\\ [Yx], & [Yy], & [Yz]\\ [Zx], & [Zy], & [Zz]\end{vmatrix} - \eta\left\{\begin{vmatrix}[Yy], & [Yz]\\ [Zy], & [Zz]\end{vmatrix} + \begin{vmatrix}[Zz], & [Zx]\\ [Xz], & [Xx]\end{vmatrix} + \begin{vmatrix}[Xx], & [Xy]\\ [Yx], & [Yy]\end{vmatrix}\right\}$$
$$+\eta^2\{[Xx]+[Yy]+[Zz]\} - \eta^3 = 0 \quad\ldots\ldots\ldots\ldots(3)$$

In the particular case of no strain, since $[Xx]$, etc., are then *equal*, not merely *proportional*, to the direction cosines of three mutually perpendicular lines, we have by well-known geometrical theorems

$$\begin{vmatrix}[Xx], & [Xy], & [Xz]\\ [Yx], & [Yy], & [Yz]\\ [Zx], & [Zy], & [Zz]\end{vmatrix} = 1, \text{ and } \begin{vmatrix}[Yy], & [Yz]\\ [Zy], & [Zz]\end{vmatrix} = [Xx], \text{ etc.}$$

Hence the cubic becomes

$$1 - (\eta - \eta^2)\{[Xx] + [Yy] + [Zz]\} - \eta^3 = 0,$$

of which one root is evidently $\eta = 1$. This leads to the above explained rotational solution, the line determined by the value 1 of η being the axis of rotation. Dividing out the factor $1 - \eta$, we get for the two remaining roots the equation

$$1 + (1 - [Xx] - [Yy] - [Zz])\,\eta + \eta^2 = 0,$$

whose roots are imaginary if the coefficient of η lies between $+2$ and -2. Now -2 is evidently its *least* value, and for that case the roots are real, each being unity. Here there is no rotation. Also $+2$ is its *greatest* value, and this gives us a pair of values each $= -1$, of which the interpretation is, that there is rotation through two right angles. In this case, as in general, one line (the axis of rotation) is determined by the equations (2) with the value $+1$ for η; but with $\eta = -1$ these equations are satisfied by any line perpendicular to the former.

The limiting case of two equal roots, when there is strain, is an interesting subject which may be left as an exercise. It separates the cases in which there is only one axis of directional identity from those in which there are three.

Let it next be proposed to find those lines of the body whose elongations are greatest or least. For this purpose we must find the equations expressing that $x_1^2 + y_1^2 + z_1^2$ is a maximum, when $x^2 + y^2 + z^2 = r^2$, a constant. First, we have

$$x_1^2 + y_1^2 + z_1^2 = Ax^2 + By^2 + Cz^2 + 2(ayz + bzx + cxy) \quad\ldots\ldots\ldots(4),$$

Displacement of a body, rigid or not, one point of which is held fixed.

where

$$\left.\begin{array}{l} A=[Xx]^2+[Yx]^2+[Zx]^2 \\ B=[Xy]^2+[Yy]^2+[Zy]^2 \\ C=[Xz]^2+[Yz]^2+[Zz]^2 \\ a=[Xy][Xz]+[Yy][Yz]+[Zy][Zz] \\ b=[Xz][Xx]+[Yz][Yx]+[Zz][Zx] \\ c=[Xx][Xy]+[Yx][Yy]+[Zx][Zy] \end{array}\right\}\ldots\ldots\ldots\ldots(5).$$

The equation

$$Ax^2+By^2+Cz^2+2(ayz+bzx+cxy)=r_1^{\,2}\ldots\ldots\ldots\ldots(6),$$

where r_1 is any constant, represents clearly the ellipsoid which a spherical surface, radius r_1, of the altered body, would become if the body were restored to its primitive condition. The problem of making r_1 a maximum when r is a given constant, leads to the following equations:—

$$x^2+y^2+z^2=r^2\ldots\ldots\ldots\ldots\ldots\ldots\ldots\ldots(7),$$

$$\left.\begin{array}{r} xdx+ydy+zdz=0, \\ (Ax+cy+bz)dx+(cx+By+az)dy+(bx+ay+Cz)dz=0. \end{array}\right\}\ (8)$$

On the other hand, the problem of making r a maximum or minimum when r_1 is given, that is to say, the problem of finding maximum and minimum diameters, or principal axes, of the ellipsoid (6), leads to these same two differential equations (8), and only differs in having equation (6) instead of (7) to complete the determination of the absolute values of x, y, and z. Hence the ratios $x:y:z$ will be the same in one problem as in the other; and therefore the *directions* determined are those of the principal axes of the ellipsoid (6). We know, therefore, by the properties of the ellipsoid, that there are three real solutions, and that the directions of the three radii so determined are mutually rectangular. The ordinary method (Lagrange's) for dealing with the differential equations, being to multiply one of them by an arbitrary multiplier, then add, and equate the coefficients of the separate differentials to zero, gives, if we take $-\eta$ as the arbitrary multiplier, and the first of the two equations the one multiplied by it,

$$\left.\begin{array}{rrr} (A-\eta)x & +cy & +bz=0, \\ cx & +(B-\eta)y & +az=0, \\ bx & +ay & +(C-\eta)z=0. \end{array}\right\}\quad(9)$$

We may find what η means if we multiply the first of these by x, the second by y, and the third by z, and add; because we thus obtain

Displacement of a body, rigid or not, one point of which is held fixed.

$$Ax^2 + By^2 + Cz^2 + 2(ayz + bzx + cxy) - \eta(x^2 + y^2 + z^2) = 0,$$

or
$$r_1^2 - \eta r^2 = 0,$$

which gives
$$\eta = \left(\frac{r_1}{r}\right)^2 \quad\ldots\ldots\ldots\ldots\ldots\ldots\ldots\ldots\ldots (10).$$

Eliminating the ratios $x : y : z$ from (9), by the usual method, we have the well-known determinant cubic

$$(A-\eta)(B-\eta)(C-\eta) - a^2(A-\eta) - b^2(B-\eta) - c^2(C-\eta) + 2abc = 0 \ldots (11),$$

of which the three roots are known to be all real. Any one of the three roots if used for η, in (9), harmonizes these three equations for the true ratios $x : y : z$; and, making the coefficients of x, y, z in them all known, allows us to determine the required ratios by any two of the equations, or symmetrically from the three, by the proper algebraic processes. Thus we have only to determine the absolute magnitudes of x, y, and z, which (7) enables us to do when their ratios are known.

It is to be remarked, that when $[Yz] = [Zy]$, $[Zx] = [Xz]$, and $[Xy] = [Yx]$, equation (3) becomes a cubic, the squares of whose roots are the roots of (11), and that the three lines determined by (2) in this case are identical with those determined by (9). The reader will find it a good analytical exercise to prove this directly from the equations. It is a necessary consequence of § 183, below.

We have precisely the same problem to solve when the question proposed is, to find what radii of a sphere remain perpendicular to the surface of the altered figure. This is obvious when viewed geometrically. The tangent plane is perpendicular to the radius when the radius is a maximum or minimum. Therefore, every plane of the body parallel to such tangent plane is perpendicular to the radius in the altered, as it was in the initial condition.

The analytical investigation of the problem, presented in the second way, is as follows:—

Let
$$l_1x_1 + m_1y_1 + n_1z_1 = 0 \quad\ldots\ldots\ldots\ldots\ldots\ldots\ldots (12)$$

be the equation of any plane of the altered substance, through the origin of co-ordinates, the axes of co-ordinates being the same fixed axes, OX, OY, OZ, which we have used of late. The direction cosines of a perpendicular to it are, of course, proportional to l_1, m_1, n_1. If, now, for x_1, y_1, z_1, we substitute their

9—2

Displacement of a body, rigid or not, one point of which is held fixed.

values, as in (1), in terms of the co-ordinates which the same point of the substance had initially, we find the equation of the same plane of the body in its initial position, which, when the terms are grouped properly, is this—

$$\{l_1[Xx] + m_1[Yx] + n_1[Zx]\}x + \{l_1[Xy] + m_1[Yy] + n_1[Zy]\}y + \{l_1[Xz] + m_1[Yz] + n_1[Zz]\}z = 0 \quad\ldots\ldots(13).$$

The direction cosines of the perpendicular to the plane are proportional to the co-efficients of x, y, z. Now these are to be the direction cosines of the same line of the substance as was altered into the line $l_1 : m_1 : n_1$. Hence, if $l : m : n$ are quantities proportional to the direction cosines of this line in its initial position, we must have

$$\left.\begin{array}{l} l_1[Xx] + m_1[Yx] + n_1[Zx] = \eta l \\ l_1[Xy] + m_1[Yy] + n_1[Zy] = \eta m \\ l_1[Xz] + m_1[Yz] + n_1[Zz] = \eta n \end{array}\right\} \quad\ldots\ldots(14),$$

where η is arbitrary. Suppose, to fix the ideas, that l_1, m_1, n_1 are the co-ordinates of a certain point of the substance in its altered state, and that l, m, n are proportional to the initial co-ordinates of the same point of the substance. Then we shall have, by the fundamental equations, the expressions for l_1, m_1, n_1 in terms of l, m, n. Using these in the first members of (14), and taking advantage of the abbreviated notation (5), we have precisely the same equations for l, m, n as (9) for x, y, z above.

Analysis of a strain into distortion and rotation

182. From the preceding analysis it follows that any homogeneous strain whatever applied to a body generally changes a sphere of the body into an ellipsoid, and causes the latter to rotate about a definite axis through a definite angle. In particular cases the sphere may remain a sphere. Also there may be no rotation. In the general case, when there is no rotation, there are three directions in the body (the axes of the ellipsoid) which remain fixed; when there *is* rotation, there are generally three such directions, but not rectangular. Sometimes, however, there is but one.

Pure strain.

183. When the axes of the ellipsoid are lines of the body whose directions do not change, the strain is said to be *pure*, or unaccompanied by rotation. The strains we have already considered were more general than this, being pure strains accompanied by rotation. We proceed to find the analytical conditions of the existence of a pure strain.

Pure strain.

Let $O\Xi$, $O\Xi'$, $O\Xi''$ be the three principal axes of the strain, and let $l, m, n,\quad l', m', n',\quad l'', m'', n''$, be their direction cosines. Let $\alpha, \alpha', \alpha''$ be the principal elongations. Then, if ξ, ξ', ξ'' be the position of a point of the unaltered body, with reference to $O\Xi$, $O\Xi'$, $O\Xi''$, its position in the body when altered will be $\alpha\xi, \alpha'\xi', \alpha''\xi''$. But if x, y, z be its initial, and x_1, y_1, z_1 its final, positions with reference to OX, OY, OZ, we have

$$\xi = lx + my + nz,\quad \xi' = \text{etc.},\quad \xi'' = \text{etc.} \quad\ldots\ldots\ldots\ldots\ldots (15),$$

and $x_1 = l\alpha\xi + l'\alpha'\xi' + l''\alpha''\xi'',\quad y_1 = \text{etc.},\quad z_1 = \text{etc.}$

For ξ, ξ', ξ'' substitute their values (15), and we have x_1, y_1, z_1 in terms of x, y, z, expressed by the following equations:—

$$\left.\begin{aligned} x_1 &= (\alpha l^2 + \alpha' l'^2 + \alpha'' l''^2)x + (\alpha lm + \alpha' l'm' + \alpha'' l''m'')y + (\alpha ln + \alpha' l'n' + \alpha'' l''n'')z \\ y_1 &= (\alpha ml + \alpha' m'l' + \alpha'' m''l'')x + (\alpha m^2 + \alpha' m'^2 + \alpha'' m''^2)y + (\alpha mn + \alpha' m'n' + \alpha'' m''n'')z \\ z_1 &= (\alpha nl + \alpha' n'l' + \alpha'' n''l'')x + (\alpha nm + \alpha' n'm' + \alpha'' n''m'')y + (\alpha n^2 + \alpha' n'^2 + \alpha'' n''^2)z \end{aligned}\right\} .(16).$$

Hence, comparing with (1) of § 181, we have

$$\left.\begin{aligned} [Xx] &= \alpha l^2 + \alpha' l'^2 + \alpha'' l''^2, \text{ etc.}; \\ [Zy] &= [Yz] = \alpha mn + \alpha' m'n' + \alpha'' m''n'', \text{ etc.} \end{aligned}\right\} \ldots\ldots\ldots (17).$$

In these equations, $l, l', l'', m, m', m'', n, n', n''$, are deducible from three independent elements, the three angular co-ordinates (§ 100, above) of a rigid body, of which one point is held fixed; and therefore, along with $\alpha, \alpha', \alpha''$, constituting in all six independent elements, may be determined so as to make the six members of these equations have any six prescribed values. Hence the conditions necessary and sufficient to insure no rotation are

$$[Zy] = [Yz],\quad [Xz] = [Zx],\quad [Xy] = [Yx] \ldots\ldots\ldots\ldots (18).$$

184. If a body experience a succession of strains, each unaccompanied by rotation, its resulting condition will generally be producible by a strain and a rotation. From this follows the remarkable corollary that three pure strains produced one after another, in any piece of matter, each without rotation, may be so adjusted as to leave the body unstrained, but rotated through some angle about some axis. We shall have, later, most important and interesting applications to fluid motion,

Composition of pure strains.

Composition of pure strains.

which (Chap. II.) will be proved to be instantaneously, or differentially, irrotational; but which may result in leaving a whole fluid mass merely turned round from its primitive position, as if it had been a rigid body. The following elementary geometrical investigation, though not bringing out a thoroughly comprehensive view of the subject, affords a rigorous demonstration of the proposition, by proving it for a particular case.

Let us consider, as above (§ 171), a simple shearing motion. A point O being held fixed, suppose the matter of the body in a plane, cutting that of the diagram perpendicularly in CD, to move in this plane from right to left parallel to DC; and in other planes parallel to it let there be motions proportional to their distances from O. Consider first a shear from P to P_1; then from P_1 on to P_2; and let O be taken in a line through P_1, perpendicular to CD. During the shear from P to P_1 a point Q moves of course to Q_1 through a distance $QQ_1 = PP_1$. Choose Q midway between P and P_1, so that $P_1Q = QP = \frac{1}{2}P_1P$. Now, as we have seen above (§ 152), the line of the body, which is the principal axis of contraction in the shear from Q to Q_1, is OA, bisecting the angle QOE at the beginning, and OA_1, bisecting Q_1OE at the end, of the whole motion considered. The angle between these two lines is half the angle Q_1OQ, that is to say, is equal to P_1OQ. Hence, if the plane CD is rotated through an angle equal to P_1OQ, in the plane of the diagram, in the same way as the hands of a watch, during the shear from Q to Q_1, or, which is the same thing, the shear from P to P_1, this shear will be effected without final rotation of its principal axes. (Imagine the diagram turned round till OA_1 lies along OA. The actual and the newly imagined position of CD will show how this plane of the body has moved during such non-rotational shear.)

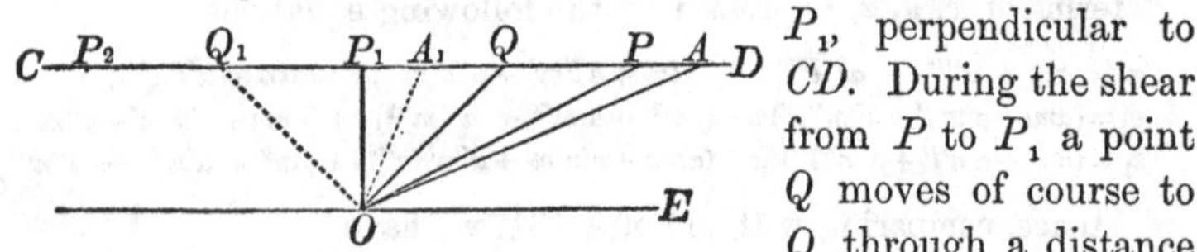

Now, let the second step, P_1 to P_2, be made so as to complete the whole shear, P to P_2, which we have proposed to consider. Such second partial shear may be made by the common shearing process parallel to the new position (imagined in the preceding parenthesis) of CD, and to make itself also non-rotational, as its predecessor has been made, we must turn further round, in the same direction, through an angle equal to Q_1OP_1. Thus in these two steps, each made non-rotational, we have turned the plane CD round through an angle equal to Q_1OQ. But now, we have a whole shear PP_2; and to make this as one non-rotational shear, we must turn CD through an angle P_1OP only, which is less than Q_1OQ by the excess of P_1OQ above QOP. Hence the resultant of the two shears, PP_1, P_1P_2, each separately deprived of rotation, is a single shear PP_2, and a rotation of its principal axes, in the direction of the hands of a watch, through an angle equal to $QOP_1 - POQ$.

Composition of pure strains.

185. Make the two partial shears each non-rotationally. Return from their resultant in a single non-rotational shear: we conclude with the body unstrained, but turned through the angle $QOP_1 - POQ$, in the same direction as the hands of a watch.

$$\begin{aligned} x_1 &= Ax + cy + bz \\ y_1 &= cx + By + az \\ z_1 &= bx + ay + Cz \end{aligned}$$

is (§ 183) the most general possible expression for the displacement of any point of a body of which one point is held fixed, strained according to any three lines at right angles to one another, as principal axes, which are kept fixed in direction, relatively to the lines of reference OX, OY, OZ.

Similarly, if the body thus strained be again non-rotationally strained, the most general possible expressions for x_2, y_2, z_2, the co-ordinates of the position to which x_1, y_1, z_1, will be brought, are

$$\begin{aligned} x_2 &= A_1x_1 + c_1y_1 + b_1z_1 \\ y_2 &= c_1x_1 + B_1y_1 + a_1z_1 \\ z_2 &= b_1x_1 + a_1y_1 + C_1z_1. \end{aligned}$$

Substituting in these, for x_1, y_1, z_1, their preceding expressions, in terms of the primitive co-ordinates, x, y, z, we have the following expressions for the co-ordinates of the position to which the point in question is brought by the two strains:—

$$\begin{aligned} x_2 &= (A_1A + c_1c + b_1b)\,x + (A_1c + c_1B + b_1a)\,y + (A_1b + c_1a + b_1C)\,z \\ y_2 &= (c_1A + B_1c + a_1b)\,x + (c_1c + B_1B + a_1a)\,y + (c_1b + B_1a + a_1C)\,z \\ z_2 &= (b_1A + a_1c + C_1b)\,x + (b_1c + a_1B + C_1a)\,y + (b_1b + a_1a + C_1C)\,z. \end{aligned}$$

Composition of pure strains.

The resultant displacement thus represented is not generally of the non-rotational character, the conditions (18) of § 183 not being fulfilled, as we see immediately. Thus, for instance, we see that the coefficient of y in the expression for x_2 is not necessarily equal to the coefficient of x in the expression for y_2.

Cor.—If both strains are infinitely small, the resultant displacement is a pure strain without rotation. For A, B, C, A_1, B_1, C_1 are each infinitely nearly unity, and a, b, etc., each infinitely small. Hence, neglecting the products of these infinitely small quantities among one another, and of any of them with the differences between the former and unity, we have a resultant displacement

$$x_2 = A_1Ax \quad + (c + c_1)\, y + (b + b_1) z$$
$$y_2 = (c_1 + c)\, x + \quad B_1By \quad + (a + a_1) z$$
$$z_2 = (b_1 + b)\, x + (a_1 + a)\, y + C_1Cz,$$

which represents a pure strain unaccompanied by rotation.

Displacement of a curve.

186. The measurement of rotation in a strained elastic solid, or in a moving fluid, is much facilitated by considering separately the displacement of any line of the substance. We are therefore led now to a short digression on the displacement of a curve, which may either belong to a continuous solid or fluid mass, or may be an elastic cord, given in any position. The propositions at which we shall arrive are, of course, applicable to a flexible but inextensible cord (§ 14, above) as a particular case.

It must be remarked, that the displacements to be considered do not depend merely on the curves occupied by the given line in its successive positions, but on the corresponding points of these curves.

Tangential displacement.

What we shall call tangential displacement is to be thus reckoned:—Divide the undisplaced curve into an infinite number of infinitely small equal parts. The sum of the tangential components of the displacements from all the points of division, multiplied by the length of each of the infinitely small parts, is *the entire tangential displacement of the curve reckoned along the undisplaced curve.* The same reckoning carried out in the displaced curve is *the entire tangential displacement reckoned on the displaced curve.*

187. The whole tangential displacement of a curve reckoned along the displaced curve, exceeds the whole tangential displacement reckoned along the undisplaced curve by half the rectangle under the sum and difference of the absolute terminal displacements, taken as positive when the displacement of the end towards which the tangential components are if positive exceeds that at the other. This theorem may be proved by a geometrical demonstration which the reader may easily supply.

Two reckonings of tangential displacement compared.

Analytically thus:—Let x, y, z be the co-ordinates of any point, P, in the undisplaced curve; x_1, y_1, z_1, those of P_1 the point to which the same point of the curve is displaced. Let dx, dy, dz be the increments of the three co-ordinates corresponding to any infinitely small arc, ds, of the first; so that

$$ds = (dx^2 + dy^2 + dz^2)^{\frac{1}{2}},$$

and let corresponding notation apply to the corresponding element of the displaced curve. Let θ denote the angle between the line PP_1 and the tangent to the undisplaced curve through P; so that we have

$$\cos\theta = \frac{x_1 - x}{D}\frac{dx}{ds} + \frac{y_1 - y}{D}\frac{dy}{ds} + \frac{z_1 - z}{D}\frac{dz}{ds},$$

where for brevity

$$D = \{(x_1 - x)^2 + (y_1 - y)^2 + (z_1 - z)^2\}^{\frac{1}{2}},$$

being the absolute space of displacement. Hence

$$D\cos\theta ds = (x_1 - x)\,dx + (y_1 - y)\,dy + (z_1 - z)\,dz.$$

Similarly we have

$$D\cos\theta_1 ds_1 = (x_1 - x)\,dx_1 + (y_1 - y)\,dy_1 + (z_1 - z)\,dz_1,$$

and therefore

$$D\cos\theta_1 ds_1 - D\cos\theta ds = (x_1 - x)\,d(x_1 - x) + (y_1 - y)\,d(y_1 - y) + (z_1 - z)\,d(z_1 - z),$$

or

$$D\cos\theta_1 ds_1 - D\cos\theta ds = \tfrac{1}{2}\,d(D^2).$$

To find the difference of the tangential displacements reckoned the two ways, we have only to integrate this expression. Thus we obtain

$$\int D\cos\theta_1 ds_1 - \int D\cos\theta ds = \tfrac{1}{2}(D''^2 - D'^2) = \tfrac{1}{2}(D'' + D')(D'' - D'),$$

where D'' and D' denote the displacements of the two ends.

Tangential displacement of a closed curve.

188. The entire tangential displacement of a closed curve is the same whether reckoned along the undisplaced or the displaced curve.

189. The entire tangential displacement from one to another of two conterminous arcs, is the same reckoned along either as along the other.

Rotation of a rigid closed curve.

190. The entire tangential displacement of a rigid closed curve when rotated through any angle about any axis, is equal to twice the area of its projection on a plane perpendicular to the axis, multiplied by the sine of the angle.

Tangential displacement in a solid, in terms of components of strain.

(*a*) *Prop.*—The entire tangential displacement round a closed curve of a homogeneously strained solid, is equal to

$$2(P\varpi + Q\rho + R\sigma),$$

where P, Q, R denote, for its initial position, the areas of its projections on the planes YOZ, ZOX, XOY respectively, and ϖ, ρ, σ are as follows :—

$$\varpi = \tfrac{1}{2}\{[Zy] - [Yz]\}$$
$$\rho = \tfrac{1}{2}\{[Xz] - [Zx]\}$$
$$\sigma = \tfrac{1}{2}\{[Yx] - [Xy]\}.$$

To prove this, let, farther,

$$a = \tfrac{1}{2}\{[Zy] + [Yz]\}$$
$$b = \tfrac{1}{2}\{[Xz] + [Zx]\}$$
$$c = \tfrac{1}{2}\{[Yx] + [Xy]\}.$$

Thus we have

$$x_1 = Ax + cy + bz + \sigma y - \rho z$$
$$y_1 = cx + By + az + \varpi z - \sigma x$$
$$z_1 = bx + ay + Cz + \rho x - \varpi y.$$

Hence, according to the previously investigated expression, we have, for the tangential displacement, reckoned along the undisplaced curve,

$$\int\{(x_1 - x)\,dx + (y_1 - y)\,dy + (z_1 - z)\,dz\}$$
$$= \int[\tfrac{1}{2}d\{(A-1)\,x^2 + (B-1)\,y^2 + (C-1)\,z^2 + 2\,(ayz + bzx + cxy)\}$$
$$+ \varpi\,(ydz - zdy) + \rho\,(zdx - xdz) + \sigma\,(xdy - ydx)].$$

The first part, $\int\frac{1}{2}d\{\ \}$, vanishes for a closed curve.

The remainder of the expression is

Tangential displacement in a solid, in terms of components of strain.

$$\varpi\int(ydz-zdy)+\rho\int(zdx-xdz)+\sigma\int(xdy-ydx),$$

which, according to the formulæ for projection of areas, is equal to

$$2P\varpi+2Q\rho+2R\sigma.$$

For, as in § 36 (*a*), we have in the plane of xy

$$\int(xdy-ydx)=\int r^2d\theta,$$

double the area of the orthogonal projection of the curve on that plane; and similarly for the other integrals.

(*b*) From this and § 190, it follows that if the body is rigid, and therefore only rotationally displaced, if at all, $[Zy]-[Yz]$ is equal to twice the sine of the angle of rotation multiplied by the cosine of the inclination of the axis of rotation to the line of reference OX.

(*c*) And in general $[Zy]-[Yz]$ measures the entire tangential displacement, divided by the area on ZOY, of any closed curve given, if a plane curve, in the plane YOZ, or, if a tortuous curve, given so as to have zero area projections on ZOX and XOY. The entire tangential displacement of any closed curve given in a plane, A, perpendicular to a line whose direction cosines are proportional to ϖ, ρ, σ, is equal to twice its area multiplied by $\sqrt{(\varpi^2+\rho^2+\sigma^2)}$. And the entire tangential displacement of any closed curve whatever is equal to twice the area of its projection on A, multiplied by $\sqrt{(\varpi^2+\rho^2+\sigma^2)}$.

In the transformation of co-ordinates, ϖ, ρ, σ transform by the elementary cosine law, and of course $\varpi^2+\rho^2+\sigma^2$ is an invariant; that is to say, its value is unchanged by transformation from one set of rectangular axes to another.

(*d*) In non-rotational homogeneous strain, the entire tangential displacement along any curve from the fixed point to (x, y, z), reckoned along the undisplaced curve, is equal to

$$\tfrac{1}{2}\{(A-1)x^2+(B-1)y^2+(C-1)z^2+2(ayz+bzx+cxy)\}.$$

Reckoned along displaced curve, it is, from this and § 187,

$$\tfrac{1}{2}\{(A-1)x^2+(B-1)y^2+(C-1)z^2+2(ayz+bzx+cxy)\}$$
$$+\tfrac{1}{2}\{[(A-1)x+cy+bz]^2+[cx+(B-1)y+az]^2$$
$$+[bx+ay+(C-1)z]^2\}.$$

And the entire tangential displacement from one point along any curve to another point, is independent of the curve, *i.e.*, is the same along any number of conterminous curves, this of course whether reckoned in each case along the undisplaced or along the displaced curve.

Hetero-geneous strain.

(*e*) Given the absolute displacement of every point, to find the strain. Let α, β, γ, be the components, relative to fixed axes, OX, OY, OZ, of the displacement of a particle, P, initially in the position x, y, z. That is to say, let $x+\alpha$, $y+\beta$, $z+\gamma$ be the co-ordinates, in the strained body, of the point of it which was initially at x, y, z.

Consider the matter all round this point in its first and second positions. Taking this point P as moveable origin, let ξ, η, ζ be the initial co-ordinates of any other point near it, and ξ_1, η_1, ζ_1 the final co-ordinates of the same.

The initial and final co-ordinates of the last-mentioned point, with reference to the fixed axes OX, OY, OZ, will be

$$x+\xi,\ y+\eta,\ z+\zeta,$$

and $$x+\alpha+\xi_1,\ y+\beta+\eta_1,\ z+\gamma+\zeta_1,$$

respectively; that is to say,

$$\alpha+\xi_1-\xi,\ \beta+\eta_1-\eta,\ \gamma+\zeta_1-\zeta$$

are the components of the displacement of the point which had initially the co-ordinates $x+\xi$, $y+\eta$, $z+\zeta$, or, which is the same thing, are the values of α, β, γ, when x, y, z are changed into

$$x+\xi,\ y+\eta,\ z+\zeta.$$

Hence, by Taylor's theorem,

$$\xi_1-\xi=\frac{d\alpha}{dx}\xi+\frac{d\alpha}{dy}\eta+\frac{d\alpha}{dz}\zeta$$

$$\eta_1-\eta=\frac{d\beta}{dx}\xi+\frac{d\beta}{dy}\eta+\frac{d\beta}{dz}\zeta$$

$$\zeta_1-\zeta=\frac{d\gamma}{dx}\xi+\frac{d\gamma}{dy}\eta+\frac{d\gamma}{dz}\zeta,$$

the higher powers and products of ξ, η, ζ being neglected. Comparing these expressions with (1) of § 181, we see that they express the changes in the co-ordinates of any displaced point of a body relatively to three rectangular axes in fixed directions through one point of it, when all other points of it are displaced relatively to this one, in any manner subject only to the condition of giving a homogeneous strain. Hence we perceive that at distances all round any point, so small that the first terms only of the expressions by Taylor's theorem for the differences of displacement are sensible, the strain is sensibly homogeneous, and we conclude that the directions of the principal axes of the strain at any point (x, y, z), and the amounts of the elongations of the matter along them, and the tangential displacements in closed curves, are to be found according to the general methods described above, by taking

Heterogeneous strain.

$$[Xx] = \frac{d\alpha}{dx} + 1, \quad [Xy] = \frac{d\alpha}{dy}, \quad [Xz] = \frac{d\alpha}{dz},$$

$$[Yx] = \frac{d\beta}{dx}, \quad [Yy] = \frac{d\beta}{dy} + 1, \quad [Yz] = \frac{d\beta}{dz},$$

$$[Zx] = \frac{d\gamma}{dx}, \quad [Zy] = \frac{d\gamma}{dy}, \quad [Zz] = \frac{d\gamma}{dz} + 1.$$

If each of these nine quantities is constant (*i.e.*, the same for all values of x, y, z), the strain is homogeneous : not unless.

Homogeneous strain.

(*f*) The condition that the strain may be infinitely small is that

Infinitely small strain.

$$\frac{d\alpha}{dx}, \quad \frac{d\alpha}{dy}, \quad \frac{d\alpha}{dz},$$

$$\frac{d\beta}{dx}, \quad \frac{d\beta}{dy}, \quad \frac{d\beta}{dz},$$

$$\frac{d\gamma}{dx}, \quad \frac{d\gamma}{dy}, \quad \frac{d\gamma}{dz},$$

must be each infinitely small.

(*g*) These formulæ apply to the most general possible motion of any substance, and they may be considered as the fundamental equations of kinematics. If we introduce time as independent variable, we have for component velocities u, v, w, parallel to the fixed axes OX, OY, OZ, the following expressions; x, y, z, t being independent variables, and α, β, γ functions of them :—

Most general motion of matter.

$$u = \frac{d\alpha}{dt}, \quad v = \frac{d\beta}{dt}, \quad w = \frac{d\gamma}{dt}.$$

(*h*) If we introduce the condition that no line of the body experiences any elongation, we have the general equations for the kinematics of a rigid body, of which, however, we have had enough already. The equations of condition to express this will be six in number, among the nine quantities $\frac{d\alpha}{dx}$, etc., which (*g*) are, in this case, each constant relatively to x, y, z. There are left three independent arbitrary elements to express any angular motion of a rigid body.

Change of position of a rigid body.

Non-rotational strain.

(*i*) If the disturbed condition is so related to the initial condition that every portion of the body can pass from its initial to its disturbed position and strain, by a translation and a strain without rotation; *i.e.*, if the three principal axes of the strain at any point are lines of the substance which retain their parallelism, we must have, § 183 (18),

$$\frac{d\beta}{dz}=\frac{d\gamma}{dy},\quad \frac{d\gamma}{dx}=\frac{d\alpha}{dz},\quad \frac{d\alpha}{dy}=\frac{d\beta}{dx};$$

and if these equations are fulfilled, the strain is non-rotational, as specified. But these three equations express neither more nor less than that

$$\alpha dx+\beta dy+\gamma dz$$

is the differential of a function of three independent variables. Hence we have the remarkable proposition, and its converse, that if $F(x, y, z)$ denote any function of the co-ordinates of any point of a body, and if every such point be displaced from its given position (x, y, z) to the point whose co-ordinates are

$$x_1=x+\frac{dF}{dx},\quad y_1=y+\frac{dF}{dy},\quad z_1=z+\frac{dF}{dz}\quad\ldots\ldots\ldots\ldots(1),$$

the principal axes of the strain at every point are lines of the substance which have retained their parallelism. The displacement back from (x_1, y_1, z_1) to (x, y, z) fulfils the same condition, and therefore we must have

Theorem of pure analysis.

$$x=x_1+\frac{dF_1}{dx_1},\quad y=y_1+\frac{dF_1}{dy_1},\quad z=z_1+\frac{dF_1}{dz_1}\ldots\ldots\ldots\ldots(2),$$

where F_1 denotes a function of x_1, y_1, z_1, and $\frac{dF_1}{dx_1}$, etc., its partial differential coefficients with reference to this system of variables. The relation between F and F_1 is clearly

$$F+F_1=-\tfrac{1}{2}D^2\ldots\ldots\ldots\ldots\ldots\ldots\ldots\ldots(3),$$

where $D^2=\frac{dF^2}{dx^2}+\frac{dF^2}{dy^2}+\frac{dF^2}{dz^2}=\frac{dF_1^{\,2}}{dx_1^{\,2}}+\frac{dF_1^{\,2}}{dy_1^{\,2}}+\frac{dF_1^{\,2}}{dz_1^{\,2}}\ldots\ldots\ldots(4).$

This, of course, may be proved by ordinary analytical methods, applied to find x, y, z in terms of x_1, y_1, z_1, when the latter are given by (1) in terms of the former.

(*j*) Let α, β, γ be any three functions of x, y, z. Let dS be any element of a surface; l, m, n the direction cosines of its normal.

Heterogeneous strain.

Then $$\iint dS\left\{l\left(\frac{d\gamma}{dy}-\frac{d\beta}{dz}\right)+m\left(\frac{d\alpha}{dz}-\frac{d\gamma}{dx}\right)+n\left(\frac{d\beta}{dx}-\frac{d\alpha}{dy}\right)\right\}$$
$$=\int(\alpha dx+\beta dy+\gamma dz)\ldots\ldots\ldots\ldots\ldots\ldots\ldots(5),$$

the former integral being over any curvilinear area bounded by a closed curve; and the latter, which may be written

$$\int ds\left(\alpha\frac{dx}{ds}+\beta\frac{dy}{ds}+\gamma\frac{dz}{ds}\right),$$

being round the periphery of this curve line*. To demonstrate this, begin with the part of the first member of (5) depending on α; that is

$$\iint dS\left(m\frac{d\alpha}{dz}-n\frac{d\alpha}{dy}\right);$$

and to evaluate it divide S into bands by planes parallel to ZOY, and each of these bands into rectangles. The breadth at x, y, z, of the band between the planes $x-\frac{1}{2}dx$ and $x+\frac{1}{2}dx$ is $\frac{dx}{\sin\theta}$, if θ denote the inclination of the tangent plane of S to the plane x. Hence if ds denote an element of the curve in which the plane x cuts the surface S, we may take

$$dS=\frac{1}{\sin\theta}dx\,ds.$$

And we have $l=\cos\theta$, and therefore may put

$$m=\sin\theta\cos\phi,\ n=\sin\theta\sin\phi.$$

Hence

$$\iint dS\left(m\frac{d\alpha}{dz}-n\frac{d\alpha}{dy}\right)=\iint dx\,ds\left(\cos\phi\frac{d\alpha}{dz}-\sin\phi\frac{d\alpha}{dy}\right)$$
$$=\iint dx\,ds\frac{d\alpha}{ds}=\int\alpha dx.$$

The limits of the s integration being properly attended to we see that the remaining integration, $\int\alpha dx$, must be performed round the periphery of the curve bounding S. By this, and corresponding evaluations of the parts of the first member of (5) depending on β and γ, the equation is proved.

* This theorem was given by Stokes in his Smith's Prize paper for 1854 (*Cambridge University Calendar*, 1854). The demonstration in the text is an expansion of that indicated in our first edition. A more synthetical proof is given in § 69 (q) of Sir W. Thomson's paper on "Vortex Motion," *Trans. R. S. E.* 1869. A thoroughly analytical proof is given by Prof. Clerk Maxwell in his *Electricity and Magnetism* (§ 24).

Hetero-geneous strain.

(*k*) It is remarkable that

$$\iint dS \left\{ l\left(\frac{d\gamma}{dy} - \frac{d\beta}{dz}\right) + m\left(\frac{d\alpha}{dz} - \frac{d\gamma}{dx}\right) + n\left(\frac{d\beta}{dx} - \frac{d\alpha}{dy}\right) \right\}$$

is the same for all surfaces having common curvilinear boundary; and when α, β, γ are the components of a displacement from x, y, z, it is the entire tangential displacement round the said curvilinear boundary, being a closed curve. It is therefore this that is nothing when the displacement of every part is non-rotational. And when it is not nothing, we see by the above propositions and corollaries precisely what the measure of the rotation is.

Displace-ment function.

(*l*) *Lastly*, We see what the meaning, for the case of no rotation, of $\int(\alpha dx + \beta dy + \gamma dz)$, or, as it has been called, "the displacement function," is. It is, the entire tangential displacement along any curve from the fixed point O, to the point P (x, y, z). And the entire tangential displacement, being in this case the same along all different curves proceeding from one to another of any two points, is equal to the difference of the values of the displacement functions at those points.

"Equation of con-tinuity."

191. As there can be neither annihilation nor generation of matter in any natural motion or action, the whole quantity of a fluid within any space at any time must be equal to the quantity originally in that space, increased by the whole quantity that has entered it and diminished by the whole quantity that has left it. This idea when expressed in a perfectly comprehensive manner for every portion of a fluid in motion constitutes what is called the "*equation of continuity*," an unhappily chosen expression.

Integral equation of continuity.

192. Two ways of proceeding to express this idea present themselves, each affording instructive views regarding the properties of fluids. In one we consider a definite portion of the fluid; follow it in its motions; and declare that the average density of the substance varies inversely as its volume. We thus obtain the equation of continuity in an integral form.

Let a, b, c be the co-ordinates of any point of a moving fluid, at a particular era of reckoning, and let x, y, z be the co-ordinates of the position it has reached at any time t from that era. To specify completely the motion, is to give each of these three varying co-ordinates as a function of a, b, c, t.

Integral equation of continuity.

Let δa, δb, δc denote the edges, parallel to the axes of co-ordinates, of a very small rectangular parallelepiped of the fluid, when $t=0$. Any portion of the fluid, if only small enough in all its dimensions, must (§ 190, e), in the motion, approximately fulfil the condition of a body uniformly strained throughout its volume. Hence if δa, δb, δc are taken infinitely small, the corresponding portion of fluid must (§ 156) remain a parallelepiped during the motion.

If a, b, c be the initial co-ordinates of one angular point of this parallelepiped: and $a+\delta a, b, c$; $a, b+\delta b, c$; $a, b, c+\delta c$; those of the other extremities of the three edges that meet in it: the co-ordinates of the same points of the fluid at time t, will be

$$x, y, z;$$

$$x+\frac{dx}{da}\delta a,\ y+\frac{dy}{da}\delta a,\ z+\frac{dz}{da}\delta a;$$

$$x+\frac{dx}{db}\delta b,\ y+\frac{dy}{db}\delta b,\ z+\frac{dz}{db}\delta b;$$

$$x+\frac{dx}{dc}\delta c,\ y+\frac{dy}{dc}\delta c,\ z+\frac{dz}{dc}\delta c.$$

Hence the lengths and direction cosines of the edges are respectively—

$$\left(\frac{dx^2}{da^2}+\frac{dy^2}{da^2}+\frac{dz^2}{da^2}\right)^{\frac{1}{2}}\delta a,\quad \frac{\frac{dx}{da}}{\left(\frac{dx^2}{da^2}+\frac{dy^2}{da^2}+\frac{dz^2}{da^2}\right)^{\frac{1}{2}}},\ \text{etc.}$$

$$\left(\frac{dx^2}{db^2}+\frac{dy^2}{db^2}+\frac{dz^2}{db^2}\right)^{\frac{1}{2}}\delta b,\quad \frac{\frac{dx}{db}}{\left(\frac{dx^2}{db^2}+\frac{dy^2}{db^2}+\frac{dz^2}{db^2}\right)^{\frac{1}{2}}},\ \text{etc.}$$

$$\left(\frac{dx^2}{dc^2}+\frac{dy^2}{dc^2}+\frac{dz^2}{dc^2}\right)^{\frac{1}{2}}\delta c,\quad \frac{\frac{dx}{dc}}{\left(\frac{dx^2}{dc^2}+\frac{dy^2}{dc^2}+\frac{dz^2}{dc^2}\right)^{\frac{1}{2}}},\ \text{etc.}$$

The volume of this parallelepiped is therefore

$$\left(\frac{dx}{da}\frac{dy}{db}\frac{dz}{dc}-\frac{dx}{da}\frac{dy}{dc}\frac{dz}{db}+\frac{dx}{db}\frac{dy}{dc}\frac{dz}{da}-\frac{dx}{db}\frac{dy}{da}\frac{dz}{dc}+\frac{dx}{dc}\frac{dy}{da}\frac{dz}{db}-\frac{dx}{dc}\frac{dy}{db}\frac{dz}{da}\right)\delta a\delta b\delta c$$

Integral equation of continuity.

or, as it is now usually written,

$$\begin{vmatrix} \frac{dx}{da}, & \frac{dy}{da}, & \frac{dz}{da} \\ \frac{dx}{db}, & \frac{dy}{db}, & \frac{dz}{db} \\ \frac{dx}{dc}, & \frac{dy}{dc}, & \frac{dz}{dc} \end{vmatrix} \delta a \delta b \delta c.$$

Now as there can be neither increase nor diminution of the quantity of matter in any portion of the fluid, the density, or the quantity of matter per unit of volume, in the infinitely small portion we have been considering, must vary inversely as its volume if this varies. Hence, if ρ denote the density of the fluid in the neighbourhood of (x, y, z) at time t, and ρ_0 the initial density, we have

$$\rho \begin{vmatrix} \frac{dx}{da}, & \frac{dy}{da}, & \frac{dz}{da} \\ \frac{dx}{db}, & \frac{dy}{db}, & \frac{dz}{db} \\ \frac{dx}{dc}, & \frac{dy}{dc}, & \frac{dz}{dc} \end{vmatrix} = \rho_0 \quad \text{..............} \quad (1),$$

which is the integral "equation of continuity."

Differential equation of continuity.

193. The form under which the equation of continuity is most commonly given, or the *differential equation of continuity*, as we may call it, expresses that the rate of diminution of the density bears to the density, at any instant, the same ratio as the rate of increase of the volume of an infinitely small portion bears to the volume of this portion at the same instant.

To find it, let a, b, c denote the co-ordinates, not when $t = 0$, but at any time $t - dt$, of the point of fluid whose co-ordinates are x, y, z at t; so that we have

$$x - a = \frac{dx}{dt} dt, \quad y - b = \frac{dy}{dt} dt, \quad z - c = \frac{dz}{dt} dt,$$

according to the ordinary notation for partial differential coefficients; or, if we denote by u, v, w, the components of the velocity of this point of the fluid, parallel to the axes of co-ordinates,

$$x - a = udt, \quad y - b = vdt, \quad z - c = wdt.$$

Differential equation of continuity.

Hence

$$\frac{dx}{da} = 1 + \frac{du}{da}dt, \quad \frac{dy}{da} = \frac{dv}{da}dt, \quad \frac{dz}{da} = \frac{dw}{da}dt;$$

$$\frac{dx}{db} = \frac{du}{db}dt, \quad \frac{dy}{db} = 1 + \frac{dv}{db}dt, \quad \frac{dz}{db} = \frac{dw}{db}dt;$$

$$\frac{dx}{dc} = \frac{du}{dc}dt, \quad \frac{dy}{dc} = \frac{dv}{dc}dt, \quad \frac{dz}{dc} = 1 + \frac{dw}{dc}dt;$$

and, as we must reject all terms involving higher powers of dt than the first, the determinant becomes simply

$$1 + \left(\frac{du}{da} + \frac{dv}{db} + \frac{dw}{dc}\right) dt.$$

This therefore expresses the ratio in which the volume is augmented in time dt. The corresponding ratio of variation of density is

$$1 + \frac{D\rho}{\rho}$$

if $D\rho$ denote the differential of ρ, the density of one and the same portion of fluid as it moves from the position (a, b, c) to (x, y, z) in the interval of time from $t - dt$ to t. Hence

$$\frac{1}{\rho}\frac{D\rho}{dt} + \frac{du}{da} + \frac{dv}{db} + \frac{dw}{dc} = 0. \quad \text{................} (1).$$

Here ρ, u, v, w are regarded as functions of a, b, c, and t, and the variation of ρ implied in $\frac{D\rho}{dt}$ is the rate of the actual variation of the density of an indefinitely small portion of the fluid as it moves away from a fixed position (a, b, c). If we alter the principle of the notation, and consider ρ as the density of whatever portion of the fluid is at time t in the neighbourhood of the fixed point (a, b, c), and u, v, w the component velocities of the fluid passing the same point at the same time, we shall have

$$\frac{D\rho}{dt} = \frac{d_t\rho}{dt} + u\frac{d_a\rho}{da} + v\frac{d_b\rho}{db} + w\frac{d_c\rho}{dc} \quad \text{................} (2).$$

Omitting again the suffixes, according to the usual imperfect notation for partial differential co-efficients, which on our new understanding can cause no embarrassment, we thus have, in virtue of the preceding equation,

$$\frac{1}{\rho}\left(\frac{d\rho}{dt} + u\frac{d\rho}{da} + v\frac{d\rho}{db} + w\frac{d\rho}{dc}\right) + \frac{du}{da} + \frac{dv}{db} + \frac{dw}{dc} = 0;$$

or,

$$\frac{d\rho}{dt} + \frac{d(\rho u)}{da} + \frac{d(\rho v)}{db} + \frac{d(\rho w)}{dc} = 0 \quad \text{..........} (3),$$

Differential equation of continuity.

which is the differential equation of continuity, in the form in which it is most commonly given.

194. The other way referred to above (§ 192) leads immediately to the differential equation of continuity.

Imagine a space fixed in the interior of a fluid, and consider the fluid which flows into this space, and the fluid which flows out of it, across different parts of its bounding surface, in any time. If the fluid is of the same density and incompressible, the whole quantity of matter in the space in question must remain constant at all times, and therefore the quantity flowing in must be equal to the quantity flowing out in any time. If, on the contrary, during any period of motion, more fluid enters than leaves the fixed space, there will be condensation of matter in that space; or if more fluid leaves than enters, there will be dilatation. The rate of augmentation of the average density of the fluid, per unit of time, in the fixed space in question, bears to the actual density, at any instant, the same ratio that the rate of acquisition of matter into that space bears to the whole matter in that space.

Let the space S be an infinitely small parallelepiped, of which the edges α, β, γ are parallel to the axes of co-ordinates, and let x, y, z be the co-ordinates of its centre; so that $x \pm \frac{1}{2}\alpha$, $y \pm \frac{1}{2}\beta$, $z \pm \frac{1}{2}\gamma$ are the co-ordinates of its angular points. Let ρ be the density of the fluid at (x, y, z), or the mean density through the space S, at the time t. The density at the time $t+dt$ will be $\rho + \frac{d\rho}{dt}dt$; and hence the quantities of fluid contained in the space S, at the times t, and $t+dt$, are respectively $\rho\alpha\beta\gamma$ and $\left(\rho + \frac{d\rho}{dt}dt\right)\alpha\beta\gamma$. Hence the quantity of fluid lost (there will of course be an absolute gain if $\frac{d\rho}{dt}$ be positive) in the time dt is

$$-\frac{d\rho}{dt}\alpha\beta\gamma dt \qquad\qquad (a).$$

Now let u, v, w be the three components of the velocity of the fluid (or of a fluid particle) at P. These quantities will be functions of x, y, z (involving also t, except in the case of "steady motion"), and will in general vary gradually from point to point of the fluid; although the analysis which follows is not restricted

"Steady motion" defined.

by this consideration, but holds even in cases where in certain places of the fluid there are abrupt transitions in the velocity, as may be seen by considering them as limiting cases of motions in which there are very sudden continuous transitions of velocity. If ω be a small plane area, perpendicular to the axis of x, and having its centre of gravity at P, the volume of fluid which flows across it in the time dt will be equal to $u\omega dt$, and the mass or quantity will be $\rho u\omega dt$. If we substitute $\beta\gamma$ for ω, the quantity which flows across either of the faces β, γ of the parallelepiped S, will differ from this only on account of the variation in the value of ρu; and therefore the quantities which flow across the two sides $\beta\gamma$ are respectively

$$\left\{\rho u - \tfrac{1}{2}\alpha\frac{d(\rho u)}{dx}\right\}\beta\gamma dt,$$

and
$$\left\{\rho u + \tfrac{1}{2}\alpha\frac{d(\rho u)}{dx}\right\}\beta\gamma dt.$$

Hence $\alpha\frac{d(\rho u)}{dx}\beta\gamma dt$, or $\frac{d(\rho u)}{dx}\alpha\beta\gamma dt$, is the excess of the quantity of fluid which leaves the parallelepiped across one of the faces $\beta\gamma$ above that which enters it across the other. By considering in addition the effect of the motion across the other faces of the parallelepiped, we find for the total quantity of fluid lost from the space S, in the time dt,

$$\left\{\frac{d(\rho u)}{dx} + \frac{d(\rho v)}{dy} + \frac{d(\rho w)}{dz}\right\}\alpha\beta\gamma dt \;\ldots\ldots\ldots\ldots\ldots\ldots\ldots (b).$$

Equating this to the expression (a), previously found, we have

$$\left\{\frac{d(\rho u)}{dx} + \frac{d(\rho v)}{dy} + \frac{d(\rho w)}{dz}\right\}\alpha\beta\gamma dt = -\frac{d\rho}{dt}\alpha\beta\gamma dt;$$

and we deduce

$$\frac{d(\rho u)}{dx} + \frac{d(\rho v)}{dy} + \frac{d(\rho w)}{dz} + \frac{d\rho}{dt} = 0 \;\ldots\ldots\ldots\ldots\ldots (4),$$

which is the required equation.

195. Several references have been made in preceding sections to the number of independent variables in a displacement, or to the degrees of *freedom* or *constraint* under which the displacement takes place. It may be well, therefore, to take a general view of this part of the subject by itself.

Freedom and constraint.

Of a point.

196. A free point has *three* degrees of freedom, inasmuch as the most general displacement which it can take is resolvable into three, parallel respectively to any three directions, and independent of each other. It is generally convenient to choose these three directions of resolution at right angles to one another.

If the point be constrained to remain always on a given surface, *one* degree of constraint is introduced, or there are left but *two* degrees of freedom. For we may take the normal to the surface as one of three rectangular directions of resolution. No displacement can be effected parallel to it: and the other two displacements, at right angles to each other, in the tangent plane to the surface, are independent.

If the point be constrained to remain on *each* of two surfaces, it loses two degrees of freedom, and there is left but one. In fact, it is constrained to remain on the curve which is common to both surfaces, and along a curve there is at each point but one direction of displacement.

Of a rigid body.

197. Taking next the case of a free rigid body, we have evidently *six* degrees of freedom to consider—*three* independent translations in rectangular directions as a point has, and three independent rotations about three mutually rectangular axes.

Freedom and constraint of a rigid body.

If it have one point fixed, it loses *three* degrees of freedom; in fact, it has now only the rotations just mentioned.

If a second point be fixed, the body loses *two* more degrees of freedom, and keeps only one freedom to rotate about the line joining the two fixed points.

If a third point, not in a line with the other two, be fixed, the body is fixed.

198. If a rigid body is forced to touch a smooth surface, *one* degree of freedom is lost; there remain *five*, two displacements parallel to the tangent plane to the surface, and three rotations. As a degree of freedom is lost by a constraint of the body to touch a smooth surface, *six* such conditions completely determine the position of the body. Thus if six points on the barrel and stock of a rifle rest on six convex portions of the surface of a fixed rigid body, the rifle may be placed, and replaced any number of times, in precisely the same position, and always left quite free to recoil when fired, for the purpose of testing its accuracy.

Freedom and constraint of a rigid body.

A fixed V under the barrel near the muzzle, and another under the swell of the stock close in front of the trigger-guard, give four of the contacts, bearing the weight of the rifle. A fifth (the one to be broken by the recoil) is supplied by a nearly vertical fixed plane close behind the second V, to be touched by the trigger-guard, the rifle being pressed forward in its V's as far as this obstruction allows it to go. This contact may be dispensed with and nothing sensible of accuracy lost, by having a mark on the second V, and a corresponding mark on barrel or stock, and sliding the barrel backwards or forwards in the V's till the two marks are, as nearly as can be judged by eye, in the same plane perpendicular to the barrel's axis. The sixth contact may be dispensed with by adjusting two marks on the heel and toe of the butt to be as nearly as need be in one vertical plane judged by aid of a plummet. This method requires less of costly apparatus, and is no doubt more accurate and trustworthy, and more quickly and easily executed, than the ordinary method of clamping the rifle in a massive metal cradle set on a heavy mechanical slide.

A geometrical clamp is a means of applying and maintaining six mutual pressures between two bodies touching one another at six points.

Geometrical clamp.

A "geometrical slide" is any arrangement to apply five degrees of constraint, and leave one degree of freedom, to the relative motion of two rigid bodies by keeping them pressed together at just five points of their surfaces.

Geometrical slide.

Ex. 1. The transit instrument would be an instance if one end of one pivot, made slightly convex, were pressed against a fixed vertical end-plate, by a spring pushing at the other end of the axis. The other four guiding points are the points, or small areas, of contact of the pivots on the Y's.

Examples of geometrical slide.

Ex. 2. Let two rounded ends of legs of a three-legged stool rest in a straight, smooth, V-shaped canal, and the third on a smooth horizontal plane*. Gravity maintains positive determinate pressures on the five bearing points; and there is a determinate distribution and amount of friction to be overcome, to produce the rectilineal translational motion thus accurately provided for.

Example of geometrical clamp.

Ex. 3. Let only one of the feet rest in a V canal, and let another rest in a trihedral hollow† in line with the canal, the third still resting on a horizontal plane. There are thus six bearing points, one on the horizontal plane, two on the sides of the canal, and three on the sides of the trihedral hollow: and the stool is fixed in a determinate position as long as all these six contacts are unbroken. Substitute for gravity a spring, or a screw and nut (of not infinitely rigid material), binding the stool to the rigid body to which these six planes belong. Thus we have a "geometrical clamp," which clamps two bodies together with perfect firmness in a perfectly definite position,

* Thomson's reprint of *Electrostatics and Magnetism*, § 346.

† A conical hollow is more easily made (as it can be bored out at once by an ordinary drill), and fulfils nearly enough for most practical applications the geometrical principle. A conical, or otherwise rounded, hollow is touched at three points by knobs or ribs projecting from a round foot resting in it, and thus again the geometrical principle is rigorously fulfilled. The virtue of the geometrical principle is well illustrated by its possible violation in this very case. Suppose the hollow to have been drilled out not quite "true," and instead of being a circular cone to have slightly elliptic horizontal sections:—A hemispherical foot will not rest steadily in it, but will be liable to a slight horizontal displacement in the direction parallel to the major axes of the elliptic sections, besides the legitimate rotation round any axis through the centre of the hemispherical surface: in fact, on this supposition there are just two points of contact of the foot in the hollow instead of three. When the foot and hollow are large enough in any particular case to allow the possibility of this defect to be of moment, it is to be obviated, not by any vain attempt to turn the hollow and the foot each perfectly "true:"—even if this could be done the desired result would be lost by the smallest particle of matter such as a chip of wood, or a fragment of paper, or a hair, getting into the hollow when, at any time in the use of the instrument, the foot is taken out and put in again. On the contrary, the true geometrical method, (of which the general principle was taught to one of us by the late Professor Willis thirty years ago,) is to alter one or other of the two surfaces so as to render it manifestly not a figure of revolution, thus:—Roughly file three round notches in the hollow so as to render it something between a trihedral pyramid and a circular cone, leaving the foot approximately round; or else roughly file at three places of the rounded foot so that horizontal sections through and a little above and below the points of contact may be (roughly) equilateral triangles with rounded corners.

without the aid of friction (except in the screw, if a screw is used); and in various practical applications gives very readily and conveniently a more securely firm connexion by one screw slightly pressed, than a clamp such as those commonly made hitherto by mechanicians can give with three strong screws forced to the utmost.

Example of geometrical clamp.

Do away with the canal and let two feet (instead of only one) rest on the plane, the other still resting in the conical hollow. The number of contacts is thus reduced to five (three in the hollow and two on the plane), and instead of a "clamp" we have again a slide. This form of slide,—a three-legged stool with two feet resting on a plane and one in a hollow,—will be found very useful in a large variety of applications, in which motion about an axis is desired when a material axis is not conveniently attainable. Its first application was to the "azimuth mirror," an instrument placed on the glass cover of a mariner's compass and used for taking azimuths of sun or stars to correct the compass, or of landmarks or other terrestrial objects to find the ship's position. It has also been applied to the "Deflector," an adjustible magnet laid on the glass of the compass bowl and used, according to a principle first we believe given by Sir Edward Sabine, to discover the "semicircular" error produced by the ship's iron. The movement may be made very frictionless when the plane is horizontal, by weighting the moveable body so that its centre of gravity is very nearly over the foot that rests in the hollow. One or two guard feet, not to touch the plane except in case of accident, ought to be added to give a broad enough base for safety.

Example of geometrical slide.

The geometrical slide and the geometrical clamp have both been found very useful in electrometers, in the "siphon recorder," and in an instrument recently brought into use for automatic signalling through submarine cables. An infinite variety of forms may be given to the geometrical slide to suit varieties of application of the general principle on which its definition is founded.

An old form of the geometrical clamp, with the six pressures produced by gravity, is the three V grooves on a stone slab bearing the three legs of an astronomical or magnetic instru-

Example of geometrical slide.

ment. It is not generally however so "well-conditioned" as the trihedral hole, the V groove, and the horizontal plane contact, described above.

For investigation of the pressures on the contact surfaces of a geometrical slide or a geometrical clamp, see § 551, below.

There is much room for improvement by the introduction of geometrical slides and geometrical clamps, in the mechanism of mathematical, optical, geodetic, and astronomical instruments: which as made at present are remarkable for disregard of geometrical and dynamical principles in their slides, micrometer screws, and clamps. Good workmanship cannot compensate for bad design, whether in the safety-valve of an iron-clad, or the movements and adjustments of a theodolite.

199. If one point be constrained to remain in a curve, there remain four degrees of freedom.

If two points be constrained to remain in given curves, there are four degrees of constraint, and we have left two degrees of freedom. One of these may be regarded as being a simple rotation about the line joining the constrained points, a motion which, it is clear, the body is free to receive. It may be shown that the other possible motion is of the most general character for one degree of freedom; that is to say, translation and rotation in any fixed proportions as of the nut of a screw.

If one line of a rigid system be constrained to remain parallel to itself, as, for instance, if the body be a three-legged stool standing on a perfectly smooth board fixed to a common window, sliding in its frame with perfect freedom, there remain *three* translations and one rotation.

But we need not further pursue this subject, as the number of combinations that might be considered is endless; and those already given suffice to show how simple is the determination of the degrees of freedom or constraint in any case that may present itself.

One degree of constraint of the most general character.

200. One degree of constraint, of the most general character, is not producible by constraining one point of the body to a curve surface; but it consists in stopping one line of the body from longitudinal motion, except accompanied by rotation round this line, in fixed proportion to the longitudinal motion, and leaving unimpeded every other motion: that is to say, free rotation about any axis perpendicular to this line (two degrees of freedom); and translation in any direction perpendicular to the same line (two degrees of freedom). These four, with the one degree of freedom to screw, constitute the five degrees of freedom, which, with one degree of constraint, make up the six elements. Remark that it is only in case (*b*) below (§ 201) that there is any point of the body which cannot move in every direction.

201. Let a screw be cut on one shaft, A, of a Hooke's joint, and let the other shaft, L, be joined to a fixed shaft, B, by a second Hooke's joint. A nut, N, turning on A, has the most general kind of motion admitted by one degree of constraint; or it is subjected to just one degree of constraint of the most general character. It has five degrees of freedom; for it may move, 1*st*, by screwing on A, the two Hooke's joints being at rest; 2*d*, it may rotate about either axis of the first Hooke's joint, or any axis in their plane (two more degrees of freedom: being freedom to rotate about two axes through one point); 3*d*, it may, by the two Hooke's joints, each bending, have irrotational translation in any direction perpendicular to the link, L, which connects the joints (two more degrees of freedom). But it cannot have a translation parallel to the line of the shafts and link without a definite proportion of rotation round this line; nor can it have rotation round this line without a definite proportion of translation parallel to it. The same statements apply to the motion of B if N is held fixed; but it is now a fixed axis, not as before a moveable one round which the screwing takes place.

Mechanical illustration.

No simpler mechanism can be easily imagined for producing one degree of constraint, of the most general kind.

Particular case (*a*).—Step of screw infinite (straight rifling), *i.e.*, the nut may slide freely, but cannot turn. Thus the one degree of constraint is, that there shall be no rotation about a certain axis, a fixed axis if we take the case of N fixed and B moveable. This is the kind and degree of freedom enjoyed by the outer ring of a gyroscope with its fly-wheel revolving infinitely fast. The outer ring, supposed taken off its stand, and held in the hand, cannot revolve about an axis perpen-

Mechanical illustration.

dicular to the plane of the inner ring*, but it may revolve freely about either of two axes at right angles to this, namely, the axis of the fly-wheel, and the axis of the inner ring relative to the outer; and it is of course perfectly free to translation in any direction.

Particular case (*b*).—Step of the screw = 0. In this case the nut may run round freely, but cannot move along the axis of the shaft. Hence the constraint is simply that the body can have no translation parallel to the line of shafts, but may have every other motion. This is the same as if any point of the body in this line were held to a fixed surface. This constraint may be produced less frictionally by not using a guiding surface, but the link and second Hooke's joint of the present arrangement, the first Hooke's joint being removed, and by pivoting one point of the body in a cup on the end of the link. Otherwise, let the end of the link be a continuous surface, and let a continuous surface of the body press on it, rolling or spinning when required, but not permitted to slide.

One degree of constraint expressed analytically.

A single degree of constraint is expressed by a single equation among the six co-ordinates specifying the position of one rigid body, relatively to another considered fixed. The effect of this on the body in any particular position is to prevent it from getting out of this position, except by means of component velocities (or infinitely small motions) fulfilling a certain linear equation among themselves.

Thus if ϖ_1, ϖ_2, ϖ_3, ϖ_4, ϖ_5, ϖ_6, be the six co-ordinates, and $F(\varpi_1 \ldots\ldots\ldots) = 0$ the condition; then

$$\frac{dF}{d\varpi_1}\delta\varpi_1 + \ldots\ldots\ldots\ldots = 0$$

is the linear equation which guides the motion through any particular position, the special values of ϖ_1, ϖ_2, ϖ_3, etc., for the particular position, being used in $\frac{dF}{d\varpi_1}$, $\frac{dF}{d\varpi_2}$, &c.

Now, whatever may be the co-ordinate system adopted, we may, if we please, reduce this equation to one between three velocities of translation u, v, w, and three angular velocities ϖ, ρ, σ.

* "The plane of the inner ring" is the plane of the axis of the fly-wheel and of the axis of the inner ring by which it is pivoted on the outer ring.

Let this equation be

One degree of constraint expressed analytically.

$$Au + Bv + Cw + A'\varpi + B'\rho + C'\sigma = 0.$$

This is equivalent to the following:—

$$q + a\omega = 0,$$

if q denote the component velocity along or parallel to the line whose direction cosines are proportional to

$$A,\ B,\ C,$$

ω the component angular velocity round an axis through the origin and in the direction whose direction cosines are proportional to $A',\ B',\ C',$

and lastly, $$a = \sqrt{\frac{A'^2 + B'^2 + C'^2}{A^2 + B^2 + C^2}}.$$

It might be supposed that by altering the origin of co-ordinates we could do away with the angular velocities, and leave only a linear equation among the components of translational velocity. It is not so; for let the origin be shifted to a point whose co-ordinates are ξ, η, ζ. The angular velocities about the new axes, parallel to the old, will be unchanged; but the linear velocities which, in composition with these angular velocities about the new axes, give $\varpi, \rho, \sigma, u, v, w$, with reference to the old, are (§ 89)

$$u - \sigma\eta + \rho\zeta = u',$$
$$v - \varpi\zeta + \sigma\xi = v',$$
$$w - \rho\xi + \varpi\eta = w'.$$

Hence the equation of constraint becomes

$$Au' + Bv' + Cw' + (A' + B\zeta - C\eta)\,\varpi + \text{etc.} = 0.$$

Now we cannot generally determine ξ, η, ζ, so as to make ϖ, etc., disappear, because this would require three conditions, whereas their coefficients, as functions of ξ, η, ζ, are not independent, since there exists the relation

$$A\,(B\zeta - C\eta) + B\,(C\xi - A\zeta) + C\,(A\eta - B\xi) = 0.$$

The simplest form we can reduce to is

$$lu' + mv' + nw' + a\,(l\varpi + m\rho + n\sigma) = 0,$$

that is to say, every longitudinal motion of a certain axis must be accompanied by a definite proportion of rotation about it.

202. These principles constitute in reality part of the general theory of "co-ordinates" in geometry. The three co-ordinates

Generalized co-ordinates. Of a point.

of either of the ordinary systems, rectangular or polar, required to specify the position of a point, correspond to the three degrees of freedom enjoyed by an unconstrained point. The most general system of co-ordinates of a point consists of three sets of surfaces, on one of each of which it lies. When one of these surfaces only is given, the point may be anywhere on it, or, in the language we have been using above, it enjoys two degrees of freedom. If a second and a third surface, on each of which also it must lie, it has, as we have seen, no freedom left: in other words, its position is completely specified, being the point in which the three surfaces meet. The analytical ambiguities, and their interpretation, in cases in which the specifying surfaces meet in more than one point, need not occupy us here.

To express this analytically, let $\psi = \alpha$, $\phi = \beta$, $\theta = \gamma$, where ψ, ϕ, θ are functions of the position of the point, and α, β, γ constants, be the equations of the three sets of surfaces, different values of each constant giving the different surfaces of the corresponding set. Any one value, for instance, of α, will determine one surface of the first set, and so for the others: and three particular values of the three constants specify a particular point, P, being the intersection of the three surfaces which they determine. Thus α, β, γ are the "co-ordinates" of P; which may be referred to as "the point (α, β, γ)." The form of the co-ordinate surfaces of the (ψ, ϕ, θ) system is defined in terms of co-ordinates (x, y, z) on any other system, plane rectangular co-ordinates for instance, if ψ, ϕ, θ are given each as a function of (x, y, z).

Origin of the differential calculus.

203. Component velocities of a moving point, parallel to the three axes of co-ordinates of the ordinary plane rectangular system, are, as we have seen, the rates of augmentation of the corresponding co-ordinates. These, according to the Newtonian fluxional notation, are written $\dot{x}$, $\dot{y}$, $\dot{z}$; or, according to Leibnitz's notation, which we have used above, $\frac{dx}{dt}$, $\frac{dy}{dt}$, $\frac{dz}{dt}$. Lagrange has combined the two notations with admirable skill and taste in the first edition* of his *Mécanique Analytique*, as we shall

* In later editions the Newtonian notation is very unhappily altered by the

see in Chap. II. In specifying the motion of a point according to the generalized system of co-ordinates, ψ, ϕ, θ must be considered as varying with the time: $\dot{\psi}$, $\dot{\phi}$, $\dot{\theta}$, or $\frac{d\psi}{dt}$, $\frac{d\phi}{dt}$, $\frac{d\theta}{dt}$, will then be the generalized components of velocity: and $\ddot{\psi}$, $\ddot{\phi}$, $\ddot{\theta}$, or $\frac{d\dot{\psi}}{dt}$, $\frac{d\dot{\phi}}{dt}$, $\frac{d\dot{\theta}}{dt}$, or $\frac{d^2\psi}{dt^2}$, $\frac{d^2\phi}{dt^2}$, $\frac{d^2\theta}{dt^2}$, will be the generalized components of acceleration.

204. On precisely the same principles we may arrange sets of co-ordinates for specifying the position and motion of a material system consisting of any finite number of rigid bodies, or material points, connected together in any way. Thus if ψ, ϕ, θ, etc., denote any number of elements, independently variable, which, when all given, fully specify its position and configuration, being of course equal in number to the degrees of freedom to move enjoyed by the system, these elements are its *co-ordinates.* When it is actually moving, their rates of variation per unit of time, or $\dot{\psi}$, $\dot{\phi}$, etc., express what we shall call its generalized component velocities; and the rates at which $\dot{\psi}$, $\dot{\phi}$, etc., augment per unit of time, or $\ddot{\psi}$, $\ddot{\phi}$, etc., its component accelerations. Thus, for example, if the system consists of a single rigid body quite free, ψ, ϕ, etc., in number six, may be three common co-ordinates of one point of the body, and three angular co-ordinates (§ 101, above) fixing its position relatively to axes in a given direction through this point. Then $\dot{\psi}$, $\dot{\phi}$, etc., will be the three components of the velocity of this point, and the velocities of the three angular motions explained in § 101, as corresponding to variations in the angular co-ordinates. Or, again, the system may consist of one rigid body supported on a fixed axis; a second, on an axis fixed relatively to the first; a third, on an axis fixed relatively to the second, and so on. There will be in this case only as many co-ordinates as there are of rigid bodies. These co-ordinates might be, for instance, the angle between a plane of the first body and a fixed plane, through the first axis; the angle between planes through the

Co-ordinates of any system.

Generalized components of velocity.

Examples.

substitution of accents, ′ and ″, for the · and ·· signifying velocities and accelerations.

Generalized components of velocity.

Examples.

second axis, fixed relatively to the first and second bodies, and so on; and the component velocities, $\dot{\psi}$, $\dot{\phi}$, etc. would then be the angular velocity of the first body relatively to directions fixed in space; the angular velocity of the second body relatively to the first; of the third relatively to the second, and so on. Or if the system be a set, i in number, of material points perfectly free, one of its $3i$ co-ordinates may be the sum of the squares of their distances from a certain point, either fixed or moving in any way relatively to the system, and the remaining $3i-1$ may be angles, or may be mere ratios of distances between individual points of the system. But it is needless to multiply examples here. We shall have illustrations enough of the principle of generalized co-ordinates, by actual use of it in Chap. II., and other parts of this book.

APPENDIX TO CHAPTER I.

A_0.—Expression in Generalized Co-ordinates for Poisson's extension of Laplace's equation.

(*a*) In § 491 (*c*) below is to be found Poisson's extension of Laplace's equation, expressed in rectilineal rectangular co-ordinates; and in § 492 an equivalent in a form quite independent of the particular kind of co-ordinates chosen: all with reference to the theory of attraction according to the Newtonian law. The same analysis is largely applicable through a great range of physical mathematics, including hydro-kinematics (the "equation of continuity" § 192), the equilibrium of elastic solids (§ 734), the vibrations of elastic solids and fluids (Vol. II.), Fourier's theory of heat, &c. Hence detaching the analytical subject from particular physical applications, consider the equation

$$\frac{d^2U}{dx^2}+\frac{d^2U}{dy^2}+\frac{d^2U}{dz^2}=-4\pi\rho\ldots\ldots\ldots\ldots\ldots(1)$$

where ρ is a given function of x, y, z, (arbitrary and discontinuous it may be). Let it be required to express in terms of generalized co-ordinates ξ, ξ', ξ'', the property of U which this equation expresses in terms of rectangular rectilinear co-ordinates. This may be done of course directly [§ (m) below] by analytical transformation, finding the expression in terms of ξ, ξ', ξ'', for the operation $\frac{d^2}{dx^2}+\frac{d^2}{dy^2}+\frac{d^2}{dz^2}$. But it is done in the form most convenient for physical applications much more easily as follows, by taking advantage of the formula of § 492 which expresses the same property of U independently of any particular system of co-ordinates. This expression is

Laplace's equation in generalized co-ordinates.

$$\iint \delta U dS = -4\pi \iiint \rho dB \text{.............................}(2),$$

where $\iint dS$ denotes integration over the whole of a closed surface S, $\iiint dB$ integration throughout the volume B enclosed by it, and δU the rate of variation of U at any point of S, per unit of length in the direction of the normal outwards.

(b) For B take an infinitely small curvilineal parallelepiped having its centre at (ξ, ξ', ξ''), and angular points at

$$(\xi \pm \tfrac{1}{2}\delta\xi,\ \xi' \pm \tfrac{1}{2}\delta\xi',\ \xi'' \pm \tfrac{1}{2}\delta\xi'').$$

Let $R\delta\xi$, $R'\delta\xi'$, $R''\delta\xi''$ be the lengths of the edges of the parallelepiped, and a, a', a'' the angles between them in order of symmetry, so that $R'R''\sin a\,\delta\xi'\delta\xi''$, &c., are the areas of its faces.

Let DU, $D'U$, $D''U$ denote the rates of variation of U, per unit of length, perpendicular to the three surfaces $\xi=$ const., $\xi'=$ const., $\xi''=$ const., intersecting in (ξ, ξ', ξ'') the centre of the parallelepiped. The value of $\iint \delta U dS$ for a section of the parallelepiped by the surface $\xi=$ const. through (ξ, ξ', ξ'') will be

$$R'R''\sin a\,\delta\xi'\,\delta\xi''\,DU.$$

Hence the values of $\iint \delta U\, dS$ for the two corresponding sides of the parallelepiped are

$$R'R''\sin a\,\delta\xi'\,\delta\xi''\,DU \pm \frac{d}{d\xi}(R'R''\sin a\,\delta\xi'\,\delta\xi''\,DU)\,.\,\tfrac{1}{2}\delta\xi.$$

Hence the value of $\iint \delta U\, dS$ for the pair of sides is

$$\frac{d}{d\xi}(R'R''\sin a\,\delta\xi'\,\delta\xi''\,DU)\,.\,\delta\xi,$$

or

$$\frac{d}{d\xi}(R'R''\sin a\,DU)\,\delta\xi\,\delta\xi'\,\delta\xi''.$$

Dealing similarly with the two other pairs of sides of the parallelepiped and adding we find the first member of (2). Its

Laplace's equation in generalized co-ordinates.

second member is $-4\pi\rho \,.\, Q\,.\,RR'R''\,\delta\xi\,\delta\xi'\,\delta\xi''$, if Q denote the ratio of the bulk of the parallelepiped to a rectangular one of equal edges. Hence equating and dividing both sides by the bulk of the parallelepiped we find

$$\frac{1}{QRR'R''}\left\{\frac{d}{d\xi}(R'R''\sin a\,DU)+\frac{d}{d\xi'}(R''R\sin a'\,D'U)\right.$$
$$\left.+\frac{d}{d\xi''}(RR'\sin a''\,D''U)\right\}=-4\pi\rho\,\ldots\,(3).$$

(c) It remains to express DU, $D'U$, $D''U$ in terms of the co-ordinates ξ, ξ', ξ''.

Denote by K, L the two points (ξ, ξ', ξ'') and $(\xi+\delta\xi, \xi', \xi'')$. From L (not shown in the diagram) draw LM perpendicular to the surface $\xi=\text{const.}$ through K. Taking an infinitely small portion of this surface for the plane of our diagram, let $K\Xi'$, $K\Xi''$ be the lines in which it is cut respectively by the surfaces $\xi''=\text{const.}$ and $\xi'=\text{const.}$ through K. Draw MN parallel to $\Xi''K$, and MG perpendicular to $K\Xi'$.

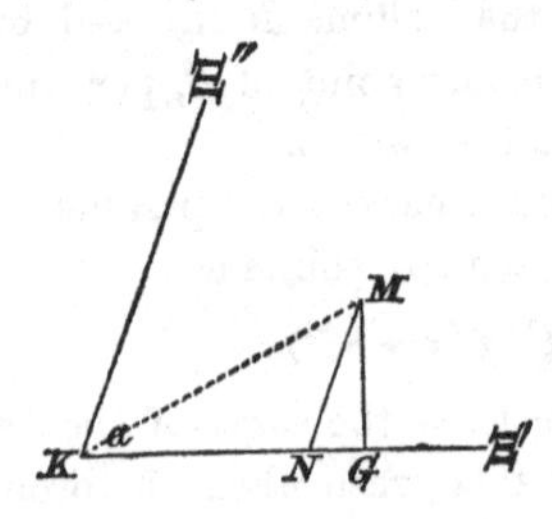

Let now p denote the angle LKM,

A' „ „ „ LGM.

We have

$ML = KL\sin p = R\sin p\,\delta\xi$,

$NM = GM\operatorname{cosec} a = ML\operatorname{cosec} a\cot A' = R\sin p\operatorname{cosec} a\cot A'\,\delta\xi$.

Similarly $KN = R\sin p\operatorname{cosec} a\cot A''\,\delta\xi$,

if A'' denotes an angle corresponding to A'; so that A' and A'' are respectively the angles at which the surfaces $\xi''=\text{const.}$ and $\xi'=\text{const.}$ cut the plane of the diagram in the lines $K\Xi'$ and $K\Xi''$.

Now the difference of values of ξ' for K and N is $\dfrac{KN}{R'}$,

and „ „ „ „ „ ξ'' „ N „ M „ $\dfrac{MN}{R''}$.

Hence if $U(K)$, $U(M)$, $U(L)$ denote the values of U respectively at the points K, M, L, we have

$$U(M)=U(K)+\frac{dU}{d\xi'}\cdot\frac{KN}{R'}+\frac{dU}{d\xi''}\cdot\frac{NM}{R''},$$

and
$$U(L)=U(K)+\frac{dU}{d\xi}\delta\xi.$$

Laplace's equation in generalized co-ordinates.

But
$$DU = \frac{U(L) - U(M)}{ML},$$
and so using the preceding expressions in the terms involved we find
$$DU = \frac{1}{R\sin p}\frac{dU}{d\xi} - \frac{1}{R'\sin a\tan A''}\frac{dU}{d\xi'} - \frac{1}{R''\sin a\tan A'}\frac{dU}{d\xi''} \quad \ldots(4).$$

Using this and the symmetrical expressions for $D'U$ and $D''U$, in (3), we have the required equation.

(d) It is to be remarked that a, a', a'' are the three sides of a spherical triangle of which A, A', A'' are the angles, and p the perpendicular from the angle A to the opposite side.

Hence by spherical trigonometry
$$\cos A = \frac{\cos a - \cos a \cos a'}{\sin a \sin a'};$$
$$\sin A = \frac{\sqrt{(1 - \cos^2 a - \cos^2 a' - \cos^2 a'' + 2\cos a\cos a'\cos a'')}}{\sin a\sin a'} \quad \ldots\ldots(5):$$
$$\sin p = \sin A' \sin a''$$
$$= \frac{\sqrt{(1 - \cos^2 a - \cos^2 a' - \cos^2 a'' + 2\cos a\cos a'\cos a'')}}{\sin a} \quad \ldots\ldots\ldots\ldots(6).$$

To find Q remark that the volume of the parallelepiped is equal to $f\sin p \,.\, gh\sin a$ if f, g, h be its edges: therefore
$$Q = \sin p \sin a \quad \ldots\ldots\ldots\ldots\ldots\ldots(7),$$
whence by (6)
$$Q = \sqrt{(1 - \cos^2 a - \cos^2 a' - \cos^2 a'' + 2\cos a\cos a'\cos a'')} \quad \ldots\ldots(8).$$

Lastly by (5) and (8) we have
$$\tan A = \frac{Q}{\cos a - \cos a'\cos a''} \quad \ldots\ldots\ldots\ldots(9).$$

(e) Using these in (4) we find
$$DU = \frac{1}{Q\sin a}\left(\frac{\sin^2 a}{R}\frac{dU}{d\xi} - \frac{\cos a'' - \cos a\cos a'}{R'}\frac{dU}{d\xi'} - \frac{\cos a' - \cos a\cos a''}{R''}\frac{dU}{d\xi''}\right) \quad \ldots\ldots(10).$$

Using this and the two symmetrical expressions in (3) and adopting a common notation [App. B (g), § 491 (c), &c. &c.], according to which Poisson's equation is written
$$\nabla^2 U = -4\pi\rho \quad \ldots\ldots\ldots\ldots\ldots\ldots(11),$$

Laplace's equation in generalized co-ordinates.

we find for the symbol ∇^2 in terms of the generalized co-ordinates ξ, ξ', ξ'',

$$\nabla^2 = \frac{1}{QRR'R''}\left\{\frac{d}{d\xi}\frac{1}{Q}\left[\frac{R'R''\sin^2 a}{R}\frac{d}{d\xi} + R''(\cos a\cos a' - \cos a'')\frac{d}{d\xi'} + R'(\cos a''\cos a - \cos a')\frac{d}{d\xi''}\right]\right.$$
$$+\frac{d}{d\xi'}\frac{1}{Q}\left[\frac{R''R\sin^2 a'}{R'}\frac{d}{d\xi'} + R(\cos a'\cos a'' - \cos a)\frac{d}{d\xi''} + R''(\cos a\cos a' - \cos a'')\frac{d}{d\xi}\right],$$
$$\left.+\frac{d}{d\xi''}\frac{1}{Q}\left[\frac{RR'\sin^2 a''}{R''}\frac{d}{d\xi''} + R'(\cos a''\cos a - \cos a')\frac{d}{d\xi} + R(\cos a'\cos a'' - \cos a)\frac{d}{d\xi'}\right]\right\} \ldots (12),$$

where for Q, its value by (8) in terms of a, a', a'' is to be used, and a, a', a'', R, R', R'' are all known functions of ξ, ξ', ξ'' when the system of co-ordinates is completely defined.

Case of rectangular co-ordinates, curved or plane.

(*f*) For the case of rectangular co-ordinates whether plane or curved $a = a' = a'' = A = A' = A'' = 90^\circ$ and $Q = 1$, and therefore we have

$$\nabla^2 = \frac{1}{RR'R''}\left\{\frac{d}{d\xi}\left(\frac{R'R''}{R}\frac{d}{d\xi}\right) + \frac{d}{d\xi'}\left(\frac{R''R}{R'}\frac{d}{d\xi'}\right) + \frac{d}{d\xi''}\left(\frac{RR'}{R''}\frac{d}{d\xi''}\right)\right\} \ldots (13),$$

which is the formula originally given by Lamé for expressing in terms of his orthogonal curved co-ordinate system the Fourier equations of the conduction of heat. The proof of the more general formula (12) given above is an extension, in purely analytical form, of a demonstration of Lamé's formula (13) which was given in terms relating to thermal conduction in an article "On the equations of Motion of Heat referred to curvilinear co-ordinates" in the *Cambridge Mathematical Journal* (1843).

(*g*) For the particular case of polar co-ordinates, r, θ, ϕ, considering the rectangular parallelepiped corresponding to δr, $\delta\theta$, $\delta\phi$ we see in a moment that the lengths of its edges are δr, $r\delta\theta$, $r\sin\theta\delta\phi$. Hence in the preceding notation $R = 1$, $R' = r$, $R'' = r\sin\theta$, and Lamé's formula (13) gives

$$\nabla^2 = \frac{1}{r^2\sin\theta}\left\{\sin\theta\frac{d}{dr}\left(r^2\frac{d}{dr}\right) + \frac{d}{d\theta}\left(\sin\theta\frac{d}{d\theta}\right) + \frac{1}{\sin\theta}\frac{d^2}{d\phi^2}\right\} \ldots (14).$$

(h) Again let the co-ordinates be of the kind which has been called "columnar"; that is to say, distance from an axis (r), angle from a plane of reference through this axis to a plane through the axis and the specified point (ϕ), and distance from a plane of reference perpendicular to the axis (z). The co-ordinate surfaces here are

Laplace's equation in columnar co-ordinates.

coaxal circular cylinders (r = const.),

planes through the axis (ϕ = const.),

planes perpendicular to the axis (z = const.).

The three edges of the infinitesimal rectangular parallelepiped are now dr, $rd\phi$, and dz. Hence $R=1$, $R'=r$, $R''=1$, and Lamé's formula gives

$$\nabla^2 = \frac{1}{r}\frac{d}{dr}\left(r\frac{d}{dr}\right) + \left(\frac{d}{rd\phi}\right)^2 + \left(\frac{d}{dz}\right)^2 \quad \text{..............(15)},$$

which is very useful for many physical problems, such as the conduction of heat in a solid circular column, the magnetization of a round bar or wire, the vibrations of air in a closed circular cylinder, the vibrations of a vortex column, &c. &c.

(i) For plane rectangular co-ordinates we have $R=R'=R''$; so in this case (13) becomes (with x, y, z for ξ, ξ', ξ''),

Algebraic transformation from plane rectangular to generalized co-ordinates.

$$\nabla^2 = \frac{d^2}{dx^2} + \frac{d^2}{dy^2} + \frac{d^2}{dz^2} \quad \text{..................(16)},$$

which is Laplace's and Fourier's original form.

(j) Suppose now it be desired to pass from plane rectangular co-ordinates to the generalized co-ordinates.

Let x, y, z be expressed as functions of ξ, ξ', ξ''; then putting for brevity

$$\frac{dx}{d\xi} = X,\quad \frac{dy}{d\xi} = Y,\quad \frac{dz}{d\xi} = Z;\quad \frac{dx}{d\xi'} = X', \text{ \&c.};\quad \frac{dx}{d\xi''} = X'', \text{ \&c. ...(17)};$$

we have

$$\left.\begin{aligned} \delta x &= X\delta\xi + X'\delta\xi' + X''\delta\xi'', \\ \delta y &= Y\delta\xi + Y'\delta\xi' + Y''\delta\xi'', \\ \delta z &= Z\delta\xi + Z'\delta\xi' + Z''\delta\xi'', \end{aligned}\right\} \quad \text{.............(18)};$$

whence

$$R = \sqrt{(X^2 + Y^2 + Z^2)},\qquad R' = \sqrt{(X'^2 + Y'^2 + Z'^2)},$$
$$R'' = \sqrt{(X''^2 + Y''^2 + Z''^2)} \quad \text{.................(19)},$$

Algebraic transformation from plane rectangular to generalized co-ordinates.

and the direction cosines of the three edges of the infinitesimal parallelepiped corresponding to $\delta\xi$, $\delta\xi'$, $\delta\xi''$ are

$$\left(\frac{X}{R}, \frac{Y}{R}, \frac{Z}{R}\right), \left(\frac{X'}{R'}, \frac{Y'}{R'}, \frac{Z'}{R'}\right), \left(\frac{X''}{R''}, \frac{Y''}{R''}, \frac{Z''}{R''}\right) \quad \text{....(20)}.$$

Hence

$$\cos a = \frac{X'X'' + Y'Y'' + Z'Z''}{R'R''}, \quad \cos a' = \frac{X''X + Y''Y + Z''Z}{R''R},$$

$$\cos a'' = \frac{XX' + YY' + ZZ'}{RR'} \quad \text{................(21)}.$$

(*k*) It is important to remark that when these expressions for $\cos a$, $\cos a'$, $\cos a''$, R, R', R'', in terms of X, &c. are used in (8), Q^2 becomes a complete square, so that $QRR'R''$ is a rational homogeneous function of the 3rd degree of X, Y, Z, X', &c.

For the ordinary process of finding from the direction cosines (20) of three lines, the sine of the angle between one of them and the plane of the other two gives

$$\sin p = \begin{vmatrix} X, & Y, & Z \\ X', & Y', & Z' \\ X'', & Y'', & Z'' \end{vmatrix} \div RR'R'' \sin a \quad \text{.........(21)};$$

from this and (7) we see that $QRR'R''$ is equal to the determinant. From this and (8) we see that

Square of a determinant.

$$\begin{aligned}(X^2 + Y^2 + Z^2)&(X'^2 + Y'^2 + Z'^2)(X''^2 + Y''^2 + Z''^2) \\ -(X^2+Y^2+Z^2)&(X'X''+Y'Y''+Z'Z'')^2-(X'^2+Y'^2+Z'^2)(X''X+Y''Y+Z''Z)^2 \\ &- (X''^2 + Y''^2 + Z''^2)(XX' + YY' + ZZ')^2 \\ + 2\,(X'X''+Y'Y''&+Z'Z'')(X''X+Y''Y+Z''Z)(XX'+YY'+ZZ') \\ &= \begin{vmatrix} X, & Y, & Z, \\ X', & Y', & Z', \\ X'', & Y'', & Z'', \end{vmatrix}^2 \quad \text{..................(22)},\end{aligned}$$

an algebraic identity which may be verified by expanding both members and comparing.

(*l*) Denoting now by T the complete determinant, we have

$$Q = \frac{T}{RR'R''} \quad \text{.......................(23)},$$

and using this for Q in (12) we have a formula for ∇^2 in which only rational functions of X, Y, Z, X', &c. appear, and which is readily verified by comparing with the following derived from (16) by direct transformation.

Algebraic transformation from plane rectangular to generalized co-ordinates.

(m) Go back to (18) and resolve for $\delta\xi$, $\delta\xi'$, $\delta\xi''$. We find

$$\delta\xi = \frac{L}{T}\delta x + \frac{M}{T}\delta y + \frac{N}{T}\delta z, \ \delta\xi' = \&c., \ \delta\xi'' = \&c.,$$

where

$$\left.\begin{array}{lll} L = Y'Z'' - Y''Z', & M = Z'X'' - Z''X', & N = X'Y'' - X''Y', \\ L' = Y''Z - YZ'', & M' = Z''X - ZX'', & N' = X''Y - XY'', \\ L'' = YZ' - Y'Z, & M'' = ZX' - Z'X, & N'' = XY' - X'Y, \end{array}\right\} \ldots\ldots(24).$$

Hence

$$\frac{d}{dx} = \frac{L}{T}\frac{d}{d\xi} + \frac{L'}{T}\frac{d}{d\xi'} + \frac{L''}{T}\frac{d}{d\xi''}, \ \frac{d}{dy} = \&c., \ \frac{d}{dz} = \&c.,$$

and thus we have

$$\nabla^2 = \left(\frac{L}{T}\frac{d}{d\xi} + \frac{L'}{T}\frac{d}{d\xi'} + \frac{L''}{T}\frac{d}{d\xi''}\right)^2 + \left(\frac{M}{T}\frac{d}{d\xi} + \frac{M'}{T}\frac{d}{d\xi'} + \frac{M''}{T}\frac{d}{d\xi''}\right)^2$$
$$+ \left(\frac{N}{T}\frac{d}{d\xi} + \frac{N'}{T}\frac{d}{d\xi'} + \frac{N''}{T}\frac{d}{d\xi''}\right)^2 \ \ldots\ldots\ldots\ldots(25).$$

Expanding this and comparing the coefficients of $\frac{d^2}{d\xi^2}$, $\frac{d^2}{d\xi d\xi'}$, $\frac{d}{d\xi}$, &c. with those of the corresponding terms of (12) with (21) and (23) we find the two formulas, (12) and (25), identical.

A.—Extension of Green's Theorem.

It is convenient that we should here give the demonstration of a few theorems of pure analysis, of which we shall have many and most important applications, not only in the subject of spherical harmonics, which follows immediately, but in the general theories of attraction, of fluid motion, and of the conduction of heat, and in the most practical investigations regarding electricity, and magnetic and electro-magnetic force.

(a) Let U and U' denote two functions of three independent variables, x, y, z, which we may conveniently regard as rectangular co-ordinates of a point P, and let α denote a quantity which may be either constant, or any arbitrary function of the variables. Let $\iiint dxdydz$ denote integration throughout a finite *singly continuous* space bounded by a close surface S; let $\iint dS$ denote integration over the whole surface S; and let δ, prefixed to any function, denote its rate of variation at any point of S, per unit of length in the direction perpendicular to S outwards.

Then

a constant gives a theorem of Green's.

$$\iiint \alpha^2 \left(\frac{dU}{dx}\frac{dU'}{dx} + \frac{dU}{dy}\frac{dU'}{dy} + \frac{dU}{dz}\frac{dU'}{dz}\right) dxdydz$$

$$\left.\begin{aligned} &= \iint dS.\, U'\alpha^2 \delta U - \iiint U' \left\{ \frac{d\left(\alpha^2 \frac{dU}{dx}\right)}{dx} + \frac{d\left(\alpha^2 \frac{dU}{dy}\right)}{dy} + \frac{d\left(\alpha^2 \frac{dU}{dz}\right)}{dz} \right\} dxdydz \\ &= \iint dS.\, U\alpha^2 \delta U' - \iiint U \left\{ \frac{d\left(\alpha^2 \frac{dU'}{dx}\right)}{dx} + \frac{d\left(\alpha^2 \frac{dU'}{dy}\right)}{dy} + \frac{d\left(\alpha^2 \frac{dU'}{dz}\right)}{dz} \right\} dxdydz \end{aligned}\right\} \quad \text{..............(1).}$$

For, taking one term of the first member alone, and integrating "by parts," we have

$$\iiint \alpha^2 \frac{dU}{dx}\frac{dU'}{dx} dxdydz = \iint U'\alpha^2 \frac{dU}{dx} dydz - \iiint U' \frac{d\left(\alpha^2 \frac{dU}{dx}\right)}{dx} dxdydz,$$

the first integral being between limits corresponding to the surface S; that is to say, being from the negative to the positive end of the portion within S, or of each portion within S, of the line x through the point $(0, y, z)$. Now if A_2 and A_1 denote the inclination of the outward normal of the surface to this line, at points where it enters and emerges from S respectively, and if dS_2 and dS_1 denote the elements of the surface in which it is cut at these points by the rectangular prism standing on $dydz$, we have

$$dydz = -\cos A_2 dS_2 = \cos A_1 dS_1.$$

Thus the first integral, between the proper limits, involves the elements $U'\alpha^2 \frac{dU}{dx} \cos A_1 dS_1$, and $-U'\alpha^2 \frac{dU}{dx} \cos A_2 dS_2$; the latter of which, as corresponding to the lower limit, is subtracted. Hence, there being in the whole of S an element dS_2 for each element dS_1, the first integral is simply

$$\iint U'\alpha^2 \frac{dU}{dx} \cos A\, dS,$$

for the whole surface. Adding the corresponding terms for y and z, and remarking that

a constant gives a theorem of Green's.

$$\frac{dU}{dx}\cos A+\frac{dU}{dy}\cos B+\frac{dU}{dz}\cos C=\delta U,$$

where B and C denote the inclinations of the outward normal through dS to lines drawn through dS in the positive directions parallel to y and z respectively, we perceive the truth of (1).

(b) Again, let U and U' denote two functions of x, y, z, which have equal values at every point of S, and of which the first fulfils the equation

Equation of the conduction of heat.

$$\frac{d\left(\alpha^2\frac{dU}{dx}\right)}{dx}+\frac{d\left(\alpha^2\frac{dU}{dy}\right)}{dy}+\frac{d\left(\alpha^2\frac{dU}{dz}\right)}{dz}=0\ldots\ldots\ldots\ldots\ldots(2),$$

for every point within S.

Then if $U'-U=u$, we have

$$\iiint\left\{\left(\alpha\frac{dU'}{dx}\right)^2+\left(\alpha\frac{dU'}{dy}\right)^2+\left(\alpha\frac{dU'}{dz}\right)^2\right\}dxdydz$$

$$=\iiint\left\{\left(\alpha\frac{dU}{dx}\right)^2+\left(\alpha\frac{dU}{dy}\right)^2+\left(\alpha\frac{dU}{dz}\right)^2\right\}dxdydz$$

$$+\iiint\left\{\left(\alpha\frac{du}{dx}\right)^2+\left(\alpha\frac{du}{dy}\right)^2+\left(\alpha\frac{du}{dz}\right)^2\right\}dxdydz\ldots\ldots\ldots\ldots(3).$$

For the first member is equal identically to the second member with the addition of

$$2\iiint\alpha^2\left(\frac{dU}{dx}\frac{du}{dx}+\frac{dU}{dy}\frac{du}{dy}+\frac{dU}{dz}\frac{du}{dz}\right)dxdydz.$$

But, by (1), this is equal to

$$2\iint dS.u\alpha^2\delta U-2\iiint u\left\{\frac{d\left(\alpha^2\frac{dU}{dx}\right)}{dx}+\frac{d\left(\alpha^2\frac{dU}{dy}\right)}{dy}+\frac{d\left(\alpha^2\frac{dU}{dz}\right)}{dz}\right\}dxdydz,$$

of which each term vanishes; the first, or the double integral, because, by hypothesis, u is equal to nothing at every point of S, and the second, or the triple integral, because of (2).

(c) The second term of the second member of (3) is essentially positive, provided α has a real value, whether positive, zero, or negative, for every point (x, y, z) within S. Hence the first member of (3) necessarily exceeds the first term of the second member. But the sole characteristic of U is that it satisfies (2). Hence U' cannot also satisfy (2). That is to say, U being any

Property of solution with U given over S.

Solution proved to

be determinate; one solution of (2), there can be no other solution agreeing with it at every point of S, but differing from it for some part of the space within S.

proved to be possible. (*d*) One solution of (2) exists, satisfying the condition that U has an arbitrary value for every point of the surface S. For let U denote any function whatever which has the given arbitrary value at each point of S; let u be any function whatever which is equal to nothing at each point of S, and which is of any real finite or infinitely small value, of the same sign as the value of

$$\frac{d\left(a^2\frac{dU}{dx}\right)}{dx}+\frac{d\left(a^2\frac{dU}{dy}\right)}{dy}+\frac{d\left(a^2\frac{dU}{dz}\right)}{dz}$$

at each internal point, and therefore, of course, equal to nothing at every internal point, if any, for which the value of this expression is nothing; and let $U' = U + \theta u$, where θ denotes any constant. Then, using the formulæ of (*b*), modified to suit the altered circumstances, and taking Q and Q' for brevity to denote

$$\iiint\left\{\left(a\frac{dU}{dx}\right)^2+\left(a\frac{dU}{dy}\right)^2+\left(a\frac{dU}{dz}\right)^2\right\}dxdydz,$$

and the corresponding integral for U', we have

$$Q' = Q - 2\theta\iiint u\left\{\frac{d}{dx}\left(a^2\frac{dU}{dx}\right)+\frac{d}{dy}\left(a^2\frac{dU}{dy}\right)+\frac{d}{dz}\left(a^2\frac{dU}{dz}\right)\right\}dxdydz$$
$$+\theta^2\iiint\left\{\left(a\frac{du}{dx}\right)^2+\left(a\frac{du}{dy}\right)^2+\left(a\frac{du}{dz}\right)^2\right\}dxdydz.$$

The coefficient of -2θ here is essentially positive, in consequence of the condition under which u is chosen, unless (2) is satisfied, in which case it is nothing; and the coefficient of θ^2 is essentially positive, if not zero, because all the quantities involved are real. Hence the equation may be written thus:—

$$Q' = Q - m\theta(n-\theta),$$

where m and n are each positive. This shows that if any positive value less than n is assigned to θ, Q' is made smaller than Q; that is to say, unless (2) is satisfied, a function, having the same value at S as U, may be found which shall make the Q integral smaller than for U. In other words, a function U, which, having any prescribed value over the surface S, makes the integral Q for the interior as small as possible, must satisfy equation (2). But the Q integral is essentially positive, and therefore there is a limit than which it cannot be made smaller.

Hence there is a solution of (2) subject to the prescribed surface condition.

Solution proved possible.

(e) We have seen (c) that there is, if one, only one, solution of (2) subject to the prescribed surface condition, and now we see that there is one. To recapitulate,—we conclude that, if the value of U be given arbitrarily at every point of any closed surface, the equation

a constant gives "Green's theorem."

$$\frac{d}{dx}\left(a^2\frac{dU}{dx}\right)+\frac{d}{dy}\left(a^2\frac{dU}{dy}\right)+\frac{d}{dz}\left(a^2\frac{dU}{dz}\right)=0$$

determines its value without ambiguity for every point within that surface. That this important proposition holds also for the whole infinite space without the surface S, follows from the preceding demonstration, with only the precaution, that the different functions dealt with must be so taken as to render all the triple integrals convergent. S need not be merely a single closed surface, but it may be any number of surfaces enclosing isolated portions of space. The extreme case, too, of S, or any detached part of S, an open shell, that is a finite unclosed surface, is clearly included. Or lastly, S, or any detached part of S, may be an infinitely extended surface, provided the value of U arbitrarily assigned over it be so assigned as to render the triple and double integrals involved all convergent.

B.—Spherical Harmonic Analysis.

The mathematical method which has been commonly referred to by English writers as that of "Laplace's Coefficients," but which is here called *spherical harmonic analysis*, has for its object the expression of an arbitrary periodic function of two independent variables in the proper form for a large class of physical problems involving arbitrary data over a spherical surface, and the deduction of solutions for every point of space.

Object of spherical harmonic analysis.

(a) A *spherical harmonic function* is defined as a homogeneous function, V, of x, y, z, which satisfies the equation

Definition of spherical harmonic functions.

$$\frac{d^2V}{dx^2}+\frac{d^2V}{dy^2}+\frac{d^2V}{dz^2}=0\text{........................}(4).$$

Its degree may be any positive or negative integer; or it may be fractional; or it may be imaginary.

Examples of spherical harmonics.

EXAMPLES. The functions written below are spherical harmonics of the degrees noted; r representing $(x^2+y^2+z^2)^{\frac{1}{2}}$:—

Degree Zero.

I. $$1; \qquad \log\frac{r+z}{r-z}.$$

$$\tan^{-1}\frac{y}{x}; \quad \tan^{-1}\frac{y}{x}\log\frac{r+z}{r-z}; \quad \frac{rz(x^2-y^2)}{(x^2+y^2)^2}; \quad \frac{2rzxy}{(x^2+y^2)^2}.$$

II. Generally, in virtue of (*g*) (15) and (13) below,

$$r\frac{dV_0}{dx}, \; r\frac{dV_0}{dy}, \; r\frac{dV_0}{dz},$$

if V_0 denote any harmonic of degree 0: for instance, group III. below.

III. $$\frac{rx}{x^2+y^2}; \quad \frac{zx}{x^2+y^2}; \quad \frac{x}{r+z}\left(=\frac{rx-zx}{x^2+y^2}\right); \quad \frac{2zy}{x^2+y^2}\tan^{-1}\frac{y}{x}-\frac{xr}{x^2+y^2}\log\frac{r+z}{r-z}.$$

$$\frac{ry}{x^2+y^2}; \quad \frac{zy}{x^2+y^2}; \quad \frac{y}{r+z}; \qquad \frac{2zx}{x^2+y^2}\tan^{-1}\frac{y}{x}+\frac{yr}{x^2+y^2}\log\frac{r+z}{r-z}.$$

IV. Generally, in virtue of (*g*) (15), (13), below,

$$\delta_{n-j-1}(r^{2(n-j)-1}\delta_n V_j),$$

where V_j denotes any spherical harmonic of integral degree, j, and δ_n, δ_{n-j-1} homogeneous integral functions of $\frac{d}{dx}, \frac{d}{dy}, \frac{d}{dz}$, of degrees n and $n-j-1$ respectively: for instance, some of group II. above, and groups V. and VI. below.

V. $$\frac{d^{n-1}(r^{2n-1})}{dz^{n-1}} \cdot \frac{d^n\tan^{-1}\frac{y}{x}}{dx^n};$$

$$\frac{d^{n-1}(r^{2n-1})}{dz^{n-1}} \cdot \frac{d^n\tan^{-1}\frac{y}{x}}{dy^n}.$$

Remark that

$$\tan^{-1}\frac{y}{x}=\frac{1}{2\sqrt{-1}}\log\frac{x+y\sqrt{-1}}{x-y\sqrt{-1}},$$

and therefore

$$\frac{d^n\tan^{-1}\frac{y}{x}}{dx^n}=(-1)^n 1.2\ldots(n-1)\frac{\sin n\phi}{(x^2+y^2)^{\frac{n}{2}}};$$

Examples of spherical harmonics.

so the preceding yields

$$\frac{d^{n-1}(r^{2n-1})}{dz^{n-1}} \frac{\begin{matrix}\sin\\ \cos\end{matrix} n\phi}{(x^2+y^2)^{\frac{n}{2}}},$$

where ϕ denotes $\tan^{-1}\frac{y}{x}$.

VI.

Taking, in IV., $j=-1$,

$$V_j = \frac{1}{r}\log\frac{r+z}{r-z},$$

$$\delta_n = \tfrac{1}{2}\left\{\left(\frac{d}{dx}+\frac{d}{dy}\sqrt{-1}\right)^n + \left(\frac{d}{dx}-\frac{d}{dy}\sqrt{-1}\right)^n\right\},$$

or $$\delta_n = \frac{1}{2\sqrt{-1}}\left\{\left(\frac{d}{dx}+\frac{d}{dy}\sqrt{-1}\right)^n - \left(\frac{d}{dx}-\frac{d}{dy}\sqrt{-1}\right)^n\right\},$$

$$\delta_{n-j-1} = \left(\frac{d}{dz}\right)^n,$$

we find

$$\left(\frac{d}{dz}\right)^n\left[r^{2n+1}\left(\frac{d}{rdr}\right)^n\left(\frac{1}{r}\log\frac{r+z}{r-z}\right)\right](x^2+y^2)^{\frac{n}{2}} \begin{matrix}\cos\\ \sin\end{matrix} n\phi,$$

where $\frac{d}{dr}$ denotes differentiation with reference to r on the supposition of z constant, and $\frac{d}{dz}$ differentiation with reference to z on supposition of x and y constant.

Degree $-i-1$, *or* $+i$, *and type* $H\{z, \sqrt{(x^2+y^2)}\} \begin{matrix}\sin\\ \cos\end{matrix} n\phi$.

H denoting a homogeneous function; n any integer; and i any positive integer.

Let $U_0^{(n)}$ and $V_0^{(n)}$ denote functions yielded by V. and VI. preceding. The following are the two* distinct functions of the degrees and types now sought, and found in virtue of (*g*) (15) below :—

$$U_{-i-1}^{(n)} = \frac{d^{i+1}}{dz^{i+1}} U_0^{(n)}, \quad V_{-i-1}^{(n)} = \frac{d^{i+1}}{dz^{i+1}} V_0 ;$$

* See § (*l*) below.

Examples of spherical harmonics.

or explicitly

$$\text{I.}\begin{cases} U^{(n)}_{-i-1} = \dfrac{d^{n+i}(r^{2n-1})}{dz^{n+i}} \dfrac{\begin{matrix}\cos\\ \sin\end{matrix} n\phi}{(x^2+y^2)^{\frac{n}{2}}}, \\ V^{(n)}_{-i-1} = \left(\dfrac{d}{dz}\right)^{n+i+1}\left[r^{2n+1}\left(\dfrac{d}{rdr}\right)^n\left(\dfrac{1}{r}\log\dfrac{r+z}{r-z}\right)\right](x^2+y^2)^{\frac{n}{2}} \begin{matrix}\cos\\ \sin\end{matrix} n\phi. \end{cases}$$

In the particular case of $n=0$, these two are not distinct. Either of them yields

$$U^{(0)}_{-i-1} = \frac{d^i\left(\dfrac{1}{r}\right)}{dz^i}.$$

II. The other harmonic of the same degree and type is

$$V^{(0)}_{-i-1} = \frac{d^i\left(\dfrac{1}{r}\log\dfrac{r+z}{r-z}\right)}{dz^i}.$$

III. To obtain the harmonics of the same types, but of degree i, multiply each of the preceding groups I. and II. by r^{2i+1}, in virtue of (g) (13) below.

Degree -1.

I. Generally, in virtue of (g) (13) below, any of the preceding functions of degree zero divided by r; or, in virtue of (g) (15), the differential coefficient of any of them with reference to x, or y, or z. For instance,

$$\text{II.}\ \frac{1}{r};$$

$$\text{III.}\ \frac{1}{r}\tan^{-1}\frac{y}{x};\quad \frac{1}{r}\log\frac{r+z}{r-z};\quad \frac{1}{r}\tan^{-1}\frac{y}{x}\log\frac{r+z}{r-z}.$$

$$\text{IV.}\begin{cases} \dfrac{x}{x^2+y^2};\quad \dfrac{xz}{r(x^2+y^2)};\quad \dfrac{x}{r(r+z)}; \\ \dfrac{y}{x^2+y^2};\quad \dfrac{yz}{r(x^2+y^2)};\quad \dfrac{y}{r(r+z)}. \end{cases}$$

Examples of spherical harmonics.

Degrees -2 *and* $+1$.

I. $\left\{\dfrac{x}{r^3},\quad \dfrac{y}{r^3},\quad \dfrac{z}{r^3};\quad x,\ y,\ z.\right.$

II. $\left\{\dfrac{z\tan^{-1}\dfrac{y}{x}}{r^3};\quad z\tan^{-1}\dfrac{y}{x}.\right.$

III. $\left\{\dfrac{z}{r^3}\log\dfrac{r+z}{r-z}-\dfrac{2}{r^3};\quad z\log\dfrac{r+z}{r-z}-2r.\right.$

IV. $\left\{\begin{array}{l} \dfrac{x^2-y^2}{(x^2+y^2)^2},\quad \dfrac{2xy}{(x^2+y^2)^2};\quad \dfrac{r^3(x^2-y^2)}{(x^2+y^2)^2},\quad \dfrac{2r^3xy}{(x^2+y^2)^2};\\ \text{or}\quad \dfrac{\cos 2\phi}{x^2+y^2},\quad \dfrac{\sin 2\phi}{x^2+y^2};\quad \dfrac{r^3\cos 2\phi}{x^2+y^2},\quad \dfrac{r^3\sin 2\phi}{x^2+y^2}.\end{array}\right.$

V. $\left\{\dfrac{1}{r^3}\left(\log\dfrac{r+z}{r-z}+\dfrac{2rz}{x^2+y^2}\right)x;\quad \left(\log\dfrac{r+z}{r-z}+\dfrac{2rz}{x^2+y^2}\right)x\right.$

(the former being $\dfrac{d}{dx}$ of III. 2 degree -1, and the latter being $-\int dz$ of VI. degree 0 with $n=1$).

The Rational Integral Harmonics of Degree 2.

I. Five distinct functions, for instance,

$$2z^2-x^2-y^2;\quad x^2-y^2;\quad yz;\quad xz;\quad xy.$$

Or one function with five arbitrary constants.

II. $\left\{\begin{array}{ll} & ax^2+by^2+cz^2+eyz+fzx+gxy,\\ \text{where} & a+b+c=0.\end{array}\right.$

Degrees $-n-1$, *and* $+n$ (n *any integer*).

With same notation and same references for proof as above for Degree 0, group IV.

I. $\delta_{n+1}V_0,\quad \delta_n V_{-1},\quad \text{or}\quad \delta_{n+j}V_{j-1}.$

II. $\delta_{n+i-j+1}(r^{2(i-j)+1}\delta_i V_{j-1}),\quad \text{and}\quad r^{2n+1}\delta_{n+i-j+1}(r^{2(i-j)+1}\delta_i V_{j-1}).$

Examples of spherical harmonics.

Degrees e + vf, *and* − e − 1 − vf.

(v denoting $\sqrt{-1}$, and e and f any real quantities.)

I.

$$\tfrac{1}{2}\left[(x+vy)^{e+vf}+(x-vy)^{e+vf}\right]; \quad \frac{1}{2v}\left[(x+vy)^{e+vf}-(x-vy)^{e+vf}\right]:$$

or $$q^{e+vf}\cos\left[(e+vf)\phi\right]; \quad q^{e+vf}\sin\left[(e+vf)\phi\right],$$

where $$q=\sqrt{(x^2+y^2)} \text{ and } \phi=\tan^{-1}\frac{y}{x}:$$

or $$\tfrac{1}{2}q^{e+vf}\left[\epsilon^{v(e+vf)\phi}+\epsilon^{-v(e+vf)\phi}\right]; \quad \frac{1}{2v}q^{e+vf}\left[\epsilon^{v(e+vf)\phi}-\epsilon^{-v(e+vf)\phi}\right]:$$

or $$\tfrac{1}{2}q^{e}\{\epsilon^{f\phi}\left[\cos(f\log q-e\phi)+v\sin(f\log q-e\phi)\right] + \epsilon^{-f\phi}\left[\cos(f\log q+e\phi)+v\sin(f\log q+e\phi)\right]\};$$

II. the same with $\frac{3\pi}{2}+e\phi$ instead of $e\phi$.

III.

$$\frac{\tfrac{1}{2}\left[(x+vy)^{e+vf}+(x-vy)^{e+vf}\right]}{r^{2(e+vf)+1}};$$

or $$\tfrac{1}{2}r^{-2e-1}q^{e}\left[\epsilon^{f\phi}\epsilon^{v(f\log q-2f\log r-e\phi)}+\epsilon^{-f\phi}\epsilon^{v(f\log q-2f\log r+e\phi)}\right];$$

or $$\tfrac{1}{2}r^{-2e-1}q^{e}\left\{\epsilon^{f\phi}\left[\cos\left(f\log\frac{q}{r^2}-e\phi\right)+v\sin\left(f\log\frac{q}{r^2}-e\phi\right)\right] + \epsilon^{-f\phi}\left[\cos\left(f\log\frac{q}{r^2}+e\phi\right)+v\sin\left(f\log\frac{q}{r^2}+e\phi\right)\right]\right\}.$$

(*b*) A *spherical surface harmonic* is the function of two angular co-ordinates, or spherical surface co-ordinates, which a spherical harmonic becomes at any spherical surface described from O, the origin of co-ordinates, as centre. Sometimes a function which, according to the definition (*a*), is simply a spherical harmonic, will be called a *spherical solid harmonic*, when it is desired to call attention to its not being confined to a spherical surface.

(*c*) A *complete spherical harmonic* is one which is finite and of single value for all finite values of the co-ordinates.

Partial Harmonics.

A *partial harmonic* is a spherical harmonic which either does not continuously satisfy the fundamental equation (4) for space completely surrounding the centre, or does not return to the same value in going once round every closed curve. The "partial" harmonic is as it were a harmonic for a part of the spherical surface: but it may be for a part which is greater than the whole, or a part of which portions jointly and independently occupy the same space.

(*d*) It will be shown, later, § (*h*), that a complete spherical harmonic is necessarily either a rational integral function of the co-ordinates, or reducible to one by a factor of the form

Algebraic quality of complete harmonics.

$$(x^2+y^2+z^2)^{\frac{m}{2}},$$

m being an integer.

(*e*) The general problem of finding harmonic functions is most concisely stated thus :—

Differential equations of a harmonic of degree n.

To find the most general integral of the equation

$$\frac{d^2u}{dx^2}+\frac{d^2u}{dy^2}+\frac{d^2u}{dz^2}=0 \quad\ldots\ldots\ldots\ldots\ldots\ldots\ldots\ldots(4')$$

subject to the condition

$$x\frac{du}{dx}+y\frac{du}{dy}+z\frac{du}{dz}=nu \quad\ldots\ldots\ldots\ldots\ldots\ldots\ldots(5),$$

the second of these equations being merely the analytical expression of the condition that u is a homogeneous function of x, y, z of the degree n, which may be any whole number positive or negative, any fraction, or any imaginary quantity.

Let $P+vQ$ be a harmonic of degree $e+vf$, P, Q, e, f being real. We have

Differential equation for real constituents of a homogeneous function of imaginary degree.

$$\left(x\frac{d}{dx}+y\frac{d}{dy}+z\frac{d}{dz}\right)(P+vQ)=(e+vf)(P+vQ);$$

and therefore

$$\left.\begin{array}{l} x\dfrac{dP}{dx}+y\dfrac{dP}{dy}+z\dfrac{dP}{dz}=eP-fQ \\ x\dfrac{dQ}{dx}+y\dfrac{dQ}{dy}+z\dfrac{dQ}{dz}=fP+eQ \end{array}\right\}\ldots\ldots\ldots\ldots\ldots(5');$$

whence

$$\left.\begin{array}{l} \left[\left(x\dfrac{d}{dx}+y\dfrac{d}{dy}+z\dfrac{d}{dz}-e\right)^2+f^2\right]P=0, \\ \text{and}\quad \left[\left(x\dfrac{d}{dx}+y\dfrac{d}{dy}+z\dfrac{d}{dz}-e\right)^2+f^2\right]Q=0, \end{array}\right\}\ldots\ldots\ldots\ldots(5'').$$

(*f*) Analytical expressions in various forms for an absolutely general integration of these equations, may be found without much difficulty ; but with us the only value or interest which any such investigation can have, depends on the availability of

Value of general symbolical expressions.

Use of complete spherical harmonics in physical problems.

its results for solutions fulfilling the conditions at bounding surfaces presented by physical problems. In a very large and most important class of physical problems regarding space bounded by a complete spherical surface, or by two complete concentric spherical surfaces, or by closed surfaces differing very little from spherical surfaces, the case of n any positive or negative integer, integrated particularly under the restriction stated in (d), is of paramount importance. It will be worked out thoroughly below.

Uses of incomplete spherical harmonics in physical problems.

Again, in similar problems regarding sections cut out of spherical spaces by two diametral planes making any angle with one another *not a sub-multiple of two right angles*, or regarding spaces bounded by two circular cones having a common vertex and axis, and by the included portion of two spherical surfaces described from their vertex as centre, solutions for cases of fractional and imaginary values of n are useful. Lastly, when the subject is a solid or fluid, shaped as a section cut from the last-mentioned spaces by two planes through the axis of the cones, inclined to one another at any angle, whether a submultiple of π or not, we meet with the case of n either integral or not, but to be integrated under a restriction differing from that specified in (d). We shall accordingly, after investigating general expressions for complete spherical harmonics, give some indications as to the determination of the incomplete harmonics, whether of fractional, of imaginary, or of integral degrees, which are required for the solution of problems regarding such portions of spherical spaces as we have just described.

A few formulæ, which will be of constant use in what follows, are brought together in the first place.

Working formulæ.

(g) Calling O the origin of co-ordinates, and P the point x, y, z, let $OP = r$, so that $x^2 + y^2 + z^2 = r^2$. Let δ, prefixed to any function, denote its rate of variation per unit of space in the direction OP; so that

$$\delta = \frac{x}{r}\frac{d}{dx} + \frac{y}{r}\frac{d}{dy} + \frac{z}{r}\frac{d}{dz} \dots\dots\dots\dots\dots\dots\dots\dots (6).$$

If H_n denote any homogeneous function of x, y, z of order n, we have clearly

$$\delta H_n = \frac{n}{r} H_n \dots\dots\dots\dots\dots\dots\dots\dots (7);$$

whence $$x\frac{dH_n}{dx} + y\frac{dH_n}{dy} + z\frac{dH_n}{dz} = nH_n \dots\dots\dots (5) \text{ or } (8),$$

the well-known differential equation of a homogeneous function; in which, of course, n may have any value, positive, integral, negative, fractional, or imaginary. Again, denoting, for brevity,

Working formulæ.

$\frac{d^2}{dx^2}+\frac{d^2}{dy^2}+\frac{d^2}{dz^2}$ by ∇^2, we have, by differentiation,

$$\nabla^2(r^m)=m(m+1)r^{m-2} \qquad (9).$$

Also, if u, u' denote any two functions,

$$\nabla^2(uu')=u'\nabla^2u+2\left(\frac{du}{dx}\frac{du'}{dx}+\frac{du}{dy}\frac{du'}{dy}+\frac{du}{dz}\frac{du'}{dz}\right)+u\nabla^2u' \qquad (10);$$

whence, if u and u' are both solutions of (4),

$$\nabla^2(uu')=2\left(\frac{du}{dx}\frac{du'}{dx}+\frac{du}{dy}\frac{du'}{dy}+\frac{du}{dz}\frac{du'}{dz}\right) \qquad (11);$$

or, by taking $u=V_n$, a harmonic of degree n, and $u'=r^m$,

$$\nabla^2(r^mV_n)=2mr^{m-2}\left(x\frac{dV_n}{dx}+y\frac{dV_n}{dy}+z\frac{dV_n}{dz}\right)+V_n\nabla^2(r^m),$$

or, by (8) and (9),

$$\nabla^2(r^mV_n)=m(2n+m+1)r^{m-2}V_n \qquad (12).$$

From this last it follows that $r^{-2n-1}V_n$ is a harmonic; which, being of degree $-n-1$, may be denoted by V_{-n-1}, so that we have

$$\left.\begin{array}{ll} & V_{-n-1}=r^{-2n-1}V_n \\ \text{or} & \frac{V_{n'}}{r^{n'}}=\frac{V_n}{r^n} \\ \text{if} & n+n'=-1 \end{array}\right\} \qquad (13),$$

a formula showing a reciprocal relation between two solid harmonics which give the same form of surface harmonic at any spherical surface described from O as centre. Again, by taking $m=-1$, in (9), we have

$$\nabla^2\frac{1}{r}=0 \qquad (14).$$

Hence $\frac{1}{r}$ is a harmonic of degree -1. We shall see later § (h), that it is the only *complete harmonic*, of this degree.

If u be any solution of the equation $\nabla^2u=0$, we have also

$$\nabla^2\frac{du}{dx}=0,$$

Working formulæ.

and so on for any number of differentiations. Hence if V_ι is a harmonic of any degree ι, $\frac{d^{j+k+l}V_\iota}{dx^j dy^k dz^l}$ is a harmonic of degree $\iota - j - k - l$; or, as we may write it,

$$\frac{d^{j+k+l}V_\iota}{dx^j dy^k dz^l} = V_{\iota-j-k-l} \quad \text{.......................(15).}$$

Theorem due to Laplace.

Again, we have a most important theorem expressed by the following equations:—

$$\iint S_i S_{i'} d\varpi = 0 \quad \text{..............................(16),}$$

where $d\varpi$ denotes an element of a spherical surface, described from O as centre with radius unity; $\iint$ an integration over the whole of this surface; and S_i, $S_{i'}$ two complete surface harmonics, of which the degrees, i and i', are neither equal to one another, nor such that $i + i' = -1$. For, denoting the solid harmonics $r^i S_i$ and $r^{i'} S_{i'}$ by V_i and $V_{i'}$ for any point (x, y, z), we have, by the general theorem (1) of A (a), above, applied to the space between any two spherical surfaces having O for their common centre, and a and a_1 their radii;—

$$\iiint \left(\frac{dV_i}{dx}\frac{dV_{i'}}{dx} + \frac{dV_i}{dy}\frac{dV_{i'}}{dy} + \frac{dV_i}{dz}\frac{dV_{i'}}{dz} \right) dxdydz$$
$$= \iint V_i \delta V_{i'} d\sigma = \iint V_{i'} \delta V_i d\sigma.$$

But, according to (7), $\delta V_{i'} = \frac{i'}{r} V_{i'}$, and $\delta V_i = \frac{i}{r} V_i$. And for the portions of the bounding surface constituted by the two spherical surfaces respectively, $d\sigma = a^2 d\varpi$, and $d\sigma = a_1{}^2 d\varpi$. Hence the two last equal members of the preceding double equations become

$$i(a^{i+i'+1} - a_1{}^{i+i'+1}) \iint S_i S_{i'} d\varpi = i'(a^{i+i'+1} - a_1{}^{i+i'+1}) \iint S_i S_{i'} d\varpi,$$

to satisfy which, when i differs from i', and $a^{i+i'+1}$ from $a_1{}^{i+i'+1}$, (16) must hold.

The corresponding theorem for partial harmonics is this:—

Extension of theorem of Laplace to partial harmonics.

Let S_i, $S_{i'}$ denote any two different partial surface harmonics of degrees i, i', having their sum different from -1; and further, fulfilling the condition that, at every point of the boundary of some one part of the spherical surface either each of them vanishes, or the rate of variation of each of them perpendicular to this boundary vanishes, and that each is finite and single in its value at every point of the enclosed portion of surface; then, with the integration $\iint$ limited to the portion of surface in question, equation (16) holds. The proof differs from the preceding only in this, that instead of taking the whole space between two concentric spherical surfaces, we must now take only the part of it enclosed by the cone having O for vertex, and containing the boundary of the spherical area considered.

Extension of theorem of Laplace to partial harmonics.

(*h*) Proceeding now to the investigation of complete harmonics, we shall first prove that every such function is either rational and integral in terms of the co-ordinates x, y, z, or is made so by a factor of the form r^m.

Investigation of complete harmonics.

Let V be any function of x, y, z, satisfying the equation

$$\nabla^2 V = 0 \quad \text{.................................(17)}$$

at every point within a spherical surface, S, described from O as centre, with any radius a. Its value at this surface, if a known function of any arbitrary character, may be expanded according to the general theorem of § 51, below, in the following series:—

$$(r=a),\quad V = S_0 + S_1 + S_2 + \ldots\ldots + S_i + \text{etc.} \quad \text{............(18)}$$

where S_1, S_2,...S_i denote the surface values of solid spherical harmonics of degrees 1, 2,...i, each a rational integral function for every point within S. But

$$S_0 + S_1\frac{r}{a} + S_2\frac{r^2}{a^2} + \ldots + S_i\frac{r^i}{a^i} + \text{etc.} \quad \text{..................(19)}$$

Harmonic solution of Green's problem for the space within a spherical surface.

is a function fulfilling these conditions, and therefore, as was proved above, A (*c*), V cannot differ from it. Now, as a particular case, let V be a harmonic function of positive degree ι, which may be denoted by $S_\iota\dfrac{r^\iota}{a^\iota}$: we must have

$$S_\iota\frac{r^\iota}{a^\iota} = S_0 + S_1\frac{r}{a} + S_2\frac{r^2}{a^2} + \ldots + S_i\frac{r^i}{a^i} + \text{etc.}$$

This cannot be unless $\iota = i$, $S_\iota = S_i$, and all the other functions S_0, S_1, S_2, etc., vanish. Hence there can be no complete spherical harmonic of positive degree, which is not, as $S_i\dfrac{r^i}{a^i}$, of integral degree and an integral rational function of the co-ordinates.

Complete harmonics of positive degrees, proved rational and integral.

Again, let V be any function satisfying (17) for every point without the spherical surface S, and vanishing at an infinite distance in every direction; and let, as before, (18) express its surface value at S. We similarly prove that it cannot differ from

$$\frac{aS_0}{r} + \frac{a^2S_1}{r^2} + \frac{a^3S_2}{r^3} + \ldots\ldots + \frac{a^{i+1}S_i}{r^{i+1}} + \text{etc.} \quad \text{..............(20).}$$

Harmonic solution of Green's problem for space external to a spherical surface.

Harmonic solution of Green's problem for space external to a spherical surface.

Hence if, as a particular case, V be any complete harmonic $\frac{r^\kappa S_\kappa}{a^\kappa}$, of negative degree κ, we must have, for all points outside S,

$$\frac{r^\kappa S_\kappa}{a^\kappa} = \frac{aS_0}{r} + \frac{a^2 S_1}{r^2} + \frac{a^3 S_2}{r^3} + \dots\dots + \frac{a^{i+1} S_i}{r^{i+1}} + \text{etc.},$$

which requires that $\kappa = -(i+1)$, $S_\kappa = S_i$, and that all the other functions S_0, S_1, S_2, etc., vanish. Hence a complete spherical harmonic of negative degree cannot be other than $\frac{a^{i+1} S_i}{r^{i+1}}$, or $\frac{a^{i+1}}{r^{2i+1}} S_i r^i$, where $S_i r^i$ is not only a rational integral function of the co-ordinates, as asserted in the enunciation, but is itself a spherical harmonic.

Complete harmonics of negative degree.

(*i*) Thus we have proved that a complete spherical harmonic, if of positive, is necessarily of integral, degree, and is, besides, a rational integral function of the co-ordinates, or if of negative degree, $-(i+1)$, is necessarily of the form $\frac{V_i}{r^{2i+1}}$, where V_i is a harmonic of positive degree, i. We shall therefore call the *order* of a complete spherical harmonic of negative degree, the *degree or order* of the complete harmonic of positive degree allied to it; and we shall call the *order* of a surface harmonic, the degree or order of the solid harmonic of positive degree, or the order of the solid harmonic of negative degree, which agrees with it at the spherical surface.

Orders and degrees of complete harmonics.

General expressions for complete harmonics.

(*j*) To obtain general expressions for complete spherical harmonics of all orders, we may first remark that, inasmuch as a constant is the only rational integral function of degree 0, a complete harmonic of degree 0 is necessarily constant. Hence, by what we have just seen, a complete harmonic of the degree -1 is necessarily of the form $\frac{A}{r}$. That this function is a harmonic we knew before, by (14).

By differentiation of harmonic of degree -1.

Hence, by (15), we see that

$$\left.\begin{array}{l} V_{-i-1} = \dfrac{d^{j+k+l}}{dx^j dy^k dz^l} \dfrac{1}{(x^2+y^2+z^2)^{\frac{1}{2}}} \\ \text{if} \qquad j+k+l = i \end{array}\right\} \dots\dots\dots\dots\dots\dots\dots (21),$$

where V_{-i-1} denotes a harmonic, which is clearly a complete harmonic, of degree $-(i+1)$. The differential coefficient here indicated, when worked out, is easily found to be a fraction, of which the numerator is a rational integral function of degree i, and the denominator is r^{2i+1}. By what we have just seen, the numerator must be a harmonic; and, denoting it by V_i, we thus have

By differentiation of harmonic of degree −1.

$$V_i = r^{2i+1} \frac{d^{j+k+l}}{dx^j dy^k dz^l} \frac{1}{r} \text{ ....(22).}$$

The number of independent harmonics of order i, which we can thus derive by differentiation from $\frac{1}{r}$, is $2i+1$. For, although there are $\frac{(i+2)(i+1)}{2}$ differential coefficients $\frac{d^{j+k+l}}{dx^j dy^k dz^l}$, for which $j+k+l=i$, only $2i+1$ of these are independent when $\frac{1}{r}$ is the subject of differentiation, inasmuch as

Number of independent harmonics of any order.

$$\left(\frac{d^2}{dx^2} + \frac{d^2}{dy^2} + \frac{d^2}{dz^2}\right)\frac{1}{r} = 0 \text{ ....(14),}$$

which gives
$$\frac{d^{2n}}{dz^{2n}} \frac{1}{r} = (-1)^n \left(\frac{d^2}{dx^2} + \frac{d^2}{dy^2}\right)^n \frac{1}{r} \text{ ....(23),}$$

Relation between differential coefficients of harmonics.

n being any integer, and shows that

$$\left.\begin{aligned} \frac{d^{j+k+l}}{dx^j dy^k dz^l} \frac{1}{r} &= (-1)^{\frac{l}{2}} \frac{d^{j+k}}{dx^j dy^k} \left(\frac{d^2}{dx^2} + \frac{d^2}{dy^2}\right)^{\frac{l}{2}} \frac{1}{r}, \text{ if } l \text{ is even,} \\ \text{or} \quad &= (-1)^{\frac{l-1}{2}} \frac{d^j d^k}{dx^j dy^k} \left(\frac{d^2}{dx^2} + \frac{d^2}{dy^2}\right)^{\frac{l-1}{2}} \frac{d}{dz} \frac{1}{r}, \text{ if } l \text{ is odd.} \end{aligned}\right\} \text{...(24).}$$

Hence, by taking $l = 0$, and $j + k = i$, in the first place, we have $i+1$ differential coefficients $\frac{d^{j+k}}{dx^j dy^k}$; and by taking next $l = 1$, and $j+k=i-1$, we have i varieties of $\frac{d^{j+k}}{dx^j dy^k}$; that is to say, we have in all $2i+1$ varieties, and no more, when $\frac{1}{r}$ is the subject. It is easily seen that these $2i+1$ varieties are in reality independent. We need not stop at present to show this, as it will be apparent in the actual expansions given below.

Now if $H_i(x, y, z)$ denote any rational integral function of x, y, z of degree i, $\nabla^2 H_i(x, y, z)$ is of degree $i-2$. Hence since in H_i there are $\frac{(i+2)(i+1)}{2}$ terms, in $\nabla^2 H_i$ there are $\frac{i(i-1)}{2}$.

Complete harmonic of any degree investigated algebraically.

Hence if $\nabla^2 H_i = 0$, we have $\frac{i(i-1)}{2}$ equations among the constant coefficients, and the number of independent constants remaining is $\frac{(i+2)(i+1)}{2} - \frac{i(i-1)}{2}$, or $2i+1$; that is to say, there are $2i+1$ constants in the general rational integral harmonic of degree i. But we have seen that there are $2i+1$ distinct varieties of differential coefficients of $\frac{1}{r}$ of order i, and that the numerator of each is a harmonic of degree i. Hence every complete harmonic of order i is expressible in terms of differential coefficients of $\frac{1}{r}$. It is impossible to form $2i+1$ functions symmetrically among three variables, except when $2i+1$ is divisible by 3; that is to say, when $i = 3n+1$, n being any integer. This class of cases does not seem particularly interesting or important, but here are two examples of it.

Example 1. $i = 1$, $2i+1 = 3$.

The harmonics are obviously

$$\frac{d}{dx}\frac{1}{r}, \quad \frac{d}{dy}\frac{1}{r}, \quad \frac{d}{dz}\frac{1}{r}.$$

Formula (25) involves z singularly, and x and y symmetrically, for every value of i greater than unity, but for the case of $i=1$ it is essentially symmetrical in respect to x, y, and z, as in this case it becomes

$$\frac{V_1}{r^3} = \left(A_0 \frac{d}{dx} + A_1 \frac{d}{dy} + B_0 \frac{d}{dz}\right)\frac{1}{r}.$$

Example 2. $i = 4$, $2i+1 = 9$.

Looking first for three differential coefficients of the 4th order, singular with respect to x, and symmetrical with respect to y and z; and thence changing cyclically to yzx and zxy, we find

$$\frac{d^4}{dy^2 dz^2}, \quad \frac{d^4}{dx\,dy^3}, \quad \frac{d^4}{dx\,dz^3},$$

$$\frac{d^4}{dz^2 dx^2}, \quad \frac{d^4}{dy\,dz^3}, \quad \frac{d^4}{dy\,dx^3},$$

$$\frac{d^4}{dx^2 dy^2}, \quad \frac{d^4}{dz\,dx^3}, \quad \frac{d^4}{dz\,dy^3}.$$

These nine differentiations of $\frac{1}{r}$ are essentially distinct and give us therefore nine distinct harmonics of the 4th order formed symmetrically among x, y, z. By putting in them for $\frac{d^2}{dz^2}$, wherever it occurs, its equivalent $-\left(\frac{d^2}{dx^2}+\frac{d^2}{dy^2}\right)$, considering that it is $\frac{1}{r}$ which is differentiated, and for $\frac{d^3}{dz^3}$, its equivalent $-\frac{d}{dz}\left(\frac{d^2}{dx^2}+\frac{d^2}{dy^2}\right)$, we may pass from them to (25).

Complete harmonic of any degree investigated algebraically.

But for every value of i the general harmonic may be exhibited as a function, with $2i+1$ constants, involving two out of the three variables symmetrically. This may be done in a variety of ways, of which we choose the two following, as being the most useful :—First,

General expression for complete harmonic of order i.

$$\left.\begin{aligned}\frac{V_i}{r^{2i+1}}=&\left\{A_0\left(\frac{d}{dx}\right)^i+A_1\left(\frac{d}{dx}\right)^{i-1}\frac{d}{dy}+A_2\left(\frac{d}{dx}\right)^{i-2}\left(\frac{d}{dy}\right)^2+\ldots+A^i\left(\frac{d}{dy}\right)^i\right\}\frac{1}{r}\\&+\left\{B_0\left(\frac{d}{dx}\right)^{i-1}+B_1\left(\frac{d}{dx}\right)^{i-2}\frac{d}{dy}+B_2\left(\frac{d}{dx}\right)^{i-3}\left(\frac{d}{dy}\right)^2+\ldots\right.\\&\qquad\left.+B_{i-1}\left(\frac{d}{dy}\right)^{i-1}\right\}\frac{d}{dz}\frac{1}{r}\end{aligned}\right\}\text{(25)}.$$

Secondly, let $\quad x+yv=\xi,\quad x-yv=\eta$ .... (26),

where, as formerly, v is taken to denote $\sqrt{-1}$.

This gives

$$\left.\begin{aligned}x=\tfrac{1}{2}(\xi+\eta),\quad y=\frac{1}{2v}(\xi-\eta),\\ \frac{1}{r}=\frac{1}{(\xi\eta+z^2)^{\frac{1}{2}}},\end{aligned}\right\}\ldots\ldots\ldots\ldots\text{(27)};$$

Imaginary linear transformation.

$$\left.\begin{aligned}&\frac{d}{dx}[x,y]=\left(\frac{d}{d\xi}+\frac{d}{d\eta}\right)[\xi,\eta],\quad \frac{d}{dy}[x,y]=v\left(\frac{d}{d\xi}-\frac{d}{d\eta}\right)[\xi,\eta],\\&\frac{d}{d\xi}[\xi,\eta]=\tfrac{1}{2}\left(\frac{d}{dx}-v\frac{d}{dy}\right)[x,y],\quad \frac{d}{d\eta}[\xi,\eta]=\tfrac{1}{2}\left(\frac{d}{dx}+v\frac{d}{dy}\right)[x,y],\end{aligned}\right\}\ldots\text{(28)},$$

where $[x, y]$ and $[\xi, \eta]$ denote the same quantity, expressed in terms of x, y, and of ξ, η respectively. From these we have, further,

Imaginary linear transformation.

$$\left.\begin{aligned}&\left(\frac{d^2}{dx^2}+\frac{d^2}{dy^2}+\frac{d^2}{dz^2}\right)[x,y,z]=\left(4\frac{d^2}{d\xi d\eta}+\frac{d^2}{dz^2}\right)[\xi,\eta,z],\\&\text{or, according to our abbreviated notation,}\\&\nabla^2=4\frac{d^2}{d\xi d\eta}+\frac{d^2}{dz^2}.\end{aligned}\right\}\ldots(29).$$

Hence, as $\nabla^2 V=0$, if V denote $\frac{1}{r}$ or any other solid harmonic,

$$\frac{d^2}{dz^2}V=-4\frac{d^2}{d\xi d\eta}V \ldots\ldots\ldots\ldots(29').$$

Using (28) in (25) and taking $\mathfrak{A}_0$, $\mathfrak{A}_1$, $\mathfrak{B}_0$, $\mathfrak{B}_1$, ... to denote another set of coefficients readily expressible in terms of A_0, A_1, B_0, B_1, ... we find

$$\left.\begin{aligned}\frac{V_i}{r^{2i+1}}=&\left\{\mathfrak{A}_0\left(\frac{d}{d\xi}\right)^i+\mathfrak{A}_1\left(\frac{d}{d\xi}\right)^{i-1}\frac{d}{d\eta}+\mathfrak{A}_2\left(\frac{d}{d\xi}\right)^{i-2}\left(\frac{d}{d\eta}\right)^2+\ldots+\mathfrak{A}_i\left(\frac{d}{d\eta}\right)^i\right\}\frac{1}{r}\\&+\left\{\mathfrak{B}_0\left(\frac{d}{d\xi}\right)^{i-1}+\mathfrak{B}_1\left(\frac{d}{d\xi}\right)^{i-2}\frac{d}{d\eta}+\mathfrak{B}_2\left(\frac{d}{d\xi}\right)^{i-3}\left(\frac{d}{d\eta}\right)^2+\ldots+\mathfrak{B}_{i-1}\left(\frac{d}{d\eta}\right)^{i-1}\right\}\frac{d}{dz}\frac{1}{r}\end{aligned}\right\}\ldots(30).$$

Expansion of elementary term.

The differentiations here are performed with great ease, by the aid of Leibnitz's theorem. Thus we have

$$\left.\begin{aligned}&r^{2(m+n)+1}\frac{d^{m+n}}{d\xi^m d\eta^n}\frac{1}{r}=(-)^{m+n}\tfrac{1}{2}.\tfrac{3}{2}.\tfrac{5}{2}\ldots(m+n-\tfrac{1}{2})\\&\left[\eta^m\xi^n-\frac{mn}{1.(m+n-\frac{1}{2})}\eta^{m-1}\xi^{n-1}r^2+\frac{m(m-1).n(n-1)}{1.2.(m+n-\frac{1}{2})(m+n-\frac{3}{2})}\eta^{m-2}\xi^{n-2}r^4-\text{etc.}\right]\\&\text{and}\\&r^{2(m+n)+3}\frac{d^{m+n+1}}{d\xi^m d\eta^n dz}\frac{1}{r}=(-)^{m+n+1}\tfrac{1}{2}.\tfrac{3}{2}.\tfrac{5}{2}\ldots(m+n+\tfrac{1}{2}).2z\\&\left[\eta^m\xi^n-\frac{mn}{1.(m+n+\frac{1}{2})}\eta^{m-1}\xi^{n-1}r^2+\frac{m(m-1).n(n-1)}{1.2.(m+n+\frac{1}{2})(m+n-\frac{1}{2})}\eta^{m-2}\xi^{n-2}r^4-\text{etc.}\right].\end{aligned}\right\}\ldots(31)$$

Polar transformation.

This expression leads at once to a real development, in terms of polar co-ordinates, thus :—Let

$$z=r\cos\theta,\quad x=r\sin\theta\cos\phi,\quad y=r\sin\theta\sin\phi \ldots\ldots\ldots(32);$$

so that $\xi=r\sin\theta\epsilon^{v\phi},\quad \eta=r\sin\theta\epsilon^{-v\phi}$ .... (33).

Then, since $\xi\eta=x^2+y^2=r^2\sin^2\theta$,

and

$$\xi^n\eta^m=(\xi\eta)^m\xi^s=(\xi\eta)^m(r\sin\theta)^s(\cos\phi+v\sin\phi)^s=(r\sin\theta)^{m+n}(\cos s\phi+v\sin s\phi),$$

where $s=n-m$; and if, further, we take

$$\left.\begin{aligned}\mathfrak{A}_n+\mathfrak{A}_m=\mathbf{A}_s,\quad (\mathfrak{A}_n-\mathfrak{A}_m)v=\mathbf{A}_s',\\ \mathfrak{B}_n+\mathfrak{B}_m=\mathbf{B}_s,\quad (\mathfrak{B}_n-\mathfrak{B}_m)v=\mathbf{B}_s',\end{aligned}\right\}\ldots\ldots\ldots\ldots(34)$$

Trigono-metrical expansions.

we have

$$\left.\begin{array}{l}
\mathfrak{A}_n \dfrac{d^{m+n}}{d\xi^m d\eta^n}\dfrac{1}{r}+\mathfrak{A}_m \dfrac{d^{m+n}}{d\xi^n d\eta^m}\dfrac{1}{r} \\
=(-)^{m+n}\tfrac{1}{2}\cdot\tfrac{3}{2}\cdot\tfrac{5}{2}\ldots(m+n-\tfrac{1}{2})r^{-(m+n+1)}(\mathbf{A}_s\cos s\phi+\mathbf{A}_s'\sin s\phi) \\
\left[\sin^{m+n}\theta-\dfrac{mn}{1.(m+n-\frac{1}{2})}\sin^{m+n-2}\theta+\dfrac{m(m-1).n(n-1)}{1.2.(m+n-\frac{1}{2})(m+n-\frac{3}{2})}\right. \\
\qquad\qquad\qquad\qquad \left.\sin^{m+n-4}\theta-\text{etc.}\right] \\
\mathfrak{B}_n \dfrac{d^{m+n+1}}{d\xi^m d\eta^n dz}\dfrac{1}{r}+\mathfrak{B}_m \dfrac{d^{m+n+1}}{d\xi^n d\eta^m dz}\dfrac{1}{r} \\
=(-)^{m+n+1}\tfrac{1}{2}\cdot\tfrac{3}{2}\cdot\tfrac{5}{2}\ldots(m+n+\tfrac{1}{2})r^{-(m+n+2)}(\mathbf{B}_s\cos s\phi+\mathbf{B}_s'\sin s\phi)2\cos\theta \\
\left[\sin^{m+n}\theta-\dfrac{mn}{1.(m+n+\frac{1}{2})}\sin^{m+n-2}\theta+\dfrac{m(m-1).n(n-1)}{1.2.(m+n+\frac{1}{2})(m+n-\frac{1}{2})}\right. \\
\qquad\qquad\qquad\qquad \left.\sin^{m+n-4}\theta-\text{etc.}\right].
\end{array}\right\} \quad (35)$$

Setting aside now constant factors, which have been retained hitherto to show the relations of the expressions we have investigated, to differential coefficients of $\frac{1}{r}$; taking Σ to denote summation with respect to the arbitrary constants, **A**, **A**$'$, **B**, **B**$'$; and putting $\sin\theta=\nu$, $\cos\theta=\mu$; we have the following perfectly general expression for a complete surface harmonic of order i:—

$$S_i=\sum^{m+n=i}(\mathbf{A}_s\cos s\phi+\mathbf{A}_s'\sin s\phi)\Theta_{(m,n)}+\sum^{m+n+1=i}(\mathbf{B}_s\cos s\phi+\mathbf{B}_s'\sin s\phi)\mu Z_{(m,n)}\ldots(36)$$

where $s=m\sim n$, and

$$\Theta_{(m,n)}=\nu^{m+n}-\frac{mn}{1.(m+n-\frac{1}{2})}\nu^{m+n-2}+\frac{m(m-1).n(n-1)}{1.2.(m+n-\frac{1}{2})(m+n-\frac{3}{2})}\nu^{m+n-4}-\text{etc.}$$

while $Z_{(m,n)}$ differs from $\Theta_{(m,n)}$ only in having $m+n+1$ in place of $m+n$, in the denominators.

The formula most commonly given for a spherical harmonic of order i (Laplace, *Mécanique Celeste*, livre III. chap. II., or Murphy's *Electricity*, Preliminary Prop. XI.) is somewhat simpler, being as follows:—

$$S_i=\sum_{s=0}^{s=i}(\mathbf{A}_s\cos s\phi+\mathbf{B}_s\sin s\phi)\Theta_i^{(s)}\ \ldots\ldots\ldots\ldots\ldots\ldots\ (37).$$

$$\Theta_i^{(s)}=\nu^s\left[\mu^{i-s}-\frac{(i-s)(i-s-1)}{2.(2i-1)}\mu^{i-s-2}+\frac{(i-s)(i-s-1)(i-s-2)(i-s-3)}{2.4.(2i-1)(2i-3)}\mu^{i-s-4}-\text{etc.}\right]$$
$$\ldots\ldots\ldots\ldots\ldots\ldots\ldots\ldots(38),$$

Trigonometrical expansions.

where it may be remarked that $\Theta_i^{(s)}$ means the same as $(-1)^{\frac{i-s}{2}}\Theta_{(m,n)}$ if $m+n=i$ and $m\sim n=s$, or as $(-1)^{\frac{i-s-1}{2}}\mu Z_{(m,n)}$ if $m+n+1=i$ and $m\sim n=s$. Formula (38) may be derived algebraically from (36) by putting $\sqrt{(1-\mu^2)}$ for ν in $\Theta_{(m,n)}\div\nu^s$ and in $Z_{(m,n)}\div\nu^s\mu$: or it may be obtained directly by the method of differentiation followed above, varied suitably. But it may also be obtained by assuming (with a_s and b_s as arbitrary constants)

$$V_i=S_ir^i=\Sigma(a_s\xi^s+b_s\eta^s)(z^{i-s}+pr^2z^{i-s-2}+qr^4z^{i-s-4}+\text{etc.}),$$

which is obviously a proper form; and determining p, q, etc., by the differential equation $\nabla^2V_i=0$, with (29).

Another form may be obtained with even greater ease, thus: Assuming

$$V_i=\Sigma(a_s\xi^s+b_s\eta^s)(z^{i-s}+p_1z^{i-s-2}\xi\eta+p_2z^{i-s-4}\xi^2\eta^2+\text{etc.}),$$

and determining p_1, p_2, etc., by the differential equation, we have

$$\left.\begin{aligned}V_i=\Sigma(a_s\xi^s+\beta_s\eta^s)\bigg[z^{i-s}&-\frac{(i-s)(i-s-1)}{4.(s+1).1}z^{i-s-2}\xi\eta\\&+\frac{(i-s)(i-s-1)(i-s-2)(i-s-3)}{4^2.(s+1)(s+2).1.2}z^{i-s-4}\xi^2\eta^2-\text{etc.}\bigg]\end{aligned}\right\}(39)$$

which might also have been found easily by the differentiation of $\frac{1}{r}$. Hence, eliminating imaginary symbols, and retaining the notation of (37) and (38), we have

$$\left.\begin{aligned}&\Theta_i^{(s)}=C\sin^s\theta\left[\mu^{i-s}-\frac{(i-s)(i-s-1)}{4.(s+1).1}\mu^{i-s-2}\nu^2\right.\\&\qquad\left.+\frac{(i-s)(i-s-1)(i-s-2)(i-s-3)}{4^2.(s+1)(s+2).1.2}\mu^{i-s-4}\nu^4-\text{etc.}\right],\\&\text{where}\qquad C=\frac{(2s+1)(2s+2)\dots(i+s)}{(2s+1)(2s+3)\dots(2i-1)}.\end{aligned}\right\}(40)$$

This value of C is found by comparing with (35). Thus we see that C must be equal to the numerical coefficient of the last term of (35), irrespectively of sign. Or C is found by comparing (40) with (38): it is equal to the coefficient of the last term of (38) divided by the coefficient of the last term within the brackets of (40). Or it is found directly (that is to say, independently of other equivalent formulas) thus:—We have, by (29′), **Trigonometrical expansions.**

$$\left.\begin{aligned}\frac{d^i}{dz^{i-s}d\eta^s}\frac{1}{r} &= (-)^{\frac{i-s}{2}}2^{i-s}\frac{d^i}{d\xi^{\frac{i-s}{2}}d\eta^{\frac{i+s}{2}}}\frac{1}{r}, \text{ if } i-s \text{ is even,}\\ \text{or}\qquad &= (-)^{\frac{i-s-1}{2}}2^{i-s-1}\frac{d^i}{dz\,d\xi^{\frac{i-s-1}{2}}d\eta^{\frac{i+s-1}{2}}}\frac{1}{r}, \text{ if } i-s \text{ is odd.}\end{aligned}\right\}(41)$$

Expanding the first member in terms of z, ξ, η, by successive differentiation, with reference first to η, s times, and then z, $i-s$ times, we find

$$(-)^i\tfrac{1}{2}\cdot\tfrac{3}{2}\ldots(s-\tfrac{1}{2})(2s+1)(2s+2)(2s+3)\ldots(i+s)z^{i-s}\xi^s\ldots\ldots(42),$$

for a term in its numerator: comparing this with (39) and (40), and the second number of (41) with (35), we find C.

(*k*) It is very important to remark, first, that **Important properties of elementary terms and auxiliary functions.**

$$\iint U_i U_{i'}' d\sigma = 0\ldots\ldots\ldots\ldots\ldots\ldots\ldots\ldots\ldots(43),$$

where U_i and $U_{i'}'$ denote any two of the elements of which V is composed in one of the preceding expressions; and secondly, that

$$\int_0^\pi \overset{(s)}{\Theta}_i \overset{(s)}{\Theta}_{i'} \sin\theta d\theta = 0\ldots\ldots\ldots\ldots\ldots\ldots\ldots(44),$$

the case of $i=i'$ being of course excluded. For, taking $r=a$, the radius of the spherical surface; and $d\sigma = a^2 d\varpi$, as above; we have $d\varpi = \sin\theta d\theta d\phi$, etc., the limits of θ and ϕ, in the integration for the whole spherical surface, being 0 to π, and 0 to 2π, respectively. Thus, since $\int_0^{2\pi}\cos s\phi\cos s'\phi\, d\phi = 0$, we see the truth of the first remark; and from (16) and (36) we infer the second, which the reader may verify algebraically, as an exercise.

(*l*) Each one of the preceding series may be taken by either end, and used with i or s, either or both of them negative or fractional or imaginary. Whether finite or infinite in its number of terms, any series thus obtained expresses when multiplied by r^i a harmonic of degree i; since it is of degree i, and satisfies $\nabla^2 V_i = 0$. In any case in which one of the preceding series is not finite, the formula taken by one end gives a converging series; taken by the other end a diverging series. Thus (40) taken in the order shown above, converges when θ is between 0 and 45°, or between 135° and 180°: and taken with the last term of that order first it converges when $\theta > 45°$ and **Expansions of partial harmonics for cones and wedges.**

Dismissal of series which diverge when θ is real.

$<135^\circ$. Thus, again, $\Theta_{(m,n)}$ and $Z_{(m,n)}$ of (36), being each of a finite number of terms when either m or n is a positive integer, become when neither is so, infinite series, which diverge when $\nu<1$ and converge when $\nu>1$. These two series, whether both infinite or one finite and the other infinite, when convergent are so related that

$$\mu Z_{(m-\frac{1}{2},\, n-\frac{1}{2})} = \sqrt{-1}\,\Theta_{(m,n)} \ldots\ldots\ldots\ldots\ldots\ldots\ldots (36'),$$

as is easily verified for a few terms by multiplying $Z_{(m-\frac{1}{2},\, n-\frac{1}{2})}$ by the expansion of $\left(1-\frac{1}{\nu^2}\right)^{\frac{1}{2}}$ in ascending powers of $\frac{1}{\nu^2}$. But expansions in ascending powers of $\frac{1}{\nu^2}$ are of comparatively little interest, as they are divergent for real values of θ, and therefore not available for the proposed physical applications. To find expansions which converge when $\nu<1$ take the last terms of (36) first. Thus, if we put

Finding of convergent expansions.

$$K=(-)^n \frac{m(m-1)\ldots(m-n+2)(m-n+1).n(n-1)\ldots2.1}{1.2\ldots.(n-1)n.(m+n-\frac{1}{2})(m+n-\frac{3}{2})\ldots(m+\frac{3}{2})(m+\frac{1}{2})} \ldots\ldots (36'');$$

supposing m to be $>n$, and n to be a positive integer, we find

$$\Theta_{(m,n)} = K.\nu^{m-n}\left[1-\frac{n(m+\frac{1}{2})}{(m-n+1).1}\nu^2+\frac{n(n-1).(m+\frac{1}{2})(m+\frac{3}{2})}{(m-n+1)(m-n+2).1.2}\nu^4-\text{etc.}\right] \ldots (36''').$$

Writing down the corresponding expression for $Z_{(m-\frac{1}{2},\, n-\frac{1}{2})}$ from (36), and using (36′), we find

$$\Theta_{(m,n)} = K\mu\nu^{m-n}\left[1-\frac{(n-\frac{1}{2})(m+1)}{(m-n+1).1}\nu^2+\frac{(n-\frac{1}{2})(n-\frac{3}{2}).(m+1)(m+2)}{(m-n+1)(m-n+2).1.2}\nu^4+\&\text{c.}\right] \ldots (36^{\text{iv}}).$$

This expansion of $\Theta_{(m,n)}$ is derivable algebraically from (36‴) by multiplying the second member of (36‴) by

$$\mu\left(1+\tfrac{1}{2}\nu^2+\frac{1.3}{2.4}\nu^4+\text{etc.}\right)$$

(which is equal to unity). Both expansions converge when $\nu^2<1$, or, for all real values of θ; just failing when $\theta=\frac{1}{2}\pi$. In choosing between the two expansions (36‴) and (36^{iv}), prefer (36^{iv}) when n differs by less than $\frac{1}{4}$ from zero or some positive integer, otherwise choose (36‴); but it is chiefly important to have them both, because (36^{iv}) is finite, but (36‴) infinite, when $n=\frac{2j-1}{2}$; and (36‴) is finite, but (36^{iv}) infinite, when $n=j-1$; j being any positive integer.

Complete expressions for spherical harmonics of any tesseral type;

$$\left.\begin{array}{lll} \text{Put now} & m+n=i, & m-n=s, \\ \text{or} & m=\tfrac{1}{2}(i+s), & n=\tfrac{1}{2}(i-s), \\ \text{and denote by} & Ku_i^{(s)} & \end{array}\right\}\ldots\ldots\ldots\ldots\ldots(36^{\text{v}})$$

what the second members of (36‴) and (36^{iv}) become with these values for m and n. Again, put

$$\left.\begin{array}{lll} & m+n=i, & n-m=s, \\ \text{or} & m=\tfrac{1}{2}(i-s), & n=\tfrac{1}{2}(i+s), \\ \text{and denote by} & Kv_i^{(s)} & \end{array}\right\}\ldots\ldots\ldots\ldots(36^{\text{vi}})$$

what the second members of (36‴) and (36^{iv}) become with these values for m and n. We thus have two equal convergent series for $u_i^{(s)}$ and two equal convergent series for $v_i^{(s)}$, and $u_i^{(s)}$, $v_i^{(s)}$ are functions of ν (or of θ) such that

in ascending powers of ν.

$$\left.\begin{array}{ll} & u_i^{(s)}(A\cos s\phi+B\sin s\phi) \\ \text{and} & v_i^{(s)}(A\cos s\phi+B\sin s\phi) \end{array}\right\}\ldots\ldots\ldots\ldots\ldots\ldots(36^{\text{vii}})$$

are surface harmonics of order i.

The first terms of $u_i^{(s)}$ and $v_i^{(s)}$ are ν^s and ν^{-s}, or $\mu\nu^s$ and $\mu\nu^{-s}$, according as they are taken from (36‴) or (36^{iv}), and in general $u_i^{(s)}$ and $v_i^{(s)}$ are distinct from one another.

Two distinct solutions are clearly needed for the physical problems. But in the particular case of s an integer, $u_i^{(s)}$ and $v_i^{(s)}$ are not distinct. For in this case each term of $v_i^{(s)}$ after the first s terms has the infinite factor $\frac{1}{s-s}$; thus if C_j denote the coefficient of the $(j+1)^{\text{th}}$ term of $v_i^{(s)}$, the first s terms of $\frac{v_i^{(s)}}{C_s}$ vanish when s is an integer, and those that follow constitute the same series as that expressing $u_i^{(s)}$, whether we take (36‴) or (36^{iv}). For the case of s an integer the wanting solution is to be found by putting

$$\left.\begin{array}{l} w_i^{(j)}=\dfrac{u_i^{(j+\sigma)}-\dfrac{v_i^{(j+\sigma)}}{C_j}}{2\sigma},\ \text{when } \sigma=0: \\ w_i^{(s)} \text{ thus found is such that} \\ \qquad w_i^{(s)}(A\cos s\phi+B\sin s\phi) \end{array}\right\}\ldots\ldots(36^{\text{viii}})$$

Complete expressions for spherical harmonics of any tesseral type;

is a surface harmonic of order i distinct from $u_i^{(s)}$. The first term of $w_i^{(s)}$, according to (36'''), is $\nu^s \log \nu$, or $\mu\nu^s \log \nu$ according to (36^{iv}), and subsequent terms are of the form $(a + b \log \nu)\, \nu^{2j}$, or $(a + b \log \nu)\, \mu\nu^{2j}$, j being an integer. The circumstances belong to a well-known class of cases in the solution of linear differential equations of the second order (see § (f') below).

Again, lastly, remark that (38), unless it is finite (which it is if and only if $i - s$ is a positive integer), diverges when $\mu < 1$ and converges when $\mu > 1$, if taken in the order in which it is given above. To obtain series which converge when $\mu < 1$ (that is to say, for real values of θ), reverse the order of (38) for the case of $i - s$ a positive integer. Thus, according as $i - s$ is even or odd, we find

in ascending powers of μ.

$$\left.\begin{aligned}
&\Theta_i^{(s)} = H\nu^s\left\{1 - \frac{(i-s).(i+s+1)}{1.2}\mu^2 + \frac{(i-s)(i-s-2).(i+s+1)(i+s+3)}{1.2.3.4}\mu^4 - \text{etc.}\right.\\
&\text{where, } i-s \text{ being even,}\\
&H = (-)^{\frac{1}{2}(i-s)}\frac{(i-s)(i-s-1)(i-s-2)(i-s-3)\ldots 4.3.2.1}{2.4\ldots(i-s-2)(i-s).(2i-1)(2i-3)\ldots(i+s+3)(i+s+1)}
\end{aligned}\right\}\ldots(38'),$$

and

$$\left.\begin{aligned}
&\Theta_i^{(s)} = H'\nu^s\left\{\mu - \frac{(i-s-1)(i+s+2)}{2.3}\mu^3 + \frac{(i-s-1)(i-s-3).(i+s+2)(i+s+4)}{2.3.4.5}\mu^5 - \text{etc.}\right.\\
&\text{where, } i-s \text{ being odd,}\\
&H' = (-)^{\frac{1}{2}(i-s-1)}\frac{(i-s)(i-s-1)(i-s-2)(i-s-3)\ldots 5.4.3.2}{2.4\ldots(i-s-3)(i-s-1).(2i-1)(2i-3)\ldots(i+s+4)(i+s+2)}
\end{aligned}\right\}(38'').$$

Then, whatever be $i - s$, or i, or s, integral or fractional, positive or negative, real or imaginary, the formulas within the brackets { } are convergent series when they are not finite integral functions of μ. Hence we see that if we put

$$\left.\begin{aligned}
&p_i^{(s)} = 1 - \frac{(i-s).(i+s+1)}{1.2}\mu^2 + \frac{(i-s)(i-s-2).(i+s+1)(i+s+3)}{1.2.3.4}\mu^4 - \text{etc.}\\
&\text{and}\\
&q_i^{(s)} = \mu - \frac{(i-s-1).(i+s+2)}{2.3}\mu^3 + \frac{(i-s-1)(i-s-3).(i+s+2)(i+s+4)}{2.3.4.5}\mu^5 - \text{etc.}
\end{aligned}\right\}(38''')$$

$$\left.\begin{aligned}
&\text{or} \quad p_i^{(s)} = A_0 + A_2\mu^2 + A_4\mu^4 + \&\text{c.},\\
&\text{and} \quad q_i^{(s)} = A_1\mu + A_3\mu^3 + A_5\mu^5 + \&\text{c.},\\
&\text{where } A_0 = 1,\ A_1 = 1, \text{ and } A_{n+2} = \frac{(n-i+s)(n+1+i+s)}{(n+1)(n+2)}A_n,
\end{aligned}\right\}(38^{iv}),$$

the functions $p_i^{(s)}$, $q_i^{(s)}$ thus expressed, whether they be algebraic or transcendental, are such that

Complete expressions for spherical harmonics of any tesseral type in ascending powers of μ.

$$\left.\begin{array}{l} p_i^{(s)}(A\cos s\phi + B\sin s\phi)\,\nu^s, \\ \text{and} \quad q_i^{(s)}(A\cos s\phi + B\sin s\phi)\,\nu^s, \end{array}\right\} \ldots\ldots\ldots\ldots\ldots\ldots (38^{\mathrm{v}})$$

are the two surface harmonics of order i, and of the form $f(\theta)\begin{smallmatrix}\sin\\ \cos\end{smallmatrix} s\phi$. For example, if $i-s$ be an even integer, $p_i^{(s)}$ is the finite function with which we are familiar as giving a rational integral solution of the form (38^{v}), and $q_i^{(s)}$ gives the solution of the same form which is not integral or rational. And if $i-s$ is odd, $q_i^{(s)}$ gives the familiar rational integral solution, and $p_i^{(s)}$ the other solution of the same form but not integral or rational.

The corresponding solid harmonics of degrees i and $-i-1$ are obtained by multiplying (38^{v}) by r^i and r^{-i-1}. Reducing the latter from polar to rectangular co-ordinates, we find them of the form

Corresponding solid harmonics.

$$\left.\begin{array}{l} \left[r^{-i-s-1} - \dfrac{(i-s)(i+s+1)}{1\,.\,2} r^{-i-s-3} z^2 + \text{etc.}\right] H_s(x, y) \\ \text{and} \\ \left[r^{-i-s-2} z - \text{etc.}\right] H_s(x, y) \end{array}\right\} \ldots\ldots (38^{\mathrm{vi}}),$$

where H_s denotes a homogeneous function of degree s. Now (15) $\dfrac{d}{dz}$ of any solid harmonic of degree $-i$ is a solid harmonic of degree $-i-1$. Hence

Successive derivation from lower orders.

$$r^{i+1} r^s \nu^s \begin{smallmatrix}\sin\\ \cos\end{smallmatrix} s\phi \frac{d}{dz}\left[r^{-i-s} q_{i-1}^{(s)}\right],$$

and

$$r^{i+1} r^s \nu^s \begin{smallmatrix}\sin\\ \cos\end{smallmatrix} s\phi \frac{d}{dz}\left[r^{-i-s} p_{i-1}^{(s)}\right],$$

are surface harmonics of order i, and they are clearly of the first and second forms of (38^{v}). Hence, putting into the forms shown in (38^{vi}) and performing the indicated differentiation for the first term of the q function and the first and second terms of the p function, so as to find the numerical coefficients of r^{-i-s-1} and $r^{-i-s-2}z$ in the immediate results of the differentiation, and then putting μr for z, we find

Successive derivation from lower orders.

$$\left.\begin{aligned} r^{-i-s-1}p_i^{(s)} &= \frac{d}{dz}\left[r^{-i-s}q_{i-1}^{(s)}\right] \\ \text{and}\qquad r^{-i-s-1}q_i^{(s)} &= -\frac{1}{i^2-s^2}\frac{d}{dz}\left[r^{-i-s}p_{i-1}^{(s)}\right] \end{aligned}\right\}\text{.........}(38^{vii}).$$

To reduce back to polar co-ordinates put for a moment $x^2+y^2=a^2$. Then we have

$$r = \frac{a}{\sqrt{(1-\mu^2)}} = \frac{a}{\nu},$$

$$z = \frac{a\mu}{\sqrt{(1-\mu^2)}} = \frac{a\mu}{\nu},$$

and
$$dz = \frac{a d\mu}{(1-\mu^2)^{\frac{3}{2}}} = \frac{a d\mu}{\nu^3}.$$

Hence, instead of (38^{vii}), we have

$$\left.\begin{aligned} p_i^{(s)} &= \nu^{-i-s+2}\frac{d}{d\mu}\{\nu^{i+s}q_{i-1}^{(s)}\}, \\ \text{and}\qquad q_i^{(s)} &= -\frac{1}{i^2-s^2}\nu^{-i-s+2}\frac{d}{d\mu}\{\nu^{i+s}p_{i-1}^{(s)}\} \end{aligned}\right\}\text{....}\ (38^{viii}).$$

[Compare § 782 (5) below.]

Supposing now s and i to be real quantities, and going back to (38^{iv}), to investigate the convergency of the series for $p_i^{(s)}$ and $q_i^{(s)}$, we see that, when n is infinitely great,

$$\frac{A_{n+2}}{A_n} = 1 + \frac{2(s-1)}{n}.$$

Now if
$$(1-\mu^2)^{-\kappa} = \Sigma B_n \mu^n,$$

we have, by the binomial theorem,

$$B_0 = 1,\quad B_1 = 0,\quad \text{and}\quad \frac{B_{n+2}}{B_n} = 1 + \frac{2(\kappa-1)}{n}.$$

Hence, when $\mu = \pm(1-e)$, where e is an infinitely small positive quantity,

$$\left.\begin{aligned} &p_i^{(s)}\nu^{2\kappa} = 0 \ \text{ or } = \infty, \\ \text{and}\qquad &q_i^{(s)}\nu^{2\kappa} = 0 \ \text{ or } = \infty, \\ \text{according as}\qquad &\kappa > s \ \text{ or } \ \kappa < s. \end{aligned}\right\}\text{......................}(38^{ix}).$$

Hence if $i > s$, the quantities within the brackets under $\frac{d}{d\mu}$ in (38^{viii}) vanish when $\mu = \pm 1$; and as they vary continuously, and within finite limits, when μ is continuously increased from -1 to $+1$, it follows that $p_i^{(s)}$ vanishes one time more than does $q_{i-1}^{(s)}$, and $q_i^{(s)}$ one time more than does $p_{i-1}^{(s)}$. Now looking to (38‴), and supposing (as we clearly may without loss of generality) that s is positive, we see that every term of $p_{i-1}^{(s)}$ is positive if $i < s+1$. Hence if i is any quantity between s and $s+1$, $\nu^{i+s} p_{i-1}^{(s)}$ vanishes when $\mu = \pm 1$, and is finite and positive for every intermediate value of μ.

Acquisition of roots with rise of order.

The rootless form of lowest order.

Hence and from the second formula of (38^{viii}), $q_i^{(s)}$ vanishes just once as μ is increased continuously from -1 to $+1$: thence and from the first of (38^{viii}), $p_{i+1}^{(s)}$ vanishes twice: hence and from the second again, $q_{i+2}^{(s)}$ vanishes thrice, and so on. Again, as the coefficient of every term of the series (38‴) for $q_i^{(s)}$ is positive when $i < s+1$, this is the case for $q_{i-1}^{(s)}$, and therefore this function vanishes only for $\mu = 0$, as μ is increased from -1 to $+1$. Hence $p_i^{(s)}$ vanishes twice; and, then, continuing alternate applications of the second and first of (38″), we see that $q_{i+1}^{(s)}$ vanishes thrice, $p_{i+2}^{(s)}$ four times, and so on. Thus, putting all together, we see that $q_{i+j-1}^{(s)}$ has j or $j+1$ roots, and $p_{i+j-1}^{(s)}$ has $j+1$ or j roots, according as j is odd or even; j being any integer and i, as defined above, any quantity between s and $s+1$. In other words, the number of roots of $p_i^{(s)}$ is the even number next above $i-s$; and the number of roots of $q_i^{(s)}$ is the odd number next above $i-s$. Farther, from (38^{viii}) we see that the roots of $p_i^{(s)}$ lie in order between those of $q_{i-1}^{(s)}$, and the roots of $q_i^{(s)}$ between those of $p_{i-1}^{(s)}$. [Compare § (p) below.] These properties of the p and q functions are of paramount importance, not only in the theory of the development of arbitrary functions by aid of them, but in the physical applications of the fractional harmonic analysis. In each case of physical application they belong to the foundation of the theory of the simple and nodal modes of the action investigated. They afford the principles for the determination of values of $i-s$, which shall make $\Theta_i^{(s)}$ or $\frac{d}{d\theta}\Theta_i^{(s)}$ vanish for each of two stated

The other form of lowest order has one root,—zero.

Census of roots of tesseral harmonics of any order.

Electric induction, motion of water, etc., in space between two coaxal cones.

values of θ. This is an analytical problem of high interest in connexion with these extensions of spherical harmonic analysis: it is essentially involved in the physical application referred to above where the spaces concerned are bounded partly by coaxal cones. When the boundary is completed by the intercepted portions of two concentric spherical surfaces, functions of the class described in (o) below also enter into the solution. When prepared to take advantage of physical applications we shall return to the subject; but it is necessary at present to restrict ourselves to these few observations.

Electric induction, motion of water, etc., in space between spherical surface and two planes meeting in a diameter.

(m) If, in physical problems such as those already referred to, the space considered is bounded by two planes meeting, at any angle $\frac{\pi}{s}$, in a diameter, and the portion of spherical surface in the angle between them (the case of $s<1$, that is to say, the case of angle exceeding two right angles, not being excluded) the harmonics required are all of fractional degrees, but each a finite algebraic function of the co-ordinates ξ, η, z if s is any incommensurable number. Thus, for instance, if the problem be to find the internal temperature at any point of a solid of the shape in question, when each point of the curved portion of its surface is maintained permanently at any arbitrarily given temperature, and its plane sides at one constant temperature, the forms and the degrees of the harmonics referred to are as follows:—

Degree.	Harmonic.	Degree.	Harmonic.	Degree.	Harmonic.
s,	ξ^s	$2s$,	ξ^{2s}	$3s$,	ξ^{3s}
$s+1$,	$r^{2s+3}\frac{d}{dz}\frac{\xi^s}{r^{2s+1}}$	$2s+1$,	$r^{4s+3}\frac{d}{dz}\frac{\xi^{2s}}{r^{4s+1}}$	$3s+1$,	$r^{6s+1}\frac{d}{dz}\frac{\xi^{3s}}{r^{6s+1}}$
$s+2$,	$r^{2s+5}\frac{d^2}{dz^2}\frac{\xi^s}{r^{2s+1}}$	$2s+2$,	$r^{4s+5}\frac{d^2}{dz^2}\frac{\xi^{2s}}{r^{4s+1}}$	......	....
$s+3$,	$r^{2s+7}\frac{d^3}{dz^3}\frac{\xi^s}{r^{2s+1}}$	$2s+3$,	$r^{4s+7}\frac{d^3}{dz^3}\frac{\xi^{2s}}{r^{4s+1}}$	......	....
......	....	......	....	......	....
......	....	......	....	......	....

These harmonics are expressed, by various formulæ (36)...(40), etc., in terms of real co-ordinates, in what precedes.

(n) It is worthy of remark that these, and every other spherical harmonic, of whatever degree, integral, real but fractional, or imaginary, are derivable by a general form of process, which includes differentiation as a particular case. Thus if $\left(\frac{d}{d\eta}\right)^s$ denotes an operation which, when s is an integer, constitutes taking the sth differential coefficient, we have clearly

Harmonic functions of all degrees derived from that of degree −1 by generalized differentiation.

$$\xi^s = r^{2s+1} P_s \left(\frac{d}{d\eta}\right)^s \frac{1}{(\xi\eta + z^2)^{\frac{1}{2}}},$$

where P_s denotes a function of s, which, when s is a real integer, becomes $(-)^s \frac{1}{2} \cdot \frac{3}{2} \cdot \frac{5}{2} \ldots (s - \frac{1}{2})$.
The investigation of this generalized differentiation presents difficulties which are confined to the evaluation of P_s, and which have formed the subject of highly interesting mathematical investigations by Liouville, Gregory, Kelland, and others.

If we set aside the factor P_s, and satisfy ourselves with determinations of *forms* of spherical harmonics, we have only to apply Leibnitz's and other obvious formulæ for differentiation with any fractional or imaginary number as index, to see that the equivalent expressions above given for a complete spherical harmonic of any degree, are derivable from $\frac{1}{r}$ by the process of generalized differentiation now indicated, so as to include every possible partial harmonic, of whatever degree, whether integral, or fractional and real, or imaginary. But, as stated above, those expressions may be used, in the manner explained, for partial harmonics, whether finite algebraic functions of ξ, η, z, or transcendents expressed by converging infinite series; quite irrespectively of the manner of derivation now remarked.

Expansions of partial harmonics obtained by common formulæ, with generalized indices.

(*o*) To illustrate the use of spherical harmonics of imaginary degrees, the problem regarding the conduction of heat specified above may be varied thus:—Let the solid be bounded by two concentric spherical surfaces, of radii a and a', and by two cones or planes, and let every point of each of these flat or conical sides be maintained with any arbitrarily given distribution of temperature, and the whole spherical portion of the boundary at one constant temperature. Harmonics will enter into the solution, of degree

Imaginary degrees useful when arbitrary functions of r are to be expressed.

$$-\frac{1}{2} + \frac{j\pi\sqrt{-1}}{\log \frac{a}{a'}},$$

where j denotes any integer. [Compare § (d') below.] Converging series for these and the others required for the solution are included in our general formulas (36)...(40), etc.

Derivation of any harmonic from that of degree −1 indicates the character and number of its nodes.

(p) The method of finding complete spherical harmonics by the differentiation of $\frac{1}{r}$, investigated above, has this great advantage, that it shows immediately very important properties which they possess with reference to the values of the variables for which they vanish. Thus, inasmuch as $\frac{1}{r}$ and all its differential coefficients vanish for $x = \pm\infty$, and for $y = \pm\infty$, and for $z = \pm\infty$, it follows that

$$\frac{d^{j+k+l}}{dx^j dy^k dz^l}\frac{1}{r}$$

vanishes j times when x is increased from $-\infty$ to $+\infty$
" k " y " " " "
and " l " z " " " "

[Compare with the investigation of the roots of $p_i^{(s)}$ and $q_i^{(s)}$ in § (l) above.]

The reader who is not familiar with Fourier's theory of equations will have no difficulty in verifying for himself the present application of the principles developed in that admirable work. Its interpretation for fractional or imaginary values of j, k, l is wonderfully interesting, and of obvious value for the physical applications of partial harmonics.

Expression of an arbitrary function in a series of surface harmonics.

Thus it appears that spherical harmonics of large real degrees, integral or fractional, or of imaginary degrees with large real parts ($\alpha + \beta\sqrt{-1}$, with α large), belong to the general class, to which Sir William R. Hamilton has applied the designation "Fluctuating Functions." This property is essentially involved in their capacity for expressing arbitrary functions, to the demonstration of which for the case of complete harmonics we now proceed, in conclusion.

Preliminary proposition.

(r) Let C be the centre and a the radius of a spherical surface, which we shall denote by S. Let P be any external or internal point, and let f denote its distance from C. Let $d\sigma$ denote an element of S, at a point E, and let $EP = D$. Then, $\iint$ denoting an integration extended over S, it is easily proved that

Preliminary proposition.

$$\left.\begin{array}{ll} & \displaystyle\iint \frac{d\sigma}{D^3} = \frac{a}{f}\frac{4\pi a}{f^2 - a^2} \text{ when } P \text{ is external to } S \\ \text{and} & \displaystyle\iint \frac{d\sigma}{D^3} = \frac{4\pi a}{a^2 - f^2} \text{ when } P \text{ is within } S \end{array}\right\} \ldots\ldots\ldots\ldots(45).$$

This is merely a particular case of a very general theorem of Green's, included in that of A (*a*), above, as will be shown when we shall be particularly occupied, later, with the general theory of Attraction: a geometrical proof of a special theorem, of which it is a case, (§ 474, fig. 2, with P infinitely distant,) will occur in connexion with elementary investigations regarding the distribution of electricity on spherical conductors: and, in the meantime, the following direct evaluation of the integral itself is given, in order that no part of the important investigation with which we are now engaged may be even temporarily incomplete.

Choosing polar co-ordinates, $\theta = ECP$, and ϕ the angle between the plane of ECP and a fixed plane through CP, we have

$$d\sigma = a^2 \sin\theta \, d\theta \, d\phi.$$

Hence, by integration from $\phi = 0$ to $\phi = 2\pi$,

$$\iint \frac{d\sigma}{D^3} = 2\pi a^2 \int_0^\pi \frac{\sin\theta d\theta}{D^3}.$$

But $$D^2 = a^2 - 2af\cos\theta + f^2;$$

and therefore $$\sin\theta d\theta = \frac{DdD}{af};$$

the limiting values of D in the integral being

$$f - a, \; f + a, \text{ when } f > a,$$

and $$a - f, \; a + f, \text{ when } f < a.$$

Hence we have

$$\iint \frac{d\sigma}{D^3} = \frac{2\pi a}{f}\left(\frac{1}{f-a} - \frac{1}{f+a}\right), \text{ or} = \frac{2\pi a}{f}\left(\frac{1}{a-f} - \frac{1}{a+f}\right),$$

in the two cases respectively, which proves (45).

(*s*) Let now $F(E)$ denote any arbitrary function of the position of E on S, and let

Solution of Green's problem for case of spherical surface, expressed by definite integral.

$$u = \iint \frac{(f^2 \sim a^2) F(E) d\sigma}{D^3} \ldots\ldots\ldots\ldots\ldots\ldots\ldots\ldots (46).$$

When f is infinitely nearly equal to a, every element of this integral will vanish except those for which D is infinitely small.

Green's problem for case of spherical surface, expressed by definite integral.

Hence the integral will have the same value as it would have if $F(E)$ had everywhere the same value as it has at the part of S nearest to P; and, therefore, denoting this value of the arbitrary function by $F(P)$, we have

$$u = F(P)\iint \frac{(f^2 \sim a^2)\,d\sigma}{D^3},$$

when f differs infinitely little from a; or, by (45),

$$u = 4\pi a F(P) \quad\ldots\ldots\ldots\ldots(46').$$

Its expansion in harmonic series.

Now, if e denote any positive quantity less than unity, we have, by expansion in a convergent series,

$$\frac{1}{(1-2e\cos\theta+e^2)^{\frac{1}{2}}} = 1 + Q_1 e + Q_2 e^2 + \text{etc.} \quad\ldots\ldots\ldots(47),$$

Q_1, Q_2, etc., denoting functions of θ, for which expressions will be investigated below. Each of them is equal to $+1$, when $\theta = 0$, and they are alternately equal to -1 and $+1$, when $\theta = \pi$. It is easily proved that each is >-1 and $<+1$, for all values of θ between 0 and π. Hence the series, which becomes the geometrical series $1 \pm e + e^2 \pm$ etc., in the extreme cases, converges more rapidly than the geometrical series, except in those extreme cases of $\theta = 0$ and $\theta = \pi$.

$$\left.\begin{aligned} &\text{Hence}\quad \frac{1}{D} = \frac{1}{f}\left(1 + \frac{Q_1 a}{f} + \frac{Q_2 a^2}{f^2} + \text{etc.}\right) \text{ when } f > a \\ &\text{and}\quad \frac{1}{D} = \frac{1}{a}\left(1 + \frac{Q_1 f}{a} + \frac{Q_2 f^2}{a^2} + \text{etc.}\right) \text{ when } f < a \end{aligned}\right\}\ldots\ldots\ldots(48).$$

Now we have
$$\frac{d\frac{1}{D}}{df} = \frac{a\cos\theta - f}{D^3},$$

and therefore
$$\frac{f^2 - a^2}{D^3} = -\left(2f\frac{d\frac{1}{D}}{df} + \frac{1}{D}\right).$$

Hence by (48),

$$\left.\begin{aligned} &\frac{f^2-a^2}{D^3} = \frac{1}{f}\left(1 + \frac{3Q_1 a}{f} + \frac{5Q_2 a^2}{f^2} + \ldots\ldots\right) \text{ when } f > a \\ &\text{and}\quad \frac{a^2-f^2}{D^3} = \frac{1}{a}\left(1 + \frac{3Q_1 f}{a} + \frac{5Q_2 f^2}{a^2} + \ldots\ldots\right) \text{ when } f < a \end{aligned}\right\}(49).$$

Hence, for u (46), we have the following expansions:—

Green's problem for case of spherical surface, solved explicitly in harmonic series.

$$u=\frac{1}{f}\left\{\iint F(E)d\sigma+\frac{3a}{f}\iint Q_1F(E)d\sigma+\frac{5a^2}{f^2}\iint Q_2F(E)d\sigma+\ldots\right\}, \text{ when } f>a,$$

and

$$u=\frac{1}{a}\left\{\iint F(E)d\sigma+\frac{3f}{a}\iint Q_1F(E)d\sigma+\frac{5f^2}{a^2}\iint Q_2F(E)d\sigma+\ldots\right\}, \text{ when } f<a \qquad \ldots\ldots\ldots(51).$$

These series being clearly convergent, except in the case of $f=a$, and, in this limiting case, the unexpanded value of u having been proved (46′) to be finite and equal to $4\pi aF(P)$, it follows that the sum of each series approaches more and more nearly to this value when f approaches to equality with a. Hence, in the limit,

$$F(P)=\frac{1}{4\pi a^2}\left\{\iint F(E)d\sigma+3\iint Q_1F(E)d\sigma+5\iint Q_2F(E)d\sigma+\text{etc.},\right\}\ldots(52),$$

Laplace's spherical harmonic expansion of an arbitrary function.

which is the celebrated development of an arbitrary function in a series of "Laplace's coefficients," or, as we now call them, *spherical harmonics.*

(*t*) The preceding investigation shows that when there is one determinate value of the arbitrary function F for every point of S, the series (52) converges to the value of this function at P. The same reason shows that when there is an abrupt transition in the value of F, across any line on S, the series cannot converge when P is *exactly on,* but must still converge, however near it may be to, this line. [Compare with last two paragraphs of § 77 above.] The degree of non-convergence is so slight that, as we see from (51), the introduction of factors e, e^2, e^3, &c. to the successive terms e being <1 by a very small difference, produces decided convergence for every position of P, and the value of the series differs very little from $F(P)$, passing very rapidly through the finite difference when P is moved across the line of abrupt change in the value of $F(P)$.

Convergence of series never lost except at abrupt changes in value of the function expressed.

(*u*) In the development (47) of

$$\frac{1}{(1-2e\cos\theta+e^2)^{\frac{1}{2}}},$$

the coefficients of $e, e^2, \ldots e^i$, are clearly rational integral functions of $\cos\theta$, of degrees $1, 2\ldots i$, respectively. They are given explicitly below in (60) and (61), with $\theta'=0$. But, if x, y, z and

Expansion of $\frac{1}{D}$ in symmetrical harmonic functions of the co-ordinates of the two points.

x', y', z' denote rectangular co-ordinates of P and of E respectively, we have

$$\cos\theta = \frac{xx' + yy' + zz'}{rr'},$$

where $r = (x^2 + y^2 + z^2)^{\frac{1}{2}}$, and $r' = (x'^2 + y'^2 + z'^2)^{\frac{1}{2}}$. Hence, denoting, as above, by Q_i the coefficient of e^i in the development, we have

$$Q_i = \frac{H_i[(x, y, z), (x', y', z')]}{r^i r'^i} \text{ ....(53)},$$

$H_i[(x, y, z), (x', y', z')]$ denoting a symmetrical function of (x, y, z) and (x', y', z'), which is homogeneous with reference to either set alone. An explicit expression for this function is of course found from the expression for Q_i in terms of $\cos\theta$.

Biaxal harmonic.

Viewed as a function of (x, y, z), $Q_i r^i r^{i'}$ is symmetrical round OE; and as a function of (x', y', z') it is symmetrical round OP. We shall therefore call it the biaxal harmonic of (x, y, z) (x', y', z') of degree i; and Q_i the biaxal surface harmonic of order i.

Expansion of $\frac{1}{D}$ by Taylor's theorem.

(v) But it is important to remark, that the coefficient of any term, such as $x'^j y'^k z'^l$, in it may be obtained alone, by means of Taylor's theorem, applied to a function of three variables, thus:—

$$\frac{1}{(1 - 2e\cos\theta + e^2)^{\frac{1}{2}}} = \frac{r}{(r^2 - 2rr'\cos\theta + r'^2)^{\frac{1}{2}}} = \frac{r}{[(x-x')^2 + (y-y')^2 + (z-z')^2]^{\frac{1}{2}}}.$$

Now if $F(x, y, z)$ denote any function of x, y, and z, we have

$$F(x+f, y+g, z+h) = \sum_{j=0}^{j=\infty}\sum_{k=0}^{k=\infty}\sum_{l=0}^{l=\infty} \frac{f^j g^k h^l}{1.2\ldots j.1.2\ldots k.1.2\ldots l}\frac{d^{j+k+l}F(x,y,z)}{dx^j dy^k dz^l};$$

where it must be remarked that the interpretation of $1.2\ldots j$, when $j = 0$, is unity, and so for k and l also. Hence, by taking $F(x, y, z) = \frac{1}{(x^2 + y^2 + z^2)^{\frac{1}{2}}}$, we have

$$\frac{1}{[(x-x')^2 + (y-y')^2 + (z-z')^2]^{\frac{1}{2}}}$$
$$= \Sigma\Sigma\Sigma \frac{(-1)^{j+k+l} x'^j y'^k z'^l}{1.2\ldots j.1.2\ldots k.1.2\ldots l}\frac{d^{j+k+l}}{dx^j dy^k dz^l}\frac{1}{(x^2 + y^2 + z^2)^{\frac{1}{2}}},$$

a development which, by comparing it with (48), above, we see to be convergent whenever

$$x'^2 + y'^2 + z'^2 < x^2 + y^2 + z^2.$$

Hence

Expression for biaxal harmonic deduced.

$$(rr')^i Q_i = r^{2i+1} \sum\sum\sum^{(j+k+l=i)} \frac{(-1)^{j+k+l} x'^j y'^k z'^l}{1.2\ldots j.1.2\ldots k.1.2\ldots l} \frac{d^{j+k+l}}{dx^j dy^k dz^l} \frac{1}{(x^2+y^2+z^2)^{\frac{1}{2}}} \ldots (54),$$

the summation including all terms which fulfil the indicated condition $(j+k+l=i)$. It is easy to verify that the second member is not only integral and homogeneous of the degree i, in x, y, z, as it is expressly in x', y', z'; but that it is symmetrical with reference to these two sets of variables. Arriving thus at the conclusion expressed above by (53), we have now, for the function there indicated, an explicit expression in terms of differential coefficients, which, further, may be immediately expanded into an algebraic form with ease.

(v') In the particular case of $x'=0$ and $y'=0$, (54) becomes reduced to a single term, a function of x, y, z symmetrical about the axis OZ; and, dividing each member by r'^i, or its equal, z'^i, we have

Axial harmonic of order i.

$$r^i Q_i = \frac{(-1)^i r^{2i+1}}{1.2.3\ldots i} \frac{d^i}{dz^i} \frac{1}{(x^2+y^2+z^2)^{\frac{1}{2}}} \ldots\ldots\ldots\ldots\ldots (55).$$

Axial harmonic with its co-ordinates transformed becomes biaxal.

By actual differentiation it is easy to find the law of successive derivation of the numerators; and thus we find, with about equal ease, either of the expansions (31), (40), or (41), above, for the case $m=n$, or the trigonometrical formulæ, which are of course obtained by putting $z = r\cos\theta$ and $x^2+y^2 = r^2\sin^2\theta$.

(w) If now we put in these, $\cos\theta = \dfrac{xx'+yy'+zz'}{rr'}$, introducing again, as in (u) above, the notation (x, y, z), (x', y', z'), we arrive at expansions of Q_i in the terms indicated in (53).

Expansions of the biaxal harmonic, of order i.

(x) Some of the most useful expansions of Q_i are very readily obtained by introducing, as before, the imaginary co-ordinates (ξ, η) instead of (x, y), according to equations (26) of (j), and similarly, (ξ', η') instead of (x', y'). Thus we have

$$D^2 = (\xi-\xi')(\eta-\eta') + (z-z')^2.$$

Hence, as above,

$$\frac{1}{[(\xi-\xi')(\eta-\eta')+(z-z')^2]^{\frac{1}{2}}}$$
$$= \sum\sum\sum \frac{(-1)^{j+k+l}\xi'^j \eta'^k z'^l}{1.2\ldots j.1.2\ldots k.1.2\ldots l} \frac{d^{j+k+l}}{d\xi^j d\eta^k dz^l} \frac{1}{(\xi\eta+z^2)^{\frac{1}{2}}}.$$

Expansions of the biaxal harmonic, of order i.

Hence

$$(rr')^i Q_i = r^{2i+1} \sum^{j+k+l=i} \sum \sum \frac{(-1)^{j+k+l} \xi'^j \eta'^k z'^l}{1.2\ldots j.1.2\ldots k.1.2\ldots l} \frac{d^{j+k+l}}{d\xi^j d\eta^k dz^l} \frac{1}{(\xi\eta + z^2)^{\frac{1}{2}}} \ldots (56).$$

Of course we have in this case

$$r^2 = \xi\eta + z^2, \quad r'^2 = \xi'\eta' + z'^2,$$

and $$\cos\theta = \frac{\xi\eta' + \xi'\eta + zz'}{rr'}.$$

And, just as above, we see that this expression, obviously a homogeneous function of ξ', η', z', of degree i, and also of η, ξ, z, involves these two systems of variables symmetrically.

Now, as we have seen above, all the ith differential coefficients of $\frac{1}{r}$ are reducible to the $2i+1$ independent forms

$$\left(\frac{d}{dz}\right)^i \frac{1}{r}, \quad \begin{array}{l} \left(\frac{d}{dz}\right)^{i-1} \frac{d}{d\eta}\frac{1}{r}, \; \left(\frac{d}{dz}\right)^{i-2} \left(\frac{d}{d\eta}\right)^2 \frac{1}{r}, \ldots \left(\frac{d}{d\eta}\right)^i \frac{1}{r}, \\ \left(\frac{d}{dz}\right)^{i-1} \frac{d}{d\xi}\frac{1}{r}, \; \left(\frac{d}{dz}\right)^{i-2} \left(\frac{d}{d\xi}\right)^2 \frac{1}{r}, \ldots \left(\frac{d}{d\xi}\right)^i \frac{1}{r}. \end{array}$$

Hence $r^i Q_i$, viewed as a function of z, ξ, η, is expressed by these $2i+1$ terms, each with a coefficient involving z', ξ', η'. And because of the symmetry we see that this coefficient must be the same function of z', η', ξ', into some factor involving none of these variables (z, ξ, η), (z', η', ξ'). Also, by the symmetry with reference to ξ, η' and η, ξ', we see that the numerical factor must be the same for the terms similarly involving ξ, η' on the one hand, and η, ξ' on the other. Hence,

Biaxal harmonic expressed in symmetrical series of differential coefficients.

$$\left.\begin{array}{l} Q_i = (rr')^{i+1} \left[E_0 \left(\frac{d}{dz'}\right)^i \frac{1}{r'} \left(\frac{d}{dz}\right)^i \frac{1}{r} \right. \\ \left. \quad + \sum_{s=1}^{s=i} E_i^{(s)} \left\{ \frac{d^i}{dz'^{i-s} d\xi'^s} \frac{1}{r'} \frac{d^i}{dz^{i-s} d\eta^s} \frac{1}{r} + \frac{d^i}{dz'^{i-s} d\eta'^s} \frac{1}{r'} \frac{d^i}{dz^{i-s} d\xi^s} \frac{1}{r} \right\} \right] \\ \text{where} \\ E_i^{(s)} = \frac{1}{1.2\ldots s.1.2\ldots(i-s).\frac{1}{2}.\frac{3}{2}\ldots(s-\frac{1}{2}).(2s+1)(2s+2)\ldots(i+s)} \end{array}\right\} \ldots (57).$$

The value of $E_i^{(s)}$ is obtained thus:—Comparing the coefficient of the term $(zz')^{i-s}(\xi\eta')^s$ in the numerator of the expression which (56) becomes when the differential coefficient is expanded, with the coefficient of the same term in (57), we have

$$\frac{(-)^i M}{1.2\ldots(i-s).1.2\ldots s} = E_i^{(s)} M^2 \ldots\ldots\ldots (58),$$

where M denotes the coefficient of $z^{i-s}\xi^s$ in $r^{2i+1}\frac{d_i}{dz^{i-s}d\eta^s}\frac{1}{r}$, or, which is the same, the coefficient of $z'^{i-s}\eta'^s$ in $r'^{2i+1}\frac{d^i}{dz'^{i-s}d\xi'^s}\frac{1}{r}$. From this, with the value (42) for M, we find $E_i^{(s)}$ as above.

Biaxal harmonic expressed in symmetrical series of differential coefficients.

(y) We are now ready to reduce the expansion of Q_n to a real trigonometrical form. First, we have, by (33),

$$(\xi\eta')^s + (\xi'\eta)^s = 2(rr'\sin\theta\sin\theta')^s\cos s(\phi-\phi') \ldots\ldots\ldots (59).$$

Let now

$$\vartheta_i^{(s)} = \sin^s\theta\left[\cos^{i-s}\theta - \frac{(i-s)(i-s-1)}{4(s+1).1}\cos^{i-s-2}\theta\sin^2\theta\right.$$
$$\left. + \frac{(i-s)(i-s-1)(i-s-2)(i-s-3)}{4^2(s+1)(s+2).1.2}\cos^{i-s-4}\theta\sin^4\theta - \text{etc.}\right] \ldots (60);$$

(that is to say, $C\vartheta_i^{(s)} = \Theta_i^{(s)}$, in accordance with the previous notation,) and let the corresponding notation with accents apply to θ'. Then, by the aid of (57), (58), and (59), we have

$$Q_i = 2\sum_{s=0}^{s=i}\frac{\frac{1}{2}.\frac{3}{2}\ldots(s-\frac{1}{2})}{1.2\ldots s}.\frac{(2s+1)(2s+2)\ldots(2s+i-s)}{1.2\ldots(i-s)}\cos s(\phi-\phi')\vartheta_i^{(s)}\vartheta_i'^{(s)} \ldots (61),$$

Trigonometrical expansion of biaxal surface harmonic.

of which, however, the first term (that for which $s=0$) must be halved.

(z) As a supplement to the fundamental proposition $\iint S_i S_{i'} d\varpi = 0$, (16) of ($g$), and the corresponding propositions, (43) and (44), regarding elementary terms of harmonics, we are now prepared to evaluate $\iint S_i^2 d\varpi$.

First, using the general expression (37) investigated above for S_i, and modifying the arbitrary constants to suit our present notation, we have

Fundamental definite integral investigated.

$$S_i = \sum_{s=0}^{s=i} A_s \cos(s\phi + a_s)\vartheta_i^{(s)} \ldots\ldots\ldots\ldots\ldots\ldots\ldots\ldots\ldots (62).$$

Hence

$$\iint S_i^2\, d\varpi = \pi\sum_0^i A_s^2 \int_0^\pi (\vartheta_i^{(s)})^2 \sin\theta\, d\theta \ldots\ldots\ldots\ldots\ldots\ldots (63).$$

To evaluate the definite integral in the second member, we have only to apply the general theorem (52) for expansion, in terms of surface harmonics, to the particular case in which the arbitrary function $F(E)$ is itself the harmonic, $\cos s\phi\vartheta_i^{(s)}$. Thus, remembering (16), we have

$$\cos s\phi\vartheta_i^{(s)} = \frac{2i+1}{4\pi}\int_0^\pi \sin\theta' d\theta' \int_0^{2\pi} d\phi' \cos s\phi' \vartheta_i'^{(s)} Q_i \ldots\ldots\ldots (64).$$

Fundamental definite integral evaluated.

Using here for Q_i its trigonometrical expansion just investigated, and performing the integration for ϕ' between the stated limits, we find that $\cos s\phi\, \vartheta_i^{(s)}$ may be divided out, and (omitting the accents in the residual definite integral) we conclude,

$$\int_0^\pi \sin\theta (\vartheta_i^{(s)})^2 d\theta = \frac{2}{2i+1} \cdot \frac{1.2...s}{\frac{1}{2}.\frac{3}{2}...(s-\frac{1}{2})} \cdot \frac{1.2...(i-s)}{(2s+1)(2s+2)...(2s+i-s)} \;...(65).$$

This holds without exception for the case $s=0$, in which the second member becomes $\frac{2}{2i+1}$. It is convenient here to recal equation (44), which, when expressed in terms of $\vartheta_i^{(s)}$ instead of $\Theta_{(m,n)}$, becomes

$$\int_0^\pi \sin\theta\, \vartheta_i^{(s)} \vartheta_{i'}^{(s)} d\theta = 0 \;..................... (66),$$

where i and i' must be different. The properties expressed by these two equations, (65) and (66), may be verified by direct integration, from the explicit expression (60) for $\vartheta_i^{(s)}$; and to do so will be a good analytical exercise on the subject.

(a') Denote for brevity the second member of (65) by (i, s), so that

$$\int_0^\pi \sin\theta (\vartheta_i^{(s)})^2 d\theta = (i, s) \;.....................(67).$$

Suppose the co-ordinates θ, ϕ to be used in (52); so that a, θ, ϕ are the three co-ordinates of P, and we may take $d\sigma = a^2 \sin\theta d\theta d\phi$. Working out by aid of (61), (65), the processes indicated symbolically in (52), we find

Spherical harmonic synthesis of arbitrary function concluded.

$$F(\theta, \phi) = \sum_{i=0}^{i=\infty} \{A_i \vartheta_i^{(0)} + \sum_{s=1}^{s=i} (A_i^{(s)} \cos s\phi + B_i^{(s)} \sin s\phi) \vartheta_i^{(s)}\} \;........(68),$$

where

$$\left.\begin{aligned} A_i &= \frac{2i+1}{4\pi} \int_0^\pi \vartheta_i^{(0)} \sin\theta d\theta \int_0^{2\pi} F(\theta,\phi) d\phi \\ A_i^{(s)} &= \frac{1}{(i,s)\pi} \int_0^\pi \vartheta_i^{(s)} \sin\theta d\theta \int_0^{2\pi} \cos s\phi\, F(\theta,\phi) d\phi \\ B_i^{(s)} &= \frac{1}{(i,s)\pi} \int_0^\pi \vartheta_i^{(s)} \sin\theta d\theta \int_0^{2\pi} \sin s\phi\, F(\theta,\phi) d\phi \end{aligned}\right\} \;...... (69),$$

which is the explicit form most convenient for general use, of the expansion of an arbitrary function of the co-ordinates θ, ϕ in spherical surface harmonics. It is most easily proved, [when once the general theorem expressed by (66) and (65) has been in any way established,] by assuming the form of expansion (68), and then determining the coefficients by multiplying both members by $\vartheta_i^{(s)} \cos s\phi \sin\theta \, d\theta \, d\phi$, and again by $\vartheta_i^{(s)} \sin s\phi \sin\theta \, d\theta \, d\phi$, and integrating in each case over the whole spherical surface.

Spherical harmonic analysis of arbitrary function.

(b′) In what precedes the expansions of surface harmonics, whether complete or not, have been obtained solely by the differentiation of $\frac{1}{r}$ with reference to rectilineal rectangular co-ordinates x, y, z. The expansions of the complete harmonics have been found simply as expressions for differential coefficients, or for linear functions of differential coefficients of $\frac{1}{r}$. The expansions of harmonics of fractional and imaginary orders have been inferred from the expansions of the complete harmonics merely by generalizing their algebraic forms. The properties of the harmonics have been investigated solely from the differential equation

Review of preceding expansions and investigations of properties.

$$\frac{d^2 V}{dx^2} + \frac{d^2 V}{dy^2} + \frac{d^2 V}{dz^2} = 0 \quad \text{.....................(70)},$$

in terms of the rectilineal rectangular co-ordinates. The original investigations of Laplace, on the other hand, were founded exclusively on the transformation of this equation into polar co-ordinates. In our first edition this transformation was not given—we now supply the omission, not only on account of the historical interest attached to "Laplace's equation" in terms of polar co-ordinates, but also because in this form it leads directly by the ordinary methods of treating differential equations, to every possible expansion of surface harmonics in polar co-ordinates.

(c′) By App. Z (g) (14) we find for Laplace's equation (20) transformed to polar co-ordinates,

$$\frac{d}{dr}\left(\frac{r^2 \, dV}{dr}\right) + \frac{1}{\sin\theta}\frac{d}{d\theta}\left(\sin\theta \frac{dV}{d\theta}\right) + \frac{1}{\sin^2\theta}\frac{d^2 V}{d\phi^2} = 0 \quad \text{........(71)}.$$

In this put

$$V = S_i r^i, \text{ or } V = S_i r^{-i-1} \quad \text{..................(72)}.$$

We find

$$i(i+1) S_i + \frac{1}{\sin\theta}\frac{d}{d\theta}\left(\sin\theta \frac{dS_i}{d\theta}\right) + \frac{1}{\sin^2\theta}\frac{d^2 S_i}{d\phi^2} = 0 \quad \text{.......(73)},$$

Laplace's equation for surface harmonic in polar coordinates.

which is the celebrated formula commonly known in England as "Laplace's Equation" for determining S_i, the "Laplace's coefficient" of order i; i being an integer, and the solutions admitted or sought for being restricted to rational integral functions of $\cos\theta$, $\sin\theta\cos\phi$ and $\sin\theta\sin\phi$.

(*d′*) Doing away now with all such restrictions, suppose i to be any number, integral or fractional, real or imaginary, only if imaginary let it be such as to make $i(i-1)$ real [compare § (*o*)] above. On the supposition that S_i is a rational integral function of $\cos\theta$, $\sin\theta\cos\phi$ and $\sin\theta\sin\phi$, it would be the sum of terms such as $\Theta_i^{(s)}\,{}^{\sin}_{\cos}\,s\phi$. Now, allowing s to have any value integral or fractional, real or imaginary, assume

$$S = \Theta_i^{(s)}\,{}^{\sin}_{\cos}\,s\phi \;\ldots\ldots\ldots\ldots\ldots\ldots\ldots\ldots(74).$$

This will be a form of particular solution adapted for application to problems such as those referred to in §§ (*l*), (*m*) above; and (73) gives, for the determination of $\Theta_i^{(s)}$,

$$\frac{1}{\sin\theta}\frac{d}{d\theta}\left(\sin\theta\frac{d\Theta_i^{(s)}}{d\theta}\right)+\left[\frac{-s^2}{\sin^2\theta}+i(i+1)\right]\Theta_i^{(s)}=0\;\ldots\ldots\ldots(75).$$

Definition of "Laplace's functions."

(*e′*) When i and s are both integers we know from § (*h*) above, and we shall verify presently, by regular treatment of it in its present form, that the differential equation (75) has for one solution a rational integral function of $\sin\theta$ and $\cos\theta$. It is this solution that gives the "Laplace's Function," or the "complete surface harmonic" of the form $\Theta_i^{(s)}\,{}^{\sin}_{\cos}\,s\phi$. But being a differential equation of the second order, (75) must have another distinct solution, and from § (*h*) above it follows that this second solution cannot be a rational integral function of $\sin\theta$, $\cos\theta$. It may of course be found by quadratures from the rational integral solution according to the regular process for finding the second particular solution of a differential equation of the second order when one particular solution is known. Thus denoting by $\Theta_i^{(s)}$ any solution, as for example the known rational integral solution expressed by equation (38), or (36) or (40) above, or § 782 (*e*) or (*f*) with (5) below, we have for the complete solution,

$$\Theta'^{(s)}_i = \Theta^{(s)}_i \int \frac{d\mu}{(1-\mu^2)\,[\Theta^{(s)}_i]^2} \quad \text{.................(76).}$$

Definition of "Laplace's functions."

For a direct investigation of the complete solution in finite terms for the case $i-s$ a positive integer, see below § (n'), Example 2; and for the case i an integer, and s either not an integer or not $<i$, see § (o') (111).

The rational integral solution alone can enter, and it alone suffices, when the problem deals with the complete spherical surface. When there are boundaries, whether by two planes meeting in a diameter at an angle equal to a submultiple of four right angles, or by coaxal cones corresponding to certain particular values of θ, or by planes and cones, both the rational integral solution and the other are required. But when there are coaxal cones for boundaries, the values of i required by the boundary conditions [§ (l)] are not generally integral, and it is only when $i-s$ is integral that either solution is a rational and integral function of $\sin\theta$ and $\cos\theta$. Hence, in general, for the class of problems referred to, two solutions are required and neither is a rational integral function of $\sin\theta$ and $\cos\theta$.

(f') The ordinary process for the solution of linear differential equations in series of powers of the independent variable when the multipliers of the differential coefficients are rational algebraic functions of the independent variable leads easily from the equation (75) to any of the forms of rational integral solutions referred to above, as well as to the second solution in a form corresponding to each of them, when i and s are integers; and, quite generally, to the two particular solutions in every case, whether i and s be integral or fractional, real or imaginary. Thus, putting as above, § (k),

$$\cos\theta = \mu, \quad \sin\theta = \nu \quad \text{.....................(77),}$$

make μ the independent variable in the first place, in order to find expansions in powers of μ: thus (75) becomes

Differential equation with ν independent variable omitted here for brevity.

$$\frac{d}{d\mu}\left[(1-\mu^2)\frac{d\Theta^{(s)}_i}{d\mu}\right] + \left[\frac{-s^2}{1-\mu^2} + i\,(i+1)\right]\Theta^{(s)}_i \quad \text{.........(78).}$$

This is the form in which "Laplace's equation" has been most commonly presented. To avoid the appearance of supposing

Commonest form of "Laplace's equation"

i and s to be integers or even real, put

$$\Theta_i^{(s)} = w, \quad i(i+1) = a, \quad s^2 = b, \ldots\ldots\ldots\ldots\ldots (79).$$

Using this notation, and multiplying both members by $(1-\mu^2)$, we have, instead of (78),

generalized.

$$(1-\mu^2)\frac{d}{d\mu}\left[(1-\mu^2)\frac{dw}{d\mu}\right] + [a(1-\mu^2) - b]\, w = 0 \ldots\ldots\ldots (80).$$

To integrate this equation, assume

$$w = \Sigma K_n \mu^n,$$

Obvious solution in ascending powers of μ;

and in the series so found for its first member equate to zero the coefficient of μ^n. Thus we find

$$(n+1)(n+2)K_{n+2} = [2n^2 - a + b]K_n - [(n-1)(n-2) - a]K_{n-2} \ldots (81).$$

The first member of this vanishes for $n = -1$, and for $n = -2$, if K_1 and K_0 be finite. Hence, we may put $K_n = 0$ for all negative values of n, give arbitrary values to K_0 and K_1, and then find K_2, K_3, K_4, &c., by applications of (81) with $n = 0$, $n = 1$, $n = 2$,... successively. Thus if we first put $K_0 = 1$, and $K_1 = 0$; then again $K_0 = 0$, $K_1 = 1$; we find two series of the forms

$$1 + K_2\mu^2 + K_4\mu^4 + \&c.$$

and $$\mu + K_3\mu^3 + K_5\mu^5 + \&c.,$$

each of which satisfies (80); and therefore the complete solution is

$$w = C(1 + K_2\mu^2 + K_4\mu^4 + \&c.) + C'(\mu + K_3\mu^3 + K_5\mu^5 + \&c.) \ldots (82).$$

From the form of (81) we see that for very great values of n we have

$$K_{n+2} = 2K_n - K_{n-2} \text{ approximately,}$$

and therefore

$$K_{n+2} - K_n = K_n - K_{n-2} \text{ approximately.}$$

Hence each of the series in (82) converges for every value of μ less than unity.

why dismissed.

(g') But this is a very unsatisfactory form of solution. It gives in the form of an infinite series $1 + K_2\mu^2 + K_4\mu^4 + \&c.$ or $\mu + K_3\mu^3 + K_5\mu^5 + \&c.$, the finite solution which we know exists in the form

Geometrical antecedents suggest modified form of solution;

$$(1-\mu^2)^{\frac{s}{2}}(A_0 + A_2\mu^2 + \dots A_{i-s}\mu^{i-s})$$

or

$$(1-\mu^2)^{\frac{s}{2}}(A_1\mu + A_3\mu^3 + \dots A_{i-s}\mu^{i-s}),$$

when b is the square of an odd integer (s), and when $a = i(i+1)$, i being an odd integer or an even integer; and, a minor defect, but still a serious one, it does not show without elaborate verification that one or other of its constituents $1 + K_2\mu^2 + \text{\&c.}$ or $\mu + K_3\mu^3 + \text{\&c.}$ consists of a finite number, $\frac{1}{2}i$ or $\frac{1}{2}(i+1)$, of terms when b is the square of an even integer and $a = i(i+1)$, i being an even integer or an odd integer.

(h') A form of solution which turns out to be much simpler in every case is suggested by our primary knowledge [§ (j) above] of integral solutions. Put

$$w = (1-\mu^2)^{\frac{\sqrt{b}}{2}} v \qquad \text{.......................(83)},$$

in (80) and divide the first member by $(1-\mu^2)^{\frac{\sqrt{b}}{2}}$. Thus we find

correspondingly modified differential equation:

$$(1-\mu^2)\frac{d^2v}{d\mu^2} - 2(\sqrt{b}+1)\mu\frac{dv}{d\mu} + [a - \sqrt{b}(\sqrt{b}+1)]v = 0 \qquad \text{......(84)}.$$

Assume now

$$v = \Sigma A_n \mu^n \qquad \text{......................(85)};$$

equating to zero the coefficient of μ^n in the first member of (84) gives

its solution in powers of μ:

$$(n+1)(n+2)A_{n+2} - [(n-1)n + 2(\sqrt{b}+1)n - a + \sqrt{b}(\sqrt{b}+1)]A_n = 0 \qquad \text{...(86)},$$

or $$(n+1)(n+2)A_{n+2} = (n + \tfrac{1}{2} + s + \alpha)(n + \tfrac{1}{2} + s - \alpha)A_n \qquad \text{.....(87)},$$

if we put $$\alpha = \sqrt{(a + \tfrac{1}{4})}, \qquad s = \sqrt{b} \qquad \text{......................(88)},$$

and with this notation (84) becomes

$$(1-\mu^2)\frac{d^2v}{d\mu^2} - 2(s+1)\mu\frac{dv}{d\mu} + [\alpha^2 - (s+\tfrac{1}{2})^2]v = 0 \qquad \text{....... (84')}.$$

The second member of (87) shows that if the series (85) is in descending powers of μ its first term must have either

$$n = -\tfrac{1}{2} - s + \alpha, \quad \text{or} \quad n = -\tfrac{1}{2} - s - \alpha:$$

descending order dismissed, ascending chosen.

the expansion thus obtained would, if not finite, be convergent when $\mu > 1$ and divergent when $\mu < 1$, and they are therefore not suited for the physical applications. On the other hand, the first member of (87) shows that if the series (85) is in ascending powers of μ, its first term must have either $n = 0$ or

Complete solution for tesseral harmonics.

$n=1$: the expansions thus obtained are necessarily convergent when $\mu < 1$, and it is therefore these that are suited for our purposes. Taking then $A_0 = 1$ and $A_1 = 0$, and denoting by p the series so found, and again $A_0 = 0$ and $A_1 = 1$, and q the series; so that we have

$$\left.\begin{aligned} p &= 1 + A_2\mu^2 + A_4\mu^4 + \text{etc.} \\ \text{and} \quad q &= \mu + A_3\mu^3 + A_5\mu^5 + \text{etc.} \end{aligned}\right\} \quad (89),$$

A_2, A_4, etc. and A_3, A_5, etc. being found by two sets of successive applications of (87); then the complete solution of (84) is

$$v = Cp + C'q \quad (90).$$

This solution is identical with (38^{iv}) of § (l) above, as we see by (88) and (79), which give

$$a = i + \tfrac{1}{2} \quad (91).$$

Alternative solution by changing sign of s:

(i') The sign of either a or s may be changed, in virtue of (88). No variation however is made in the solution by changing the sign of a [which corresponds to changing i into $-i-1$, and verifies (13) (g) above]: but a very remarkable variation is made by changing the sign of s, from which, looking to (88), (83), (87), we infer that if $\mathfrak{p}$ and $\mathfrak{q}$ denote what p and q become when $-s$ is substituted for s in (89), we have

$$\left.\begin{aligned} \mathfrak{p} &= (1-\mu^2)^s p \\ \text{and} \quad \mathfrak{q} &= (1-\mu^2)^s q \end{aligned}\right\} \quad (92);$$

and the prescribed modification of (89) gives

$$\left.\begin{aligned} \mathfrak{p} &= 1 + \mathfrak{A}_2\mu^2 + \mathfrak{A}_4\mu^4 + \text{etc.} \\ \mathfrak{q} &= \mu + \mathfrak{A}_3\mu^3 + \mathfrak{A}_5\mu^5 + \text{etc.} \end{aligned}\right\} \quad (93),$$

$\mathfrak{A}_2$, $\mathfrak{A}_4$, etc., and $\mathfrak{A}_3$, $\mathfrak{A}_5$, etc. being found by successive applications of

$$\mathfrak{A}_{n+2} = \frac{(n+\frac{1}{2}-s+a)(n+\frac{1}{2}-s-a)}{(n+1)(n+2)}\mathfrak{A}_n \quad (94).$$

(j') In the case of "complete harmonics" s is zero or an integer, and the $\mathfrak{p}$ or $\mathfrak{q}$ solution expressing the result of multiplying the already finite and integral p or q solution by the integral polynomial $(1-\mu^2)^s$, is only interesting on account of the way of obtaining it from (87), etc. in virtue of (88). But when either $a - \frac{1}{2}$ or s is not an integer, the possession of the alternative solutions, p or $\mathfrak{p}$, q or $\mathfrak{q}$ may come to be of great intrinsic importance, in respect to obtaining results in finite form. For, supposing a and s to be both positive, it is impossible that both p and q can be finite polynomials, but one or both of $\mathfrak{p}$ and $\mathfrak{q}$ may be so; or

its importance.

one of the p, q forms and the other of the $\mathfrak{p}$, $\mathfrak{q}$ forms may be finite. This we see from (87) and (94), which show as follows:—

Cases of finite algebraic solution.

1. If $\frac{1}{2}+s-a$ is positive, p and q must each be an infinite series; but $\mathfrak{p}$ *or* $\mathfrak{q}$ will be finite if either $\frac{1}{2}+s-a$ or $\frac{1}{2}+s+a$ is a positive integer*; and *both* $\mathfrak{p}$ and $\mathfrak{q}$ will be finite if $\frac{1}{2}+s-a$ and $\frac{1}{2}+s+a$ are positive integers differing by unity or any odd number.

2. If $a \overline{\gtrless} s+\frac{1}{2}$, one of the two series $\mathfrak{p}$, $\mathfrak{q}$ must be infinite; and if $a-s-\frac{1}{2}$ is zero or a positive integer, one of the two series p, q is finite. If, lastly, $a+s-\frac{1}{2}$ is zero or a positive integer, one of the two $\mathfrak{p}$, $\mathfrak{q}$ is finite. It is p that is finite if $a-s-\frac{1}{2}$ is zero or even, q if it is odd: and $\mathfrak{p}$ that is finite if $a+s-\frac{1}{2}$ is zero or even, $\mathfrak{q}$ if it is odd. Hence it is p and $\mathfrak{p}$, or q and $\mathfrak{q}$ that are finite if $2s$ be zero or even; but it is p and $\mathfrak{q}$, or q and $\mathfrak{p}$ that are finite if $2s$ be odd. Hence in this latter case the complete solution is a finite algebraic function of μ.

(k') Remembering that by a and s we denote the positive values of the square roots indicated in (88), we collect from (j') 1 and 2, that, if $\mathbf{F}$ denote a rational integral function of μ and $(1-\mu^2)^{\pm\frac{1}{2}s}$, the character of the solution of (80) is as follows in the several cases indicated:—

I. $\begin{cases} \mathbf{A}; \; a < s+\frac{1}{2}; \; \text{if } s \text{ and } a-\frac{1}{2} \text{ are integers.} \\ \mathbf{B}; \; a \overline{\gtrless} s+\frac{1}{2}; \; \text{if } s+\frac{1}{2} \text{ and } a \text{ are integers.} \\ \qquad \text{The complete solution is } \mathbf{F}. \end{cases}$

II. $\begin{cases} \mathbf{A}; \; a < s+\frac{1}{2}; \; \text{if } s \pm (a-\frac{1}{2}) \text{ is an integer, but } a-\frac{1}{2} \text{ not an integer.} \\ \mathbf{B}; \; a \overline{\gtrless} s+\frac{1}{2}; \; \text{if } a-\frac{1}{2} \pm s \text{ is an integer, but } s+\frac{1}{2} \text{ not an integer.} \\ \text{A particular solution } \textit{is } \mathbf{F}; \text{ but the complete solution } \textit{is not } \mathbf{F}. \end{cases}$

(l') "Complete Spherical Harmonics," or "Laplace's Coefficients," are included in the particular solution $\mathbf{F}$ of Case II. **B.**

(m') Differentiate (84′) and put

$$\frac{dv}{d\mu} = u \text{ .... } (95).$$

* Unity being understood as included in the class of "positive integers."

Three modes of derivation with change of type.

We find immediately

$$(1-\mu^2)\frac{d^2u}{d\mu^2}-2(s+2)\mu\frac{du}{d\mu}+[a^2-(s+\tfrac{3}{2})^2]u=0 \quad\text{.........(96).}$$

Let
$$u'=\mu\frac{dv}{d\mu}+(\pm a+s+\tfrac{1}{2})v \quad\text{....................(97).}$$

We have, as will be proved presently,

$$(1-\mu^2)\frac{d^2u'}{d\mu^2}-2(s+2)\mu\frac{du'}{d\mu}+[(a\pm 1)^2-(s+\tfrac{3}{2})^2]u'=0 \quad\text{...(98).}$$

Lastly, let
$$u''=(1-\mu^2)\frac{dv}{d\mu}-(\pm a+s+\tfrac{1}{2})\mu v \quad\text{.............. (99).}$$

We have, as will be proved presently,

$$(1-\mu^2)\frac{d^2u''}{d\mu^2}-2(s+1)\mu\frac{du''}{d\mu}+[(a\pm 1)^2-(s+\tfrac{1}{2})^2]u''=0 \quad\text{...(100).}$$

The operation $\frac{d}{dz}$ performed on a solid harmonic of degree $-a-\frac{1}{2}$, and type $H\{z, \surd(x^2+y^2)\}\,{}^{\sin}_{\cos}\,s\phi$, and transformed to polar co-ordinates r, μ, ϕ, with attention to (83), gives the transition from v to u'', as expressed in (99), and thus (100) is proved by (g) (15).

Similarly the operation

$$\left(\frac{d}{dx}+v\frac{d}{dy}\right)H[z, \surd(x^2+y^2)](x+vy)^s+\left(\frac{d}{dx}-v\frac{d}{dy}\right)H[z, \surd(x^2+y^2)](x-vy)^s,$$

transformed to co-ordinates r, μ, ϕ, gives (97), and thus (98) is proved by (g) (15).

Thus it was that (97) (98), and (99) (100) were found. But, assuming (97) and (99) arbitrarily as it were, we prove (98) and (100) most easily as follows. Let

$$u'=\Sigma B'_n\mu^n, \quad\text{and}\quad u''=\Sigma B''_n\mu^n \quad\text{................(101).}$$

Then, by (97) and (99), with (85), we find

$$\left.\begin{aligned} B'_{n+2}&=(n+2\pm a+s+\tfrac{1}{2})A_{n+2}\\ \text{and}\quad B''_{n+2}&=(\mp a+s-\tfrac{1}{2})\frac{n+1+\frac{1}{2}+s\pm a}{n+2}A_{n+1}\end{aligned}\right\}\quad\text{......... (102).}$$

Lastly, applying (87), we find that the corresponding equation is satisfied by $B'_{n+2}\div B'_n$, with $a\pm 1$ and $s+1$ instead of a and s; and by $B''_{n+2}\div B''_n$, with $a\pm 1$ instead of a, but with s unchanged.

As to (95) and (96), they merely express for the generalized surface harmonics the transition from s to $s+1$ without change of i shown for complete harmonics by Murphy's formula, § 782 (6) below.

Examples of derivation.

(n') *Examples of* (95) (96), *and* (99) (100).

Example 1. Let $a = s + \frac{1}{2}$.

$$\left.\begin{array}{l} (84') \text{ becomes } \quad (1-\mu^2)\dfrac{d^2v}{d\mu^2} - 2(s+1)\mu\dfrac{dv}{d\mu} = 0, \\ \text{of which the complete solution is} \\ \qquad v = C\displaystyle\int\frac{d\mu}{(1-\mu^2)^{s+1}} + C', \end{array}\right\} \ldots\ldots\ldots\ldots (103).$$

Tesserals from sectorial by increase of s, with order i unchanged.

By (95) (96) we find

$$\left.\begin{array}{l} u = C\dfrac{d^{n-1}[(1-\mu^2)^{-s-1}]}{d\mu^{n-1}} \\ \text{as a solution of} \\ (1-\mu^2)\dfrac{d^2u}{d\mu^2} - 2(n+s+1)\mu\dfrac{du}{d\mu} - n(n+2s+1)v = 0 \end{array}\right\} \ldots\ldots\ldots (104).$$

This is the particular finite solution indicated in § (k') II. **A.**

The liberty we now have to let a be negative as well as positive allows us now to include in our formula for u the cases represented by the double sign $\pm$ in II. **A** of (k').

Example 2. By m successive applications of (99) (100), with the upper sign, to v of (103), we find for the complete integral of

$$\left.\begin{array}{l} (1-\mu^2)\dfrac{d^2u'}{d\mu^2} - 2(s+1)\mu\dfrac{du'}{d\mu} + m(m+2s+1)u' = 0 \\ u' = C\left\{f(\mu)\displaystyle\int\frac{d\mu}{(1-\mu^2)^{s+1}} + F(\mu)\right\} + C'\mathbf{F}(\mu) \end{array}\right\} \ldots (105),$$

Tesserals from sectorial by increase of i, with s unchanged.

where $f(\mu)$, $F(\mu)$, $\mathbf{F}(\mu)$ denote rational integral algebraic functions of μ.

Of this solution the part $C'\mathbf{F}(\mu)$ is the particular finite solution indicated in § (k') II. **B**. We now see that the complete solution involves no other transcendent than $\displaystyle\int\frac{d\mu}{(1-\mu^2)^{s+1}}$. When s is an integer, this is reducible to the form

$$a \log\frac{1+\mu}{1-\mu} + \mathbf{f}(\mu),$$

Examples of derivation continued.

a being a constant and $\mathbf{f}(\mu)$ a rational integral algebraic function of μ. In this case, remembering that (105) is what (84') becomes when $m+s+\frac{1}{2}$ is put for a, we may recur to our notation of §§ (g) (j), by putting i for $m+s$, which is now an integer: and going back, by (83) to (80) or (78), put

$$w=(1-\mu^2)^{\frac{s}{2}}u' \qquad\qquad (83');$$

thus (105) is equivalent to

$$\frac{d}{d\mu}\left[(1-\mu^2)\frac{dw}{d\mu}\right]+\left[\frac{-s^2}{1-\mu^2}+i(i+1)\right]w=0 \qquad\qquad (78').$$

The process of Example 2, § (n'), gives the complete integral of this equation when $i-s$ is a positive integer. When also s, and therefore also i, is an integer, the transcendent involved becomes $\log\frac{1+\mu}{1-\mu}$: in this case the algebraic part of the solution [or $C'\mathbf{F}(\mu)(1-\mu^2)^{\frac{s}{2}}$ according to the notation of (105) and (78')] is the ordinary "Laplace's Function" of order and type (i, s); the $\Theta_i^{(s)}$, $\vartheta_i^{(s)}$, &c. of our previous notations of §§ (j), (y). It is interesting to know that the other particular solution which we now have, completing the solution of the differential equation for these functions, involves nothing of transcendent but $\log\frac{1+\mu}{1-\mu}$.

(o') *Examples of* (99) (100), *and* (95) (96) *continued.*

Algebraic case of last example.

Example 3. Returning to (n'), Example 2, let $s+\frac{1}{2}$ be an integer: the integral $\int\frac{d\mu}{(1-\mu^2)^{s+1}}$ is algebraic. Thus we have the case of (k') I. **B**, in which the complete solution is algebraic.

(p') Returning to (n'), Example 1: let $a=\frac{1}{2}$ and $s=0$; (103) becomes

Zonal of order zero:

$$\left.\begin{aligned}&(1-\mu^2)\frac{d^2v}{d\mu^2}-2\mu\frac{dv}{d\mu}=0,\\ &\text{of which the complete integral is}\\ &v=\tfrac{1}{2}C\log\frac{1+\mu}{1-\mu}+C'\end{aligned}\right\} \qquad (103').$$

growing into one sectorial by augmentation of s, with order still zero:

As before, apply (95) (96) n times successively: we find

$$u_1=\tfrac{1}{2}.1.2\ldots(n-1)C\left[\left(\frac{1}{1-\mu}\right)^n-\left(\frac{-1}{1+\mu}\right)^n\right] \qquad (106)$$

as one solution of

$$(1-\mu^2)\frac{d^2u}{d\mu^2}-2(n+1)\mu\frac{du}{d\mu}-n(n+1)u=0\text{ ....}(96').$$

the other derived from this by lowering order to -1, equivalent to zero.

To find the other: treat (106) by (99) (100) with the lower sign; the effect is to diminish a from $\frac{1}{2}$ to $-\frac{1}{2}$, and therefore to make no change in the differential equation, but to derive from (106) another particular solution, which is as follows:

$$u_2=\tfrac{1}{2}.1.2\ldots(n-1).n.C\left[\left(\frac{1}{1-\mu}\right)^n+\left(\frac{-1}{1+\mu}\right)^n\right]\text{ ......}(106').$$

Complete solution for tesserals of order zero:

Giving any different values to C in (106) and (106′), and, using K, K' to denote two arbitrary constants, adding we have the complete solution of (96′), which we may write as follows:

$$u=\frac{K}{(1-\mu)^n}+\frac{K'}{(1+\mu)^n}\text{ .... }(107).$$

(q') That (107) is the solution of (96′) we verify in a moment by trial, and in so doing we see farther that it is the complete solution, whether n be integral or not.

derivation from it of both tesserals of every integral order;

(r') Example 4. Apply (99) (100) with upper sign i times to (107) and successive results. We get thus the complete solution of (84′) for $a-\frac{1}{2}=i$ *any* integer, if n is *not* an integer. But if n is an integer we get the complete solution only provided $i<n$: this is case I. **A** of § (k'). If we take $i=n-1$, the result, algebraic as it is, may be proved to be expressible in the form

$$u=\frac{C+C'\int d\mu(1-\mu^2)^{n-1}}{(1-\mu^2)^n},$$

which is therefore for n an integer the complete integral of

$$(1-\mu^2)\frac{d^2u}{d\mu^2}-2(n+1)\mu\frac{du}{d\mu}-2nu=0,$$

$\Bigg\}$ .... (108):

except case of s an integer and $i \gtreqless s$, when only Laplace's function is found.

being the case of (84′) for which $a=s-\frac{1}{2}$, and $s=n$ an integer: applying to this (99) (100) with upper sign, the constant C disappears, and we find $u'=C'$ as *a* solution of

$$(1-\mu^2)\frac{d^2u'}{d\mu^2}-2(n+1)\mu\frac{du'}{d\mu}=0\text{ .... }(109).$$

Hence, for $i\gtreqless n$ one solution is lost. The other, found by

Examples of derivation continued.

continued applications of (99) (100) with upper sign, is the regular "Laplace's function" growing from $C' \sin^n \theta \begin{smallmatrix}\sin\\ \cos\end{smallmatrix} n\phi$, which is the case represented by $u'=C'$ in (109). But in this continuation we are only doing for the case of n an integer, part of what was done in § (n'), Example 2, where the other part, from the other part of the solution of (109) now lost, gives the other part of the complete solution of Laplace's equation subject to the limitation $i-n$ (or $i-s$) a positive integer, but not to the limitation of i an integer or n an integer.

(s') Returning to the commencement of § (r'), with s put for n, we find a complete solution growing in the form

$$\frac{Kf_i(\mu)}{(1-\mu)^s} + (-)^i \frac{K'f_i(-\mu)}{(1+\mu)^s} \quad\ldots\ldots\ldots\ldots\ldots\ldots\ldots (110);$$

which may be immediately reduced to

$$\frac{Kf_i(\mu)(1+\mu)^s + (-)^i K'f_i(-\mu)(1-\mu)^s}{(1-\mu^2)^s} \quad\ldots\ldots\ldots\ldots (110');$$

f_i denoting an integral algebraic function of the ith degree, readily found by the proper successive applications of (99) (100). Hence, by (83) (79), we have

$$w = \frac{Kf_i(\mu)(1+\mu)^s + (-)^i K'f_i(-\mu)(1-\mu)^s}{(1-\mu^2)^{\frac{s}{2}}} \quad\ldots\ldots\ldots (111),$$

as the complete solution of Laplace's equation

$$\frac{d}{d\mu}\left[(1-\mu^2)\frac{dw}{d\mu}\right] + \left[\frac{-s^2}{1-\mu^2} + i(i+1)\right] w = 0 \quad\ldots\ldots\ldots (112),$$

Finite algebraic expression of complete solution for tesseral harmonics of *integral order*.

for the case of i an integer without any restriction as to the value of s, which may be integral or fractional, real or imaginary, with no failure except the case of s an integer and $i > s$, of which the complete treatment is included in § (m'), Example 2, above.

CHAPTER II.

DYNAMICAL LAWS AND PRINCIPLES.

205. In the preceding chapter we considered as a subject of pure geometry the motion of points, lines, surfaces, and volumes, whether taking place with or without change of dimensions and form; and the results we there arrived at are of course altogether independent of the idea of *matter*, and of the *forces* which matter exerts. We have heretofore assumed the *existence* merely of motion, distortion, etc.; we now come to the consideration, not of how we *might* consider such motions, etc., to be produced, but of the *actual* causes which in the material world *do* produce them. The axioms of the present chapter must therefore be considered to be due to actual experience, in the shape either of observation or experiment. How this experience is to be obtained will form the subject of a subsequent chapter.

Ideas of matter and force introduced.

206. We cannot do better, at all events in commencing, than follow Newton somewhat closely. Indeed the introduction to the *Principia* contains in a most lucid form the general foundations of Dynamics. The *Definitiones* and *Axiomata sive Leges Motûs*, there laid down, require only a few amplifications and additional illustrations, suggested by subsequent developments, to suit them to the present state of science, and to make a much better introduction to dynamics than we find in even some of the best modern treatises.

207. We cannot, of course, give a definition of *Matter* which will satisfy the metaphysician, but the naturalist may be content to know matter as *that which can be perceived by the senses*, or as *that which can be acted upon by, or can exert, force.* The

Matter.

latter, and indeed the former also, of these definitions involves the idea of *Force*, which, in point of fact, is a direct object of sense; probably of all our senses, and certainly of the "muscular sense." To our chapter on Properties of Matter we must refer for further discussion of the question, *What is matter?* And we shall then be in a position to discuss the question of the subjectivity of *Force*.

Force.

208. *The Quantity of Matter* in a body, or, as we now call it, the *Mass* of a body, is proportional, according to Newton, to the *Volume* and the *Density* conjointly. In reality, the definition gives us the meaning of density rather than of mass; for it shows us that if twice the original quantity of matter, air for example, be forced into a vessel of given capacity, the density will be doubled, and so on. But it also shows us that, of matter of uniform density, the mass or quantity is proportional to the volume or space it occupies.

Mass. Density.

Let M be the mass, ρ the density, and V the volume, of a homogeneous body. Then

$$M = V\rho;$$

if we so take our units that unit of mass is that of unit volume of a body of unit density.

If the density vary from point to point of the body, we have evidently, by the above formula and the elementary notation of the integral calculus,

$$M = \iiint \rho \, dxdydz,$$

where ρ is supposed to be a known function of x, y, z, and the integration extends to the whole space occupied by the matter of the body whether this be continuous or not.

It is worthy of particular notice that, in this definition, Newton says, if there be anything which *freely* pervades the interstices of all bodies, this is *not* taken account of in estimating their Mass or Density.

209. Newton further states, that a practical measure of the mass of a body is its *Weight*. His experiments on pendulums, by which he establishes this most important result, will be described later, in our chapter on Properties of Matter.

Measurement of mass.

As will be presently explained, the unit mass most convenient for British measurements is an imperial pound of matter.

210. The *Quantity of Motion,* or the *Momentum,* of a rigid body moving without rotation is proportional to its mass and velocity conjointly. The whole motion is the sum of the motions of its several parts. Thus a doubled mass, or a doubled velocity, would correspond to a double quantity of motion; and so on.

Momentum.

Hence, if we take as unit of momentum the momentum of a unit of matter moving with unit velocity, the momentum of a mass M moving with velocity v is Mv.

211. *Change of Quantity of Motion,* or *Change of Momentum,* is proportional to the mass moving and the change of its velocity conjointly.

Change of momentum.

Change of velocity is to be understood in the general sense of § 27. Thus, in the figure of that section, if a velocity represented by OA be changed to another represented by OC, the change of velocity is represented in magnitude and direction by AC.

212. *Rate of Change of Momentum* is proportional to the mass moving and the acceleration of its velocity conjointly. Thus (§ 35, *b*) the rate of change of momentum of a falling body is constant, and in the vertical direction. Again (§ 35, *a*) the rate of change of momentum of a mass M, describing a circle of radius R, with uniform velocity V, is $\frac{MV^2}{R}$, and is directed to the centre of the circle; that is to say, it is a change of direction, not a change of speed, of the motion. Hence if the mass be compelled to keep in the circle by a cord attached to it and held fixed at the centre of the circle, the force with which the cord is stretched is equal to $\frac{MV^2}{R}$: this is called the centrifugal force of the mass M moving with velocity V in a circle of radius R.

Rate of change of momentum.

Generally (§ 29), for a body of mass M moving anyhow in space there is change of momentum, at the rate, $M\frac{d^2s}{dt^2}$ in the direc-

Rate of change of momentum.

tion of motion, and $M\frac{v^2}{\rho}$ towards the centre of curvature of the path; and, if we choose, we may exhibit the whole acceleration of momentum by its three rectangular components $M\frac{d^2x}{dt^2}$, $M\frac{d^2y}{dt^2}$, $M\frac{d^2z}{dt^2}$, or, according to the Newtonian notation, $M\ddot{x}$, $M\ddot{y}$, $M\ddot{z}$.

Kinetic energy.

213. The *Vis Viva,* or *Kinetic Energy,* of a moving body is proportional to the mass and the square of the velocity, conjointly. If we adopt the same units of mass and velocity as before, there is particular advantage in defining kinetic energy as *half* the product of the mass and the square of its velocity.

214. *Rate of Change of Kinetic Energy* (when defined as above) is the product of the velocity into the component of rate of change of momentum in the direction of motion.

For $$\frac{d}{dt}\left(\frac{Mv^2}{2}\right) = v\,\frac{d(Mv)}{dt}.$$

Particle and point.

215. It is to be observed that, in what precedes, with the exception of the definition of mass, we have taken no account of the dimensions of the moving body. This is of no consequence so long as it does not rotate, and so long as its parts preserve the same relative positions amongst one another. In this case we may suppose the whole of the matter in it to be condensed in one point or particle. We thus speak of a *material particle,* as distinguished from a *geometrical point.* If the body rotate, or if its parts change their relative positions, then we cannot choose any one point by whose motions alone we may determine those of the other points. In such cases the momentum and change of momentum of the whole body in any direction are, the sums of the momenta, and of the changes of momentum, of its parts, in these directions; while the kinetic energy of the whole, being non-directional, is simply the sum of the kinetic energies of the several parts or particles.

Inertia.

216. Matter has an innate power of resisting external influences, so that every body, as far as it can, remains at rest, or moves uniformly in a straight line.

This, the *Inertia* of matter, is proportional to the quantity of matter in the body. And it follows that some *cause* is requisite to disturb a body's uniformity of motion, or to change its direction from the natural rectilinear path. **Inertia.**

217. Force is any cause which tends to alter a body's natural state of rest, or of uniform motion in a straight line. **Force.**

Force is wholly expended in the *Action* it produces; and the body, after the force ceases to act, retains by its inertia the direction of motion and the velocity which were given to it. Force may be of divers kinds, as pressure, or gravity, or friction, or any of the attractive or repulsive actions of electricity, magnetism, etc.

218. The three elements specifying a force, or the three elements which must be known, before a clear notion of the force under consideration can be formed, are, its place of application, its direction, and its magnitude. **Specification of a force.**

(*a*) The place of application of a force. The first case to be considered is that in which the place of application is a point. It has been shown already in what sense the term "point" is to be taken, and, therefore, in what way a force may be imagined as acting at a point. In reality, however, the place of application of a force is always either a surface or a space of three dimensions occupied by matter. The point of the finest needle, or the edge of the sharpest knife, is still a surface, and acts by pressing over a finite area on bodies to which it may be applied. Even the most rigid substances, when brought together, do not touch at a point merely, but mould each other so as to produce a surface of application. On the other hand, gravity is a force of which the place of application is the whole matter of the body whose weight is considered; and the smallest particle of matter that has weight occupies some finite portion of space. Thus it is to be remarked, that there are two kinds of force, distinguishable by their place of application—force, whose place of application is a surface, and force, whose place of application is a solid. When a heavy body rests on the ground, or on a table, force of the second character, acting downwards, is balanced by force of the first character acting upwards. **Place of application.**

Direction. (*b*) The second element in the specification of a force is its direction. The direction of a force is the line in which it acts. If the place of application of a force be regarded as a point, a line through that point, in the direction in which the force tends to move the body, is the direction of the force. In the case of a force distributed over a surface, it is frequently possible and convenient to assume a single point and a single line, such that a certain force acting at that point in that line would produce sensibly the same effect as is really produced.

Magnitude. (*c*) The third element in the specification of a force is its magnitude. This involves a consideration of the method followed in dynamics for measuring forces. Before measuring anything, it is necessary to have a unit of measurement, or a standard to which to refer, and a principle of numerical specification, or a mode of referring to the standard. These will be supplied presently. See also § 258, below.

Accelerative effect. **219.** The *Accelerative Effect of a Force* is proportional to the velocity which it produces in a given time, and is measured by that which is, or would be, produced in unit of time; in other words, the *rate of change of velocity* which it produces. This is simply what we have already defined as acceleration, § 28.

Measure of force. **220.** The *Measure of a Force* is the quantity of motion which it produces per unit of time.

The reader, who has been accustomed to speak of a force of so many pounds, or so many tons, may be startled when he finds that such expressions are not definite unless it be specified at what part of the earth's surface the pound, or other definite quantity of matter named, is to be weighed; for the *heaviness* or *gravity* of a given quantity of matter differs in different latitudes. But the force required to produce a stated quantity of motion in a given time is perfectly definite, and independent of locality. Thus, let W be the mass of a body, g the velocity it would acquire in falling freely for a second, and P the force of gravity upon it, measured in kinetic or absolute units. We have

$$P = Wg.$$

221. According to the system commonly followed in mathematical treatises on dynamics till fourteen years ago, when a small instalment of the first edition of the present work was issued for the use of our students, the unit of mass was g times the mass of the standard or unit weight. This definition, giving a varying and a very unnatural unit of mass, was exceedingly inconvenient. By taking the gravity of a constant mass for the unit of force it makes the unit of force greater in high than in low latitudes. In reality, standards of weight are *masses*, not *forces*. They are employed primarily in commerce for the purpose of measuring out a definite *quantity* of matter; not an amount of matter which shall be attracted by the earth with a given force.

Inconvenient system of modern treatises.

Standards of weight are *masses*, and not primarily intended for measurement of force.

A merchant, with a balance and a set of standard weights, would give his customers the same quantity of the same kind of matter however the earth's attraction might vary, depending as he does upon *weights* for his measurement; another, using a spring-balance, would defraud his customers in high latitudes, and himself in low, if his instrument (which depends on constant forces and not on the gravity of constant masses) were correctly adjusted in London.

It is a secondary application of our standards of weight to employ them for the measurement of *forces*, such as steam pressures, muscular power, etc. In all cases where great accuracy is required, the results obtained by such a method have to be reduced to what they would have been if the measurements of force had been made by means of a perfect spring-balance, graduated so as to indicate the forces of gravity on the standard weights in some conventional locality.

It is therefore very much simpler and better to take the imperial pound, or other national or international standard weight, as, for instance, the gramme (see the chapter on Measures and Instruments), as the unit of mass, and to derive from it, according to Newton's definition above, the unit of force. This is the method which Gauss has adopted in his great improvement (§ 223 below) of the system of measurement of forces.

Clairault's formula for the amount of gravity.

222. The formula, deduced by Clairault from observation, and a certain theory regarding the figure and density of the earth, may be employed to calculate the most probable value of the apparent force of gravity, being the resultant of true gravitation and centrifugal force, in any locality where no pendulum observation of sufficient accuracy has been made. This formula, with the two coefficients which it involves, corrected according to the best modern pendulum observations (Airy, *Encyc. Metropolitana, Figure of the Earth*), is as follows:—

Let G be the apparent force of gravity on a unit mass at the equator, and g that in any latitude λ; then

$$g = G\,(1 + \cdot 005133 \sin^2\lambda).$$

The value of G, in terms of the British absolute unit, to be explained immediately, is

$$32\cdot 088.$$

According to this formula, therefore, polar gravity will be

$$g = 32\cdot 088 \times 1\cdot 005133 = 32\cdot 2527.$$

223. Gravity having failed to furnish a definite standard, independent of locality, recourse must be had to something else. The principle of measurement indicated as above by Newton, but first introduced practically by Gauss, furnishes us with what we want. According to this principle, the unit force is that force which, acting on a national standard unit of matter during the unit of time, generates the unity of velocity.

Gauss's absolute Unit of Force.

This is known as Gauss's absolute unit; absolute, because it furnishes a standard force independent of the differing amounts of gravity at different localities. It is however terrestrial and inconstant if the unit of time depends on the earth's rotation, as it does in our present system of chronometry. The period of vibration of a piece of quartz crystal of specified shape and size and at a stated temperature (a tuning-fork, or bar, as one of the bars of glass used in the "musical glasses") gives us a unit of time which is constant through all space and all time, and independent of the earth. A unit of force founded on such a unit of time would be better entitled to the designation *abso-*

lute than is the "absolute unit" now generally adopted, which is founded on the *mean solar second*. But this depends essentially on one particular piece of matter, and is therefore liable to all the accidents, etc. which affect so-called National Standards however carefully they may be preserved, as well as to the almost insuperable practical difficulties which are experienced when we attempt to make exact copies of them. Still, in the present state of science, we are really confined to such approximations. The recent discoveries due to the Kinetic theory of gases and to Spectrum analysis (especially when it is applied to the light of the heavenly bodies) indicate to us *natural standard* pieces of matter such as atoms of hydrogen, or sodium, ready made in infinite numbers, all absolutely alike in every physical property. The time of vibration of a sodium particle corresponding to any one of its modes of vibration, is known to be absolutely independent of its position in the universe, and it will probably remain the same so long as the particle itself exists. The wave-length for that particular ray, *i.e.* the space through which light is propagated *in vacuo* during the time of one complete vibration of this period, gives a perfectly invariable unit of length; and it is possible that at some not very distant day the mass of such a sodium particle may be employed as a natural standard for the remaining fundamental unit. This, the latest improvement made upon our original suggestion of a *Perennial Spring* (First edition, § 406), is due to Clerk Maxwell*; who has also communicated to us another very important and interesting suggestion for founding the unit of time upon physical properties of a substance without the necessity of specifying any particular quantity of it. It is this, water being chosen as the substance of all others known to us which is most easily obtained in perfect purity and in perfectly definite physical condition.—Call the standard density of water the maximum density of the liquid when under the pressure of its own vapour alone. The time of revolution of an infinitesimal satellite close to the surface of a globe of water at standard density (or of any kind of matter at the same density) may be taken as the unit of time; for it is independent of the size of the globe. This has

Maxwell's two suggestions for Absolute Unit of Time.

* *Electricity and Magnetism*, 1872.

Third suggestion for Absolute Unit of Time.

suggested to us still another unit, founded, however, still upon the same physical principle. The time of the gravest simple harmonic infinitesimal vibration of a globe of liquid, water at standard density, or of other perfect liquids at the same density, may be taken as the unit of time; for the time of the simple harmonic vibration of any one of the fundamental modes of a liquid sphere is independent of the size of the sphere.

Let f be the force of gravitational attraction between two units of matter at unit distance. The force of gravity at the surface of a globe of radius r, and density ρ, is $\frac{4\pi}{3}f\rho r$. Hence if ω be the angular velocity of an infinitesimal satellite, we have, by the equilibrium of centrifugal force and gravity (§§ 212, 477),

$$\omega^2 r = \frac{4\pi}{3} f\rho r.$$

Hence
$$\omega = \sqrt{\frac{4\pi f\rho}{3}},$$

and therefore if T be the satellite's period,

$$T = 2\pi\sqrt{\frac{3}{4\pi f\rho}}$$

(which is equal to the period of a simple pendulum whose length is the globe's radius, and weighted end infinitely near the surface of the globe). And it has been proved* that if a globe of liquid be distorted infinitesimally according to a spherical harmonic of order i, and left at rest, it will perform simple harmonic oscillations in a period equal to

$$2\pi\sqrt{\left\{\frac{3}{4\pi f\rho}\cdot\frac{2i+1}{2i(i-1)}\right\}}.$$

Hence if T' denote the period of the gravest, that, namely, for which $i=2$, we have

$$T' = T\sqrt{\frac{5}{4}}.$$

The semi-period of an infinitesimal satellite round the earth is equal, reckoned in seconds, to the square root of the number of metres in the earth's radius, the metre being very approximately

* "Dynamical Problems regarding Elastic Spheroidal Shells and Spheroids of Incompressible Liquid" (W. Thomson), *Phil. Trans.* Nov. 27, 1862.

the length of the seconds pendulum, whose period is two seconds. Hence taking the earth's radius as 6,370,000 metres, and its density as $5\frac{1}{2}$ times that of our standard globe,

Suggestions for Absolute Unit of Time.

$$T = 3 \text{ h. } 17 \text{ m.}$$

$$T' = 3 \text{ h. } 40 \text{ m.}$$

224. The absolute unit depends on the unit of matter, the unit of time, and the unit of velocity; and as the unit of velocity depends on the unit of space and the unit of time, there is, in the definition, a single reference to mass and space, but a *double* reference to time; and this is a point that must be particularly attended to.

225. The unit of mass may be the British imperial pound; the unit of space the British standard foot; and, accurately enough for practical purposes for a few thousand years, the unit of time may be the mean solar second.

British absolute unit.

We accordingly define the British absolute unit force as "the force which, acting on one pound of matter for one second, generates a velocity of one foot per second." Prof. James Thomson has suggested the name "Poundal" for this unit of force.

Comparison with gravity.

226. To illustrate the reckoning of force in "absolute measure," find how many absolute units will produce, in any particular locality, the same effect as the force of gravity on a given mass. To do this, measure the effect of gravity in producing acceleration on a body unresisted in any way. The most accurate method is indirect, by means of the pendulum. The result of pendulum experiments made at Leith Fort, by Captain Kater, is, that the velocity which would be acquired by a body falling unresisted for one second is at that place 32·207 feet per second. The preceding formula gives exactly 32·2, for the latitude 55° 33′, which is approximately that of Edinburgh. The variation in the force of gravity for one degree of difference of latitude about the latitude of Edinburgh is only ·0000832 of its own amount. It is nearly the same, though somewhat more, for every degree of latitude southwards, as far as the southern limits of the British Isles. On the other hand, the variation per degree is sensibly less, as far north as the Orkney and Shetland Isles. Hence

Gravity of Unit weight or mass in terms of Kinetic Unit.

the augmentation of gravity per degree from south to north throughout the British Isles is at most about $\frac{1}{12000}$ of its whole amount in any locality. The average for the whole of Great Britain and Ireland differs certainly but little from 32·2. Our present application is, that the force of gravity at Edinburgh is 32·2 times the force which, acting on a pound for a second, would generate a velocity of one foot per second; in other words, 32·2 is the number of absolute units which measures the weight of a pound in this latitude. Thus, approximately, the poundal is equal to the gravity of about half an ounce.

227. Forces (since they involve only direction and magnitude) may be represented, as velocities are, by straight lines in their directions, and of lengths proportional to their magnitudes, respectively.

Also the laws of composition and resolution of any number of forces acting at the same point, are, as we shall show later (§ 255), the same as those which we have already proved to hold for velocities; so that with the substitution of force for velocity, §§ 26, 27, are still true.

Effective component of a force

228. In rectangular resolution the *Component* of a force in any direction, (sometimes called the *Effective Component* in that direction,) is therefore found by multiplying the magnitude of the force by the cosine of the angle between the directions of the force and the component. The remaining component in this case is perpendicular to the other.

It is very generally convenient to resolve forces into components parallel to three lines at right angles to each other; each such resolution being effected by multiplying by the cosine of the angle concerned.

Geometrical Theorem preliminary to definition of centre of inertia.

229. The point whose distances from three planes at right angles to one another are respectively equal to the mean distances of any group of points from these planes, is at a distance from any plane whatever, equal to the mean distance of the group from the same plane. Hence of course, if it is in motion, its velocity perpendicular to that plane is the mean of the velocities of the several points, in the same direction.

Geometrical Theorem preliminary to definition of centre of inertia.

Let (x_1, y_1, z_1), etc., be the points of the group in number i; and $\bar{x}$, $\bar{y}$, $\bar{z}$ be the co-ordinates of a point at distances respectively equal to their mean distances from the planes of reference; that is to say, let

$$\bar{x} = \frac{x_1 + x_2 + \text{etc.}}{i}, \quad \bar{y} = \frac{y_1 + y_2 + \text{etc.}}{i}, \quad \bar{z} = \frac{z_1 + z_2 + \text{etc.}}{i}.$$

Thus, if p_1, p_2, etc., and p, denote the distances of the points in question from any plane at a distance a from the origin of co-ordinates, perpendicular to the direction (l, m, n), the sum of a and p_1 will make up the projection of the broken line x_1, y_1, z_1 on (l, m, n), and therefore

$$p_1 = lx_1 + my_1 + nz_1 - a, \text{ etc.};$$

and similarly, $\qquad p = l\bar{x} + m\bar{y} + n\bar{z} - a.$

Substituting in this last the expressions for $\bar{x}$, $\bar{y}$, $\bar{z}$, we find

$$p = \frac{p_1 + p_2 + \text{etc.}}{i},$$

which is the theorem to be proved. Hence, of course,

$$\frac{dp}{dt} = \frac{1}{i}\left(\frac{dp_1}{dt} + \frac{dp_2}{dt} + \text{etc.}\right).$$

Centre of inertia.

230. The *Centre of Inertia* of a system of equal material points (whether connected with one another or not) is the point whose distance is equal to their average distance from any plane whatever (§ 229).

A group of material points of unequal masses may always be imagined as composed of a greater number of equal material points, because we may imagine the given material points divided into different numbers of very small parts. In any case in which the magnitudes of the given masses are incommensurable, we may approach as near as we please to a rigorous fulfilment of the preceding statement, by making the parts into which we divide them sufficiently small.

On this understanding the preceding definition may be applied to define the centre of inertia of a system of material points, whether given equal or not. The result is equivalent to this:—

Centre of Inertia.

The centre of inertia of any system of material points whatever (whether rigidly connected with one another, or connected in any way, or quite detached), is a point whose distance from any plane is equal to the sum of the products of each mass into its distance from the same plane divided by the sum of the masses.

We also see, from the proposition stated above, that a point whose distance from three rectangular planes fulfils this condition, must fulfil this condition also for every other plane.

The co-ordinates of the centre of inertia, of masses w_1, w_2, etc., at points (x_1, y_1, z_1), (x_2, y_2, z_2), etc., are given by the following formulæ :—

$$\bar{x} = \frac{w_1 x_1 + w_2 x_2 + \text{etc.}}{w_1 + w_2 + \text{etc.}} = \frac{\Sigma wx}{\Sigma w}, \quad \bar{y} = \frac{\Sigma wy}{\Sigma w}, \quad \bar{z} = \frac{\Sigma wz}{\Sigma w}.$$

These formulæ are perfectly general, and can easily be put into the particular shape required for any given case. Thus, suppose that, instead of a set of detached material points, we have a continuous distribution of matter through certain definite portions of space; the density at x, y, z being ρ, the elementary principles of the integral calculus give us at once

$$\bar{x} = \frac{\iiint \rho x dx dy dz}{\iiint \rho dx dy dz}, \text{ etc.,}$$

where the integrals extend through all the space occupied by the mass in question, in which ρ has a value different from zero.

The Centre of Inertia or Mass is thus a perfectly definite point in every body, or group of bodies. The term *Centre of Gravity* is often very inconveniently used for it. The theory of the resultant action of gravity which will be given under Abstract Dynamics shows that, except in a definite class of distributions of matter, there is no one fixed point which can properly be called the Centre of Gravity of a rigid body. In ordinary cases of terrestrial gravitation, however, an approximate solution is available, according to which, in common parlance, the term "*Centre of Gravity*" may be used as equivalent to *Centre of Inertia;* but it must be carefully remembered that the fundamental ideas involved in the two definitions are essentially different.

The second proposition in § 229 may now evidently be stated thus:—The sum of the momenta of the parts of the system in any direction is equal to the momentum in the same direction of a mass equal to the sum of the masses moving with a velocity equal to the velocity of the centre of inertia.

Centre of Inertia.

231. The *Moment* of any physical agency is the numerical measure of its importance. Thus, the moment of a force round a point or round a line, signifies the measure of its importance as regards producing or balancing rotation round that point or round that line.

Moment.

232. The *Moment* of a force about a point is defined as the product of the force into its perpendicular distance from the point. It is numerically double the area of the triangle whose vertex is the point, and whose base is a line representing the force in magnitude and direction. It is often convenient to represent it by a line numerically equal to it, drawn through the vertex of the triangle perpendicular to its plane, through the front of a watch held in the plane with its centre at the point, and facing so that the force tends to turn round this point in a direction opposite to the hands. The moment of a force round any axis is the moment of its component in any plane perpendicular to the axis, round the point in which the plane is cut by the axis. Here we imagine the force resolved into two components, one parallel to the axis, which is ineffective so far as rotation round the axis is concerned; the other perpendicular to the axis (that is to say, having its line in any plane perpendicular to the axis). This latter component may be called the effective component of the force, with reference to rotation round the axis. And its moment round the axis may be defined as its moment round the nearest point of the axis, which is equivalent to the preceding definition. It is clear that the moment of a force round any axis, is equal to the area of the projection on any plane perpendicular to the axis, of the figure representing its moment round any point of the axis.

Moment of a force about a point.

Moment of a force about an axis.

233. The projection of an area, plane or curved, on any plane, is the area included in the projection of its bounding line.

Digression on projection of areas.

Digression on projection of areas.

If we imagine an area divided into any number of parts, the projections of these parts on any plane make up the projection of the whole. But in this statement it must be understood that the areas of partial projections are to be reckoned as positive if particular sides, which, for brevity, we may call the outside of the projected area and the front of the plane of projection, face the same way, and negative if they face oppositely.

Of course if the projected surface, or any part of it, be a plane area at right angles to the plane of projection, the projection vanishes. The projections of any two shells having a common edge, on any plane, are equal, but with the same, or opposite, signs as the case may be. Hence, by taking two such shells facing opposite ways, we see that the projection of a closed surface (or a shell with evanescent edge), on any plane, is nothing.

Equal areas in one plane, or in parallel planes, have equal projections on any plane, whatever may be their figures.

Hence the projection of any plane figure, or of any shell, edged by a plane figure, on another plane, is equal to its area, multiplied by the cosine of the angle at which its plane is inclined to the plane of projection. This angle is acute or obtuse, according as the outside of the projected area, and the front of plane of projection, face on the whole towards the same parts, or oppositely. Hence lines representing, as above described, moments about a point in different planes, are to be compounded as forces are.—See an analogous theorem in § 96.

Couple.

234. A *Couple* is a pair of equal forces acting in dissimilar directions in parallel lines. The *Moment* of a couple is the sum of the moments of its forces about any point in their plane, and is therefore equal to the product of either force into the shortest distance between their directions. This distance is called the *Arm* of the couple.

The *Axis of a Couple* is a line drawn from any chosen point of reference perpendicular to the plane of the couple, of such magnitude and in such direction as to represent the magnitude of the moment, and to indicate the direction in which the couple tends to turn. The most convenient rule for fulfilling the latter condition is this:—Hold a watch with its centre at the point of reference, and with its plane parallel to the plane of the couple. Then, according as the motion of the hands is contrary to or along with the direction in which the couple tends to turn, draw the axis of the couple through the face or through the back of the watch, *from* its centre. Thus a couple is completely represented by its axis; and couples are to be resolved and compounded by the same geometrical constructions performed with reference to their axes as forces or velocities, with reference to the lines directly representing them.

Couple.

235. If we substitute, for the force in § 232, a velocity, we have the moment of a velocity about a point; and by introducing the mass of the moving body as a factor, we have an important element of dynamical science, the *Moment of Momentum.* The laws of composition and resolution are the same as those already explained; but for the sake of some simple applications we give an elementary investigation.

Moment of velocity.

Moment of momentum.

The moment of a rectilineal motion is the product of its length into the distance of its line from the point.

Moment of a rectilineal displacement.

The moment of the resultant velocity of a particle about any point in the plane of the components is equal to the algebraic sum of the moments of the components, the proper sign of each moment being determined as above, § 233. The same is of course true of moments of displacements, of moments of forces and of moments of momentum.

First, consider two component motions, AB and AC, and let AD be their resultant (§ 27). Their half moments round the point O are respectively the areas OAB, OCA. Now OCA, together with half the area of the parallelogram $CABD$, is equal to OBD. Hence the sum of the two half moments together with half the area of the parallelogram, is equal to AOB together with BOD, that is to say, to the area of the whole figure $OABD$. But ABD, a part of this figure, is equal to half the area of the parallelogram; and therefore the remainder, OAD, is equal to the sum of the two half moments. But OAD is half the moment of the resultant velocity round the point O. Hence the moment of the

For two forces, motions, velocities, or momenta, in one plane, the sum of their moments proved equal to the moment of their resultant round any point in that plane.

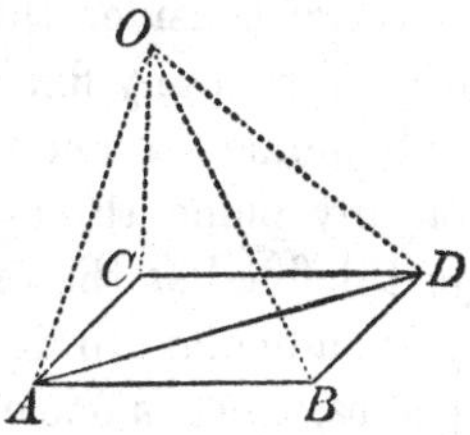

O C D A B

resultant is equal to the sum of the moments of the two components.

If there are any number of component rectilineal motions in one plane, we may compound them in order, any two taken together first, then a third, and so on; and it follows that the sum of their moments is equal to the moment of their resultant. It follows, of course, that the sum of the moments of any number of component velocities, all in one plane, into which the velocity of any point may be resolved, is equal to the moment of their resultant, round any point in their plane. It follows also, that if velocities, in different directions all in one plane, be successively given to a moving point, so that at any time its velocity is their resultant, the moment of its velocity at any time is the sum of the moments of all the velocities which have been successively given to it.

Any number of moments in one plane compounded by addition.

Cor.—If one of the components always passes through the point, its moment vanishes. This is the case of a motion in which the acceleration is directed to a fixed point, and we thus reproduce the theorem of § 36, *a*, that in this case the areas described by the radius-vector are proportional to the times; for, as we have seen, the moment of velocity is double the area traced out by the radius-vector in unit of time.

Moment round an axis.

236. The moment of the velocity of a point round any axis is the moment of the velocity of its projection on a plane perpendicular to the axis, round the point in which the plane is cut by the axis.

Moment of a whole motion, round an axis.

The moment of the whole motion of a point during any time, round any axis, is twice the area described in that time by the radius-vector of its projection on a plane perpendicular to that axis.

If we consider the conical area traced by the radius-vector drawn from any fixed point to a moving point whose motion is not confined to one plane, we see that the projection of this area on any plane through the fixed point is half of what we have just defined as the moment of the whole motion round an axis perpendicular to it through the fixed point. Of all these planes, there is one on which the projection of the area is greater than on any other; and the projection of the conical area on any plane perpendicular to this plane, is equal to nothing, the proper interpretation of positive and negative projections being used.

Moment of a whole motion, round an axis.

If any number of moving points are given, we may similarly consider the conical surface described by the radius-vector of each drawn from one fixed point. The same statement applies to the projection of the many-sheeted conical surface, thus presented. The resultant axis of the whole motion in any finite time, round the fixed point of the motions of all the moving points, is a line through the fixed point perpendicular to the plane on which the area of the whole projection is greater than on any other plane; and the moment of the whole motion round the resultant axis, is twice the area of this projection.

Resultant axis.

The resultant axis and moment of velocity, of any number of moving points, relatively to any fixed point, are respectively the resultant axis of the whole motion during an infinitely short time, and its moment, divided by the time.

The moment of the whole motion round any axis, of the motion of any number of points during any time, is equal to the moment of the whole motion round the resultant axis through any point of the former axis, multiplied into the cosine of the angle between the two axes.

The resultant axis, relatively to any fixed point, of the whole motion of any number of moving points, and the moment of the whole motion round it, are deduced by the same elementary constructions from the resultant axes and moments of the individual points, or partial groups of points of the system, as the direction and magnitude of a resultant displacement are deduced from any given lines and magnitudes of component displacements.

Moment of momentum.

Corresponding statements apply, of course, to the moments of velocity and of momentum.

237. If the point of application of a force be displaced through a small space, the resolved part of the displacement in the direction of the force has been called its *Virtual Velocity*.

Virtual velocity.

Virtual velocity.

This is positive or negative according as the virtual velocity is in the same, or in the opposite, direction to that of the force.

The product of the force, into the virtual velocity of its point of application, has been called the *Virtual Moment* of the force. These terms we have introduced since they stand in the history and developments of the science; but, as we shall show further on, they are inferior substitutes for a far more useful set of ideas clearly laid down by Newton.

Work.

238. A force is said to *do work* if its place of application has a positive component motion in its direction; and the work done by it is measured by the product of its amount into this component motion.

Thus, in lifting coals from a pit, the amount of work done is proportional to the weight of the coals lifted; that is, to the force overcome in raising them; and also to the height through which they are raised. The unit for the measurement of work adopted in practice by British engineers, is that required to overcome a force equal to the gravity of a pound through the space of a foot; and is called a *Foot-Pound.*

Practical unit.

Scientific unit.

In purely scientific measurements, the unit of work is not the foot-pound, but the kinetic unit force (§ 225) acting through unit of space. Thus, for example, as we shall show further on, this unit is adopted in measuring the work done by an electric current, the units for electric and magnetic measurements being founded upon the kinetic unit force.

If the weight be raised obliquely, as, for instance, along a smooth inclined plane, the space through which the force has to be overcome is increased in the ratio of the length to the height of the plane; but the force to be overcome is not the whole gravity of the weight, but only the component of the gravity parallel to the plane; and this is less than the gravity in the ratio of the height of the plane to its length. By multiplying these two expressions together, we find, as we might expect, that the amount of work required is unchanged by the substitution of the oblique for the vertical path.

Work of a force.

239. Generally, for any force, the work done during an infinitely small displacement of the point of application is the virtual moment of the force (§ 237), or is the product of the resolved part of the force in the direction of the displacement into the displacement.

Work of a force.

From this it appears, that if the motion of the point of application be always perpendicular to the direction in which a force acts, such a force does no work. Thus the mutual normal pressure between a fixed and moving body, as the tension of the cord to which a pendulum bob is attached, or the attraction of the sun on a planet if the planet describe a circle with the sun in the centre, is a case in which no work is done by the force.

240. The work done by a force, or by a couple, upon a body turning about an axis, is the product of the moment of the force or couple into the angle (in radians, or fraction of a radian) through which the body acted on turns, if the moment remains the same in all positions of the body. If the moment be variable, the statement is only valid for infinitely small displacements, but may be made accurate by employing the proper *average* moment of the force or of the couple. The proof is obvious.

Work of a couple.

If Q be the moment of the force or couple for a position of the body given by the angle θ, $Q(\theta_1 - \theta_0)$ if Q is constant, or $\int_{\theta_0}^{\theta_1} Qd\theta = q(\theta_1 - \theta_0)$ where q is the proper average value of Q when variable, is the work done by the couple during the rotation from θ_0 to θ_1.

241. Work done on a body by a force is always shown by a corresponding increase of vis viva, or kinetic energy, if no other forces act on the body which can do work or have work done against them. If work be done against any forces, the increase of kinetic energy is less than in the former case by the amount of work so done. In virtue of this, however, the body possesses an equivalent in the form of *Potential Energy* (§ 273), if its physical conditions are such that these forces will act equally, and in the same directions, if the motion of the system is reversed. Thus there may be no change of kinetic energy pro-

Transformation of work.

Potential energy.

Potential energy.

duced, and the work done may be wholly stored up as potential energy.

Thus a weight requires work to raise it to a height, a spring requires work to bend it, air requires work to compress it, etc.; but a raised weight, a bent spring, compressed air, etc., are *stores* of energy which can be made use of at pleasure.

Newton's Laws of Motion.

242. In what precedes we have given some of Newton's *Definitiones* nearly in his own words; others have been enunciated in a form more suitable to modern methods; and some terms have been introduced which were invented subsequent to the publication of the *Principia*. But the *Axiomata, sive Leges Motûs*, to which we now proceed, are given in Newton's own words; the two centuries which have nearly elapsed since he first gave them have not shown a necessity for any addition or modification. The first two, indeed, were discovered by Galileo, and the third, in some of its many forms, was known to Hooke, Huyghens, Wallis, Wren, and others; before the publication of the *Principia*. Of late there has been a tendency to split the second law into two, called respectively the second and third, and to ignore the third entirely, though using it *directly* in every dynamical problem; but all who have done so have been forced *indirectly* to acknowledge the completeness of Newton's system, by introducing as an axiom what is called D'Alembert's principle, which is really Newton's rejected third law in another form. Newton's own interpretation of his third law directly points out not only D'Alembert's principle, but also the modern principles of Work and Energy.

Axiom.

243. An Axiom is a proposition, the truth of which must be admitted as soon as the terms in which it is expressed are clearly understood. But, as we shall show in our chapter on "Experience," physical axioms are axiomatic to those only who have sufficient knowledge of the action of physical causes to enable them to see their truth. Without further remark we shall give Newton's Three Laws; it being remembered that, as the properties of matter *might* have been such as to render a totally different set of laws axiomatic, these laws must be considered as resting on convictions drawn from observation and experiment, *not* on intuitive perception.

244. Lex I. *Corpus omne perseverare in statu suo quiescendi vel movendi uniformiter in directum, nisi quatenus illud à viribus impressis cogitur statum suum mutare.* **Newton's first law.**

Every body continues in its state of rest or of uniform motion in a straight line, except in so far as it may be compelled by force to change that state.

245. The meaning of the term *Rest*, in physical science is essentially relative. Absolute rest is undefinable. If the universe of matter were finite, its centre of inertia might fairly be considered as absolutely at rest; or it might be imagined to be moving with any uniform velocity in any direction whatever through infinite space. But it is remarkable that the first law of motion enables us (§ 249, below) to explain what may be called *directional* rest. As will soon be shown, § 267, the plane in which the moment of momentum of the universe (if finite) round its centre of inertia is the greatest, which is clearly determinable from the actual motions at any instant, is fixed in direction in space. **Rest.**

246. We may logically convert the assertion of the first law of motion as to velocity into the following statements:—

The times during which any particular body, not compelled by force to alter the speed of its motion, passes through equal spaces, are equal. And, again—Every other body in the universe, not compelled by force to alter the speed of its motion, moves over equal spaces in successive intervals, during which the particular chosen body moves over equal spaces.

247. The first part merely expresses the convention universally adopted for the measurement of *Time*. The earth, in its rotation about its axis, presents us with a case of motion in which the condition, of not being compelled by force to alter its speed, is more nearly fulfilled than in any other which we can easily or accurately observe. And the numerical measurement of time practically rests on defining *equal intervals of time*, as *times during which the earth turns through equal* **Time.**

angles. This is, of course, a mere convention, and not a law of nature; and, as we now see it, is a part of Newton's first law.

Examples of the law.

248. The remainder of the law is not a convention, but a great truth of nature, which we may illustrate by referring to small and trivial cases as well as to the grandest phenomena we can conceive.

A curling-stone, projected along a horizontal surface of ice, travels equal distances, except in so far as it is retarded by friction and by the resistance of the air, in successive intervals of time during which the earth turns through equal angles. The sun moves through equal portions of interstellar space in times during which the earth turns through equal angles, except in so far as the resistance of interstellar matter, and the attraction of other bodies in the universe, alter his speed and that of the earth's rotation.

Directional fixedness.

249. If two material points be projected from one position, A, at the same instant with any velocities in any directions, and each left to move uninfluenced by force, the line joining them will be always parallel to a fixed direction. For the law asserts, as we have seen, that $AP : AP' :: AQ : AQ'$, if P, Q, and again P', Q' are simultaneous positions; and therefore PQ is parallel to $P'Q'$. Hence if four material points O, P, Q, R are all projected at one instant from one position, OP, OQ, OR are fixed directions of reference ever after. But, practically, the determination of fixed directions in space, § 267, is made to depend upon the rotation of groups of particles exerting forces on each other, and thus involves the Third Law of Motion.

The "Invariable Plane" of the solar system.

250. The whole law is singularly at variance with the tenets of the ancient philosophers who maintained that circular motion is perfect.

The last clause, "*nisi quatenus*," etc., admirably prepares for the introduction of the second law, by conveying the idea that *it is force alone which can produce a change of motion.* How, we naturally inquire, does the change of motion produced depend on the magnitude and direction of the force which produces it? And the answer is—

251. Lex II. *Mutationem motûs proportionalem esse vi motrici impressæ, et fieri secundum lineam rectam quâ vis illa imprimitur.*

Newton's second law

Change of motion is proportional to force applied, and takes place in the direction of the straight line in which the force acts.

252. If any force generates motion, a double force will generate double motion, and so on, whether simultaneously or successively, instantaneously, or gradually applied. And this motion, if the body was moving beforehand, is either added to the previous motion if directly conspiring with it; or is subtracted if directly opposed; or is geometrically compounded with it, according to the kinematical principles already explained, if the line of previous motion and the direction of the force are inclined to each other at an angle. (This is a paraphrase of Newton's own comments on the second law.)

253. In Chapter I. we have considered change of velocity, or acceleration, as a purely geometrical element, and have seen how it may be at once inferred from the given initial and final velocities of a body. By the definition of quantity of motion (§ 210), we see that, if we multiply the change of velocity, thus geometrically determined, by the mass of the body, we have the change of motion referred to in Newton's law as the measure of the force which produces it.

It is to be particularly noticed, that in this statement there is nothing said about the actual motion of the body before it was acted on by the force: it is only the *change* of motion that concerns us. Thus the same force will produce precisely the same change of motion in a body, whether the body be at rest, or in motion with any velocity whatever.

254. Again, it is to be noticed that nothing is said as to the body being under the action of *one* force only; so that we may logically put a part of the second law in the following (apparently) amplified form:—

When any forces whatever act on a body, then, whether the body be originally at rest or moving with any velocity and in any direction, each force produces in the body the exact change of motion which it would have produced if it had acted singly on the body originally at rest.

Composition of forces.

255. A remarkable consequence follows immediately from this view of the second law. Since forces are measured by the changes of motion they produce, and their directions assigned by the directions in which these changes are produced; and since the changes of motion of one and the same body are in the directions of, and proportional to, the changes of velocity—a single force, measured by the resultant change of velocity, and in its direction, will be the equivalent of any number of simultaneously acting forces. Hence

The resultant of any number of forces (applied at one point) is to be found by the same geometrical process as the resultant of any number of simultaneous velocities.

256. From this follows at once (§ 27) the construction of the *Parallelogram of Forces* for finding the resultant of two forces, and the *Polygon of Forces* for the resultant of any number of forces, in lines all through one point.

The case of the equilibrium of a number of forces acting at one point, is evidently deducible at once from this; for if we introduce one other force equal and opposite to their resultant, this will produce a change of motion equal and opposite to the resultant change of motion produced by the given forces; that is to say, will produce a condition in which the point experiences no change of motion, which, as we have already seen, is the only kind of rest of which we can ever be conscious.

257. Though Newton perceived that the Parallelogram of Forces, or the fundamental principle of Statics, is essentially involved in the second law of motion, and gave a proof which is virtually the same as the preceding, subsequent writers on Statics (especially in this country) have very generally ignored the fact; and the consequence has been the introduction of various unnecessary Dynamical Axioms, more or less obvious, but in reality included in or dependent upon Newton's laws of motion. We have retained Newton's method, not only on account of its admirable simplicity, but because we believe it contains the most philosophical foundation for the static as well as for the kinetic branch of the dynamic science.

258. But the second law gives us the means of measuring force, and also of measuring the mass of a body.

Measurement of force and mass.

For, if we consider the actions of various forces upon the same body for equal times, we evidently have changes of velocity produced which are *proportional to* the forces. The changes of velocity, then, give us in this case the means of comparing the magnitudes of different forces. Thus the velocities acquired in one second by the same mass (falling freely) at different parts of the earth's surface, give us the relative amounts of the earth's attraction at these places.

Again, if equal forces be exerted on different bodies, the changes of velocity produced in equal times must be *inversely* as the masses of the various bodies. This is approximately the case, for instance, with trains of various lengths started by the same locomotive: it is exactly realized in such cases as the action of an electrified body on a number of solid or hollow spheres of the same external diameter, and of different metals or of different thicknesses.

Again, if we find a case in which different bodies, each acted on by a force, acquire in the same time the same changes of velocity, the forces must be proportional to the masses of the bodies. This, when the resistance of the air is removed, is the case of falling bodies; and from it we conclude that the weight of a body in any given locality, or the force with which the earth attracts it, is proportional to its mass; a most important physical truth, which will be treated of more carefully in the chapter devoted to "Properties of Matter."

259. It appears, lastly, from this law, that every theorem of Kinematics connected with acceleration has its counterpart in Kinetics.

Translations from the kinematics of a point.

For instance, suppose X, Y, Z to be the components, parallel to fixed axes of x, y, z respectively, of the whole force acting on a particle of mass M. We see by § 212 that

$$M\frac{d^2x}{dt^2} = X, \quad M\frac{d^2y}{dt^2} = Y, \quad M\frac{d^2z}{dt^2} = Z;$$

or

$$M\ddot{x} = X, \quad M\ddot{y} = Y, \quad M\ddot{z} = Z.$$

Translations from the kinematics of a point.

Also, from these, we may evidently write,

$$M\ddot{s} = X\frac{dx}{ds} + Y\frac{dy}{ds} + Z\frac{dz}{ds} = X\frac{\dot{x}}{\dot{s}} + Y\frac{\dot{y}}{\dot{s}} + Z\frac{\dot{z}}{\dot{s}},$$

$$0 = X\frac{\dot{y}\ddot{z} - \dot{z}\ddot{y}}{\rho^{-1}\dot{s}^3} + Y\frac{\dot{z}\ddot{x} - \dot{x}\ddot{z}}{\rho^{-1}\dot{s}^3} + Z\frac{\dot{x}\ddot{y} - \dot{y}\ddot{x}}{\rho^{-1}\dot{s}^3},$$

$$\frac{M\dot{s}^2}{\rho} = X\frac{\dot{s}\ddot{x} - \dot{x}\ddot{s}}{\rho^{-1}\dot{s}^3} + Y\frac{\dot{s}\ddot{y} - \dot{y}\ddot{s}}{\rho^{-1}\dot{s}^3} + Z\frac{\dot{s}\ddot{z} - \dot{z}\ddot{s}}{\rho^{-1}\dot{s}^3}.$$

The second members of these equations are respectively the components of the impressed force, along the tangent (§ 9), perpendicular to the osculating plane (§ 9), and towards the centre of curvature, of the path described.

Measurement of force and mass.

260. We have, by means of the first two laws, arrived at a *definition* and a *measure* of force; and have also found how to compound, and therefore also how to resolve, forces; and also how to investigate the motion of a single particle subjected to given forces. But more is required before we can completely understand the more complex cases of motion, especially those in which we have mutual actions between or amongst two or more bodies; such as, for instance, attractions, or pressures, or transference of energy in any form. This is perfectly supplied by

Newton's third law.

261. LEX III. *Actioni contrariam semper et æqualem esse reactionem: sive corporum duorum actiones in se mutuò semper esse æquales et in partes contrarias dirigi.*

To every action there is always an equal and contrary reaction: or, the mutual actions of any two bodies are always equal and oppositely directed.

262. If one body presses or draws another, it is pressed or drawn by this other with an equal force in the opposite direction. If any one presses a stone with his finger, his finger is pressed with the same force in the opposite direction by the stone. A horse towing a boat on a canal is dragged backwards by a force equal to that which he impresses on the towing-rope forwards. By whatever amount, and in whatever direction, one body has its motion changed by impact upon another, this other body has its motion changed by the same amount in the opposite direction; for at each instant during the impact the force between them was equal and opposite on the two. When neither of the two bodies has any rotation, whether before or after impact, the changes of velocity which they experience are inversely as their masses.

Newton's third law.

When one body attracts another from a distance, this other attracts it with an equal and opposite force. This law holds not only for the attraction of gravitation, but also, as Newton himself remarked and verified by experiment, for magnetic attractions: also for electric forces, as tested by Otto-Guericke.

263. What precedes is founded upon Newton's own comments on the third law, and the actions and reactions contemplated are simple forces. In the scholium appended, he makes the following remarkable statement, introducing another description of actions and reactions subject to his third law, the full meaning of which seems to have escaped the notice of commentators:—

Si æstimetur agentis actio ex ejus vi et velocitate conjunctim; et similiter resistentis reactio æstimetur conjunctim ex ejus partium singularum velocitatibus et viribus resistendi ab earum attritione, cohæsione, pondere, et acceleratione oriundis; erunt actio et reactio, in omni instrumentorum usu, sibi invicem semper æquales.

In a previous discussion Newton has shown what is to be understood by the velocity of a force or resistance; *i.e.*, that it is the velocity of the point of application of the force *resolved in the direction of the force.* Bearing this in mind, we may read the above statement as follows:—

If the Activity of an agent be measured by its amount and its velocity conjointly; and if, similarly, the Counter-activity of the resistance be measured by the velocities of its several parts and their several amounts conjointly, whether these arise from friction, cohesion, weight, or acceleration;—Activity and Counter-activity, in all combinations of machines, will be equal and opposite.*

Farther on (§§ 264, 293) we shall give an account of the

* We translate Newton's word "*Actio*" here by "Activity" to avoid confusion with the word "Action" so universally used in modern dynamical treatises, according to the definition of § 326 below, in relation to Maupertuis' principle of "Least Action."

splendid dynamical theory founded by D'Alembert and Lagrange on this most important remark.

D'Alembert's principle.

264. Newton, in the passage just quoted, points out that forces of resistance against acceleration are to be reckoned as reactions equal and opposite to the actions by which the acceleration is produced. Thus, if we consider any one material point of a system, its reaction against acceleration must be equal and opposite to the resultant of the forces which that point experiences, whether by the actions of other parts of the system upon it, or by the influence of matter not belonging to the system. In other words, it must be in equilibrium with these forces. Hence Newton's view amounts to this, that all the forces of the system, with the reactions against acceleration of the material points composing it, form groups of equilibrating systems for these points considered individually. Hence, by the principle of superposition of forces in equilibrium, all the forces acting on points of the system form, with the reactions against acceleration, an equilibrating set of forces on the whole system. This is the celebrated principle first explicitly stated, and very usefully applied, by D'Alembert in 1742, and still known by his name. We have seen, however, that it is very distinctly implied in Newton's own interpretation of his third law of motion. As it is usual to investigate the general equations or conditions of equilibrium, in dynamical treatises, before entering in detail on the kinetic branch of the subject, this principle is found practically most useful in showing how we may write down at once the equations of motion for any system for which the equations of equilibrium have been investigated.

Mutual forces between particles of a rigid body.

265. Every rigid body may be imagined to be divided into indefinitely small parts. Now, in whatever form we may eventually find a *physical* explanation of the origin of the forces which act between these parts, it is certain that each such small part may be considered to be held in its position relatively to the others by mutual forces in lines joining them.

266. From this we have, as immediate consequences of the second and third laws, and of the preceding theorems relating to Centre of Inertia and Moment of Momentum, a number of important propositions such as the following:—

Motion of centre of inertia of a rigid body.

(*a*) The centre of inertia of a rigid body moving in any manner, but free from external forces, moves uniformly in a straight line.

(*b*) When any forces whatever act on the body, the motion of the centre of inertia is the same as it would have been had these forces been applied with their proper magnitudes and directions at that point itself.

Moment of momentum of a rigid body.

(*c*) Since the moment of a force acting on a particle is the same as the moment of momentum it produces in unit of time, the changes of moment of momentum in any two parts of a rigid body due to their mutual action are equal and opposite. Hence the moment of momentum of a rigid body, about any axis which is fixed in direction, and passes through a point which is either fixed in space or moves uniformly in a straight line, is unaltered by the mutual actions of the parts of the body.

(*d*) The rate of increase of moment of momentum, when the body is acted on by external forces, is the sum of the moments of these forces about the axis.

Conservation of momentum, and of moment of momentum.

267. We shall for the present take for granted, that the mutual action between two rigid bodies may in every case be imagined as composed of pairs of equal and opposite forces in straight lines. From this it follows that the sum of the quantities of motion, parallel to any fixed direction, of two rigid bodies influencing one another in any possible way, remains unchanged by their mutual action; also that the sum of the moments of momentum of all the particles of the two bodies, round any line in a fixed direction in space, and passing through any point moving uniformly in a straight line in any direction, remains constant. From the first of these propositions we infer that the centre of inertia of any number of mutually influencing bodies, if in motion, continues moving uniformly in a straight line, unless in so far as the direction or velocity of its motion is changed by forces acting mutually between them and some other matter not belonging to them; also that the centre of inertia of any body or system of bodies moves

The "Invariable Plane" is a plane through the centre of inertia, perpendicular to the resultant axis.

just as all their matter, if concentrated in a point, would move under the influence of forces equal and parallel to the forces really acting on its different parts. From the second we infer that the axis of resultant rotation through the centre of inertia of any system of bodies, or through any point either at rest or moving uniformly in a straight line, remains unchanged in direction, and the sum of moments of momenta round it remains constant if the system experiences no force from without. This principle used to be called *Conservation of Areas*, a very ill-considered designation. From this principle it follows that if by internal action such as geological upheavals or subsidences, or pressure of the winds on the water, or by evaporation and rain- or snow-fall, or by any influence not depending on the attraction of sun or moon (even though dependent on solar heat), the disposition of land and water becomes altered, the component round any fixed axis of the moment of momentum of the earth's rotation remains constant.

Terrestrial application.

Rate of doing work.

268. The foundation of the abstract theory of energy is laid by Newton in an admirably distinct and compact manner in the sentence of his scholium already quoted (§ 263), in which he points out its application to mechanics*. The *actio agentis*, as he defines it, which is evidently equivalent to the product of the effective component of the force, into the velocity of the point on which it acts, is simply, in modern English phraseology, the rate at which the agent works. The subject for measurement here is precisely the same as that for which Watt, a hundred years later, introduced the practical unit of a "*Horse-power*," or the rate at which an agent works when overcoming 33,000 times the weight of a pound through the space of a foot in a minute; that is, producing 550 foot-pounds of work per second. The unit, however, which is most generally convenient is that which Newton's definition implies, namely, the rate of doing work in which the unit of energy is produced in the unit of time.

Horse-power.

* The reader will remember that we use the word "mechanics" in its true classical sense, the science of machines, the sense in which Newton himself used it, when he dismissed the further consideration of it by saying (in the scholium referred to), *Cœterum mechanicam tractare non est hujus instituti.*

269. Looking at Newton's words (§ 263) in this light, we see that they may be logically converted into the following form :—

Energy in abstract dynamics.

Work done on any system of bodies (in Newton's statement, the parts of any machine) has its equivalent in work done against friction, molecular forces, or gravity, if there be no acceleration; but if there be acceleration, part of the work is expended in overcoming the resistance to acceleration, and the additional kinetic energy developed is equivalent to the work so spent. This is evident from § 214.

When part of the work is done against molecular forces, as in bending a spring; or against gravity, as in raising a weight; the recoil of the spring, and the fall of the weight, are capable at any future time, of reproducing the work originally expended (§ 241). But in Newton's day, and long afterwards, it was supposed that work was *absolutely lost* by friction; and, indeed, this statement is still to be found even in recent authoritative treatises. But we must defer the examination of this point till we consider in its modern form the principle of *Conservation of Energy*.

270. If a system of bodies, given either at rest or in motion, be influenced by no forces from without, the sum of the kinetic energies of all its parts is augmented in any time by an amount equal to the whole work done in that time by the mutual forces, which we may imagine as acting between its points. When the lines in which these forces act remain all unchanged in length, the forces do no work, and the sum of the kinetic energies of the whole system remains constant. If, on the other hand, one of these lines varies in length during the motion, the mutual forces in it will do work, or will consume work, according as the distance varies with or against them.

271. A limited system of bodies is said to be *dynamically conservative* (or simply *conservative*, when force is understood to be the subject), if the mutual forces between its parts always perform, or always consume, the same amount of work during any motion whatever, by which it can pass from one particular configuration to another.

Conservative system.

Foundation of the theory of energy.

272. The whole theory of energy in physical science is founded on the following proposition:—

If the mutual forces between the parts of a material system are independent of their velocities, whether relative to one another, or relative to any external matter, the system must be dynamically conservative.

Physical axiom that "the Perpetual Motion is impossible" introduced.

For if more work is done by the mutual forces on the different parts of the system in passing from one particular configuration to another, by one set of paths than by another set of paths, let the system be directed, by frictionless constraint, to pass from the first configuration to the second by one set of paths and return by the other, over and over again for ever. It will be a continual source of energy without any consumption of materials, which is impossible.

Potential energy of conservative system.

273. The *potential energy* of a conservative system, in the configuration which it has at any instant, is the amount of work required to bring it to that configuration against its mutual forces during the passage of the system from any one chosen configuration to the configuration at the time referred to. It is generally, but not always, convenient to fix the particular configuration chosen for the zero of reckoning of potential energy, so that the potential energy, in every other configuration practically considered, shall be positive.

274. The potential energy of a conservative system, at any instant, depends solely on its configuration at that instant, being, according to definition, the same at all times when the system is brought again and again to the same configuration. It is therefore, in mathematical language, said to be a function of the co-ordinates by which the positions of the different parts of the system are specified. If, for example, we have a conservative system consisting of two material points; or two rigid bodies, acting upon one another with force dependent only on the relative position of a point belonging to one of them, and a point belonging to the other; the potential energy of the system depends upon the co-ordinates of one of these points relatively to lines of reference in fixed directions through the other. It will therefore, in general, depend on three indepen-

Potential energy of conservative system.

dent co-ordinates, which we may conveniently take as the distance between the two points, and two angles specifying the absolute direction of the line joining them. Thus, for example, let the bodies be two uniform metal globes, electrified with any given quantities of electricity, and placed in an insulating medium such as air, in a region of space under the influence of a vast distant electrified body. The mutual action between these two spheres will depend solely on the relative position of their centres. It will consist partly of gravitation, depending solely on the distance between their centres, and of electric force, which will depend on the distance between them, but also, in virtue of the inductive action of the distant body, will depend on the absolute direction of the line joining their centres. In our divisions devoted to gravitation and electricity respectively, we shall investigate the portions of the mutual potential energy of the two bodies depending on these two agencies separately. The former we shall find to be the product of their masses divided by the distance between their centres; the latter a somewhat complicated function of the distance between the centres and the angle which this line makes with the direction of the resultant electric force of the distant electrified body. Or again, if the system consist of two balls of soft iron, in any locality of the earth's surface, their mutual action will be partly gravitation, and partly due to the magnetism induced in them by terrestrial magnetic force. The portion of the mutual potential energy depending on the latter cause, will be a function of the distance between their centres and the inclination of this line to the direction of the terrestrial magnetic force. It will agree in mathematical expression with the potential energy of electric action in the preceding case, so far as the inclination is concerned, but the law of variation with the distance will be less easily determined.

Inevitable loss of energy of visible motions.

275. In nature the hypothetical condition of § 271 is *apparently violated* in all circumstances of motion. A material system can never be brought through any returning cycle of motion without spending more work against the mutual forces of its parts than is gained from these forces, because no relative motion can take place without meeting with frictional or

Inevitable loss of energy of visible motions.

other forms of resistance; among which are included (1) mutual friction between solids sliding upon one another; (2) resistances due to the viscosity of fluids, or imperfect elasticity of solids; (3) resistances due to the induction of electric currents; (4) resistances due to varying magnetization under the influence of imperfect magnetic retentiveness. No motion in nature can take place without meeting resistance due to some, if not to all, of these influences. It is matter of every day experience that friction and imperfect elasticity of solids impede the action of all artificial mechanisms; and that even when bodies are detached, and left to move freely in the air, as falling bodies, or as projectiles, they experience resistance owing to the viscosity of the air.

The greater masses, planets and comets, moving in a less resisting medium, show less indications of resistance*. Indeed it cannot be said that observation upon any one of these bodies, with the exception of Encke's comet, has demonstrated resistance. But the analogies of nature, and the ascertained facts of physical science, forbid us to doubt that every one of them, every star, and every body of any kind moving in any part of space, has its relative motion impeded by the air, gas, vapour, medium, or whatever we choose to call the substance occupying the space immediately round it; just as the motion of a rifle bullet is impeded by the resistance of the air.

Effect of tidal friction.

276. There are also indirect resistances, owing to friction impeding the tidal motions, on all bodies (like the earth) partially or wholly covered by liquid, which, as long as these bodies move relatively to neighbouring bodies, must keep drawing off energy from their relative motions. Thus, if we consider, in the first place, the action of the moon alone, on the earth with its oceans, lakes, and rivers, we perceive that it must tend to equalize the periods of the earth's rotation about its axis, and of the revolution of the two bodies about their centre of inertia; because as long as these periods differ, the tidal action on the

* Newton, *Principia*. (Remarks on the first law of motion.) "Majora autem Planetarum et Cometarum corpora motus suos et progressivos et circulares, in spatiis minus resistentibus factos, conservant diutius."

earth's surface must keep subtracting energy from their motions. To view the subject more in detail, and, at the same time, to avoid unnecessary complications, let us suppose the moon to be a uniform spherical body. The mutual action and reaction of gravitation between her mass and the earth's, will be equivalent to a single force in some line through her centre; and must be such as to impede the earth's rotation as long as this is performed in a shorter period than the moon's motion round the earth. It must therefore lie in some such direction as the line MQ in the diagram, which represents, necessarily with enormous exaggeration, its deviation, OQ, from the earth's centre. Now the actual force on the moon in the line MQ, may be regarded as consisting of a force in the line MO towards the earth's centre, sensibly equal in amount to the whole force, and a comparatively very small force in the line MT perpendicular to MO. This latter is very nearly tangential to the moon's path, and is in the direction *with* her motion. Such a force, if suddenly commencing to act, would, in the first place, increase the moon's velocity; but after a certain time she would have moved so much farther from the earth, in virtue of this acceleration, as to have lost, by moving against the earth's attraction, as much velocity as she had gained by the tangential accelerating force. The effect of a continued tangential force, acting with the motion, but so small in amount as to make only a small deviation at any moment from the circular form of the orbit, is to gradually increase the distance from the central body, and to cause as much again as its own amount of work to be done against the attraction of the central mass, by the kinetic energy of motion lost. The circumstances will be readily understood, by considering this motion round the central body in a very gradual spiral path tending outwards. Provided the law of the central force is the inverse square of the distance, the tangential component of the central force against the motion will be twice as great as the disturbing tangential force in the direction with the motion; and therefore one-half of the amount of work done

Effect of tidal friction.

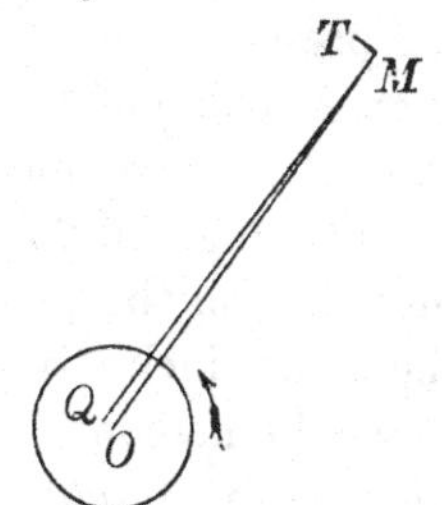

Inevitable loss of energy of visible motions. Tidal friction.

against the former, is done by the latter, and the other half by kinetic energy taken from the motion. The integral effect on the moon's motion, of the particular disturbing cause now under consideration, is most easily found by using the principle of moments of momenta. Thus we see that as much moment of momentum is gained in any time by the motions of the centres of inertia of the moon and earth relatively to their common centre of inertia, as is lost by the earth's rotation about its axis. The sum of the moments of momentum of the centres of inertia of the moon and earth as moving at present, is about 4·45 times the present moment of momentum of the earth's rotation. The average plane of the former is the ecliptic; and therefore the axes of the two momenta are inclined to one another at the average angle of 23° 27½′, which, as we are neglecting the sun's influence on the plane of the moon's motion, may be taken as the actual inclination of the two axes at present. The resultant, or whole moment of momentum, is therefore 5·38 times that of the earth's present rotation, and its axis is inclined 19° 13′ to the axis of the earth. Hence the ultimate tendency of the tides is, to reduce the earth and moon to a simple uniform rotation with this resultant moment round this resultant axis, as if they were two parts of one rigid body: in which condition the moon's distance would be increased (approximately) in the ratio 1 : 1·46, being the ratio of the square of the present moment of momentum of the centres of inertia to the square of the whole moment of momentum; and the period of revolution in the ratio 1 : 1·77, being that of the cubes of the same quantities. The distance would therefore be increased to 347,100 miles, and the period lengthened to 48·36 days. Were there no other body in the universe but the earth and the moon, these two bodies might go on moving thus for ever, in circular orbits round their common centre of inertia, and the earth rotating about its axis in the same period, so as always to turn the same face to the moon, and therefore to have all the liquids at its surface at rest relatively to the solid. But the existence of the sun would prevent any such state of things from being permanent. There would be solar tides—twice high water and twice low water—in the period of the earth's revolution relatively to the sun (that is

Inevitable loss of energy of visible motions. Tidal friction.

to say, twice in the solar day, or, which would be the same thing, the month). This could not go on without loss of energy by fluid friction. It is easy to trace the whole course of the disturbance in the earth's and moon's motions which this cause would produce*: its first effect must be to bring the moon to fall in to the earth, with compensation for loss of moment of momentum of the two round their centre of inertia in increase of its distance from the sun, and then to reduce the very rapid rotation of the compound bôdy, Earth-and-Moon, after the collision, and farther increase its distance from the Sun till ultimately, (corresponding action on liquid matter on the Sun having *its* effect also, and it being for our illustration supposed that there are no other planets,) the two bodies shall rotate round their common centre of inertia, like parts of one rigid body. It is remarkable that the whole frictional effect of the lunar and solar tides should be, first to augment the moon's distance from the earth to a maximum, and then to diminish it, till ultimately the moon falls in to the earth: and first to diminish, after that to increase, and lastly to diminish the earth's rotational velocity. We hope to return to the subject later, and to consider the general problem of the motion of any number of rigid bodies or material points acting on one another with mutual forces, under any actual physical law, and therefore, as we shall see, necessarily subject to loss of energy as long as any of their mutual distances vary; that is to say, until all subside into a state of motion in circles round an axis passing through their centre of inertia, like parts of one rigid body. It is probable

* The friction of these solar tides on the earth would cause the earth to rotate still slower; and then the moon's influence, tending to keep the earth rotating with always the same face towards herself, would resist this further reduction in the speed of the rotation. Thus (as explained above with reference to the moon) there would be from the sun a force opposing the earth's rotation, and from the moon a force promoting it. Hence according to the preceding explanation applied to the altered circumstances, the line of the earth's attraction on the moon passes now as before, not through the centre of inertia of the earth, but now in a line slightly *behind* it (instead of *before*, as formerly). It therefore now resists the moon's motion of revolution. The combined effect of this resistance and of the earth's attraction on the moon is, like that of a resisting medium, to cause the moon to fall in towards the earth in a spiral path with gradually increasing velocity.

Inevitable loss of energy of visible motions. Tidal friction.

that the moon, in ancient times liquid or viscous in its outer layer if not throughout, was thus brought to turn always the same face to the earth.

277. We have no data in the present state of science for estimating the relative importance of tidal friction, and of the resistance of the resisting medium through which the earth and moon move; but whatever it may be, there can be but one ultimate result for such a system as that of the sun and planets, if continuing long enough under existing laws, and not disturbed by meeting with other moving masses in space. That result is the falling together of all into one mass, which, although rotating for a time, must in the end come to rest relatively to the surrounding medium.

Ultimate tendency of the solar system.

Conservation of energy.

278. The theory of energy cannot be completed until we are able to examine the physical influences which accompany loss of energy in each of the classes of resistance mentioned above, § 275. We shall then see that in every case in which energy is lost by resistance, heat is generated; and we shall learn from Joule's investigations that the quantity of heat so generated is a perfectly definite equivalent for the energy lost. Also that in no natural action is there ever a development of energy which cannot be accounted for by the disappearance of an equal amount elsewhere by means of some known physical agency. Thus we shall conclude, that if any limited portion of the material universe could be perfectly isolated, so as to be prevented from either giving energy to, or taking energy from, matter external to it, the sum of its potential and kinetic energies would be the same at all times: in other words, that every material system subject to no other forces than actions and reactions between its parts, is a dynamically conservative system, as defined above, § 271. But it is only when the inscrutably minute motions among small parts, possibly the ultimate molecules of matter, which constitute light, heat, and magnetism; and the intermolecular forces of chemical affinity; are taken into account, along with the palpable motions and measurable forces of which we become cognizant by direct observation, that we can recognise the universally conservative character of all natural dynamic action, and perceive the bearing of the principle of reversibility on the whole class of natural actions involving resistance, which seem to violate it. In the meantime, in our studies of abstract dynamics, it will be sufficient to introduce a special reckoning for energy lost in working against, or gained from work done by, forces not belonging palpably to the conservative class.

Conservation of energy.

279. As of great importance in farther developments, we prove a few propositions intimately connected with energy.

280. The kinetic energy of any system is equal to the sum of the kinetic energies of a mass equal to the sum of the masses of the system, moving with a velocity equal to that of its centre of inertia, and of the motions of the separate parts relatively to the centre of inertia.

Kinetic energy of a system.

For if x, y, z be the co-ordinates of any particle, m, of the system; ξ, η, ζ its co-ordinates relative to the centre of inertia; and $\bar{x}$, $\bar{y}$, $\bar{z}$, the co-ordinates of the centre of inertia itself; we have for the whole kinetic energy

$$\tfrac{1}{2}\Sigma m\left\{\left(\frac{dx}{dt}\right)^2+\left(\frac{dy}{dt}\right)^2+\left(\frac{dz}{dt}\right)^2\right\}=\tfrac{1}{2}\Sigma m\left\{\left(\frac{d(\bar{x}+\xi)}{dt}\right)^2+\left(\frac{d(\bar{y}+\eta)}{dt}\right)^2+\left(\frac{d(\bar{z}+\zeta)}{dt}\right)^2\right\}.$$

But by the properties of the centre of inertia, we have

$$\Sigma m\frac{d\bar{x}}{dt}\frac{d\xi}{dt}=\frac{d\bar{x}}{dt}\Sigma m\frac{d\xi}{dt}=0, \text{ etc. etc.}$$

Hence the preceding is equal to

$$\tfrac{1}{2}\Sigma m\left\{\left(\frac{d\bar{x}}{dt}\right)^2+\left(\frac{d\bar{y}}{dt}\right)^2+\left(\frac{d\bar{z}}{dt}\right)^2\right\}+\tfrac{1}{2}\Sigma m\left\{\left(\frac{d\xi}{dt}\right)^2+\left(\frac{d\eta}{dt}\right)^2+\left(\frac{d\zeta}{dt}\right)^2\right\},$$

which proves the proposition.

281. The kinetic energy of rotation of a rigid system about any axis is (§ 95) expressed by $\tfrac{1}{2}\Sigma mr^2\omega^2$, where m is the mass of any part, r its distance from the axis, and ω the angular velocity of rotation. It may evidently be written in the form $\tfrac{1}{2}\omega^2\Sigma mr^2$. The factor Σmr^2 is of very great importance in kinetic investigations, and has been called the *Moment of Inertia* of the system about the axis in question. The moment of inertia about any axis is therefore found by summing the

Moment of inertia.

Moment of inertia.

products of the masses of all the particles each into the square of its distance from the axis.

Moment of momentum of a rotating rigid body.

It is important to notice that the moment of momentum of any rigid system about an axis, being $\Sigma mvr = \Sigma mr^2\omega$, is the product of the angular velocity into the moment of inertia.

If we take a quantity k, such that

$$k^2\Sigma m = \Sigma mr^2$$

Radius of gyration.

k is called the *Radius of Gyration* about the axis from which r is measured. The radius of gyration about any axis is therefore the distance from that axis at which, if the whole mass were placed, it would have the same moment of inertia as before. In a fly-wheel, where it is desirable to have as great a moment of inertia with as small a mass as possible, within certain limits of dimensions, the greater part of the mass is formed into a ring of the largest admissible diameter, and the radius of this ring is then approximately the radius of gyration of the whole.

Fly-wheel.

Moment of inertia about any axis.

A rigid body being referred to rectangular axes passing through any point, it is required to find the moment of inertia about an axis through the origin making given angles with the co-ordinate axes.

Let λ, μ, ν be its direction-cosines. Then the distance (r) of the point x, y, z from it is, by § 95,

$$r^2 = (\mu z - \nu y)^2 + (\nu x - \lambda z)^2 + (\lambda y - \mu x)^2,$$

and therefore

$$Mk^2 = \Sigma mr^2 = \Sigma m\left[\lambda^2(y^2+z^2) + \mu^2(z^2+x^2) + \nu^2(x^2+y^2) - 2\mu\nu yz - 2\nu\lambda zx - 2\lambda\mu xy\right]$$

which may be written

$$A\lambda^2 + B\mu^2 + C\nu^2 - 2\alpha\mu\nu - 2\beta\nu\lambda - 2\gamma\lambda\mu,$$

where A, B, C are the moments of inertia about the axes, and $\alpha = \Sigma myz$, $\beta = \Sigma mzx$, $\gamma = \Sigma mxy$. From its derivation we see that this quantity is *essentially positive.* Hence when, by a proper linear transformation, it is deprived of the terms containing the products of λ, μ, ν, it will be brought to the form

$$Mk^2 = A\lambda^2 + B\mu^2 + C\nu^2 = Q,$$

where A, B, C are essentially positive. They are evidently the moments of inertia about the new rectangular axes of co-ordinates, and λ, μ, ν the corresponding direction-cosines of the axis round which the moment of inertia is to be found.

Moment of inertia about any axis.

Let $A > B > C$, if they are unequal. Then

$$A\lambda^2 + B\mu^2 + C\nu^2 = Q(\lambda^2 + \mu^2 + \nu^2)$$

shows that Q cannot be greater than A, nor less than C. Also, if A, B, C be equal, Q is equal to each.

If a, b, c be the radii of gyration about the new axes of x, y, z,

$$A = Ma^2, \quad B = Mb^2, \quad C = Mc^2,$$

and the above equation gives

$$k^2 = a^2\lambda^2 + b^2\mu^2 + c^2\nu^2.$$

But if x, y, z be any point in the line whose direction-cosines are λ, μ, ν, and r its distance from the origin, we have

$$\frac{x}{\lambda} = \frac{y}{\mu} = \frac{z}{\nu} = r, \text{ and therefore}$$

$$k^2r^2 = a^2x^2 + b^2y^2 + c^2z^2.$$

If, therefore, we consider the ellipsoid whose equation is

$$a^2x^2 + b^2y^2 + c^2z^2 = \epsilon^4,$$

we see that it intercepts on the line whose direction-cosines are λ, μ, ν—and about which the radius of gyration is k, a length r which is given by the equation

$$k^2r^2 = \epsilon^4;$$

or the rectangle under any radius-vector of this ellipsoid and the radius of gyration about it is constant. Its semi-axes are evidently $\frac{\epsilon^2}{a}$, $\frac{\epsilon^2}{b}$, $\frac{\epsilon^2}{c}$ where ϵ may have any value we may assign. Thus it is evident that

282. For every rigid body there may be described about any point as centre, an ellipsoid (called *Poinsot's Momental Ellipsoid**) which is such that the length of any radius-vector is

Momental ellipsoid.

* The definition is not Poinsot's, but ours. The momental ellipsoid as we define it is fairly called Poinsot's, because of the splendid use he has made of it in his well-known kinematic representation of the solution of the problem —to find the motion of a rigid body with one point held fixed but otherwise influenced by no forces—which, with Sylvester's beautiful theorem completing it so as to give a purely kinematical mechanism to show the time which the body takes to attain any particular position, we reluctantly keep back for our Second Volume.

Momental ellipsoid.

inversely proportional to the radius of gyration of the body about that radius-vector as axis.

Principal axes.

The axes of this ellipsoid are, and might be defined as, the *Principal Axes* of inertia of the body for the point in question: but the best definition of principal axes of inertia is given below. First take two preliminary lemmas :—

Equilibration of Centrifugal Forces.

(1) If a rigid body rotate round any axis, the centrifugal forces are reducible to a single force perpendicular to the axis of rotation, and to a couple (§ 234 above) having its axis parallel to the line of this force.

(2) But in particular cases the couple may vanish, or both couple and force may vanish and the centrifugal forces be in equilibrium. The force vanishes if, and only if, the axis of rotation passes through the body's centre of inertia.

Definition of Principal Axes of Inertia.

DEF. (1). Any axis is called a principal axis of a body's inertia, or simply a principal axis of the body, if when the body rotates round it the centrifugal forces either balance or are reducible to a single force.

DEF. (2). A principal axis not through the centre of inertia is called a principal axis of inertia for the point of itself through which the resultant of centrifugal forces passes.

DEF. (3). A principal axis which passes through the centre of inertia is a principal axis for every point of itself.

The proofs of the lemmas may be safely left to the student as exercises on § 559 below; and from the proof the identification of the principal axes as now defined with the principal axes of Poinsot's momental ellipsoid is seen immediately by aid of the analysis of § 281.

283. The proposition of § 280 shows that the moment of inertia of a rigid body about any axis is equal to that which the mass, if collected at the centre of inertia, would have about this axis, together with that of the body about a parallel axis through its centre of inertia. It leads us naturally to investigate the relation between principal axes for any point and principal axes for the centre of inertia. The following investigation proves the remarkable theorem of § 284, which was first given in 1811 by Binet in the *Journal de l'École Polytechnique.*

Let the origin, O, be the centre of inertia, and the axes the principal axes at that point. Then, by §§ 280, 281, we have for the moment of inertia about a line through the point P (ξ, η, ζ), whose direction-cosines are λ, μ, ν;

Principal axes.

$$Q = A\lambda^2 + B\mu^2 + C\nu^2 + M\{(\mu\zeta - \nu\eta)^2 + (\nu\xi - \lambda\zeta)^2 + (\lambda\eta - \mu\xi)^2\}$$
$$= \{A + M(\eta^2 + \zeta^2)\}\lambda^2 + \{B + M(\zeta^2 + \xi^2)\}\mu^2 + \{C + M(\xi^2 + \eta^2)\}\nu^2$$
$$- 2M(\mu\nu\eta\zeta + \nu\lambda\zeta\xi + \lambda\mu\xi\eta).$$

Substituting for Q, A, B, C their values, and dividing by M, we have

$$k^2 = (a^2 + \eta^2 + \zeta^2)\lambda^2 + (b^2 + \zeta^2 + \xi^2)\mu^2 + (c^2 + \xi^2 + \eta^2)\nu^2$$
$$- 2(\eta\zeta\mu\nu + \zeta\xi\nu\lambda + \xi\eta\lambda\mu).$$

Let it be required to find λ, μ, ν so that the direction specified by them may be a principal axis. Let $s = \lambda\xi + \mu\eta + \nu\zeta$, *i.e.* let s represent the projection of OP on the axis sought.

The axes of the ellipsoid

$$(a^2 + \eta^2 + \zeta^2)x^2 + \ldots\ldots - 2(\eta\zeta yz + \ldots\ldots) = H \;\ldots\ldots(a),$$

are found by means of the equations

$$\left.\begin{array}{r}(a^2 + \eta^2 + \zeta^2 - p)\lambda - \xi\eta\mu - \zeta\xi\nu = 0\\ -\xi\eta\lambda + (b^2 + \zeta^2 + \xi^2 - p)\mu - \eta\zeta\nu = 0\\ -\zeta\xi\lambda - \eta\zeta\mu + (c^2 + \xi^2 + \eta^2 - p)\nu = 0\end{array}\right\}\ldots\ldots\ldots\ldots\ldots(b).$$

If, now, we take f to denote OP, or $(\xi^2 + \eta^2 + \zeta^2)^{\frac{1}{2}}$, these equations, where p is clearly the square of the radius of gyration about the axis to be found, may be written

$$(a^2 + f^2 - p)\lambda - \xi(\xi\lambda + \eta\mu + \zeta\nu) = 0,$$
$$\text{etc.} = \text{etc.},$$

or

$$(a^2 + f^2 - p)\lambda - \xi s = 0,$$
$$\text{etc.} = \text{etc.},$$

or

$$\left.\begin{array}{r}(a^2 - K)\lambda - \xi s = 0\\ (b^2 - K)\mu - \eta s = 0\\ (c^2 - K)\nu - \zeta s = 0\end{array}\right\}\ldots\ldots\ldots\ldots\ldots\ldots(c)$$

where $K = p - f^2$. Hence

$$\lambda = \frac{\xi s}{a^2 - K}, \text{ etc.}$$

Multiply, in order, by ξ, η, ζ, add, and divide by s, and we get

$$\frac{\xi^2}{a^2 - K} + \frac{\eta^2}{b^2 - K} + \frac{\zeta^2}{c^2 - K} = 1 \;\ldots\ldots\ldots\ldots\ldots\ldots(d).$$

Principal axes.

By (c) we see that (λ, μ, ν) is the direction of the normal through the point P, (ξ, η, ζ) of the surface represented by the equation

$$\frac{x^2}{a^2-K}+\frac{y^2}{b^2-K}+\frac{z^2}{c^2-K}=1 \quad \ldots\ldots\ldots\ldots\ldots\ldots (e),$$

which is obviously a surface of the second degree confocal with the ellipsoid

$$\frac{x^2}{a^2}+\frac{y^2}{b^2}+\frac{z^2}{c^2}=1 \quad \ldots\ldots\ldots\ldots\ldots\ldots (f),$$

and passing through P in virtue of (d), which determines K accordingly. The three roots of this cubic are clearly all real; one of them is less than the least of a^2, b^2, c^2, and positive or negative according as P is within or without the ellipsoid (f). And if $a > b > c$, the two others are between c^2 and b^2, and between b^2 and a^2, respectively. The addition of f^2 to each gives the square of the radius of gyration round the corresponding principal axis. Hence

Binet's Theorem.

284. The principal axes for any point of a rigid body are normals to the three surfaces of the second order through that point, confocal with the ellipsoid, which has its centre at the centre of inertia, and its three principal diameters co-incident with the three principal axes for that point, and equal respectively to the doubles of the radii of gyration round them. This ellipsoid is called the *Central Ellipsoid.*

Central ellipsoid.

Kinetic symmetry round a point;

285. A rigid body is said to be kinetically symmetrical about its centre of inertia when its moments of inertia about three principal axes through that point are equal; and therefore necessarily the moments of inertia about *all* axes through that point equal, § 281, and all these axes principal axes. About it uniform spheres, cubes, and in general any complete crystalline solid of the first system (see chapter on Properties of Matter), are kinetically symmetrical.

round an axis.

A rigid body is kinetically symmetrical about an *axis* when this axis is one of the principal axes through the centre of inertia, and the moments of inertia about the other two, and therefore about any line in their plane, are equal. A spheroid, a square or equilateral triangular prism or plate, a circular ring, disc, or cylinder, or any complete crystal of the second or fourth system, is kinetically symmetrical about its axis.

286. The only actions and reactions between the parts of a system, not belonging palpably to the conservative class, which we shall consider in abstract dynamics, are those of friction between solids sliding on solids, except in a few instances in which we shall consider the general character and ultimate results of effects produced by viscosity of fluids, imperfect elasticity of solids, imperfect electric conduction, or imperfect magnetic retentiveness. We shall also, in abstract dynamics, consider forces as applied to parts of a limited system arbitrarily from without. These we shall call, for brevity, the applied forces.

Energy in abstract dynamics.

287. The law of energy may then, in abstract dynamics, be expressed as follows:—

The whole work done in any time, on any limited material system, by applied forces, is equal to the whole effect in the forms of potential and kinetic energy produced in the system, together with the work lost in friction.

288. This principle may be regarded as comprehending the whole of abstract dynamics, because, as we now proceed to show, the conditions of equilibrium and of motion, in every possible case, may be immediately derived from it.

289. A material system, whose relative motions are unresisted by friction, is in equilibrium in any particular configuration if, and is not in equilibrium unless, the work done by the applied forces is equal to the potential energy gained, in any possible infinitely small displacement from that configuration. This is the celebrated principle of "virtual velocities" which Lagrange made the basis of his *Mécanique Analytique*. The ill-chosen name "virtual velocities" is now falling into disuse.

Equilibrium.

290. To prove it, we have first to remark that the system cannot possibly move away from any particular configuration except by work being done upon it by the forces to which it is subject: it is therefore in equilibrium if the stated condition is fulfilled. To ascertain that nothing less than this condition can secure its equilibrium, let us first consider a system having only one degree of freedom to move. Whatever forces act on the whole system, we may always hold it in equilibrium by a single force applied to any one point of the system in its line

Principle of virtual velocities.

Principle of virtual velocities.

of motion, opposite to the direction in which it tends to move, and of such magnitude that, in any infinitely small motion in either direction, it shall resist, or shall do, as much work as the other forces, whether applied or internal, altogether do or resist. Now, by the principle of superposition of forces in equilibrium, we might, without altering their effect, apply to any one point of the system such a force as we have just seen would hold the system in equilibrium, and another force equal and opposite to it. All the other forces being balanced by one of these two, they and it might again, by the principle of superposition of forces in equilibrium, be removed; and therefore the whole set of given forces would produce the same effect, whether for equilibrium or for motion, as the single force which is left acting alone. This single force, since it is in a line in which the point of its application is free to move, must move the system. Hence the given forces, to which this single force has been proved equivalent, cannot possibly be in equilibrium unless their whole work for an infinitely small motion is nothing, in which case the single equivalent force is reduced to nothing. But whatever amount of freedom to move the whole system may have, we may always, by the application of frictionless constraint, limit it to one degree of freedom only; —and this may be freedom to execute any particular motion whatever, possible under the given conditions of the system. If, therefore, in any such infinitely small motion, there is variation of potential energy uncompensated by work of the applied forces, constraint limiting the freedom of the system to only this motion will bring us to the case in which we have just demonstrated there cannot be equilibrium. But the application of constraints limiting motion cannot possibly disturb equilibrium, and therefore the given system under the actual conditions cannot be in equilibrium in any particular configuration if there is more work done than resisted in any possible infinitely small motion from that configuration by all the forces to which it is subject.

Neutral equilibrium.

291. If a material system, under the influence of internal and applied forces, varying according to some definite law, is balanced by them in any position in which it may be placed, its equilibrium is said to be neutral. This is the case with any spherical body of uniform material resting on a horizontal plane. A right cylinder or cone, bounded by plane ends perpendicular to the axis, is also in neutral equilibrium on a horizontal plane. Practically, any mass of moderate dimensions is in neutral equilibrium when its centre of inertia only is fixed, since, when its longest dimension is small in comparison with the earth's radius, gravity is, as we shall see, approximately equivalent to a single force through this point.

Neutral equilibrium.

But if, when displaced infinitely little in any direction from a particular position of equilibrium, and left to itself, it commences and continues vibrating, without ever experiencing more than infinitely small deviation in any of its parts, from the position of equilibrium, the equilibrium in this position is said to be stable. A weight suspended by a string, a uniform sphere in a hollow bowl, a loaded sphere resting on a horizontal plane with the loaded side lowest, an oblate body resting with one end of its shortest diameter on a horizontal plane, a plank, whose thickness is small compared with its length and breadth, floating on water, etc. etc., are all cases of stable equilibrium; if we neglect the motions of rotation about a vertical axis in the second, third, and fourth cases, and horizontal motion in general, in the fifth, for all of which the equilibrium is neutral.

Stable equilibrium.

If, on the other hand, the system can be displaced in any way from a position of equilibrium, so that when left to itself it will not vibrate within infinitely small limits about the position of equilibrium, but will move farther and farther away from it, the equilibrium in this position is said to be unstable. Thus a loaded sphere resting on a horizontal plane with its load as high as possible, an egg-shaped body standing on one end, a board floating edgeways in water, etc. etc., would present, if they could be realised in practice, cases of unstable equilibrium.

Unstable equilibrium.

When, as in many cases, the nature of the equilibrium varies with the direction of displacement, if unstable for any possible displacement it is practically unstable on the whole. Thus a coin standing on its edge, though in neutral equilibrium for displacements in its plane, yet being in unstable equilibrium

Unstable equilibrium.

for those perpendicular to its plane, is practically unstable. A sphere resting in equilibrium on a saddle presents a case in which there is stable, neutral, or unstable equilibrium, according to the direction in which it may be displaced by rolling, but, practically, it would be unstable.

Test of the nature of equilibrium.

292. The theory of energy shows a very clear and simple test for discriminating these characters, or determining whether the equilibrium is neutral, stable, or unstable, in any case. If there is just as much work resisted as performed by the applied and internal forces in any possible displacement the equilibrium is neutral, but not unless. If in every possible infinitely small displacement from a position of equilibrium they do less work among them than they resist, the equilibrium is thoroughly stable, and not unless. If in any or in every infinitely small displacement from a position of equilibrium they do more work than they resist, the equilibrium is unstable. It follows that if the system is influenced only by internal forces, or if the applied forces follow the law of doing always the same amount of work upon the system passing from one configuration to another by all possible paths, the whole potential energy must be constant, in all positions, for neutral equilibrium; must be a minimum for positions of thoroughly stable equilibrium; must be either an absolute maximum, or a maximum for some displacements and a minimum for others when there is unstable equilibrium.

Deduction of the equations of motion of any system.

293. We have seen that, according to D'Alembert's principle, as explained above (§ 264), forces acting on the different points of a material system, and their reactions against the accelerations which they actually experience in any case of motion, are in equilibrium with one another. Hence in any actual case of motion, not only is the actual work done by the forces equal to the kinetic energy produced in any infinitely small time, in virtue of the actual accelerations; but so also is the work which would be done by the forces, in any infinitely small time, if the velocities of the points constituting the system, were at any instant changed to any possible infinitely small velocities, and the accelerations unchanged. This statement, when put in the concise language of mathematical analysis, constitutes Lagrange's application of the "principle of virtual velocities" to express the conditions of D'Alembert's equilibrium between the forces acting, and the resistances of the masses to acceleration. It comprehends, as we have seen, every possible condition of every case of motion. The "equations of motion" in any particular case are, as Lagrange has shown, deduced from it with great ease.

Deduction of the equations of motion of any system.

Let m be the mass of any one of the material points of the system; x, y, z its rectangular co-ordinates at time t, relatively to axes fixed in direction (§ 249) through a point reckoned as fixed (§ 245); and X, Y, Z the components, parallel to the same axes, of the whole force acting on it. Thus $-m\frac{d^2x}{dt^2}$, $-m\frac{d^2y}{dt^2}$, $-m\frac{d^2z}{dt^2}$ are the components of the reaction against acceleration. And these, with X, Y, Z, for the whole system, must fulfil the conditions of equilibrium. Hence if $\delta x, \delta y, \delta z$ denote any arbitrary variations of x, y, z consistent with the conditions of the system, we have

$$\Sigma\left\{\left(X - m\frac{d^2x}{dt^2}\right)\delta x + \left(Y - m\frac{d^2y}{dt^2}\right)\delta y + \left(Z - m\frac{d^2z}{dt^2}\right)\delta z\right\} = 0..(1),$$

Indeterminate equation of motion of any system.

where Σ denotes summation to include all the particles of the system. This may be called the indeterminate, or the variational, equation of motion. Lagrange used it as the foundation of his whole kinetic system, deriving from it all the common equations of motion, and his own remarkable equations in generalized co-ordinates (presently to be given). We may write it otherwise as follows:

$$\Sigma m(\ddot{x}\delta x + \ddot{y}\delta y + \ddot{z}\delta z) = \Sigma(X\delta x + Y\delta y + Z\delta z) \quad ......(2),$$

where the first member denotes the work done by forces equal to those required to produce the real accelerations, acting through the spaces of the arbitrary displacements; and the second member the work done by the actual forces through these imagined spaces.

If the moving bodies constitute a conservative system, and if V denote its potential energy in the configuration specified by $(x, y, z, \text{etc.})$, we have of course (§§ 241, 273)

$$\delta V = -\Sigma(X\delta x + Y\delta y + Z\delta z) \quad ....(3),$$

and therefore the indeterminate equation of motion becomes

$$\Sigma m(\ddot{x}\delta x + \ddot{y}\delta y + \ddot{z}\delta z) = -\delta V \quad\text{.................(4)},$$

Of conservative system.

where δV denotes the excess of the potential energy in the configuration ($x+\delta x$, $y+\delta y$, $z+\delta z$, etc.) above that in the configuration (x, y, z, etc.).

One immediate particular result must of course be the common equation of energy, which must be obtained by supposing δx, δy, δz, etc., to be the actual variations of the co-ordinates in an infinitely small time δt. Thus if we take $\delta x = \dot{x}\delta t$, etc., and divide both members by δt, we have

$$\Sigma(X\dot{x} + Y\dot{y} + Z\dot{z}) = \Sigma m(\ddot{x}\dot{x} + \ddot{y}\dot{y} + \ddot{z}\dot{z}) \quad\text{...........(5)}.$$

Equation of energy.

Here the first member is composed of Newton's *Actiones Agentium*; with his *Reactiones Resistentium* so far as friction, gravity, and molecular forces are concerned, subtracted: and the second consists of the portion of the *Reactiones* due to acceleration. As we have seen above (§ 214), the second member is the rate of increase of $\Sigma\frac{1}{2}m(\dot{x}^2+\dot{y}^2+\dot{z}^2)$ per unit of time. Hence, denoting by v the velocity of one of the particles, and by W the integral of the first member multiplied by dt, that is to say, the integral work done by the working and resisting forces in any time, we have

$$\Sigma\tfrac{1}{2}mv^2 = W + E_0 \quad\text{.......................(6)},$$

E_0 being the initial kinetic energy. This is the integral equation of energy. In the particular case of a conservative system, W is a function of the co-ordinates, irrespectively of the time, or of the paths which have been followed. According to the previous notation, with besides V_0 to denote the potential energy of the system in its initial configuration, we have $W = V_0 - V$, and the integral equation of energy becomes

$$\Sigma\tfrac{1}{2}mv^2 = V_0 - V + E_0,$$

or, if E denote the sum of the potential and kinetic energies, a constant,

$$\Sigma\tfrac{1}{2}mv^2 = E - V \quad\text{.......................(7)}.$$

The general indeterminate equation gives immediately, for the motion of a system of free particles,

$$m_1\ddot{x}_1 = X_1,\ m_1\ddot{y}_1 = Y_1,\ m_1\ddot{z}_1 = Z_1,\ m_2\ddot{x}_2 = X_2, \text{ etc.}$$

Of these equations the three for each particle may of course be treated separately if there is no mutual influence between the particles: but when they exert force on one another, X_1, Y_1, etc., will each in general be a function of all the co-ordinates.

From the indeterminate equation (1) Lagrange, by his method of multipliers, deduces the requisite number of equations for determining the motion of a rigid body, or of any system of connected particles or rigid bodies, thus:—Let the number of the particles be i, and let the connexions between them be expressed by n equations,

Constraint introduced into the indeterminate equation.

$$\left.\begin{array}{l} F(x_1, y_1, z_1, x_2, \ldots) = 0 \\ F_{\prime}(x_1, y_1, z_1, x_2, \ldots) = 0 \\ \quad \text{etc.} \qquad \text{etc.} \end{array}\right\} \ldots\ldots\ldots\ldots\ldots\ldots(8)$$

being the *kinematical equations* of the system. By taking the variations of these we find that every possible infinitely small displacement δx_1, δy_1, δz_1, δx_2, ... must satisfy the n linear equations

$$\frac{dF}{dx_1}\delta x_1 + \frac{dF}{dy_1}\delta y_1 + \text{etc.} = 0, \quad \frac{dF_{\prime}}{dx_1}\delta x_1 + \frac{dF_{\prime}}{dy_1}\delta y_1 + \text{etc.} = 0, \quad \text{etc.} \ldots (9).$$

Multiplying the first of these by λ, the second by $\lambda_{\prime}$, etc., adding to the indeterminate equation, and then equating the coefficients of δx_1, δy_1, etc., each to zero, we have

$$\left.\begin{array}{l} \lambda\dfrac{dF}{dx_1} + \lambda_{\prime}\dfrac{dF_{\prime}}{dx_1} + \ldots + X_1 - m_1\dfrac{d^2x_1}{dt^2} = 0 \\ \lambda\dfrac{dF}{dy_1} + \lambda_{\prime}\dfrac{dF_{\prime}}{dy_1} + \ldots + Y_1 - m_1\dfrac{d^2y_1}{dt^2} = 0 \\ \qquad \text{etc.} \qquad\qquad \text{etc.} \end{array}\right\} \ldots\ldots\ldots(10).$$

These are in all $3i$ equations to determine the n unknown quantities λ, $\lambda_{\prime}$, ..., and the $3i - n$ independent variables to which x_1, y_1, ... are reduced by the kinematical equations (8). The same equations may be found synthetically in the following manner, by which also we are helped to understand the precise meaning of the terms containing the multipliers λ, $\lambda_{\prime}$, etc.

Determinate equations of motion deduced.

First let the particles be free from constraint, but acted on both by the given forces X_1, Y_1, etc., and by forces depending on mutual distances between the particles and upon their positions relatively to fixed objects subject to the law of conservation, and having for their potential energy

$$-\tfrac{1}{2}(kF^2 + k_{\prime}F_{\prime}^2 + \text{etc.}),$$

so that components of the forces actually experienced by the different particles shall be

Determinate equations of motion deduced.

$$X_1 + kF\frac{dF}{dx_1} + k_,F_,\frac{dF_,}{dx_1} + \text{etc.} + \tfrac{1}{2}\left(F^2\frac{dk}{dx_1} + F_,^2\frac{dk_,}{dx_1} + \text{etc.}\right)$$

$$\text{etc.,} \qquad \text{etc.}$$

Hence the equations of motion are

$$\left.\begin{aligned} m_1\frac{d^2x_1}{dt^2} &= X_1 + kF\frac{dF}{dx_1} + k_,F_,\frac{dF_,}{dx_1} + \text{etc.} + \tfrac{1}{2}\left(F^2\frac{dk}{dx_1} + F_,^2\frac{dk_,}{dx_1} + \text{etc.}\right) \\ m_1\frac{d^2y_1}{dt^2} &= \text{etc.} \\ \text{etc.,} &\qquad \text{etc.} \end{aligned}\right\} \quad (11).$$

Now suppose k, $k_,$, etc. to be infinitely great:—in order that the *forces on the particles* may not be infinitely great, we must have

$$F=0, \quad F_,=0, \quad \text{etc.},$$

that is to say, the equations of condition (8) must be fulfilled; and the last groups of terms in the second members of (11) now disappear because they contain the squares of the infinitely small quantities F, $F_,$, etc. Put now $kF=\lambda$, $k_,F_,=\lambda_,$, etc., and we have equations (10). This second mode of proving Lagrange's equations of motion of a constrained system corresponds precisely to the imperfect approach to the ideal case which can be made by real mechanism. The levers and bars and guide-surfaces cannot be infinitely rigid. Suppose then k, $k_,$, etc. to be finite but very great quantities, and to be some functions of the co-ordinates depending on the elastic qualities of the materials of which the guiding mechanism is composed:—equations (11) will express the motion, and by supposing k, $k_,$, etc. to be greater and greater we approach more and more nearly to the ideal case of absolutely rigid mechanism constraining the precise fulfilment of equations (8).

The problem of finding the motion of a system subject to any *unvarying* kinematical conditions whatever, under the action of any given forces, is thus reduced to a question of pure analysis. In the still more general problem of determining the motion when certain parts of the system are constrained to move in a specified manner, the equations of condition (8) involve not only the co-ordinates, but also t, the time. It is easily seen however that the equations (10) still hold, and with (8) fully determine the motion. For:—consider the equations of equilibrium of the particles acted on by any forces X_1', Y_1', etc., and constrained by proper mechanism to fulfil the equations of condition (8) with the actual values of the parameters for any particular value of t. The equations of equilibrium will be uninfluenced by the fact that some of the parameters of the conditions (8) have different values at different times. Hence, with $X_1 - m_1 \frac{d^2x_1}{dt^2}$, $Y_1 - m_1 \frac{d^2y_1}{dt^2}$, instead of X_1', Y_1', etc., according to D'Alembert's principle, the equations of motion will still be (8), (9), and (10) quite independently of whether the parameters of (8) are all constant, or have values varying in any arbitrary manner with the time.

Determinate equations of motion deduced.

To find the equation of energy multiply the first of equations (10) by $\dot{x}_1$, the second by $\dot{y}_1$, etc., and add. Then remarking that in virtue of (8) we have

Equation of energy.

$$\frac{dF}{dx_1}\dot{x}_1 + \frac{dF}{dy_1}\dot{y}_1 + \text{etc.} + \left(\frac{dF}{dt}\right) = 0,$$

$$\frac{dF_{\prime}}{dx_1}\dot{x}_1 + \frac{dF_{\prime}}{dy_1}\dot{y}_1 + \text{etc.} + \left(\frac{dF_{\prime}}{dt}\right) = 0,$$

partial differential coefficients of F, $F_{\prime}$, etc. with reference to t being denoted by $\left(\frac{dF}{dt}\right)$, $\left(\frac{dF_{\prime}}{dt}\right)$, etc.; and denoting by T the kinetic energy or $\frac{1}{2}\Sigma m(\dot{x}^2 + \dot{y}^2 + \dot{z}^2)$, we find

$$\frac{dT}{dt} = \Sigma(X\dot{x} + Y\dot{y} + Z\dot{z}) - \lambda\left(\frac{dF}{dt}\right) - \lambda_{\prime}\left(\frac{dF_{\prime}}{dt}\right) - \text{etc.} = 0 \text{....(12)}.$$

When the kinematic conditions are "*unvarying*," that is to say, when the equations of condition are equations among the co-ordinates with constant parameters, we have

$$\left(\frac{dF}{dt}\right) = 0, \quad \left(\frac{dF_{\prime}}{dt}\right) = 0, \text{ etc.,}$$

and the equation of energy becomes

$$\frac{dT}{dt} = \Sigma(X\dot{x} + Y\dot{y} + Z\dot{z}) \text{ ....(13)},$$

showing that in this case the fulfilment of the equations of condition involves neither gain nor loss of energy. On the other hand, equation (12) shows how to find the work performed or consumed in the fulfilment of the kinematical conditions when they are not unvarying.

Equation of energy.

As a simple example of varying constraint, which will be very easily worked out by equations (8) and (10), perfectly illustrating the general principle, the student may take the case of a particle acted on by any given forces and free to move anywhere in a plane which is kept moving with any given uniform or varying angular velocity round a fixed axis.

Gauss's principle of least constraint.

When there are connexions between any parts of a system, the motion is in general not the same as if all were free. If we consider any particle during any infinitely small time of the motion, and call the product of its mass into the square of the distance between its positions at the end of this time, on the two suppositions, the *constraint*: the sum of the constraints is a minimum. This follows easily from (1).

Impact.

294. When two bodies, in relative motion, come into contact, pressure begins to act between them to prevent any parts of them from jointly occupying the same space. This force commences from nothing at the first point of collision, and gradually increases per unit of area on a gradually increasing surface of contact. If, as is always the case in nature, each body possesses some degree of elasticity, and if they are not kept together after the impact by cohesion, or by some artificial appliance, the mutual pressure between them will reach a maximum, will begin to diminish, and in the end will come to nothing, by gradually diminishing in amount per unit of area on a gradually diminishing surface of contact. The whole process would occupy not greatly more or less than an hour if the bodies were of such dimensions as the earth, and such degrees of rigidity as copper, steel, or glass. It is finished, probably, within a thousandth of a second if they are globes of any of these substances not exceeding a yard in diameter.

295. The whole amount, and the direction, of the "*Impact*" experienced by either body in any such case, are reckoned according to the "change of momentum" which it experiences. The amount of the impact is measured by the amount, and its direction by the direction, of the change of momentum which is produced. The component of an impact in a direction parallel to any fixed line is similarly reckoned according to the component change of momentum in that direction.

296. If we imagine the whole time of an impact divided into a very great number of equal intervals, each so short that the force does not vary sensibly during it, the component change of momentum in any direction during any one of these intervals will (§ 220) be equal to the force multiplied by the measure of the interval. Hence the component of the impact is equal to the sum of the forces in all the intervals, multiplied by the length of each interval. **Impact.**

Let P be the component force in any direction at any instant, τ, of the interval, and let I be the amount of the corresponding component of the whole impact. Then

$$I = \int P d\tau.$$

297. Any force in a constant direction acting in any circumstances, for any time great or small, may be reckoned on the same principle; so that what we may call its whole amount during any time, or its "*time-integral*," will measure, or be measured by, the whole momentum which it generates in the time in question. But this reckoning is not often convenient or useful except when the whole operation considered is over before the position of the body, or configuration of the system of bodies, involved, has altered to such a degree as to bring any other forces into play, or alter forces previously acting, to such an extent as to produce any sensible effect on the momentum measured. Thus if a person presses gently with his hand, during a few seconds, upon a mass suspended by a cord or chain, he produces an effect which, if we know the degree of the force at each instant, may be thoroughly calculated on elementary principles. No approximation to a full determination of the motion, or to answering such a partial question as "how great will be the whole deflection produced?" can be founded on a knowledge of the "*time-integral*" alone. If, for instance, the force be at first very great and gradually diminish, the effect will be very different from what it would be if the force were to increase very gradually and to cease suddenly, even although the time-integral were the same in the two cases. But if the same body is "struck a blow," in a horizontal direction, either by the hand, or by a mallet or other somewhat **Time-integral.**

Time-integral.

hard mass, the action of the force is finished before the suspending cord has experienced any sensible deflection from the vertical. Neither gravity nor any other force sensibly alters the effect of the blow. And therefore the whole momentum at the end of the blow is sensibly equal to the "amount of the impact," which is, in this case, simply the time-integral.

Ballistic pendulum.

298. Such is the case of Robins' *Ballistic Pendulum*, a massive cylindrical block of wood cased in a cylindrical sheath of iron closed at one end and moveable about a horizontal axis at a considerable distance above it—employed to measure the velocity of a cannon or musket-shot. The shot is fired into the block in a horizontal direction along the axis of the block and perpendicular to the axis of suspension. The impulsive penetration is so nearly instantaneous, and the inertia of the block so large compared with the momentum of the shot, that the ball and pendulum are moving on as one mass before the pendulum has been sensibly deflected from the vertical. This is essential to the regular use of the apparatus. The iron sheath with its flat end must be strong enough to guard against splinters of wood flying sidewise, and to keep in the bullet.

299. Other illustrations of the cases in which the time-integral gives us the complete solution of the problem may be given without limit. They include all cases in which the direction of the force is always coincident with the direction of motion of the moving body, and those special cases in which the time of action of the force is so short that the body's motion does not, during its lapse, sensibly alter its relation to the direction of the force, or the action of any other forces to which it may be subject. Thus, in the vertical fall of a body, the time-integral gives us at once the change of momentum; and the same rule applies in most cases of forces of brief duration, as in a "drive" in cricket or golf.

Direct impact of spheres.

300. The simplest case which we can consider, and the one usually treated as an introduction to the subject, is that of the collision of two smooth spherical bodies whose centres before collision were moving in the same straight line. The force between them at each instant must be in this line, because of the symmetry of circumstances round it; and by the third law it must be equal in amount on the two bodies. Hence (LEX II.) they must experience changes of motion at equal rates in contrary directions; and at any instant of the impact the integral amounts of these changes of motion must be equal. Let us suppose, to fix the ideas, the two bodies to be moving both before and after impact in the same direction in one line: one of them gaining on the other before impact, and either following it at a less speed, or moving along with it, as the case may be, after the impact is completed. Cases in which the former is driven backwards by the force of the collision, or in which the two moving in opposite directions meet in collision, are easily reduced to dependence on the same formula by the ordinary algebraic convention with regard to positive and negative signs.

Direct impact of spheres.

In the standard case, then, the quantity of motion lost, up to any instant of the impact, by one of the bodies, is equal to that gained by the other. Hence at the instant when their velocities are equalized they move as one mass with a momentum equal to the sum of the momenta of the two before impact. That is to say, if v denote the common velocity at this instant, we have

$$(M+M')\,v = MV + M'V',$$

or

$$v = \frac{MV + M'V'}{M+M'},$$

if M, M' denote the masses of the two bodies, and V, V' their velocities before impact.

During this first period of the impact the bodies have been, on the whole, coming into closer contact with one another, through a compression or deformation experienced by each, and resulting, as remarked above, in a fitting together of the two surfaces over a finite area. No body in nature is perfectly inelastic; and hence, at the instant of closest approximation, the mutual force called into action between the two bodies continues, and tends to separate them. Unless prevented by natural surface cohesion or welding (such as is always found, as we shall see later in our chapter on Properties of Matter, however hard and well polished the surfaces may

Direct impact of spheres.

be), or by artificial appliances (such as a coating of wax, applied in one of the common illustrative experiments; or the coupling applied between two railway carriages when run together so as to push in the springs, according to the usual practice at railway stations), the two bodies are actually separated by this force, and move away from one another. Newton found that, *provided the impact is not so violent as to make any sensible permanent indentation in either body*, the relative velocity of separation after the impact bears a proportion to their previous relative velocity of approach, which is constant for the same two bodies. This proportion, always less than unity, approaches more and more nearly to it the harder the bodies are. Thus with balls of compressed wool he found it $\frac{5}{9}$, iron nearly the same, glass $\frac{15}{16}$. The results of more recent experiments on the same subject have confirmed Newton's law. These will be described later. In any case of the collision of two balls, let e denote this proportion, to which we give the name *Coefficient of Restitution*;* and, with previous notation, let in addition U, U' denote the velocities of the two bodies after the conclusion of the impact; in the standard case each being positive, but $U' > U$. Then we have

Effect of elasticity.

Newton's experiments.

$$U' - U = e(V - V')$$

and, as before, since one has lost as much momentum as the other has gained,

$$MU + M'U' = MV + M'V'.$$

From these equations we find

$$(M + M')\,U = MV + M'V' - eM'(V - V'),$$

with a similar expression for U'.

Also we have, as above,

$$(M + M')\,v = MV + M'V'.$$

Hence, by subtraction,

$$(M + M')(v - U) = eM'(V - V') = e\{M'V - (M + M')\,v + MV\}$$

* In most modern treatises this is called a "coefficient of elasticity," which is clearly a mistake; suggested, it may be, by Newton's words, but inconsistent with his facts, and utterly at variance with modern language and modern knowledge regarding elasticity.

and therefore

Direct impact of spheres.

$$v - U = e(V - v).$$

Of course we have also

$$U' - v = e(v - V').$$

These results may be put in words thus:—The *relative* velocity of either of the bodies with regard to the centre of inertia of the two is, after the completion of the impact, reversed in direction, and diminished in the ratio $e : 1$.

301. Hence the loss of kinetic energy, being, according to §§ 267, 280, due only to change of kinetic energy relative to the centre of inertia, is to this part of the whole as $1 - e^2 : 1$.

Thus

$$\text{Initial kinetic energy} = \tfrac{1}{2}(M + M')v^2 + \tfrac{1}{2}M(V - v)^2 + \tfrac{1}{2}M'(v - V')^2.$$

$$\text{Final} \quad ,, \quad ,, \quad = \tfrac{1}{2}(M + M')v^2 + \tfrac{1}{2}M(v - U)^2 + \tfrac{1}{2}M'(U' - v)^2.$$

$$\text{Loss} \quad = \tfrac{1}{2}(1 - e^2)\{M(V - v)^2 + M'(v - V')^2\}.$$

302. When two elastic bodies, the two balls supposed above for instance, impinge, some portion of their previous kinetic energy will always remain in them as vibrations. A *portion* of the loss of energy (miscalled the effect of imperfect elasticity) is necessarily due to this cause in every real case.

Distribution of energy after impact.

Later, in our chapter on Properties of Matter, it will be shown as a result of experiment, that forces of elasticity are, to a very close degree of accuracy, simply proportional to the strains (§ 154), within the limits of elasticity, in elastic solids which, like metals, glass, etc., bear but small deformations without permanent change. Hence when two such bodies come into collision, sometimes with greater and sometimes with less mutual velocity, but with all other circumstances similar, the velocities of all particles of either body, at corresponding times of the impacts, will be always in the same proportion. Hence the velocity of separation of the centres of inertia after impact will bear a constant proportion to the previous velocity of approach; which agrees with the Newtonian Law. It is therefore probable that a very sensible portion, if not the whole, of the loss of energy in the visible motions of two elastic bodies, after impact, experimented on by Newton, may have been due

Newton's experimental law consistent with *perfect* elasticity.

Distribution of energy after impact.

to vibrations; but unless some other cause also was largely operative, it is difficult to see how the loss was so much greater with iron balls than with glass.

303. In certain definite extreme cases, imaginable although not realizable, no energy will be spent in vibrations, and the two bodies will separate, each moving simply as a rigid body, and having in this simple motion the whole energy of work done on it by elastic force during the collision. For instance, let the two bodies be cylinders, or prismatic bars with flat ends, of the same kind of substance, and of equal and similar transverse sections; and let this substance have the property of compressibility with perfect elasticity, in the direction of the length of the bar, and of absolute resistance to change in every transverse dimension. Before impact, let the two bodies be placed with their lengths in one line, and their transverse sections (if not circular) similarly situated, and let one or both be set in motion in this line. The result, as regards the motions of the two bodies after the collision, will be sensibly the same if they are of any real ordinary elastic solid material, provided the greatest transverse diameter of each is very small in comparison with its length. Then, if the lengths of the two be equal, they will separate after impact with the same relative velocity as that with which they approached, and neither will retain any vibratory motion after the end of the collision.

304. If the two bars are of unequal length, the shorter will, after the impact, be exactly in the same state as if it had struck another of its own length, and it therefore will move as a rigid body after the collision. But the other will, along with a motion of its centre of gravity, calculable from the principle that its whole momentum must (§ 267) be changed by an amount equal exactly to the momentum gained or lost by the first, have also a vibratory motion, of which the whole kinetic and potential energy will make up the deficiency of energy which we shall presently calculate in the motions of the centres of inertia. For simplicity, let the longer body be supposed to be at rest before the collision. Then the shorter on striking it will be left at rest; this being clearly the result in the case of

Distribution of energy after impact.

$e=1$ in the preceding formulæ (§ 300) applied to the impact of one body striking another of equal mass previously at rest. The longer bar will move away with the same momentum, and therefore with less velocity of its centre of inertia, and less kinetic energy of this motion, than the other body had before impact, in the ratio of the smaller to the greater mass. It will also have a very remarkable vibratory motion, which, when its length is more than double of that of the other, will consist of a wave running backwards and forwards through its length, and causing the motion of its ends, and, in fact, of every particle of it, to take place by "fits and starts," not continuously. The full analysis of these circumstances, though very simple, must be reserved until we are especially occupied with waves, and the kinetics of elastic solids. It is sufficient at present to remark, that the motions of the centres of inertia of the two bodies after impact, whatever they may have been previously, are given by the preceding formulæ with for e the value $\frac{M'}{M}$, where M' and M are the smaller and the larger mass respectively.

305. The mathematical theory of the vibrations of solid elastic spheres has not yet been worked out; and its application to the case of the vibrations produced by impact presents considerable difficulty. Experiment, however, renders it certain, that but a small part of the whole kinetic energy of the previous motions can remain in the form of vibrations after the impact of two equal spheres of glass or of ivory. This is proved, for instance, by the common observation, that one of them remains nearly motionless after striking the other previously at rest; since, the velocity of the common centre of inertia of the two being necessarily unchanged by the impact, we infer that the second ball acquires a velocity nearly equal to that which the first had before striking it. But it is to be expected that unequal balls of the same substance coming into collision will, by impact, convert a very sensible proportion of the kinetic energy of their previous motions into energy of vibrations; and generally, that the same will be the case when equal or unequal masses of different substances come into colli-

Distribution of energy after impact.

sion; although for one particular proportion of their diameters, depending on their densities and elastic qualities, this effect will be a minimum, and possibly not much more sensible than it is when the substances are the same and the diameters equal.

306. It need scarcely be said that in such cases of impact as that of the tongue of a bell, or of a clock-hammer striking its bell (or spiral spring as in the American clocks), or of pianoforte hammers striking the strings, or of a drum struck with the proper implement, a large part of the kinetic energy of the blow is spent in generating vibrations.

Moment of an impact about an axis.

307. The *Moment of an impact* about any axis is derived from the line and amount of the impact in the same way as the moment of a velocity or force is determined from the line and amount of the velocity or force, §§ 235, 236. If a body is struck, the change of its moment of momentum about any axis is equal to the moment of the impact round that axis. But, without considering the measure of the impact, we see (§ 267) that the moment of momentum round any axis, lost by one body in striking another, is, as in every case of mutual action, equal to that gained by the other.

Ballistic pendulum.

Thus, to recur to the ballistic pendulum—the line of motion of the bullet at impact may be in any direction whatever, but the only part which is effective is the component in a plane perpendicular to the axis. We may therefore, for simplicity, consider the motion to be in a line perpendicular to the axis, though not necessarily horizontal. Let m be the mass of the bullet, v its velocity, and p the distance of its line of motion from the axis. Let M be the mass of the pendulum with the bullet lodged in it, and k its radius of gyration. Then if ω be the angular velocity of the pendulum when the impact is complete,

$$mvp = Mk^2\omega,$$

from which the solution of the question is easily determined.

For the kinetic energy after impact is changed (§ 241) into its equivalent in potential energy when the pendulum reaches its position of greatest deflection. Let this be given by the angle θ: then the height to which the centre of inertia is raised is $h(1-\cos\theta)$ if h be its distance from the axis. Thus

Ballistic pendulum.

$$Mgh\,(1-\cos\theta)=\tfrac{1}{2}Mk^2\omega^2=\tfrac{1}{2}\frac{m^2v^2p^2}{Mk^2},$$

or

$$2\sin\frac{\theta}{2}=\frac{mvp}{Mk\sqrt{gh}},$$

an expression for the chord of the angle of deflection. In practice the chord of the angle θ is measured by means of a light tape or cord attached to a point of the pendulum, and slipping with small friction through a clip fixed close to the position occupied by that point when the pendulum hangs at rest.

Work done by impact.

308. *Work done by an impact* is, in general, the product of the impact into half the sum of the initial and final velocities of the point at which it is applied, resolved in the direction of the impact. In the case of direct impact, such as that treated in § 300, the initial kinetic energy of the body is $\frac{1}{2}MV^2$, the final $\frac{1}{2}MU^2$, and therefore the gain, by the impact, is

$$\tfrac{1}{2}M\,(U^2-V^2),$$

or, which is the same,

$$M\,(U-V)\,.\,\tfrac{1}{2}\,(U+V).$$

But $M(U-V)$ is (§ 295) equal to the amount of the impact. Hence the proposition: the extension of which to the most general circumstances is easily seen.

Let ι be the amount of the impulse up to time τ, and I the whole amount, up to the end, T. Thus,—

$$\iota=\int_0^\tau Pd\tau,\ \ I=\int_0^T Pd\tau;\ \text{also}\ P=\frac{d\iota}{d\tau}.$$

Whatever may be the conditions to which the body struck is subjected, the change of velocity in the point struck is proportional to the amount of the impulse up to any part of its whole time, so that, if $\mathfrak{M}$ be a constant depending on the masses and conditions of constraint involved, and if U, v, V denote the component velocities of the point struck, in the direction of the impulse, at the beginning, at the time τ, and at the end, respectively, we have

$$v=U+\frac{\iota}{\mathfrak{M}},\ \ V=U+\frac{I}{\mathfrak{M}}.$$

Hence, for the rate of the doing of work by the force P, at the instant t, we have

$$Pv=PU+\frac{\iota P}{\mathfrak{M}}.$$

Work done by impact.

Hence for the whole work (W) done by it,

$$W = \int_0^T \left(PU + \frac{\iota P}{\mathfrak{M}}\right) d\tau$$

$$= UI + \frac{1}{\mathfrak{M}} \int_0^I \iota d\iota = UI + \tfrac{1}{2} \frac{I^2}{\mathfrak{M}}$$

$$= UI + \tfrac{1}{2} I (V - U) = I \cdot \tfrac{1}{2} (U + V).$$

309. It is worthy of remark, that if any number of impacts be applied to a body, their whole effect will be the same whether they be applied together or successively (provided that the whole time occupied by them be infinitely short), although the work done by each particular impact is in general different according to the order in which the several impacts are applied. The whole amount of work is the sum of the products obtained by multiplying each impact by half the sum of the components of the initial and final velocities of the point to which it is applied.

Equations of impulsive motion.

310. The effect of any stated impulses, applied to a rigid body, or to a system of material points or rigid bodies connected in any way, is to be found most readily by the aid of D'Alembert's principle; according to which the given impulses, and the impulsive reaction against the generation of motion, measured in amount by the momenta generated, are in equilibrium; and are therefore to be dealt with mathematically by applying to them the equations of equilibrium of the system.

Let P_1, Q_1, R_1 be the component impulses on the first particle, m_1, and let $\dot{x}_1$, $\dot{y}_1$, $\dot{z}_1$ be the components of the velocity instantaneously acquired by this particle. Component forces equal to $(P_1 - m_1\dot{x}_1)$, $(Q_1 - m_1\dot{y}_1)$, ... must equilibrate the system, and therefore we have (§ 290)

$$\Sigma \{(P - m\dot{x})\, \delta x + (Q - m\dot{y})\, \delta y + (R - m\dot{z})\, \delta z\} = 0 \ldots\ldots\ldots\ldots (a)$$

where δx_1, δy_1, ... denote the components of any infinitely small displacements of the particles possible under the conditions of the system. Or, which amounts to the same thing, since any possible infinitely small displacements are simply proportional to any possible velocities in the same directions,

$$\Sigma \{(P - m\dot{x})\, u + (Q - m\dot{y})\, v + (Q - m\dot{z})\, w\} = 0 \ldots\ldots\ldots\ldots (b)$$

where u_1, v_1, w_1 denote any possible component velocities of the first particle, etc.

Equations of impulsive motion.

One particular case of this equation is of course had by supposing $u_1, v_1, \ldots$ to be equal to the velocities $\dot{x}_1, \dot{y}_1, \ldots$ actually acquired; and, by halving, etc., we find

$$\Sigma\left(P.\tfrac{1}{2}\dot{x}+Q.\tfrac{1}{2}\dot{y}+R.\tfrac{1}{2}\dot{z}\right)=\tfrac{1}{2}\Sigma m(\dot{x}^2+\dot{y}^2+\dot{z}^2)\ldots\ldots.(c).$$

This agrees with § 308 above.

311. Euler discovered that the kinetic energy acquired from rest by a rigid body in virtue of an impulse fulfils a maximum-minimum condition. Lagrange* extended this proposition to a system of bodies connected by any invariable kinematic relations, and struck with any impulses. Delaunay found that it is really always a maximum *when the impulses are given, and when different motions possible under the conditions of the system, and fulfilling the law of energy* [§ 310 (*c*)], *are considered.* Farther, Bertrand shows that the energy actually acquired is not merely a "maximum," but exceeds the energy of any other motion fulfilling these conditions; and that the amount of the excess is equal to the energy of the motion which must be compounded with either to produce the other.

Theorem of Euler, extended by Lagrange.

Equation of impulsive motion.

Let $\dot{x}_1', \dot{y}_1' \ldots$ be the component velocities of any motion whatever fulfilling the equation (*c*), which becomes

$$\tfrac{1}{2}\Sigma(P\dot{x}'+Q\dot{y}'+R\dot{z}')=\tfrac{1}{2}\Sigma m(\dot{x}'^2+\dot{y}'^2+\dot{z}'^2)=T'\ldots\ldots.(d).$$

If, then, we take $\dot{x}_1'-\dot{x}_1=u_1$, $\dot{y}_1'-\dot{y}_1=v_1$, etc., we have

$$T'-T=\tfrac{1}{2}\Sigma m\{(2\dot{x}+u)u+(2\dot{y}+v)v+(2\dot{z}+w)w\}$$

$$=\Sigma m(\dot{x}u+\dot{y}v+\dot{z}w)+\tfrac{1}{2}\Sigma m(u^2+v^2+w^2)\ldots\ldots(e).$$

But, by (*b*),

$$\Sigma m(\dot{x}u+\dot{y}v+\dot{z}w)=\Sigma(Pu+Qv+Rw)\ldots\ldots\ldots.(f);$$

and, by (*c*) and (*d*),

$$\Sigma(Pu+Qv+Rw)=2T'-2T\ldots\ldots\ldots\ldots\ldots(g).$$

Hence (*e*) becomes

$$T'-T=2(T'-T)+\tfrac{1}{2}\Sigma m(u^2+v^2+w^2),$$

whence $$T-T'=\tfrac{1}{2}\Sigma m(u^2+v^2+w^2)\ldots\ldots\ldots\ldots\ldots(h),$$

which is Bertrand's result.

* *Mécanique Analytique*, 2nde partie, 3me section, § 37.

Liquid set in motion impulsively.

312. The energy of the motion generated suddenly in a mass of incompressible liquid given at rest completely filling a vessel of any shape, when the vessel is suddenly set in motion, or when it is suddenly bent out of shape in any way whatever, subject to the condition of not changing its volume, *is less than the energy of any other motion it can have with the same motion of its bounding surface.* The consideration of this theorem, which, so far as we know, was first published in the *Cambridge and Dublin Mathematical Journal* [Feb. 1849], has led us to a general *minimum* property regarding motion acquired by any system when *any prescribed velocities* are generated suddenly in any of its parts; announced in the *Proceedings of the Royal Society of Edinburgh* for April, 1863. It is, that provided impulsive forces are applied to the system only at places where the velocities to be produced are prescribed, the kinetic energy is *less* in the actual motion than in any other motion which the system can take, and which has the same values for the prescribed velocities. The excess of the energy of any possible motion above that of the actual motion is (as in Bertrand's theorem) equal to the energy of the motion which must be compounded with either to produce the other. The proof is easy:—here it is:—

Equations (d), (e), and (f) hold as in § (311). But now each velocity component, u_1, v_1, w_1, u_2, etc. vanishes for which the component impulse P_1, Q_1, R_1, P_2, etc. does not vanish (because $\dot{x}_1+u_1$, $\dot{y}_1+v_1$, etc. fulfil the prescribed velocity conditions). Hence every product P_1u_1, Q_1v_1, etc. vanishes. Hence now instead of (g) and (h) we have

$$\Sigma(\dot{x}u+\dot{y}v+\dot{z}w)=0 \qquad (g'),$$

and

$$T'-T=\tfrac{1}{2}\Sigma m(u^2+v^2+w^2) \qquad (h').$$

We return to the subject in §§ 316, 317 as an illustration of the use of Lagrange's generalized co-ordinates; to the introduction of which into Dynamics we now proceed.

Impulsive motion referred to generalized co-ordinates.

313. The method of generalized co-ordinates explained above (§ 204) is extremely useful in its application to the dynamics of a system; whether for expressing and working out the details of any particular case in which there is any finite number of degrees of freedom, or for proving general principles applicable even to cases, such as that of a liquid, as described in the preceding section, in which there may be an infinite number of degrees of freedom. It leads us to generalize the measure of inertia, and the resolution and composition of forces, impulses, and momenta, on dynamical principles corresponding with the kinematical principles explained in § 204, which gave us generalized component velocities: and, as we shall see later, the generalized equations of continuous motion are not only very convenient for the solution of problems, but most *instructive* as to the nature of relations, however complicated, between the motions of different parts of a system. In the meantime we shall consider the generalized expressions for the impulsive generation of motion. We have seen above (§ 308) that the kinetic energy acquired by a system given at rest and struck with any given impulses, is equal to half the sum of the products of the component forces multiplied each into the corresponding component of the velocity acquired by its point of application, when the ordinary system of rectangular co-ordinates is used. Precisely the same statement holds on the generalized system, and if stated as the convention agreed upon, it suffices to define the generalized components of impulse, those of velocity having been fixed on kinematical principles (§ 204). Generalized components of momentum of any specified motion are, of course, equal to the generalized components of the impulse by which it could be generated from rest.

Impulsive motion referred to generalized co-ordinates.

Generalized components of impulse or momentum.

(*a*) Let $\psi, \phi, \theta, \ldots$ be the generalized co-ordinates of a material system at any time; and let $\dot{\psi}, \dot{\phi}, \dot{\theta}, \ldots$ be the corresponding generalized velocity-components, that is to say, the rates at which $\psi, \phi, \theta, \ldots$ increase per unit of time, at any instant, in the actual motion. If x_1, y_1, z_1 denote the common rectangular co-ordinates of one particle of the system, and $\dot{x}_1, \dot{y}_1, \dot{z}_1$ its component velocities, we have

$$\left.\begin{array}{l}\dot{x}_1 = \dfrac{dx_1}{d\psi}\dot{\psi} + \dfrac{dx_1}{d\phi}\dot{\phi} + \text{etc.}\\[2ex] \dot{y}_1 = \dfrac{dy_1}{d\psi}\dot{\psi} + \dfrac{dy_1}{d\phi}\dot{\phi} + \text{etc.}\\[1ex] \quad\;\;\text{etc.} \qquad\quad \text{etc.}\end{array}\right\} \ldots\ldots\ldots\ldots\ldots\ldots(1).$$

Hence the kinetic energy, which is $\Sigma\frac{1}{2}m(\dot{x}^2+\dot{y}^2+\dot{z}^2)$, in terms of rectangular co-ordinates, becomes a quadratic function of $\dot{\psi}$, $\dot{\phi}$, etc., when expressed in terms of generalized co-ordinates, so that if we denote it by T we have

Generalized expression for kinetic energy.

$$T=\tfrac{1}{2}\{(\psi,\psi)\dot{\psi}^2+(\phi,\phi)\dot{\phi}^2+\ldots+2(\psi,\phi)\dot{\psi}\dot{\phi}+\ldots\}\ldots\ldots(2),$$

where (ψ,ψ), (ϕ,ϕ), (ψ,ϕ), etc., denote various functions of the co-ordinates, determinable according to the conditions of the system. The only condition essentially fulfilled by these coefficients is, that they must give a finite positive value to T for all values of the variables.

(b) Again let (X_1, Y_1, Z_1), (X_2, Y_2, Z_2), etc., denote component forces on the particles (x_1, y_1, z_1), (x_2, y_2, z_2), etc., respectively; and let $(\delta x_1, \delta y_1, \delta z_1)$, etc., denote the components of any infinitely small motions possible without breaking the conditions of the system. The work done by those forces, upon the system when so displaced, will be

$$\Sigma(X\delta x+Y\delta y+Z\delta z)\ldots\ldots\ldots\ldots\ldots\ldots\ldots(3).$$

To transform this into an expression in terms of generalized co-ordinates, we have

$$\left.\begin{aligned}\delta x_1&=\frac{dx_1}{d\psi}\delta\psi+\frac{dx_1}{d\phi}\delta\phi+\text{etc.}\\ \delta y_1&=\frac{dy_1}{d\psi}\delta\psi+\frac{dy_1}{d\phi}\delta\phi+\text{etc.}\\ &\quad\text{etc.}\qquad\quad\text{etc.}\end{aligned}\right\}\ldots\ldots\ldots\ldots\ldots(4),$$

and it becomes

$$\Psi\delta\psi+\Phi\delta\phi+\text{etc.}\ldots\ldots\ldots\ldots\ldots\ldots\ldots\ldots(5),$$

where

Generalized components of force,

$$\left.\begin{aligned}\Psi&=\Sigma\left(X\frac{dx}{d\psi}+Y\frac{dy}{d\psi}+Z\frac{dz}{d\psi}\right)\\ \Phi&=\Sigma\left(X\frac{dx}{d\phi}+Y\frac{dy}{d\phi}+Z\frac{dz}{d\phi}\right)\\ &\quad\text{etc.}\qquad\quad\text{etc.}\end{aligned}\right\}\ldots\ldots\ldots\ldots\ldots(6).$$

These quantities, Ψ, Φ, etc., are clearly *the generalized components of the force on the system.*

Let $\underset{\cdot}{\Psi}$, $\underset{\cdot}{\Phi}$, etc. denote component impulses, generalized on the same principle; that is to say, let

of impulse.

$$\underset{\cdot}{\Psi}=\int_0^\tau\Psi dt,\quad \underset{\cdot}{\Phi}=\int_0^\tau\Phi dt,\ \text{etc.},$$

where Ψ, Φ, ... denote generalized components of the continuous force acting at any instant of the infinitely short time τ, within which the impulse is completed.

Impulsive generation of motion referred to generalized co-ordinates.

If this impulse is applied to the system, previously in motion in the manner specified above, and if $\delta\dot\psi$, $\delta\dot\phi$, ... denote the resulting augmentations of the components of velocity, the means of the component velocities before and after the impulse will be

$$\dot\psi + \tfrac{1}{2}\delta\dot\psi,\quad \dot\phi + \tfrac{1}{2}\delta\dot\phi,\ \ldots\ldots$$

Hence, according to the general principle explained above for calculating the work done by an impulse, the whole work done in this case is

$$\underset{\cdot}{\Psi}(\dot\psi + \tfrac{1}{2}\delta\dot\psi) + \underset{\cdot}{\Phi}(\dot\phi + \tfrac{1}{2}\delta\dot\phi) + \text{etc.}$$

To avoid unnecessary complications, let us suppose $\delta\dot\psi$, $\delta\dot\phi$, etc., to be each infinitely small. The preceding expression for the work done becomes

$$\underset{\cdot}{\Psi}\dot\psi + \underset{\cdot}{\Phi}\dot\phi + \text{etc.};$$

and, as the effect produced by this work is augmentation of kinetic energy from T to $T + \delta T$, we must have

$$\delta T = \underset{\cdot}{\Psi}\dot\psi + \underset{\cdot}{\Phi}\dot\phi + \text{etc.}$$

Now let the impulses be such as to augment $\dot\psi$ to $\dot\psi + \delta\dot\psi$, and to leave the other component velocities unchanged. We shall have

$$\underset{\cdot}{\Psi}\dot\psi + \underset{\cdot}{\Phi}\dot\phi + \text{etc.} = \frac{dT}{d\dot\psi}\delta\dot\psi.$$

Dividing both members by $\delta\dot\psi$, and observing that $\frac{dT}{d\dot\psi}$ is a linear function of $\dot\psi$, $\dot\phi$, etc., we see that $\frac{\underset{\cdot}{\Psi}}{\delta\dot\psi}$, $\frac{\underset{\cdot}{\Phi}}{\delta\dot\psi}$, etc., must be equal to the coefficients of $\dot\psi$, $\dot\phi$, ... respectively in $\frac{dT}{d\dot\psi}$.

(*c*) From this we see, further, that the impulse required to produce the component velocity $\dot\psi$ from rest, or to generate it in the system moving with any other possible velocity, has for its components

$$(\psi, \psi)\,\dot\psi,\quad (\psi, \phi)\,\dot\psi,\quad (\psi, \theta)\,\dot\psi,\ \text{etc.}$$

Hence we conclude that to generate the whole resultant velocity $(\dot\psi, \dot\phi, \ldots)$ from rest, requires an impulse, of which the components, if denoted by ξ, η, ζ, ... , are expressed as follows :—

Momenta in terms of velocities.

$$\left.\begin{aligned}\xi &= (\psi, \psi)\dot{\psi} + (\phi, \psi)\dot{\phi} + (\theta, \psi)\dot{\theta} + \ldots\\ \eta &= (\psi, \phi)\dot{\psi} + (\phi, \phi)\dot{\phi} + (\theta, \phi)\dot{\theta} + \ldots\\ \zeta &= (\psi, \theta)\dot{\psi} + (\phi, \theta)\dot{\phi} + (\theta, \theta)\dot{\theta} + \ldots\\ &\text{etc.}\end{aligned}\right\}\ldots\ldots\ldots\ldots(7),$$

where it must be remembered that, as seen in the original expression for T, from which they are derived, (ϕ, ψ) means the same thing as (ψ, ϕ), and so on. The preceding expressions are the differential coefficients of T with reference to the velocities; that is to say,

$$\xi = \frac{dT}{d\dot{\psi}},\quad \eta = \frac{dT}{d\dot{\phi}},\quad \zeta = \frac{dT}{d\dot{\theta}}\ldots\ldots\ldots\ldots(8).$$

(*d*) The second members of these equations being linear functions of $\dot{\psi}, \dot{\phi}, \ldots$, we may, by ordinary elimination, find $\dot{\psi}, \dot{\phi}$, etc., in terms of ξ, η, etc., and the expressions so obtained are of course linear functions of the last-named elements. And, since T is a quadratic function of $\dot{\psi}, \dot{\phi}$, etc., we have

$$2T = \xi\dot{\psi} + \eta\dot{\phi} + \zeta\dot{\theta} + \text{etc.}\ldots\ldots\ldots\ldots(9).$$

Kinetic energy in terms of momentums and velocities.

From this, on the supposition that $T, \dot{\psi}, \dot{\phi}, \ldots$ are expressed in terms of $\xi, \eta, \ldots$, we have by differentiation

$$2\frac{dT}{d\xi} = \dot{\psi} + \xi\frac{d\dot{\psi}}{d\xi} + \eta\frac{d\dot{\phi}}{d\xi} + \zeta\frac{d\dot{\theta}}{d\xi} + \text{etc.}$$

Now the algebraic process by which $\dot{\psi}, \dot{\phi}$, etc., are obtained in terms of ξ, η, etc., shows that, inasmuch as the coefficient of $\dot{\phi}$ in the expression, (7), for ξ, is equal to the coefficient of $\dot{\psi}$, in the expression for η, and so on; the coefficient of η in the expression for $\dot{\psi}$ must be equal to the coefficient of ξ in the expression for $\dot{\phi}$, and so on; that is to say,

$$\frac{d\dot{\psi}}{d\eta} = \frac{d\dot{\phi}}{d\xi},\quad \frac{d\dot{\psi}}{d\zeta} = \frac{d\dot{\theta}}{d\xi},\ \text{etc.}$$

Hence the preceding expression becomes

$$2\frac{dT}{d\xi} = \dot{\psi} + \xi\frac{d\dot{\psi}}{d\xi} + \eta\frac{d\dot{\psi}}{d\eta} + \zeta\frac{d\dot{\psi}}{d\zeta} + \ldots = 2\dot{\psi},$$

and therefore

Velocities in terms of momentums.

$$\left.\begin{aligned}&\dot{\psi} = \frac{dT}{d\xi}.\\ \text{Similarly}\qquad &\dot{\phi} = \frac{dT}{d\eta},\ \text{etc.}\end{aligned}\right\}\ldots\ldots\ldots\ldots(10).$$

These expressions solve the direct problem,—to find the velocity produced by a given impulse (ξ, η, ...), when we have the kinetic energy, T, expressed as a quadratic function of the components of the impulse.

Velocities in terms of momentums.

(*e*) If we consider the motion simply, without reference to the impulse required to generate it from rest, or to stop it, the quantities ξ, η, ... are clearly to be regarded as the components of the momentum of the motion, according to the system of generalized co-ordinates.

(*f*) The following algebraic relation will be useful :—

Reciprocal relation between momentums and velocities in two motions.

$$\xi_{,}\dot\psi + \eta_{,}\dot\phi + \zeta_{,}\dot\theta + \text{etc.} = \xi\dot\psi_{,} + \eta\dot\phi_{,} + \zeta\dot\theta_{,} + \text{etc} \ldots\ldots\ldots\ldots(11),$$

where, ξ, η, ψ, ϕ, etc., having the same signification as before, $\xi_{,}$, $\eta_{,}$, $\zeta_{,}$, etc., denote the impulse-components corresponding to any other values, $\dot\psi_{,}$, $\dot\phi_{,}$, $\dot\theta_{,}$, etc., of the velocity-components. It is proved by observing that each member of the equation becomes a symmetrical function of $\dot\psi$, $\dot\psi_{,}$; $\dot\phi$, $\dot\phi_{,}$; etc. ; when for $\xi_{,}$, $\eta_{,}$, etc., their values in terms of $\dot\psi_{,}$, $\dot\phi_{,}$, etc., and for ξ, η, etc., their values in terms of $\dot\psi$, $\dot\phi$, etc., are substituted.

314. A material system of any kind, given at rest, and subjected to an impulse in any specified direction, and of any given magnitude, moves off so as to take the greatest amount of kinetic energy which the specified impulse can give it, subject to § 308 or § 309 (*c*).

Application of generalized co-ordinates to theorems of § 311.

Let ξ, η, ... be the components of the given impulse, and $\dot\psi$, $\dot\phi$, ... the components of the actual motion produced by it, which are determined by the equations (10) above. Now let us suppose the system be guided, by means of merely directive constraint, to take, from rest, under the influence of the given impulse, some motion ($\dot\psi_{,}$, $\dot\phi_{,}$, ...) different from the actual motion; and let $\xi_{,}$, $\eta_{,}$, ... be the impulse which, with this constraint removed, would produce the motion ($\dot\psi_{,}$, $\dot\phi_{,}$, ...). We shall have, for this case, as above,

$$T_{,} = \tfrac{1}{2}(\xi_{,}\dot\psi_{,} + \eta_{,}\dot\phi_{,} + \ldots).$$

But $\xi_{,} - \xi$, $\eta_{,} - \eta$ ... are the components of the impulse experienced in virtue of the constraint we have supposed introduced. They neither perform nor consume work on the system when moving as directed by this constraint ; that is to say,

$$(\xi_{,} - \xi)\dot\psi_{,} + (\eta_{,} - \eta)\dot\phi_{,} + (\zeta_{,} - \zeta)\dot\theta_{,} + \text{etc.} = 0 \ldots\ldots\ldots\ldots(12);$$

Application of generalized co-ordinates to theorems of § 311.

and therefore

$$2T_{,} = \xi\dot{\psi}_{,} + \eta\dot{\phi}_{,} + \zeta\dot{\theta}_{,} + \text{etc.} \quad \ldots\ldots\ldots\ldots\ldots\ldots(13).$$

Hence we have

$$2(T - T_{,}) = \xi(\dot{\psi} - \dot{\psi}_{,}) + \eta(\dot{\phi} - \dot{\phi}_{,}) + \text{etc.}$$
$$= (\xi - \xi_{,})(\dot{\psi} - \dot{\psi}_{,}) + (\eta - \eta_{,})(\dot{\phi} - \dot{\phi}_{,}) + \text{etc.}$$
$$+ \xi_{,}(\dot{\psi} - \dot{\psi}_{,}) + \eta_{,}(\dot{\phi} - \dot{\phi}_{,}) + \text{etc.}$$

But, by (11) and (12) above, we have

$$\xi_{,}(\dot{\psi} - \dot{\psi}_{,}) + \eta_{,}(\dot{\phi} - \dot{\phi}_{,}) + \text{etc.} = (\xi - \xi_{,})\dot{\psi}_{,} + (\eta - \eta_{,})\dot{\phi}_{,} + \text{etc.} = 0,$$

and therefore we have finally

$$2(T - T_{,}) = (\xi - \xi_{,})(\dot{\psi} - \dot{\psi}_{,}) + (\eta - \eta_{,})(\dot{\phi} - \dot{\phi}_{,}) + \text{etc.} \quad \ldots(14),$$

Theorems of § 311 in terms of generalized co-ordinates.

that is to say, T exceeds $T_{,}$ by the amount of the kinetic energy that would be generated by an impulse ($\xi - \xi_{,}$, $\eta - \eta_{,}$, $\zeta - \zeta_{,}$, etc.) applied simply to the system, which is essentially positive. In other words,

315. If the system is guided to take, under the action of a given impulse, any motion ($\dot{\psi}_{,}$, $\dot{\phi}_{,}$, ...) different from the natural motion ($\dot{\psi}$, $\dot{\phi}$, ...), it will have less kinetic energy than that of the natural motion, by a difference equal to the kinetic energy of the motion ($\dot{\psi} - \dot{\psi}_{,}$, $\dot{\phi} - \dot{\phi}_{,}$, ...).

COR. If a set of material points are struck independently by impulses each given in amount, more kinetic energy is generated if the points are perfectly free to move each independently of all the others, than if they are connected in any way. And the deficiency of energy in the latter case is equal to the amount of the kinetic energy of the motion which geometrically compounded with the motion of either case would give that of the other.

Problems whose data involve impulses and velocities.

(*a*) Hitherto we have either supposed the motion to be fully given, and the impulses required to produce them, to be to be found; or the impulses to be given and the motions produced by them to be to be found. A not less important class of problems is presented by supposing as many linear equations of condition between the impulses and components of motion to be given as there are degrees of freedom of the system to move (or independent co-ordinates). These equations, and as many more supplied by (8) or their equivalents (10), suffice for the complete solution of the problem, to determine the impulses and the motion.

Problems whose data involve impulses and velocities.

(*b*) A very important case of this class is presented by prescribing, among the velocities alone, a number of linear equations with constant terms, and supposing the impulses to be so directed and related as to do no work on any velocities satisfying another prescribed set of linear equations with no constant terms; the whole number of equations of course being equal to the number of independent co-ordinates of the system. The equations for solving this problem need not be written down, as they are obvious; but the following reduction is useful, as affording the easiest proof of the *minimum* property stated below.

(*c*) The given equations among the velocities may be reduced to a set, each homogeneous, except one equation with a constant term. Those homogeneous equations diminish the number of degrees of freedom; and we may transform the co-ordinates so as to have the number of independent co-ordinates diminished accordingly. Farther, we may choose the new co-ordinates, so that the linear function of the velocities in the single equation with a constant term may be one of the new velocity-components; and the linear functions of the velocities appearing in the equation connected with the prescribed conditions as to the impulses may be the remaining velocity-components. Thus the impulse will fulfil the condition of doing no work on any other component velocity than the one which is given, and the general problem—

General problem (compare § 312).

316. Given any material system at rest: let any parts of it be set in motion suddenly with any specified velocities, possible according to the conditions of the system; and let its other parts be influenced only by its connexions with those; required the motion:

takes the following very simple form :—An impulse of the character specified as a particular component, according to the generalized method of co-ordinates, acts on a material system; its amount being such as to produce a given velocity-component of the corresponding type. It is required to find the motion.

The solution of course is to be found from the equations

$$\dot{\psi} = A, \qquad \eta = 0, \qquad \zeta = 0 \ldots\ldots\ldots\ldots\ldots\ldots\ldots(15)$$

(which are the special equations of condition of the problem) and the general kinetic equations (7), or (10). Choosing the latter, and denoting by $[\xi, \xi]$, $[\xi, \eta]$, etc., the coefficients of $\frac{1}{2}\xi^2$, $\xi\eta$, etc.,

General problem (compare § 312).

in T, we have

$$\xi = \frac{A}{[\xi, \xi]}, \quad \dot{\phi} = \frac{[\xi, \eta]}{[\xi, \xi]}A, \quad \dot{\theta} = \frac{[\xi, \zeta]}{[\xi, \xi]}A, \text{ etc.} \qquad (16)$$

for the result.

This result possesses the remarkable property, that the kinetic energy of the motion expressed by it is less than that of any other motion which fulfils the prescribed condition as to velocity. For, if $\xi_,$, $\eta_,$, $\zeta_,$, etc., denote the impulses required to produce any other motion, $\dot{\psi}_,$, $\dot{\phi}_,$, $\dot{\theta}_,$, etc., and $T_,$ the corresponding kinetic energy, we have, by (9),

$$2T_, = \xi_,\dot{\psi}_, + \eta_,\dot{\phi}_, + \zeta_,\dot{\theta}_, + \text{etc.}$$

But by (11),

$$\xi_,\dot{\psi} + \eta_,\dot{\phi} + \zeta_,\dot{\theta} + \text{etc.} = \xi\dot{\psi}_,,$$

since, by (15), we have $\eta = 0$, $\xi = 0$, etc. Hence

$$2T_, = \xi\dot{\psi}_, + \xi_,(\dot{\psi}_, - \dot{\psi}) + \eta_,(\dot{\phi}_, - \dot{\phi}) + \zeta_,(\dot{\theta}_, - \dot{\theta}) + \ldots$$

Now let also this second case ($\dot{\psi}_,$, $\dot{\phi}_,$,...) of motion fulfil the prescribed velocity-condition $\dot{\psi}_, = A$. We shall have

$$\xi_,(\dot{\psi}_, - \dot{\psi}) + \eta_,(\dot{\phi}_, - \dot{\phi}) + \zeta_,(\dot{\theta}_, - \dot{\theta}) + \ldots$$
$$= (\xi_, - \xi)(\dot{\psi}_, - \dot{\psi}) + (\eta_, - \eta)(\dot{\phi}_, - \dot{\phi}) + (\zeta_, - \zeta)(\dot{\theta}_, - \dot{\theta}) + \ldots.$$

since $\dot{\psi}_, - \dot{\psi} = 0$, $\eta = 0$, $\zeta = 0$,.... Hence if $\mathfrak{T}$ denote the kinetic energy of the differential motion ($\dot{\psi}_, - \psi$, $\dot{\phi}_, - \phi$,...) we have

$$2T_, = 2T + 2\mathfrak{T} \qquad (17);$$

but $\mathfrak{T}$ is essentially positive and therefore $T_,$, the kinetic energy of any motion fulfilling the prescribed velocity-condition, but differing from the actual motion, is greater than T the kinetic energy of the actual motion; and the amount, $\mathfrak{T}$, of the difference is given by the equation

$$2\mathfrak{T} = \eta_,(\dot{\phi}_, - \dot{\phi}) + \zeta_,(\dot{\theta}_, - \dot{\theta}) + \text{etc.} \qquad (18),$$

or in words,

Kinetic energy a minimum in this case.

317. The solution of the problem is this:—The motion actually taken by the system is the motion which has less kinetic energy than any other fulfilling the prescribed velocity-conditions. And the excess of the energy of any other such motion, above that of the actual motion, is equal to the energy of the motion which must be compounded with either to produce the other.

In dealing with cases it may often happen that the use of the co-ordinate system required for the application of the solution (16) is not convenient; but in all cases, even in such as in examples (2) and (3) below, which involve an infinite number of degrees of freedom, the minimum property now proved affords an easy solution.

Kinetic energy a minimum in this case.

Example (1). Let a smooth plane, constrained to keep moving with a given normal velocity, q, come in contact with a free inelastic rigid body at rest: to find the motion produced. The velocity-condition here is, that the motion shall consist of any motion whatever giving to the point of the body which is struck a stated velocity, q, perpendicular to the impinging plane, compounded with any motion whatever giving to the same point any velocity parallel to this plane. To express this condition, let u, v, w be rectangular component linear velocities of the centre of gravity, and let ϖ, ρ, σ be component angular velocities round axes through the centre of gravity parallel to the line of reference. Thus, if x, y, z denote the co-ordinates of the point struck relatively to these axes through the centre of gravity, and if l, m, n be the direction cosines of the normal to the impinging plane, the prescribed velocity-condition becomes

Impact of a smooth rigid plane of infinite mass on a free rigid body at rest.

$$(u+\rho z-\sigma y)\,l+(v+\sigma x-\varpi z)\,m+(w+\varpi y-\rho x)\,n=-q \quad \text{.........}(a),$$

the negative sign being placed before q on the understanding that the motion of the impinging plane is obliquely, if not directly, *towards* the centre of gravity, when l, m, n are each positive. If, now, we suppose the rectangular axes through the centre of gravity to be principal axes of the body, and denote by Mf^2, Mg^2, Mh^2 the moments of inertia round them, we have

$$T=\tfrac{1}{2}M\,(u^2+v^2+w^2+f^2\varpi^2+g^2\rho^2+h^2\sigma^2) \quad \text{..............}(b).$$

This must be made a minimum subject to the equation of condition (a). Hence, by the ordinary method of indeterminate multipliers,

$$\left.\begin{array}{c} Mu+\lambda l=0,\ Mv+\lambda m=0,\ Mw+\lambda n=0 \\ Mf^2\varpi+\lambda\,(ny-mz)=0,\ Mg^2\rho+\lambda(lz-nx)=0,\ Mh^2\sigma+\lambda(mx-ly)=0 \end{array}\right\}(c).$$

These six equations give each of them explicitly the value of one of the six unknown quantities u, v, w, ϖ, ρ, σ, in terms of λ and data. Using the values thus found in (a), we have an equation to determine λ; and thus the solution is completed. The first three of equations (c) show that λ, which has entered as an indeterminate multiplier, is to be interpreted as the measure of the amount of the impulse.

Generation of motion by impulse in an inextensible cord or chain.

Example (2). A stated velocity in a stated direction is communicated impulsively to each end of a flexible inextensible cord forming any curvilineal arc: it is required to find the initial motion of the whole cord.

Let x, y, z be the co-ordinates of any point P in it, and $\dot{x}$, $\dot{y}$, $\dot{z}$ the components of the required initial velocity. Let also s be the length from one end to the point P.

If the cord were extensible, the rate per unit of time of the stretching per unit of length which it would experience at P, in virtue of the motion $\dot{x}$, $\dot{y}$, $\dot{z}$, would be

$$\frac{dx}{ds}\frac{d\dot{x}}{ds}+\frac{dy}{ds}\frac{d\dot{y}}{ds}+\frac{dz}{ds}\frac{d\dot{z}}{ds}.$$

Hence, as the cord is inextensible, by hypothesis,

$$\frac{dx}{ds}\frac{d\dot{x}}{ds}+\frac{dy}{ds}\frac{d\dot{y}}{ds}+\frac{dz}{ds}\frac{d\dot{z}}{ds}=0 \quad\ldots\ldots\ldots\ldots\ldots\ldots (a).$$

Subject to this, the kinematical condition of the system, and

$$\left.\begin{array}{l}\dot{x}=u\\ \dot{y}=v\\ \dot{z}=w\end{array}\right\}\text{ when } s=0,\quad \left.\begin{array}{l}\dot{x}=u'\\ \dot{y}=v'\\ \dot{z}=w'\end{array}\right\}\text{ when } s=l,$$

l denoting the length of the cord, and (u, v, w), (u', v', w'), the components of the given velocities at its two ends: it is required to find $\dot{x}$, $\dot{y}$, $\dot{z}$ at every point, so as to make

$$\int_0^l \tfrac{1}{2}\mu\,(\dot{x}^2+\dot{y}^2+\dot{z}^2)\,ds \quad\ldots\ldots\ldots\ldots\ldots\ldots (b)$$

a minimum, μ denoting the mass of the string per unit of length, at the point P, which need not be uniform from point to point; and of course

$$ds=(dx^2+dy^2+dz^2)^{\frac{1}{2}} \quad\ldots\ldots\ldots\ldots\ldots\ldots (c).$$

Multiplying (a) by λ, an indeterminate multiplier, and proceeding as usual according to the method of variations, we have

$$\int_0^l\left\{\mu\,(\dot{x}\delta\dot{x}+\dot{y}\delta\dot{y}+\dot{z}\delta\dot{z})+\lambda\left(\frac{dx}{ds}\frac{d\delta\dot{x}}{ds}+\frac{dy}{ds}\frac{d\delta\dot{y}}{ds}+\frac{dz}{ds}\frac{d\delta\dot{z}}{ds}\right)\right\}ds=0,$$

in which we may regard x, y, z as known functions of s, and this it is convenient we should make independent variable. Integrating "by parts" the portion of the first member which contains λ, and attending to the terminal conditions, we find, according to the regular process, for the equations containing the solution

Generation of motion by impulse in an inextensible cord or chain.

$$\mu\dot{x} = \frac{d}{ds}\left(\lambda\frac{dx}{ds}\right),\ \mu\dot{y} = \frac{d}{ds}\left(\lambda\frac{dy}{ds}\right),\ \mu\dot{z} = \frac{d}{ds}\left(\lambda\frac{dz}{ds}\right) \ldots\ldots\ldots\ldots (d).$$

These three equations with (a) suffice to determine the four unknown quantities, $\dot{x}$, $\dot{y}$, $\dot{z}$, and λ. Using (d) to eliminate $\dot{x}$, $\dot{y}$, $\dot{z}$ from (a), we have

$$0 = \frac{d\frac{1}{\mu}}{ds}\left\{\frac{dx}{ds}\frac{d}{ds}\left(\lambda\frac{dx}{ds}\right) + \ldots\right\} + \frac{1}{\mu}\left\{\frac{dx}{ds}\frac{d^2}{ds^2}\left(\lambda\frac{dx}{ds}\right) + \ldots\right\}.$$

Taking now s for independent variable, and performing the differentiation here indicated, with attention to the following relations:—

$$\frac{dx^2}{ds^2} + \ldots = 1,\ \frac{dx}{ds}\frac{d^2x}{ds^2} + \ldots = 0,$$

$$\frac{dx}{ds}\frac{d^3x}{ds^3} + \ldots + \left(\frac{d^2x}{ds^2}\right)^2 + \ldots = 0,$$

and the expression (§ 9) for ρ, the radius of curvature, we find

$$\frac{1}{\mu}\frac{d^2\lambda}{ds^2} + \frac{d\left(\frac{1}{\mu}\right)}{ds}\frac{d\lambda}{ds} - \frac{\lambda}{\mu\rho^2} = 0 \ldots\ldots\ldots\ldots (e),$$

a linear differential equation of the second order to determine λ, when μ and ρ are given functions of s.

The interpretation of (d) is very obvious. It shows that λ is the impulsive tension at the point P of the string; and that the velocity which this point acquires instantaneously is the resultant of $\frac{1}{\mu}\frac{d\lambda}{ds}$ tangential, and $\frac{\lambda}{\rho\mu}$ towards the centre of curvature. The differential equation (e) therefore shows the law of transmission of the instantaneous tension along the string, and proves that it depends solely on the mass of the cord per unit of length in each part, and the curvature from point to point, but not at all on the plane of curvature, of the initial form. Thus, for instance, it will be the same along a helix as along a circle of the same curvature.

Generation of motion by impulse in an inextensible cord or chain.

With reference to the fulfilling of the six terminal equations, a difficulty occurs inasmuch as $\dot{x}$, $\dot{y}$, $\dot{z}$ are expressed by (d) immediately, without the introduction of fresh arbitrary constants, in terms of λ, which, as the solution of a differential equation of the second degree, involves only two arbitrary constants. The explanation is, that at any point of the cord, at any instant, any velocity in any direction perpendicular to the tangent may be generated without at all altering the condition of the cord even at points infinitely near it. This, which seems clear enough without proof, may be demonstrated analytically by transforming the kinematical equation (a) thus. Let f be the component tangential velocity, q the component velocity towards the centre of curvature, and p the component velocity perpendicular to the osculating plane. Using the elementary formulas for the direction cosines of these lines (§ 9), and remembering that s is now independent variable, we have

$$\dot{x} = f\frac{dx}{ds} + q\frac{\rho d^2x}{ds^2} + p\frac{\rho\,(dzd^2y - dyd^2z)}{ds^3}, \quad \dot{y} = \text{etc.}$$

Substituting these in (a) and reducing, we find

$$\frac{df}{ds} = \frac{q}{\rho} \qquad \text{..............................} \qquad (f),$$

a form of the kinematical equation of a flexible line which will be of much use to us later.

We see, therefore, that if the tangential components of the impressed terminal velocities have any prescribed values, we may give besides, to the ends, any velocities whatever perpendicular to the tangents, without altering the motion acquired by any part of the cord. From this it is clear also, that the directions of the terminal impulses are necessarily tangential; or, in other words, that an impulse inclined to the tangent at either end, would generate an infinite transverse velocity.

To express, then, the terminal conditions, let F and F' be the tangential velocities produced at the ends, which we suppose known. We have, for any point, P, as seen above from (d),

$$f = \frac{1}{\mu}\frac{d\lambda}{ds} \qquad \text{............................} \qquad (g),$$

and hence when

Generation of motion by impulse in an inextensible cord or chain.

$$\left.\begin{array}{ll} & s=0,\ \frac{1}{\mu}\frac{d\lambda}{ds}=F \\ \text{and when} & s=l,\ \frac{1}{\mu}\frac{d\lambda}{ds}=F' \end{array}\right\} \ldots\ldots\ldots\ldots\ldots\ldots\ldots\ldots (h),$$

which suffice to determine the constants of integration of (d). Or if the data are the tangential impulses, I, I', required at the ends to produce the motion, we have

$$\left.\begin{array}{ll} \text{when} & s=0,\ \lambda=I, \\ \text{and when} & s=l,\ \lambda=I' \end{array}\right\} \ldots\ldots\ldots\ldots\ldots\ldots\ldots\ldots (i).$$

Or if either end be free, we have $\lambda=0$ at it, and any prescribed condition as to impulse applied, or velocity generated, at the other end.

The solution of this problem is very interesting, as showing how rapidly the propagation of the impulse falls off with "change of direction" along the cord. The reader will have no difficulty in illustrating this by working it out in detail for the case of a cord either uniform or such that $\mu\dfrac{d\frac{1}{\mu}}{ds}$ is constant, and given in the form of a circle or helix. When μ and ρ are constant, for instance, the impulsive tension decreases in the proportion of 1 to ϵ per space along the curve equal to ρ. The results have curious, and dynamically most interesting, bearings on the motions of a whip lash, and of the rope in harpooning a whale.

Impulsive motion of incompressible liquid.

Example (3). Let a mass of incompressible liquid be given at rest completely filling a closed vessel of any shape; and let, by suddenly commencing to change the shape of this vessel, any arbitrarily prescribed normal velocities be suddenly produced in the liquid at all points of its bounding surface, subject to the condition of not altering the volume: It is required to find the instantaneous velocity of any interior point of the fluid.

Let x, y, z be the co-ordinates of any point P of the space occupied by the fluid, and let u, v, w be the components of the required velocity of the fluid at this point. Then ρ being the density of the fluid, and $\iiint$ denoting integration throughout the space occupied by the fluid, we have

$$T=\iiint \tfrac{1}{2}\rho\,(u^2+v^2+w^2)\,dxdydz \ldots\ldots\ldots\ldots (a),$$

Impulsive motion of incompressible liquid.

which, subject to the kinematical condition (§ 193),

$$\frac{du}{dx}+\frac{dv}{dy}+\frac{dw}{dz}=0 \;\ldots\ldots\ldots\ldots\ldots\ldots\ldots\ldots (b),$$

must be the least possible, with the given surface values of the normal component velocity. By the method of variation we have

$$\iiint\left\{\rho(u\delta u+v\delta v+w\delta w)+\lambda\left(\frac{d\delta u}{dx}+\frac{d\delta v}{dy}+\frac{d\delta w}{dz}\right)\right\}dxdydz=0\ldots.(c).$$

But integrating by parts we have

$$\iiint\lambda\left(\frac{d\delta u}{dx}+\frac{d\delta v}{dy}+\frac{d\delta w}{dz}\right)dxdydz=\iint\lambda\,(\delta u dydz+\delta v dzdx+\delta w dxdy)$$

$$-\iiint\left(\delta u\frac{d\lambda}{dx}+\delta v\frac{d\lambda}{dy}+\delta w\frac{d\lambda}{dz}\right)dxdydz\,\ldots..(d),$$

and if l, m, n denote the direction cosines of the normal at any point of the surface, dS an element of the surface, and $\iint$ integration over the whole surface, we have

$$\iint\lambda\,(\delta u dydz+\delta v dzdx+\delta w dxdy)=\iint\lambda\,(l\delta u+m\delta v+n\delta w)\,dS=0,$$

since the normal component of the velocity is given, which requires that $l\delta u+m\delta v+n\delta w=0$. Using this in going back with the result to (c), (d), and equating to zero the coefficients of δu, δv, δw, we find

$$\rho u=\frac{d\lambda}{dx},\quad \rho v=\frac{d\lambda}{dy},\quad \rho w=\frac{d\lambda}{dz}\ldots\ldots\ldots\ldots\ldots\ldots\ldots(e).$$

These, used to eliminate u, v, w from (b), give

$$\frac{d}{dx}\left(\frac{1}{\rho}\frac{d\lambda}{dx}\right)+\frac{d}{dy}\left(\frac{1}{\rho}\frac{d\lambda}{dy}\right)+\frac{d}{dz}\left(\frac{1}{\rho}\frac{d\lambda}{dz}\right)=0\;\ldots\ldots\ldots.(f),$$

an equation for the determination of λ, whence by (e) the solution is completed.

The condition to be fulfilled, besides the kinematical equation (b), amounts to this merely,—that $\rho\,(udx+vdy+wdz)$ must be a complete differential. If the fluid is homogeneous, ρ is constant, and $udx+vdy+wdz$ must be a complete differential; in other words, the motion suddenly generated must be of the "non-rotational" character [§ 190, (i)] throughout the fluid mass. The equation to determine λ becomes, in this case,

$$\frac{d^2\lambda}{dx^2}+\frac{d^2\lambda}{dy^2}+\frac{d^2\lambda}{dz^2}=0\;\ldots\ldots\ldots\ldots\ldots\ldots\ldots\ldots(g).$$

Impulsive motion of incompressible liquid. From the hydrodynamical principles explained later it will appear that λ, the function of which $\rho\,(udx + vdy + wdz)$ is the differential, is the impulsive pressure at the point (x, y, z) of the fluid. Hence we may infer that the equation (f), with the condition that λ shall have a given value at every point of a certain closed surface, has a possible and a determinate solution for every point within that surface. This is precisely the same problem as the determination of the permanent temperature at any point within a heterogeneous solid of which the surface is kept permanently with any non-uniform distribution of temperature over it, (f) being Fourier's equation for the uniform conduction of heat through a solid of which the conducting power at the point (x, y, z) is $\frac{1}{\rho}$. The possibility and the determinateness of this problem (with an exception regarding multiply continuous spaces, to be fully considered in Vol. II.) were both proved above [Chap. I. App. A, (e)] by a demonstration, the comparison of which with the present is instructive. The other case of superficial condition—that with which we have commenced here—shows that the equation (f), with $l\frac{d\lambda}{dx} + m\frac{d\lambda}{dy} + n\frac{d\lambda}{dz}$ given arbitrarily for every point of the surface, has also (with like qualification respecting multiply continuous spaces) a possible and single solution for the whole interior space. This, as we shall see in examining the mathematical theory of magnetic induction, may also be inferred from the general theorem (e) of App. A above, by supposing a to be zero for all points without the given surface, and to have the value $\frac{1}{\rho}$ for any internal point (x, y, z).

Lagrange's equations of motion in terms of generalized co-ordinates **318.** The equations of continued motion of a set of free particles acted on by any forces, or of a system connected in any manner and acted on by any forces, are readily obtained in terms of Lagrange's Generalized Co-ordinates by the regular and direct process of analytical transformation, from the ordinary forms of the equations of motion in terms of Cartesian (or rectilineal rectangular) co-ordinates. It is convenient first to effect the transformation for a set of free particles acted on by any forces. The case of any system with invariable connexions, or with connexions varied in a given manner, is then to be dealt with by supposing one or more of the generalized co-ordinates to be constant: or to be given functions of the time. Thus the generalized equations of motion are merely those for the reduced number of the co-ordinates remaining un-given; and their integration determines these co-ordinates.

deduced direct by transformation from the equations of motion in terms of Cartesian co-ordinates.

Let m_1, m_2, etc. be the masses, x_1, y_1, z_1, x_2, etc. be the co-ordinates of the particles; and X_1, Y_1, Z_1, X_2, etc. the components of the forces acting upon them. Let ψ, ϕ, etc. be other variables equal in number to the Cartesian co-ordinates, and let there be the same number of relations given between the two sets of variables; so that we may either regard ψ, ϕ, etc. as known functions of x_1, y_1, etc., or x_1, y_1, etc. as known functions of ψ, ϕ, etc. Proceeding on the latter supposition we have the equations (*a*), (1), of § 313; and we have equations (*b*), (6), of the same section for the generalized components Ψ, Φ, etc. of the force on the system.

For the Cartesian equations of motion we have

$$X_1 = m_1 \frac{d^2x_1}{dt^2}, \quad Y_1 = m_1 \frac{d^2y_1}{dt^2}, \quad Z_1 = m_1 \frac{d^2z_1}{dt^2}, \quad X_2 = m_2 \frac{d^2x_2}{dt^2} \text{ etc.} \ldots (19).$$

Multiplying the first by $\frac{dx_1}{d\psi}$, the second by $\frac{dy_1}{d\psi}$, and so on, and adding all the products, we find by 313 (6)

$$\Psi = m_1 \left(\frac{d^2x_1}{dt^2} \frac{dx_1}{d\psi} + \frac{d^2y_1}{dt^2} \frac{dy_1}{d\psi} + \frac{d^2z_1}{dt^2} \frac{dz_1}{d\psi} \right) + m_2 (\text{etc.}) + \text{etc.} \ldots (20).$$

Now

$$\frac{d^2x_1}{dt^2} \frac{dx_1}{d\psi} = \frac{d}{dt}\left(\dot{x}_1 \frac{dx_1}{d\psi} \right) - \dot{x}_1 \frac{d}{dt} \frac{dx_1}{d\psi} = \frac{d}{dt}\left(\dot{x}_1 \frac{d\dot{x}_1}{d\dot{\psi}} \right) - \dot{x}_1 \frac{d\dot{x}_1}{d\psi}$$

$$= \frac{d}{dt} \left\{ \tfrac{1}{2} \frac{d(\dot{x}_1^2)}{d\dot{\psi}} \right\} - \tfrac{1}{2} \frac{d(\dot{x}_1^2)}{d\psi} \ldots\ldots (21).$$

Using this and similar expressions with reference to the other co-ordinates in (20), and remarking that

$$\tfrac{1}{2} m_1 (\dot{x}_1^2 + \dot{y}_1^2 + \dot{z}_1^2) + \tfrac{1}{2} m_2 (\text{etc.}) + \text{etc.} = T \ldots\ldots (22),$$

if, as before, we put T for the kinetic energy of the system; we find

$$\Psi = \frac{d}{dt} \frac{dT}{d\dot{\psi}} - \frac{dT}{d\psi} \ldots\ldots (23).$$

The substitutions of $\frac{d\dot{x}_1}{d\dot{\psi}}$ for $\frac{dx_1}{d\psi}$ and of $\frac{d\dot{x}_1}{d\psi}$ for $\frac{d}{dt}\frac{dx_1}{d\psi}$ used above, suppose $\dot{x}_1$ to be a function of the co-ordinates, and of the generalized velocity-components, as shown in equations (1) of § 313. It is on this supposition [which makes T a quadratic function of the generalized velocity-components with functions of the co-ordinates as coefficients as shown in § 313 (2)] that the differentiations $\frac{d}{d\dot{\psi}}$ and $\frac{d}{d\psi}$ in (23) are performed. Proceeding similarly with reference to ϕ, etc., we find expressions similar to (23) for Φ, etc., and thus we have for the equations of motion in terms of the generalized co-ordinates

Lagrange's equations of motion in terms of generalized co-ordinates deduced direct by transformation from the equations of motion in terms of Cartesian co-ordinates.

$$\left.\begin{aligned}\frac{d}{dt}\frac{dT}{d\dot{\psi}}-\frac{dT}{d\psi}&=\Psi,\\ \frac{d}{dt}\frac{dT}{d\dot{\phi}}-\frac{dT}{d\phi}&=\Phi,\\ \text{etc.}&\end{aligned}\right\}\text{.....................(24).}$$

It is to be remarked that there is nothing in the preceding transformation which would be altered by supposing t to appear in the relations between the Cartesian and the generalized co-ordinates: thus if we suppose these relations to be

$$\left.\begin{aligned}F(x_1, y_1, z_1, x_2, \ldots\ldots \psi, \phi, \theta, \ldots\ldots t)&=0\\ F_1(x_1, y_1, z_1, x_2, \ldots\ldots \psi, \phi, \theta, \ldots\ldots t)&=0\\ \text{etc.}&\end{aligned}\right\}\text{...........(25),}$$

we now, instead of § 313 (1), have

$$\left.\begin{aligned}\dot{x}_1&=\left(\frac{dx_1}{dt}\right)+\frac{dx_1}{d\psi}\dot{\psi}+\frac{dx_1}{d\phi}\dot{\phi}+\text{etc.}\\ \dot{y}_1&=\left(\frac{dy_1}{dt}\right)+\frac{dy_1}{d\psi}\dot{\psi}+\frac{dy_1}{d\phi}\dot{\phi}+\text{etc.}\\ &\text{etc.}\end{aligned}\right\}\text{.............(26),}$$

where $\left(\frac{dx_1}{dt}\right)$ denotes what the velocity-component $\dot{x}_1$ would be if ψ, ϕ, etc. were constant; being analytically the partial differential coefficient with reference to t of the formula derived from (26) to express x_1 as a function of t, ψ, ϕ, θ, etc.

Using (26) in (22) we now find instead of a homogeneous quadratic function of $\dot{\psi}$, $\dot{\phi}$, etc., as in (2) of § 313, a mixed

Lagrange's equations of motion in terms of generalized co-ordinates deduced direct by transformation from the equations of motion in terms of Cartesian co-ordinates.

function of zero degree and first and second degrees, for the kinetic energy, as follows :—

$$T = K + (\psi)\dot{\psi} + (\phi)\dot{\phi} + \ldots + \tfrac{1}{2}\{(\psi,\psi)\dot{\psi}^2 + (\phi,\phi)\dot{\phi}^2 + \ldots 2(\psi,\phi)\dot{\psi}\dot{\phi}\ldots\} \ldots (27),$$

where

$$\left.\begin{aligned} K &= \tfrac{1}{2}\Sigma m\left\{\left(\left(\frac{dx}{dt}\right)\right)^2 + \left(\left(\frac{dy}{dt}\right)\right)^2 + \left(\left(\frac{dz}{dt}\right)\right)^2\right\} \\ (\psi) &= \Sigma m\left\{\left(\frac{dx}{dt}\right)\frac{dx}{d\psi} + \left(\frac{dy}{dt}\right)\frac{dy}{d\psi} + \left(\frac{dz}{dt}\right)\frac{dz}{d\psi}\right\}, \text{ etc.} \\ (\psi,\psi) &= \Sigma m\left\{\left(\frac{dx}{d\psi}\right)^2 + \left(\frac{dy}{d\psi}\right)^2 + \left(\frac{dz}{d\psi}\right)^2\right\}, \text{ etc.} \\ (\psi,\phi) &= \Sigma m\left(\frac{dx}{d\psi}\frac{dx}{d\phi} + \frac{dy}{d\psi}\frac{dy}{d\phi} + \frac{dz}{d\psi}\frac{dz}{d\phi}\right), \text{ etc.} \\ &\text{etc.} \end{aligned}\right\} \ldots (28);$$

K, (ψ), (ϕ), (ψ, ψ), (ψ, ϕ), etc. being thus in general each a known function of t, ψ, ϕ, etc.

Equations (24) above are Lagrange's celebrated equations of motion in terms of generalized co-ordinates. It was first pointed out by Vieille* that they are applicable not only when ψ, ϕ, etc. are related to x_1, y_1, z_1, x_2, etc. by invariable relations as supposed in Lagrange's original demonstration, but also when the relations involve t in the manner shown in equations (25). Lagrange's original demonstration, to be found in the Fourth Section of the Second Part of his *Mécanique Analytique*, consisted of a transformation from Cartesian to generalized co-ordinates of the indeterminate equation of motion; and it is the same demonstration with unessential variations that has been hitherto given, so far as we know, by all subsequent writers including ourselves in our first edition (§ 329). It seems however an unnecessary complication to introduce the indeterminate variations δx, δy, etc.; and we find it much simpler to deduce Lagrange's generalized equations by direct transformation from the equations of motion (19) of a free particle.

* Sur les équations différentielles de la dynamique, *Liouville's Journal*, 1849, p. 201.

When the kinematic relations are invariable, that is to say when t does not appear in the equations of condition (25), we find from (27) and (28),

Lagrange's generalized form of the equations of motion expanded.

$$T = \tfrac{1}{2}\{(\psi, \psi)\dot{\psi}^2 + 2(\psi, \phi)\dot{\psi}\dot{\phi} + (\phi, \phi)\dot{\phi}^2 + \ldots\} \ldots\ldots (29),$$

$$\left.\begin{aligned} \frac{d}{dt}\frac{dT}{d\dot{\psi}} = (\psi, \psi)\ddot{\psi} + (\psi, \phi)\ddot{\phi} + \ldots \\ + \left\{\frac{d(\psi,\psi)}{d\psi}\dot{\psi} + \frac{d(\psi,\psi)}{d\phi}\dot{\phi} + \ldots\right\}\dot{\psi} \\ + \left\{\frac{d(\psi,\phi)}{d\psi}\dot{\psi} + \frac{d(\psi,\phi)}{d\phi}\dot{\phi} + \ldots\right\}\dot{\phi} \\ + \ldots\ldots\ldots\ldots\ldots\ldots\ldots\ldots \end{aligned}\right\} \ldots\ldots\ldots (29'),$$

and

$$\frac{dT}{d\psi} = \tfrac{1}{2}\left\{\frac{d(\psi,\psi)}{d\psi}\dot{\psi}^2 + 2\frac{d(\psi,\phi)}{d\psi}\dot{\psi}\dot{\phi} + \frac{d(\phi,\phi)}{d\psi}\dot{\phi}^2 + \ldots\right\} \; (29'').$$

Hence the ψ-equation of motion expanded in this, the most important class of cases, is as follows:

$$(\psi, \psi)\ddot{\psi} + (\psi, \phi)\ddot{\phi} + \ldots + Q_\psi(T) = \Psi,$$

where

$$Q_\psi(T) = \tfrac{1}{2}\left\{\frac{d(\psi,\psi)}{d\psi}\dot{\psi}^2 + 2\frac{d(\psi,\psi)}{d\phi}\dot{\psi}\dot{\phi} + \left[2\frac{d(\psi,\phi)}{d\phi} - \frac{d(\phi,\phi)}{d\psi}\right]\dot{\phi}^2 + \ldots\right\} \ldots\ldots\ldots\ldots (29''').$$

Remark that $Q_\psi(T)$ is a quadratic function of the velocity-components derived from that which expresses the kinetic energy (T) by the process indicated in the second of these equations, in which ψ appears singularly, and the other co-ordinates symmetrically with one another.

Multiply the ψ-equation by $\dot{\psi}$, the ϕ-equation by $\dot{\phi}$, and so on; and add. In what comes from $Q_\psi(T)$ we find terms

Equation of energy.

$$+2\frac{d(\psi,\psi)}{d\phi}\dot{\psi}\dot{\phi}\,.\,\dot{\psi}, \text{ and} - \frac{d(\psi,\psi)}{d\phi}\dot{\psi}^2\,.\,\dot{\phi};$$

which together yield $\quad + \dfrac{d(\psi,\psi)}{d\phi}\dot{\psi}^2\,.\,\dot{\phi}.$

With this, and the rest simply as shown in (29'''), we find

$$\begin{aligned} &[(\psi,\psi)\ddot{\psi} + (\psi,\phi)\ddot{\phi} + \ldots]\dot{\psi} \\ &+ [(\psi,\phi)\ddot{\psi} + (\phi,\phi)\ddot{\phi} + \ldots]\dot{\phi} \\ &+ \ldots\ldots\ldots\ldots\ldots\ldots\ldots \end{aligned}$$

Equation of energy.

$$+\frac{dT}{d\psi}\dot{\psi}+\frac{dT}{d\phi}\dot{\phi}+\ldots \qquad =\Psi\dot{\psi}+\Phi\dot{\phi}+\ldots\ldots\ldots\ldots(29^{\text{iv}}),$$

or

$$\frac{dT}{dt}=\Psi\dot{\psi}+\Phi\dot{\phi}+\ldots\ldots\ldots\ldots\ldots\ldots\ldots(29^{\text{v}}).$$

Hamilton's form.

When the kinematical relations are invariable, that is to say, when t does not appear in the equations of condition (25), the equations of motion may be put under a slightly different form first given by Hamilton, which is often convenient; thus:—Let T, $\dot{\psi}$, $\dot{\phi}$,..., be expressed in terms of ξ, η,..., the impulses required to produce the motion from rest at any instant [§ 313 (d)]; so that T will now be a homogeneous quadratic function, and $\dot{\psi}$, $\dot{\phi}$, ... each a linear function, of these elements, with coefficients—functions of ψ, ϕ, etc., depending on the kinematical conditions of the system, but not on the particular motion. Thus, denoting, as in § 322 (29), by ∂, partial differentiation with reference to ξ, η, ..., ψ, ϕ,..., considered as independent variables, we have [§ 313 (10)]

$$\dot{\psi}=\frac{\partial T}{d\xi}, \qquad \dot{\phi}=\frac{\partial T}{d\eta}, \ \ldots\ldots\ldots\ldots\ldots\ldots(30),$$

and, allowing d to denote, as in what precedes, the partial differentiations with reference to the system $\dot{\psi}$, $\dot{\phi}$, ..., ψ, ϕ, ..., we have [§ 313 (8)]

$$\xi=\frac{dT}{d\dot{\psi}}, \qquad \eta=\frac{dT}{d\dot{\phi}}, \ \ldots\ldots\ldots\ldots\ldots\ldots\ldots(31).$$

The two expressions for T being, as above, § 313,

$$T=\tfrac{1}{2}\{(\psi,\psi)\dot{\psi}^2+\ldots+2(\psi,\phi)\dot{\psi}\dot{\phi}+\ldots\}=\tfrac{1}{2}\{[\psi,\psi]\xi^2+\ldots+2[\psi,\phi]\xi\eta+\ldots\}(32),$$

the second of these is to be obtained from the first by substituting for $\dot{\psi}$, $\dot{\phi}$...., their expressions in terms of ξ, η, ... Hence

$$\frac{\partial T}{d\psi}=\frac{dT}{d\psi}+\frac{dT}{d\dot{\psi}}\frac{\partial\dot{\psi}}{d\psi}+\frac{dT}{d\dot{\phi}}\frac{\partial\dot{\phi}}{d\psi}+\ldots=\frac{dT}{d\psi}+\xi\frac{\partial}{d\psi}\frac{\partial T}{d\xi}+\eta\frac{\partial}{d\psi}\frac{\partial T}{d\eta}+\ldots$$

$$=\frac{dT}{d\psi}+\frac{\partial}{d\psi}\left(\xi\frac{\partial T}{d\xi}+\eta\frac{\partial T}{d\eta}+\ldots\right)=\frac{dT}{d\psi}+2\frac{\partial T}{d\psi}.$$

From this we conclude

$$\frac{\partial T}{d\psi}=-\frac{dT}{d\psi}\text{; and, similarly, }\frac{\partial T}{d\phi}=-\frac{dT}{d\phi},\text{ etc. }\ldots\ldots(33).$$

Hence Lagrange's equations become

$$\frac{d\xi}{dt}+\frac{\partial T}{d\psi}=\Psi,\text{ etc.}\ldots\ldots\ldots\ldots\ldots\ldots\ldots\ldots(34).$$

Hamilton's form.

In § 327 below a purely analytical proof will be given of Lagrange's generalized equations of motion, establishing them directly as a deduction from the principle of "Least Action," independently of any expression either of this principle or of the equations of motion in terms of Cartesian co-ordinates. In their Hamiltonian form they are also deduced in § 330 (33) from the principle of Least Action ultimately, but through the beautiful "Characteristic Equation" of Hamilton.

319. Hamilton's form of Lagrange's equations of motion in terms of generalized co-ordinates expresses that what is required to prevent any one of the components of momentum from varying is a corresponding component force equal in amount to the rate of change of the kinetic energy per unit increase of the corresponding co-ordinate, with all components of momentum constant: and that whatever is the amount of the component force, its excess above this value measures the rate of increase of the component momentum.

In the case of a conservative system, the same statement takes the following form:—The rate at which any component momentum increases per unit of time is equal to the rate, per unit increase of the corresponding co-ordinate, at which the sum of the potential energy, and the kinetic energy for constant momentums, diminishes. This is the celebrated "canonical form" of the equations of motion of a system, though why it has been so called it would be hard to say.

"Canonical form" of Hamilton's general equations of motion of a conservative system.

Let V denote the potential energy, so that [§ 293 (3)]

$$\Psi\delta\psi + \Phi\delta\phi + \ldots = -\delta V,$$

and therefore $$\Psi = -\frac{dV}{d\psi}, \quad \Phi = -\frac{dV}{d\phi}, \ldots$$

Let now U denote the algebraic expression for the sum of the potential energy, V, in terms of the co-ordinates, ψ, ϕ..., and the kinetic energy, T, in terms of the co-ordinates and the components of momentum, ξ, η,.... Then

$$\left.\begin{array}{ll} & \dfrac{d\xi}{dt} = -\dfrac{\partial U}{d\psi}, \text{ etc.} \\ \text{also} & \dfrac{d\psi}{dt} = \dfrac{\partial U}{d\xi}, \text{ etc.} \end{array}\right\} \ldots\ldots\ldots\ldots\ldots\ldots(35),$$

the latter being equivalent to (30), since the potential energy does not contain ξ, η, etc.

In the following examples we shall adhere to Lagrange's form (24), as the most convenient for such applications.

Examples of the use of Lagrange's generalized equations of motion;—polar co-ordinates.

Example (A).—Motion of a single point (m) referred to polar co-ordinates (r, θ, ϕ). From the well-known geometry of this case we see that δr, $r\delta\theta$, and $r\sin\theta\delta\phi$ are the amounts of linear displacement corresponding to infinitely small increments, δr, $\delta\theta$, $\delta\phi$, of the co-ordinates: also that these displacements are respectively in the direction of r, of the arc $r\delta\theta$ (of a great circle) in the plane of r and the pole, and of the arc $r\sin\theta\delta\phi$ (of a small circle in a plane perpendicular to the axis); and that they are therefore at right angles to one another. Hence if F, G, H denote the components of the force experienced by the point, in these three rectangular directions, we have

$$F = R,\quad Gr = \Theta,\quad \text{and}\quad Hr\sin\theta = \Phi\,;$$

R, Θ, Φ being what the generalized components of force (§ 313) become for this particular system of co-ordinates. We also see that $\dot r$, $r\dot\theta$, and $r\sin\theta\dot\phi$ are three components of the velocity, along the same rectangular directions. Hence

$$T = \tfrac{1}{2}m(\dot r^2 + r^2\dot\theta^2 + r^2\sin^2\theta\dot\phi^2).$$

From this we have

$$\frac{dT}{d\dot r} = m\dot r,\quad \frac{dT}{d\dot\theta} = mr^2\dot\theta,\quad \frac{dT}{d\dot\phi} = mr^2\sin^2\theta\dot\phi\,;$$

$$\frac{dT}{dr} = mr(\dot\theta^2 + \sin^2\theta\dot\phi^2),\quad \frac{dT}{d\theta} = mr^2\sin\theta\cos\theta\dot\phi^2,\quad \frac{dT}{d\phi} = 0.$$

Hence the equations of motion become

$$m\left\{\frac{d\dot r}{dt} - r(\dot\theta^2 + \sin^2\theta\dot\phi^2)\right\} = F,$$

$$m\left\{\frac{d(r^2\dot\theta)}{dt} - r^2\sin\theta\cos\theta\dot\phi^2\right\} = Gr,$$

$$m\frac{d(r^2\sin^2\theta\dot\phi)}{dt} = Hr\sin\theta\,;$$

or, according to the ordinary notation of the differential calculus,

$$m\left\{\frac{d^2r}{dt^2} - r\left(\frac{d\theta^2}{dt^2} + \sin^2\theta\frac{d\phi^2}{dt^2}\right)\right\} = F,$$

$$m\left\{\frac{d}{dt}\left(r^2\frac{d\theta}{dt}\right)-r^2\sin\theta\cos\theta\frac{d\phi^2}{dt^2}\right\}=Gr,$$

$$m\frac{d}{dt}\left(r^2\sin^2\theta\frac{d\phi}{dt}\right)=Hr\sin\theta.$$

Examples of the use of Lagrange's generalized equations of motion;—polar co-ordinates.

If the motion is confined to one plane, that of r, θ, we have $\frac{d\phi}{dt}=0$, and therefore $H=0$, and the two equations of motion which remain are

$$m\left(\frac{d^2r}{dt^2}-r\frac{d\theta^2}{dt^2}\right)=F,\quad m\frac{d}{dt}\left(r^2\frac{d\theta}{dt}\right)=Gr.$$

These equations might have been written down at once in terms of the second law of motion from the kinematical investigation of § 32, in which it was shown that $\frac{d^2r}{dt^2}-r\frac{d\theta^2}{dt^2}$, and $\frac{1}{r}\frac{d}{dt}\left(r^2\frac{d\theta}{dt}\right)$ are the components of acceleration along and perpendicular to the radius-vector, when the motion of a point in a plane is expressed according to polar co-ordinates, r, θ.

The same equations, with ϕ instead of θ, are obtained from the polar equations in three dimensions by putting $\theta=\frac{1}{2}\pi$, which implies that $G=0$, and confines the motion to the plane (r,ϕ).

Dynamical problem.

Example (B).—Two particles are connected by a string; one of them, m, moves in any way on a smooth horizontal plane, and the string, passing through a smooth infinitely small aperture in this plane, bears the other particle m', hanging vertically downwards, and only moving in this vertical line: (the string remaining always stretched in any practical illustration, but, in the problem, being of course supposed capable of transmitting negative tension with its two parts straight.) Let l be the whole length of the string, r that of the part of it from m to the aperture in the plane, and let θ be the angle between the direction of r and a fixed line in the plane. We have

$$T=\tfrac{1}{2}\{m(\dot r^2+r^2\dot\theta^2)+m'\dot r^2\},$$

$$\frac{dT}{d\dot r}=(m+m')\dot r,\qquad \frac{dT}{d\dot\theta}=mr^2\dot\theta,$$

$$\frac{dT}{dr}=mr\dot\theta^2,\qquad \frac{dT}{d\theta}=0.$$

Also, there being no other external force than gm', the weight of the second particle,

$$R=-gm',\quad \Theta=0.$$

Examples of the use of Lagrange's generalized equations of motion; dynamical problem.

Hence the equations of motion are

$$(m+m')\ddot{r} - mr\dot{\theta}^2 = -m'g, \qquad m\frac{d(r^2\dot{\theta})}{dt} = 0.$$

The motion of m' is of course that of a particle influenced only by a force towards a fixed centre; but the law of this force, P (the tension of the string), is remarkable. To find it we have (§ 32), $P = m(-\ddot{r} + r\dot{\theta}^2)$. But, by the equations of the motion,

$$\ddot{r} - r\dot{\theta}^2 = -\frac{m'}{m+m'}(g + r\dot{\theta}^2), \text{ and } \dot{\theta} = \frac{h}{mr^2},$$

where h (according to the usual notation) denotes the moment of momentum of the motion, being an arbitrary constant of integration. Hence

$$P = \frac{mm'}{m+m'}\left(g + \frac{h^2}{m^2}r^{-3}\right).$$

Case of stable equilibrium due to motion.

The particular case of projection which gives m a circular motion and leaves m' at rest is interesting, inasmuch as (§ 350, below) the motion of m is stable, and therefore m' is in stable equilibrium.

Examples continued; C (a), folding door.

Example (C).—A rigid body m is supported on a fixed axis, and another rigid body n is supported on the first, by another axis; the motion round each axis being perfectly free.

Case (a).—*The second axis parallel to the first.* At any time, t, let ϕ and ψ be the inclinations of a fixed plane through the first axis to the plane of it and the second axis, and to a plane through the second axis and the centre of inertia of the second body. These two co-ordinates, ϕ, ψ, it is clear, completely specify the configuration of the system. Now let a be the distance of the second axis from the first, and b that of the centre of inertia of the second body from the second axis. The velocity of the second axis will be $a\dot{\phi}$; and the velocity of the centre of inertia of the second body will be the resultant of two velocities

$$a\dot{\phi}, \text{ and } b\dot{\psi},$$

in lines inclined to one another at an angle equal to $\psi - \phi$, and its square will therefore be equal to

$$a^2\dot{\phi}^2 + 2ab\dot{\phi}\dot{\psi}\cos(\psi - \phi) + b^2\dot{\psi}^2.$$

Hence, if m and n denote the masses, j the radius of gyration of the first body about the fixed axis, and k that of the second body about a parallel axis through its centre of inertia; we have, according to §§ 280, 281,

Examples continued; C (a), folding door.

$$T = \tfrac{1}{2}\{mj^2\dot{\phi}^2 + n[a^2\dot{\phi}^2 + 2ab\dot{\phi}\dot{\psi}\cos(\psi - \phi) + b^2\dot{\psi}^2 + k^2\dot{\psi}^2]\}.$$

Hence we have,

$$\frac{dT}{d\dot{\phi}} = mj^2\dot{\phi} + na^2\dot{\phi} + nab\cos(\psi - \phi)\dot{\psi}; \quad \frac{dT}{d\dot{\psi}} = nab\cos(\psi - \phi)\dot{\phi} + n(b^2 + k^2)\dot{\psi};$$

$$\frac{dT}{d\phi} = -\frac{dT}{d\psi} = nab\sin(\psi - \phi)\dot{\phi}\dot{\psi}.$$

The most general supposition we can make as to the applied forces, is equivalent to assuming a couple, Φ, to act on the first body, and a couple, Ψ, on the second, each in a plane perpendicular to the axes; and these are obviously what the generalized components of stress become in this particular co-ordinate system, ϕ, ψ. Hence the equations of motion are

$$(mj^2 + na^2)\ddot{\phi} + nab\,\frac{d[\dot{\psi}\cos(\psi - \phi)]}{dt} - nab\sin(\psi - \phi)\dot{\phi}\dot{\psi} = \Phi,$$

$$nab\,\frac{d[\dot{\phi}\cos(\psi - \phi)]}{dt} + n(b^2 + k^2)\ddot{\psi} + nab\sin(\psi - \phi)\dot{\phi}\dot{\psi} = \Psi.$$

If there is no other applied force than gravity, and if, as we may suppose without losing generality, the two axes are horizontal, the potential energy of the system will be

$$gmh(1 - \cos\phi) + gn\{a[1 - \cos(\phi + A)] + b[1 - \cos(\psi + A)]\},$$

the distance of the centre of inertia of the first body from the fixed axis being denoted by h, the inclination of the plane through the fixed axis and the centre of inertia of the first body, to the plane of the two axes, being denoted by A, and the fixed plane being so taken that $\phi = 0$ when the former plane is vertical. By differentiating this, with reference to ϕ and ψ, we therefore have

$$-\Phi = gmh\sin\phi + gna\sin(\phi + A), \quad -\Psi = gnb\sin(\psi + A).$$

We shall examine this case in some detail later, in connexion with the interference of vibrations, a subject of much importance in physical science.

When there are no applied or intrinsic working forces, we have $\Phi = 0$ and $\Psi = 0$: or, if there are mutual forces between the two bodies, but no forces applied from without, $\Phi + \Psi = 0$. In

Examples of the use of Lagrange's generalized equations of motion.

either of these cases we have the following first integral:—

$$(mj^2 + na^2)\dot{\phi} + m'ab\cos(\psi - \phi)(\dot{\phi} + \dot{\psi}) + n(b^2 + k^2)\dot{\psi} = C;$$

obtained by adding the two equations of motion and integrating. This, which clearly expresses the constancy of the whole moment of momentum, gives $\dot{\phi}$ and $\dot{\psi}$ in terms of $(\dot{\psi} - \dot{\phi})$ and $(\psi - \phi)$. Using these in the integral equation of energy, provided the mutual forces are functions of $\psi - \phi$, we have a single equation between $\dfrac{d(\psi - \phi)}{dt}$, $(\psi - \phi)$, and constants, and thus the full solution of the problem is reduced to quadratures. [It is worked out fully below, as Sub-example G_1.]

C (*b*). Motion of governing masses in Watt's centrifugal governor: also of gimballed compass-bowl.

Case (b).—The second axis perpendicular to the first. For simplicity suppose the pivoted axis of the second body, n, to be a principal axis relatively [§ 282 Def. (2)] to the point, N, in which it is cut by a plane perpendicular to it through the fixed axis of the first body, m. Let NE and NF be n's two other principal axes. Denote now by

h the distance from N to m's fixed axis;

k, e, f the radii of gyration of n round its three principal axes through N;

j the radius of gyration of m round its fixed axis;

θ the inclination of NE to m's fixed axis;

ψ the inclination of the plane parallel to n's pivoted axis through m's fixed axis, to a fixed plane through the latter.

Remarking that the component angular velocities of n round NE and NF are $\dot{\psi}\cos\theta$ and $\dot{\psi}\sin\theta$, we find immediately

$$T = \tfrac{1}{2}\{[mj^2 + n(h^2 + e^2\cos^2\theta + f^2\sin^2\theta)]\dot{\psi}^2 + nk^2\dot{\theta}^2\},$$

or, if we put

$$mj^2 + n(h^2 + f^2) = G,\quad n(e^2 - f^2) = D;$$

$$T = \tfrac{1}{2}\{(G + D\cos^2\theta)\dot{\psi}^2 + nk^2\dot{\theta}^2\}.$$

The farther working out of this case we leave as a simple but most interesting exercise for the student. We may return to it later, as its application to the theory of centrifugal chronometric regulators is very important.

Example (C′). Take the case C (b) and mount a third body M upon an axis OC fixed relatively to n in any position parallel to NE. Suppose for simplicity O to be the centre of inertia of M and OC one of its principal axes; and let OA, OB be its two other principal axes relative to O. The notation being in other respects the same as in Example C (b), denote now farther by A, B, C the moments of inertia of M round OA, OB, OC; ϕ the angle between the plane AOC and the plane through the fixed axis of m perpendicular to the pivoted axis of n; ϖ, ρ, σ the component angular velocities of M round OA, OB, OC.

Motion of a rigid body pivoted on one of its principal axes mounted on a gimballed bowl.

In the annexed diagram, taken from § 101 above, ZCZ' is a

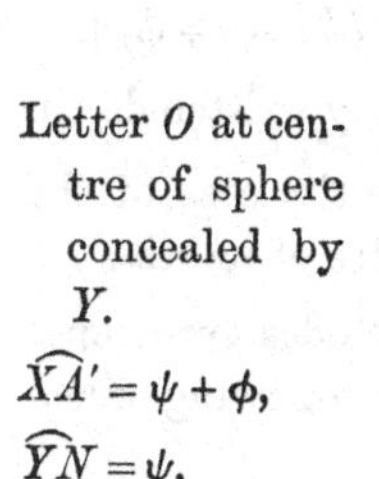

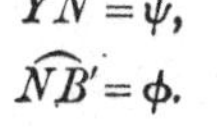

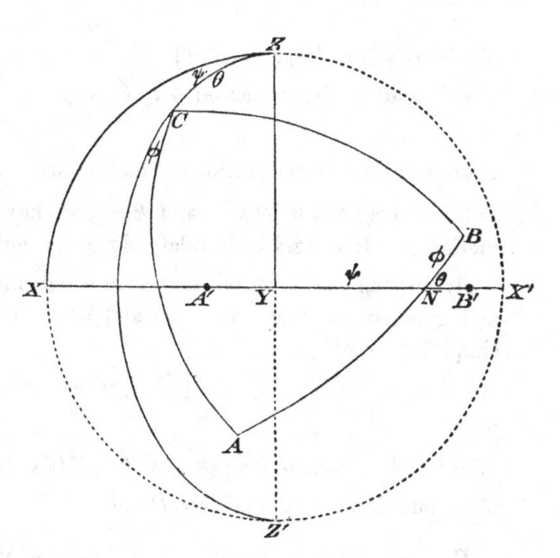

circle of unit radius having its centre at O and its plane parallel to the fixed axis of m and perpendicular to the pivoted axis of n.

The component velocities of C in the direction of the arc ZC and perpendicular to it are $\dot{\theta}$ and $\dot{\psi}\sin\theta$; and the component angular velocity of the plane ZCZ' round OC is $\dot{\psi}\cos\theta$. Hence

$$\varpi = \dot{\theta}\sin\phi - \dot{\psi}\sin\theta\cos\phi,$$

$$\rho = \dot{\theta}\cos\phi + \dot{\psi}\sin\theta\sin\phi,$$

and

$$\sigma = \dot{\psi}\cos\theta + \dot{\phi}.$$

[Compare § 101.]

Motion of a rigid body pivoted on one of its principal axes mounted on a gimballed bowl.

The kinetic energy of the motion of M relatively to O, its centre of inertia, is (§ 281)

$$\tfrac{1}{2}(A\varpi^2 + B\rho^2 + C\sigma^2);$$

and (§ 280) its whole kinetic energy is obtained by adding the kinetic energy of a material point equal to its mass moving with the velocity of its centre of inertia. This latter part of the kinetic energy of M is most simply taken into account by supposing n to include a material point equal to M placed at O; and using the previous notation k, e, f for radii of gyration of n on the understanding that n now includes this addition. Hence for the present example, with the preceding notation G, D, we have

$$T = \tfrac{1}{2}\{(G + D\cos^2\theta)\dot{\psi}^2 + nk^2\dot{\theta}^2\} \\ + A(\dot{\theta}\sin\phi - \dot{\psi}\sin\theta\cos\phi)^2 + B(\dot{\theta}\cos\phi + \dot{\psi}\sin\theta\sin\phi)^2 \\ + C(\dot{\psi}\cos\theta + \dot{\phi})^2\}.$$

Rigid body rotating freely; referred to the ψ, ϕ, θ co-ordinates (§ 101).

From this the three equations of motion are easily written down.

By putting $G = 0$, $D = 0$, and $k = 0$, we have the case of the motion of a free rigid body relatively to its centre of inertia.

Gyroscopes and gyrostats.

By putting $B = A$ we fall on a case which includes gyroscopes and gyrostats of every variety; and have the following much simplified formula:

$$T = \tfrac{1}{2}\{[G + A + (D - A)\cos^2\theta]\dot{\psi}^2 + (nk^2 + A)\dot{\theta}^2 + C(\dot{\psi}\cos\theta + \dot{\phi})^2\},$$

or

$$T = \tfrac{1}{2}\{(E + F\cos^2\theta)\dot{\psi}^2 + (nk^2 + A)\dot{\theta}^2 + C(\dot{\psi}\cos\theta + \dot{\phi})^2\},$$

if we put $E = G + A$, and $F = D - A$.

Gyroscopic pendulum.

Example (D).—*Gyroscopic pendulum.*—A rigid body, P, is attached to one axis of a universal flexure joint (§ 109), of which the other is held fixed, and a second body, Q, is supported on P by a fixed axis, in line with, or parallel to, the first-mentioned arm of the joint. For simplicity, we shall suppose Q to be kinetically symmetrical about its bearing axis, and OB to be a principal axis of an ideal rigid body, PQ, composed of P and a mass so distributed along the bearing axis of the actual body Q as to have the same centre of inertia and the same moments of inertia round axes perpendicular to it. Let AO be the fixed arm, O the joint, OB the movable arm bearing the body P, and coinciding with, or parallel to, the axis of Q. Let $BOA' = \theta$; let ϕ be the angle which the plane AOB makes with a fixed plane of reference, through OA, chosen so as to contain a second principal axis of the imagined rigid body, PQ, when OB is placed in line with AO; and let ψ be the angle between a plane of reference in Q through its axis of symmetry and the plane of the two principal axes of PQ already mentioned. These three co-ordinates (θ, ϕ, ψ) clearly specify the configuration of the system at any time, t. Let the moments of inertia of the imagined rigid body PQ, round its principal axis OB, the other principal axis referred to above, and the remaining one, be denoted by $\mathfrak{A}$, $\mathfrak{B}$, $\mathfrak{C}$ respectively; and let $\mathfrak{A}'$ be the moment of inertia of Q round its bearing axis.

Gyroscopic pendulum.

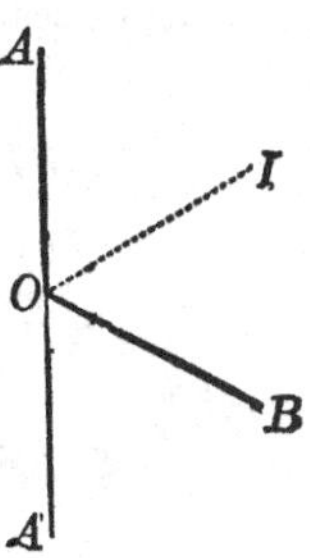

We have seen (§ 109) that, with the kind of joint we have supposed at O, every possible motion of a body rigidly connected with OB, is resolvable into a rotation round OI, the line bisecting the angle AOB, and a rotation round the line through O perpendicular to the plane AOB. The angular velocity of the latter is $\dot{\theta}$, according to our present notation. The former would give to any point in OB the same absolute velocity by rotation round OI, that it has by rotation with angular velocity $\dot{\phi}$ round AA'; and is therefore equal to

$$\frac{\sin A'OB}{\sin IOB}\dot{\phi} = \frac{\sin\theta}{\cos\frac{1}{2}\theta}\dot{\phi} = 2\dot{\phi}\sin\tfrac{1}{2}\theta.$$

This may be resolved into $2\dot{\phi}\sin^2\frac{1}{2}\theta = \dot{\phi}(1-\cos\theta)$ round OB, and $2\dot{\phi}\sin\frac{1}{2}\theta\cos\frac{1}{2}\theta = \dot{\phi}\sin\theta$ round the perpendicular to OB, in plane AOB. Again, in virtue of the symmetrical character of the joint with reference to the line OI, the angle ϕ, as defined above, will be equal to the angle between the plane of the two first-mentioned principal axes of body P, and the plane AOB. Hence the axis of the angular velocity $\dot{\phi}\sin\theta$, is inclined to the principal axis of moment $\mathfrak{B}$ at an angle equal to ϕ. Resolving therefore this angular velocity, and $\dot{\theta}$, into components round the axes of $\mathfrak{B}$ and $\mathfrak{C}$, we find, for the whole component angular velocities of the imagined rigid body PQ, round these axes, $\dot{\phi}\sin\theta\cos\phi + \dot{\theta}\sin\phi$, and $-\dot{\phi}\sin\theta\sin\phi + \dot{\theta}\cos\phi$, respectively. The whole kinetic energy, T, is composed of that of the imagined rigid body PQ, and that of Q about axes through its centre of

Gyroscopic pendulum.

inertia: we therefore have

$$2T = \mathfrak{A}(1-\cos\theta)^2\dot{\phi}^2 + \mathfrak{B}(\dot{\phi}\sin\theta\cos\phi + \dot{\theta}\sin\phi)^2 + \mathfrak{C}(\dot{\phi}\sin\theta\sin\phi - \dot{\theta}\cos\phi)^2 + \mathfrak{A}'\{\dot{\psi} - \dot{\phi}(1-\cos\theta)\}^2.$$

Hence $\dfrac{dT}{d\dot{\psi}} = \mathfrak{A}'\{\dot{\psi} - \dot{\phi}(1-\cos\theta)\}$, $\dfrac{dT}{d\psi} = 0$,

$$\frac{dT}{d\dot{\phi}} = \mathfrak{A}(1-\cos\theta)^2\dot{\phi} + \mathfrak{B}(\dot{\phi}\sin\theta\cos\phi + \dot{\theta}\sin\phi)\sin\theta\cos\phi$$
$$+ \mathfrak{C}(\dot{\phi}\sin\theta\sin\phi - \dot{\theta}\cos\phi)\sin\theta\sin\phi - \mathfrak{A}'\{\dot{\psi} - \dot{\phi}(1-\cos\theta)\}(1-\cos\theta),$$

$$\frac{dT}{d\phi} = -\mathfrak{B}(\dot{\phi}\sin\theta\cos\phi + \dot{\theta}\sin\phi)(\dot{\phi}\sin\theta\sin\phi - \dot{\theta}\cos\phi)$$
$$+ \mathfrak{C}(\dot{\phi}\sin\theta\sin\phi - \dot{\theta}\cos\phi)(\dot{\phi}\sin\theta\cos\phi + \dot{\theta}\sin\phi),$$

$$\frac{dT}{d\dot{\theta}} = \mathfrak{B}(\dot{\phi}\sin\theta\cos\phi + \dot{\theta}\sin\phi)\sin\phi - \mathfrak{C}(\dot{\phi}\sin\theta\sin\phi - \dot{\theta}\cos\phi)\cos\phi$$

and $$\frac{dT}{d\theta} = \mathfrak{A}(1-\cos\theta)\sin\theta\dot{\phi}^2 + \mathfrak{B}\cos\theta\cos\phi\dot{\phi}(\dot{\phi}\sin\theta\cos\phi + \dot{\theta}\sin\phi)$$
$$+ \mathfrak{C}\cos\theta\sin\phi\dot{\phi}(\dot{\phi}\sin\theta\sin\phi - \dot{\theta}\cos\phi) - \mathfrak{A}'\sin\theta\dot{\phi}\{\dot{\psi} - (1-\cos\theta)\dot{\phi}\}.$$

Now let a couple, G, act on the body Q, in a plane perpendicular to its axis, and let L, M, N act on P, in the plane perpendicular to OB, in the plane $A'OB$, and in the plane through OB perpendicular to the diagram. If ψ is kept constant, and ϕ varied, the couple G will do or resist work in simple addition with L. Hence, resolving $L + G$ and N into components round OI, and perpendicular to it, rejecting the latter, and remembering that $2\sin\frac{1}{2}\theta\dot{\phi}$ is the angular velocity round OI, we have

$$\Phi = 2\sin\tfrac{1}{2}\theta\{-(L+G)\sin\tfrac{1}{2}\theta + N\cos\tfrac{1}{2}\theta\} = \{-(L+G)(1-\cos\theta) + N\sin\theta\}.$$

Also, obviously

$$\Psi = G, \quad \Theta = M.$$

Using these several expressions in Lagrange's general equations (24), we have the equations of motion of the system. They will be of great use to us later, when we shall consider several particular cases of remarkable interest and of very great importance.

Example of varying relation without constraint (rotating axes).

Example (E).—*Motion of a free particle referred to rotating axes.* Let x, y, z be the co-ordinates of a moving particle referred to axes rotating with a constant or varying angular velocity round the axis OZ. Let x_1, y_1, z, be its co-ordinates referred to the same axis, OZ, and two axes OX_1, OY_1, fixed in the plane perpendicular to it. We have

Example of varying relation without constraint (rotating axes).

$$x_1 = x\cos\alpha - y\sin\alpha,\quad y_1 = x\sin\alpha + y\cos\alpha\,;$$

$$\dot{x}_1 = \dot{x}\cos\alpha - \dot{y}\sin\alpha - (x\sin\alpha + y\cos\alpha)\,\dot{\alpha},\quad \dot{y}_1 = \text{etc.}$$

where α, the angle X_1OX, must be considered as a given function of t. Hence

$$T = \tfrac{1}{2}m\,\{\dot{x}^2 + \dot{y}^2 + \dot{z}^2 + 2\,(x\dot{y} - y\dot{x})\,\dot{\alpha} + (x^2 + y^2)\,\dot{\alpha}^2\},$$

$$\frac{dT}{d\dot{x}} = m\,(\dot{x} - y\dot{\alpha}),\quad \frac{dT}{d\dot{y}} = m\,(\dot{y} + x\dot{\alpha}),\quad \frac{dT}{d\dot{z}} = m\dot{z},$$

$$\frac{dT}{dx} = m\,(\dot{y}\dot{\alpha} + x\dot{\alpha}^2),\quad \frac{dT}{dy} = m\,(-\dot{x}\dot{\alpha} + y\dot{\alpha}^2),\quad \frac{dT}{dz} = 0.$$

Also,

$$\frac{d}{dt}\frac{dT}{d\dot{x}} = m\,(\ddot{x} - \dot{y}\dot{\alpha} - y\ddot{\alpha}),\quad \frac{d}{dt}\frac{dT}{d\dot{y}} = m\,(\ddot{y} + \dot{x}\dot{\alpha} + x\ddot{\alpha}),$$

and hence the equations of motion are

$$m\,(\ddot{x} - 2\dot{y}\dot{\alpha} - x\dot{\alpha}^2 - y\ddot{\alpha}) = X,\quad m\,(\ddot{y} + 2\dot{x}\dot{\alpha} - y\dot{\alpha}^2 + x\ddot{\alpha}) = Y,\quad m\ddot{z} = Z,$$

X, Y, Z denoting simply the components of the force on the particle, parallel to the moving axes at any instant. In this example t enters into the relation between fixed rectangular axes and the co-ordinate system to which the motion is referred; but there is no constraint. The next is given as an example of varying, or kinetic, constraint.

Example of varying relation due to kinetic constraint.

Example (F).—*A particle, influenced by any forces, and attached to one end of a string of which the other is moved with any constant or varying velocity in a straight line.* Let θ be the inclination of the string at time t, to the given straight line, and ϕ the angle between two planes through this line, one containing the string at any instant, and the other fixed. These two co-ordinates (θ, ϕ) specify the position, P, of the particle at any instant, the length of the string being a given constant, a, and the distance OE, of its other end E, from a fixed point, O, of the line in which it is moved, being a given function of t, which we shall denote by u. Let x, y, z be the co-ordinates of the particle referred to three fixed rectangular axes. Choosing OX as the given straight line, and YOX the fixed plane from which ϕ is measured, we have

$$x = u + a\cos\theta,\quad y = a\sin\theta\cos\phi,\quad z = a\sin\theta\sin\phi,$$

$$\dot{x} = \dot{u} - a\sin\theta\dot{\theta}\,;$$

Example of varying relation due to kinetic constraint.

and for $\dot{y}$, $\dot{z}$ we have the same expressions as in Example (A). Hence

$$T = \mathfrak{T} + \tfrac{1}{2}m\,(\dot{u}^2 - 2\dot{u}\dot{\theta}a\sin\theta)$$

where $\mathfrak{T}$ denotes the same as the T of Example (A), with $\dot{r} = 0$, and $r = a$. Hence, denoting as there, by G and H the two components of the force on the particle, perpendicular to EP, respectively in the plane of θ and perpendicular to it, we find, for the two required equations of motion,

$$m\{a(\ddot{\theta} - \sin\theta\cos\theta\dot{\phi}^2) - \sin\theta\,\ddot{u}\} = G, \text{ and } ma\frac{d(\sin^2\theta\dot{\phi})}{dt} = H.$$

These show that the motion is the same as if E were fixed, and a force equal to $-m\ddot{u}$ were applied to the particle in a direction parallel to EX; a result that might have been arrived at at once by superimposing on the whole system an acceleration equal and opposite to that of E, to effect which on P the force $-m\ddot{u}$ is required.

Example (F′). Any case of varying relations such that in 318 (27) the coefficients (ψ, ψ), (ψ, ϕ) ... are independent of t. Let $\mathfrak{T}$ denote the quadratic part, L the linear part, and K [as in § 318 (27)] the constant part of T in respect to the velocity components, so that

$$\left.\begin{aligned} \mathfrak{T} &= \tfrac{1}{2}\{(\psi,\psi)\dot{\psi}^2 + 2(\psi,\phi)\dot{\psi}\dot{\phi} + (\phi,\phi)\dot{\phi}^2 + \ldots\} \\ L &= (\psi)\dot{\psi} + (\phi)\dot{\phi} + \ldots \\ K &= (\psi, \phi, \theta, \ldots) \end{aligned}\right\} \ldots\ldots.(a),$$

where (ψ, ψ), (ψ, ϕ), (ϕ, ϕ) ... denote functions of the co-ordinates without t, and (ψ), (ϕ), ..., $(\psi, \phi, \theta, \ldots)$ functions of the co-ordinates and, may be also, of t; and

$$T = \mathfrak{T} + L + K \ldots\ldots\ldots\ldots\ldots\ldots\ldots\ldots.(b).$$

We have

$$\frac{dK}{d\dot{\psi}} = 0.$$

Hence the contribution from K to the first member of the ψ-equation of motion is simply $-\frac{dK}{d\psi}$. Again we have

$$\frac{dL}{d\dot{\psi}} = (\psi);$$

hence

$$\frac{d}{dt}\frac{dL}{d\dot{\psi}} = \frac{d(\psi)}{d\psi}\dot{\psi} + \frac{d(\psi)}{d\phi}\dot{\phi} + \text{etc.} + \left(\frac{d(\psi)}{dt}\right).$$

Farther we have

Example of varying relation due to kinetic constraint.

$$\frac{dL}{d\psi} = \frac{d(\psi)}{d\psi}\dot{\psi} + \frac{d(\phi)}{d\psi}\dot{\phi} + \ldots.$$

Hence the whole contribution from L to the ψ-equation of motion is

$$\left(\frac{d(\psi)}{d\phi} - \frac{d(\phi)}{d\psi}\right)\dot{\phi} + \left(\frac{d(\psi)}{d\theta} - \frac{d(\theta)}{d\psi}\right)\dot{\theta} + \ldots + \left(\frac{d(\psi)}{dt}\right) \ldots\ldots (c).$$

Lastly, the contribution from $\mathfrak{T}$ is the same as the whole from T in § 318 (29‴); so that we have

$$\frac{d}{dt}\frac{d\mathfrak{T}}{d\dot{\psi}} - \frac{d\mathfrak{T}}{d\psi} = (\psi, \psi)\ddot{\psi} + (\psi, \phi)\ddot{\phi} + \ldots$$

$$+ \tfrac{1}{2}\left\{\frac{d(\psi,\psi)}{d\psi}\dot{\psi}^2 + 2\frac{d(\psi,\psi)}{d\phi}\dot{\psi}\dot{\phi} + \left[2\frac{d(\psi,\phi)}{d\phi} - \frac{d(\phi,\phi)}{d\psi}\right]\dot{\phi}^2 + \ldots\right\} \;(d),$$

and the completed ψ-equation of motion is

$$\frac{d}{dt}\frac{d\mathfrak{T}}{d\dot{\psi}} - \frac{d\mathfrak{T}}{d\psi} + \left(\frac{d(\psi)}{d\phi} - \frac{d(\phi)}{d\psi}\right)\dot{\phi} + \left(\frac{d(\psi)}{d\theta} - \frac{d(\theta)}{d\psi}\right)\dot{\theta} + \ldots$$

$$+ \left(\frac{d(\psi)}{dt}\right) - \frac{dK}{d\psi} = \Psi \ldots\ldots\ldots (e).$$

It is important to remark that the coefficient of $\dot{\phi}$ in this ψ-equation is equal but of opposite sign to the coefficient of $\dot{\psi}$ in the ϕ-equation. [Compare Example G (19) below.]

Equation of energy.

Proceeding as in § 318 (29ⁱᵛ) (29ᵛ), we have in respect to $\mathfrak{T}$ precisely the same formulas as there in respect to T. The terms involving first powers of the velocities simply, balance in the sum: and we find finally

$$\frac{d\mathfrak{T}}{dt} + \left(\frac{dL}{dt}\right) - \frac{d_{(\psi,\phi,\ldots)}K}{dt} = \Psi\dot{\psi} + \Phi\dot{\phi} + \ldots\ldots\ldots\ldots\ldots (f),$$

where $d_{(\psi,\phi,\ldots)}$ denotes differentiation on the supposition of $\psi, \phi, \ldots$ variable; and t constant, where it appears explicitly.

Now with this notation we have

$$\frac{dL}{dt} = \left(\frac{dL}{dt}\right) + \frac{d_{(\psi,\phi,\ldots)}L}{dt} + (\psi)\ddot{\psi} + (\phi)\ddot{\phi} + \ldots,$$

and

$$\frac{dK}{dt} = \left(\frac{dK}{dt}\right) + \frac{d_{(\psi,\phi,\ldots)}K}{dt}.$$

Hence from (f) we have

$$\frac{dT}{dt} = \frac{d(\mathfrak{T} + L + K)}{dt} = \Psi\dot{\psi} + \Phi\dot{\phi} + \ldots + \frac{d_{(\psi,\phi,\ldots)}L}{dt} + (\psi)\ddot{\psi} + (\phi)\ddot{\phi} + \ldots$$

$$+ 2\frac{d_{(\psi,\phi,\ldots)}K}{dt} + \left(\frac{dK}{dt}\right) \ldots\ldots\ldots\ldots (g).$$

Exercise for student.

Take, for illustration, Examples (E) and (F) from above; in which we have

$$\text{[Example (E)]} \qquad \mathfrak{T} = \tfrac{1}{2} m (\dot{x}^2 + \dot{y}^2 + \dot{z}^2),$$
$$L = m\dot{a}(x\dot{y} - y\dot{x}),$$
$$K = \tfrac{1}{2} m\dot{a}^2 (x^2 + y^2),$$

$$\text{and [Example (F)]} \qquad \mathfrak{T} = \tfrac{1}{2} ma^2 (\sin^2\theta\dot{\phi}^2 + \dot{\theta}^2),$$
$$L = - m\dot{u}a \sin\theta\dot{\theta},$$
$$K = \tfrac{1}{2} m\dot{u}^2.$$

Write out explicitly in each case equations (f) and (g), and verify them by direct work from the equations of motion forming the conclusions of the examples as treated above (remembering that $\dot{a}$ and $\dot{u}$ are to be regarded as given explicit functions of t).

Ignoration of co-ordinates.

Example (G).—*Preliminary to Gyrostatic connexions and to Fluid Motion.* Let there be one or more co-ordinates χ, χ', etc. which do not appear in the coefficients of velocities in the expression for T; that is to say let $\frac{dT}{d\chi} = 0$, $\frac{dT}{d\chi'} = 0$, etc. The equations corresponding to these co-ordinates become

$$\frac{d}{dt}\frac{dT}{d\dot{\chi}} = \text{X}, \quad \frac{d}{dt}\frac{dT}{d\dot{\chi}'} = \text{X}', \text{ etc.} \qquad \text{(1)}.$$

Farther let us suppose that the force-components X, X′, etc. corresponding to the co-ordinates χ, χ', etc. are each zero: we shall have

$$\frac{dT}{d\dot{\chi}} = C, \quad \frac{dT}{d\dot{\rho}} = C', \text{ etc.} \qquad \text{(2)};$$

or, expanded according to previous notation [318 (29)],

$$\left.\begin{array}{l} (\psi, \chi)\dot{\psi} + (\phi, \chi)\dot{\phi} + \ldots + (\chi, \chi)\dot{\chi} + (\chi, \chi')\dot{\chi}' + \ldots = C \\ (\psi, \chi')\dot{\psi} + (\phi, \chi')\dot{\phi} + \ldots + (\chi', \chi)\dot{\chi} + (\chi', \chi')\dot{\chi}' + \ldots = C' \\ \ldots\ldots\ldots\ldots\ldots\ldots\ldots\ldots \end{array}\right\} \ldots (3).$$

Hence, if we put

$$\left.\begin{array}{l} (\psi, \chi)\dot{\psi} + (\phi, \chi)\dot{\phi} + \ldots = P \\ (\psi, \chi')\dot{\psi} + (\phi, \chi')\dot{\phi} + \ldots = P' \\ \ldots\ldots\ldots\ldots \end{array}\right\} \ldots\ldots (4),$$

Ignoration of co-ordinates.

we have

$$\left.\begin{array}{l}(\chi,\chi)\,\dot\chi+(\chi,\chi')\,\dot\chi'+\ldots\;=C-P\\(\chi',\chi)\,\dot\chi+(\chi',\chi')\,\dot\chi'+\ldots=C'-P'\\\ldots\ldots\ldots\ldots\ldots\ldots\ldots\ldots\ldots\ldots\end{array}\right\}\ldots\ldots\ldots\ldots(5).$$

Resolving these for $\dot\chi, \dot\chi', \ldots$ we find

$$\dot\chi=\frac{\left|\begin{array}{l}(\chi',\chi'),\,(\chi',\chi''),\,\ldots\\(\chi'',\chi'),(\chi'',\chi''),\ldots\\\ldots\ldots\ldots\ldots\end{array}\right|(C-P)+\left|\begin{array}{l}(\chi'',\chi'),\,(\chi'',\chi''),\ldots\\(\chi''',\chi'),(\chi''',\chi''),\ldots\\\ldots\ldots\ldots\ldots\end{array}\right|(C'-P')+\ldots}{\left|\begin{array}{l}(\chi,\chi),\,(\chi,\chi'),\,(\chi,\chi''),\,\ldots\\(\chi',\chi),\,(\chi',\chi'),\,(\chi',\chi''),\,\ldots\\(\chi'',\chi),\,(\chi'',\chi'),\,(\chi'',\chi''),\ldots\\\ldots\ldots\ldots\ldots\ldots\end{array}\right|}\quad(6),$$

and symmetrical expressions for $\dot\chi', \dot\chi'', \ldots$, or, as we may write them short,

$$\left.\begin{array}{l}\dot\chi\;=(C,C)\,(C-P)+(C,C')\,(C'-P')\;+\ldots\\\dot\chi'=(C',C)\,(C-P)+(C',C')\,(C'-P')+\ldots\\\ldots\ldots\ldots\ldots\ldots\ldots\ldots\ldots\ldots\ldots\ldots\end{array}\right\}\ldots\ldots\ldots(7),$$

where (C, C), (C, C'), (C',C'), ... denote functions of the retained co-ordinates ψ, ϕ, θ, ... It is to be remembered that, because $(\chi,\chi')=(\chi',\chi)$, $(\chi,\chi'')=(\chi'',\chi)$, we see from (6) that

$(C, C')=(C', C)$, $(C, C'')=(C'', C)$, $(C', C'')=(C'', C')$, and so on...(8).

The following formulas for $\dot\chi, \dot\chi', \ldots$, condensed in respect to C, C', C'' by aid of the notation (14) below, and expanded in respect to $\dot\psi$, $\dot\phi$, ..., by (4), will also be useful.

$$\left.\begin{array}{l}\dot\chi=\dfrac{dK}{dC}-(M\dot\psi+N\dot\phi+\ldots)\\[2ex]\dot\chi'=\dfrac{dK}{dC'}-(M'\dot\psi+N'\dot\phi+\ldots)\\[1ex]\ldots\ldots\ldots\ldots\ldots\ldots\ldots\ldots\end{array}\right\}\ldots\ldots\ldots\ldots(9),$$

where

$$\left.\begin{array}{l}M\;=(C,C)\,.\,(\psi,\chi)+(C,C')\,.\,(\psi,\chi')\;+\ldots\\N\;=(C,C)\,.\,(\phi,\chi)+(C,C')\,.\,(\phi,\chi')+\ldots\\\ldots\ldots\ldots\ldots\ldots\ldots\ldots\ldots\ldots\ldots\ldots\\M'=(C',C)\,.\,(\psi,\chi)+(C',C')\,.\,(\psi,\chi')+\ldots\\\ldots\ldots\ldots\ldots\ldots\ldots\ldots\ldots\ldots\ldots\ldots\end{array}\right\}\ldots\ldots(10).$$

The elimination of $\dot\chi, \dot\chi', \ldots$ from T by these expressions for

Ignoration of co-ordinates.

them is facilitated by remarking that, as it is a quadratic function of $\dot\psi$, $\dot\phi$, ... $\dot\chi$, $\dot\chi'$, ..., we have

$$T = \tfrac{1}{2}\left\{\dot\psi \frac{dT}{d\dot\psi} + \dot\phi \frac{dT}{d\dot\phi} + \ldots + \dot\chi \frac{dT}{d\dot\chi} + \dot\chi' \frac{dT}{d\dot\chi'} + \ldots\right\}.$$

Hence by (3),

$$T = \tfrac{1}{2}\left\{\dot\psi \frac{dT}{d\dot\psi} + \dot\phi \frac{dT}{d\dot\phi} + \ldots + \dot\chi C + \dot\chi' C' + \ldots\right\},$$

so that we have now only first powers of $\dot\chi$, $\dot\chi'$, ... to eliminate. Gleaning out $\dot\chi$, $\dot\chi'$, ... from the first group of terms, and denoting by T_0 the part of T not containing $\dot\chi$, $\dot\chi'$, ..., we find

$$\begin{aligned} T = T_0 + \tfrac{1}{2}\{&[(\psi,\chi)\,\dot\psi + (\phi,\chi)\,\dot\phi + \ldots + C]\,\dot\chi \\ &+ [(\psi,\chi')\,\dot\psi + (\phi,\chi')\,\dot\phi + \ldots + C']\,\dot\chi' \\ &+ \ldots\ldots\ldots\ldots\},\end{aligned}$$

or, according to the notation of (4),

$$T = T_0 + \tfrac{1}{2}\{(C + P)\,\dot\chi + (C' + P')\,\dot\chi' + \ldots\}.$$

Eliminating now $\dot\chi$, $\dot\chi'$, ... by (7) we find

$$T = T_0 + \tfrac{1}{2}\{(C, C)\,(C^2 - P^2) + 2\,(C, C')\,(CC' - PP') + (C', C')\,(C'^2 - P'^2) + \ldots\} \quad \ldots\ldots\ldots (11).$$

It is remarkable that only second powers, and products, *not first powers*, of the velocity-components $\dot\psi$, $\dot\phi$, ... appear in this expression. We may write it thus:—

$$T = \mathfrak{T} + K \quad \ldots\ldots\ldots\ldots\ldots\ldots\ldots (12),$$

where $\mathfrak{T}$ denotes a quadratic function of $\dot\psi$, $\dot\phi$, ..., as follows:—

$$\mathfrak{T} = T_0 - \tfrac{1}{2}\{(C, C)\,P^2 + 2\,(C, C')\,PP' + (C', C')\,P'^2 + \ldots\} \ldots (13),$$

and K a quantity independent of $\dot\psi$, $\dot\phi$, ..., as follows:—

$$K = \tfrac{1}{2}\{(C, C)\,C^2 + 2\,(C, C')\,CC' + (C', C')\,C'^2 + \ldots\} \ldots\ldots (14).$$

Next, to eliminate $\dot\chi$, $\dot\chi'$, ... from the Lagrange's equations, we have, in virtue of (12) and of the constitutions of T, $\mathfrak{T}$, and K,

$$\frac{dT}{d\dot\psi} + \frac{dT}{d\dot\chi}\frac{d\dot\chi}{d\dot\psi} + \frac{dT}{d\dot\chi'}\frac{d\dot\chi'}{d\dot\psi} + \text{etc.} = \frac{d\mathfrak{T}}{d\dot\psi} \quad \ldots\ldots\ldots (15),$$

where $\dfrac{d\dot\chi}{d\dot\psi}$, $\dfrac{d\dot\chi'}{d\dot\psi}$, etc. are to be found by (7) or (9), and therefore are simply the coefficients of $\dot\psi$ in (9); so that we have

$$\frac{d\dot\chi}{d\dot\psi} = -M, \quad \frac{d\dot\chi'}{d\dot\psi} = -M' \quad \ldots\ldots\ldots\ldots\ldots (16),$$

where M, M' are functions of ψ, ϕ, ... explicitly expressed by (10). Using (16) in (15) we find

$$\frac{dT}{d\psi} = \frac{d\mathfrak{T}}{d\psi} + CM + C'M' + \text{etc.} \quad \text{.................(17).}$$

Ignoration of co-ordinates.

Again remarking that $\mathfrak{T} + K$ contains ψ, both as it appeared originally in T, and as farther introduced in the expressions (7) for $\dot{\chi}, \dot{\chi}', \ldots$, we see that

$$\frac{d}{d\psi}(\mathfrak{T} + K) = \frac{dT}{d\psi} + \frac{dT}{d\dot{\chi}}\frac{d\dot{\chi}}{d\psi} + \frac{dT}{d\dot{\chi}'}\frac{d\dot{\chi}'}{d\psi} + \ldots$$

$$= \frac{dT}{d\psi} + C\frac{d\dot{\chi}}{d\psi} + C'\frac{d\dot{\chi}'}{d\psi} + \ldots$$

And by (9) we have

$$\frac{d\dot{\chi}}{d\psi} = -\left(\dot{\psi}\frac{dM}{d\psi} + \dot{\phi}\frac{dN}{d\psi} + \ldots\right) + \frac{d}{d\psi}\frac{dK}{dC};$$

which, used in the preceding, gives

$$\frac{d}{d\psi}(\mathfrak{T} + K) = \frac{dT}{d\psi} - C\left(\dot{\psi}\frac{dM}{d\psi} + \dot{\phi}\frac{dN}{d\psi} + \ldots\right) - C'\left(\dot{\psi}\frac{dM'}{d\psi} + \dot{\phi}\frac{dN'}{d\psi} + \ldots\right) - \text{etc.} + 2\frac{dK}{d\psi}.$$

Hence

$$\frac{dT}{d\psi} = \frac{d\mathfrak{T}}{d\psi} - \frac{dK}{d\psi} + \Sigma C\left(\dot{\psi}\frac{dM}{d\psi} + \dot{\phi}\frac{dN}{d\psi} + \ldots\right) \quad \text{..........(18),}$$

where Σ denotes summation with regard to the constants C, C', etc.

Using this and (17) in the Lagrange's ψ-equation, we find finally for the ψ-equation of motion in terms of the non-ignored co-ordinates alone, and conclude the symmetrical equations for ϕ, etc., as follows,

$$\left.\begin{array}{l}
\dfrac{d}{dt}\left(\dfrac{d\mathfrak{T}}{d\dot{\psi}}\right) - \dfrac{d\mathfrak{T}}{d\psi} + \Sigma C\left\{\left(\dfrac{dM}{d\phi} - \dfrac{dN}{d\psi}\right)\dot{\phi} + \left(\dfrac{dM}{d\theta} - \dfrac{dO}{d\psi}\right)\dot{\theta} + \ldots\right\} + \dfrac{dK}{d\psi} = \Psi \\
\dfrac{d}{dt}\left(\dfrac{d\mathfrak{T}}{d\dot{\phi}}\right) - \dfrac{d\mathfrak{T}}{d\phi} + \Sigma C\left\{\left(\dfrac{dN}{d\psi} - \dfrac{dM}{d\phi}\right)\dot{\psi} + \left(\dfrac{dN}{d\theta} - \dfrac{dO}{d\phi}\right)\dot{\theta} + \ldots\right\} + \dfrac{dK}{d\phi} = \Phi \\
\dfrac{d}{dt}\left(\dfrac{d\mathfrak{T}}{d\dot{\theta}}\right) - \dfrac{d\mathfrak{T}}{d\theta} + \Sigma C\left\{\left(\dfrac{dO}{d\psi} - \dfrac{dM}{d\theta}\right)\dot{\psi} + \left(\dfrac{dO}{d\phi} - \dfrac{dN}{d\theta}\right)\dot{\phi} + \ldots\right\} + \dfrac{dK}{d\theta} = \Theta \\
\ldots\ldots\ldots\ldots\ldots\ldots\ldots\ldots\ldots\ldots\ldots\ldots\ldots\ldots
\end{array}\right\} \text{(19).}$$

[Compare Example F′ (e) above. It is important to remark that in each equation of motion the first power of the related velocity-component disappears; and the coefficient of each of the other velocity-components in this equation is equal but of opposite sign to the coefficient of the velocity-component corresponding to this equation, in the equation corresponding to that other velocity-component.]

Equation of Energy.

The equation of energy, found as above [§ 318 (29^{iv}) and (29^{v})], is

$$\frac{d(\mathfrak{T}+K)}{dt}=\Psi\dot{\psi}+\Phi\dot{\phi}+\text{etc.} \quad\ldots\ldots\ldots\ldots\ldots(20).$$

The interpretation, considering (12), is obvious. The contrast with Example F′ (*g*) is most instructive.

Sub-Example (G_1).—Take, from above, Example C, case (*a*): and put $\phi=\psi+\theta$; also, for brevity, $mj^2+na^2=B$, $n(b^2+k^2)=A$, and $nab=c$. We have*

$$T=\tfrac{1}{2}\{A\dot{\psi}^2+2c\dot{\psi}(\dot{\psi}+\dot{\theta})\cos\theta+B(\dot{\psi}+\dot{\theta})^2\};$$

and from this find

$$\frac{dT}{d\psi}=0,\quad \frac{dT}{d\dot{\psi}}=A\dot{\psi}+c(2\dot{\psi}+\dot{\theta})\cos\theta+B(\dot{\psi}+\dot{\theta});$$

$$\frac{dT}{d\theta}=-c\dot{\psi}(\dot{\psi}+\dot{\theta})\sin\theta,\quad \frac{dT}{d\dot{\theta}}=c\dot{\psi}\cos\theta+B(\dot{\psi}+\dot{\theta}).$$

Here the co-ordinate θ alone, and not the co-ordinate ψ, appears in the coefficients. Suppose now $\Psi=0$ [which is the case considered at the end of C (*a*) above]. We have $\frac{dT}{d\dot{\psi}}=C$, and deduce

$$\dot{\psi}=\frac{C-(c\cos\theta+B)\dot{\theta}}{A+B+2c\cos\theta},$$

$$T=\tfrac{1}{2}\left(\dot{\psi}\frac{dT}{d\dot{\psi}}+\dot{\theta}\frac{dT}{d\dot{\theta}}\right)=\tfrac{1}{2}\{\dot{\psi}C+\dot{\theta}[(c\cos\theta+B)\dot{\psi}+B\dot{\theta}]\}$$

$$=\tfrac{1}{2}\{\dot{\psi}[C+(c\cos\theta+B)\dot{\theta}]+B\dot{\theta}\}$$

$$=\tfrac{1}{2}\left\{\frac{C^2-(c\cos\theta+B)^2\dot{\theta}^2}{A+B+2c\cos\theta}+B\dot{\theta}^2\right\}=\tfrac{1}{2}\frac{C^2+(AB-c^2\cos^2\theta)\dot{\theta}^2}{A+B+2c\cos\theta}.$$

Hence
$$\mathfrak{T}=\tfrac{1}{2}\frac{AB-c^2\cos^2\theta}{A+B+2c\cos\theta}\dot{\theta}^2,$$

and
$$K=\tfrac{1}{2}\frac{C^2}{A+B+2c\cos\theta}:$$

* Remark that, according to the alteration from ψ, $\dot{\psi}$, ϕ, $\dot{\phi}$, to ψ, $\dot{\psi}$, θ, $\dot{\theta}$, as independent variables,

$$\frac{dT}{d\psi}=\left(\frac{dT}{d\psi}\right)+\left(\frac{dT}{d\phi}\right),\quad \frac{dT}{d\theta}=\left(\frac{dT}{d\phi}\right);$$

and
$$\frac{dT}{d\dot{\psi}}=\left(\frac{dT}{d\dot{\psi}}\right)+\left(\frac{dT}{d\dot{\phi}}\right),\quad \frac{dT}{d\dot{\theta}}=\left(\frac{dT}{d\dot{\phi}}\right);$$

where ( ) indicates the original notation of C (*a*).

and the one equation of the motion becomes

Equation of Energy.

$$\frac{d}{dt}\left(\frac{AB-c^2\cos^2\theta}{A+B+2c\cos\theta}\dot\theta\right)-\tfrac{1}{2}\dot\theta^2\frac{d}{d\theta}\left(\frac{AB-c^2\cos^2\theta}{A+B+2c\cos\theta}\right)=\Theta-\frac{dK}{d\theta};$$

which is to be fully integrated first by multiplying by $d\theta$ and integrating once; and then solving for dt and integrating again with respect to θ. The first integral, being simply the equation of energy integrated, is [Example G (20)]

$$\mathfrak{T}=\int\Theta\, d\theta-K;$$

and the final integral is

$$t=\int d\theta\sqrt{\frac{AB-\cos^2\theta}{2(A+B+2c\cos\theta)(\int\Theta d\theta-K)}}.$$

Ignoration of co-ordinates.

In the particular case in which the motion commences from rest, or is such that it can be brought to rest by proper applications of force-components, Ψ, Φ, etc. without any of the force-components X, X′, etc., we have $C=0$, $C'=0$, etc.; and the elimination of $\dot\chi$, $\dot\chi'$, etc. by (3) renders T a homogeneous quadratic function of $\dot\psi$, $\dot\phi$, etc. without C, C', etc.; and the equations of motion become

$$\left.\begin{aligned}\frac{d}{dt}\frac{dT}{d\dot\psi}-\frac{dT}{d\psi}&=\Psi\\ \frac{d}{dt}\frac{dT}{d\dot\phi}-\frac{dT}{d\phi}&=\Phi\\ \frac{d}{dt}\frac{dT}{d\dot\theta}-\frac{dT}{d\theta}&=\Theta\\ \text{etc.}\quad\text{etc.}&\end{aligned}\right\}\dots\dots\dots\dots(21).$$

We conclude that on the suppositions made, the elimination of the velocity-components corresponding to the non-appearing co-ordinates gives an expression for the kinetic energy in terms of the remaining velocity-components and corresponding co-ordinates which may be used in the generalised equations just as if these were the sole co-ordinates. The reduced number of equations of motion thus found suffices for the determination of the co-ordinates which they involve without the necessity for knowing or finding the other co-ordinates. If the farther question be put,—to determine the ignored co-ordinates, it is to be answered by a simple integration of equations (7) with $C=0$, $C'=0$, etc.

One obvious case of application for this example is a system in which any number of fly wheels, that is to say, bodies which are

Ignoration of co-ordinates.

kinetically symmetrical round an axis (§ 285), are pivoted frictionlessly on any moveable part of the system. In this case with the particular supposition $C=0$, $C'=0$, etc., the result is simply that the motion is the same as if each fly wheel were deprived of moment of inertia round its bearing axis, that is to say reduced to a line of matter fixed in the position of this axis and having unchanged moment of inertia round any axis perpendicular to it. But if C, C', etc. be not each zero we have a case embracing a very interesting class of dynamical problems in which the motion of a system having what we may call gyrostatic links or connexions is the subject. Example (D) above is an example, in which there is just one fly wheel and one moveable body on which it is pivoted. The ignored co-ordinate is ψ; and supposing now Ψ to be zero, we have

$$\dot{\psi} - \dot{\phi}(1 - \cos\theta) = C \text{ ....} (a).$$

If we suppose $C=0$ all the terms having $\mathfrak{A}'$ for a factor vanish and the motion is the same as if the fly wheel were deprived of inertia round its bearing axis, and we had simply the motion of the "ideal rigid body PQ" to consider. But when C does not vanish we eliminate $\dot{\psi}$ from the equations by means of (a). It is important to remark that in every case of Example (G) in which $C=0$, $C'=0$, etc. the motion at each instant possesses the property (§ 312 above) of having less kinetic energy than any other motion for which the velocity-components of the non-ignored co-ordinates have the same values.

Take for another example the final form of Example C′ above, putting B for C, and A for $nk^2 + A$. We have

$$T = \tfrac{1}{2}\{(E + F\cos^2\theta)\dot{\psi}^2 + B(\dot{\psi}\cos\theta + \dot{\phi})^2 + A\dot{\theta}^2\} \text{ ...(22)}.$$

Here neither ψ nor ϕ appears in the coefficients. Let us suppose $\Phi = 0$, and eliminate $\dot{\phi}$, to let us ignore ϕ. We have

$$\frac{dT}{d\dot{\phi}} = B(\dot{\psi}\cos\theta + \dot{\phi}) = C.$$

Hence
$$\dot{\phi} = \frac{C}{B} - \dot{\psi}\cos\theta \text{ ....(23)},$$

$$\mathfrak{T} = \tfrac{1}{2}\{(E + F\cos^2\theta)\dot{\psi}^2 + A\dot{\theta}^2\} \text{ ....(24)},$$

and
$$K = \tfrac{1}{2}\frac{C^2}{B} \text{ ....(25)}.$$

The place of $\dot{\chi}$ in (9) above is now taken by $\dot{\phi}$, and comparing with (23) we find

$$M = \cos\theta,\ N = 0,\ O = 0.$$

Hence, and as K is constant, the equations of motion (19) become

Ignoration of co-ordinates.

$$\left.\begin{array}{l} \dfrac{d}{dt}\dfrac{d\mathfrak{T}}{d\dot{\psi}} - \dfrac{d\mathfrak{T}}{d\psi} - C\sin\theta\dot{\theta} = \Psi \\ \text{and} \quad \dfrac{d}{dt}\dfrac{d\mathfrak{T}}{d\dot{\theta}} - \dfrac{d\mathfrak{T}}{d\theta} + C\sin\theta\dot{\psi} = \Theta \end{array}\right\} \ldots\ldots\ldots\ldots\ldots(26);$$

and, using (24) and expanding,

$$\left.\begin{array}{l} \dfrac{d\{(E+F\cos^2\theta)\dot{\psi}\}}{dt} - C\sin\theta\dot{\theta} = \Psi \\ A\ddot{\theta} + F\sin\theta\cos\theta\dot{\psi}^2 + C\sin\theta\dot{\psi} = \Theta \end{array}\right\} \ldots\ldots\ldots\ldots(27).$$

A most important case for the "ignoration of co-ordinates" is presented by a large class of problems regarding the motion of frictionless incompressible fluid in which we can ignore the infinite number of co-ordinates of individual portions of the fluid and take into account only the co-ordinates which suffice to specify the whole boundary of the fluid, including the bounding surfaces of any rigid or flexible solids immersed in the fluid. The analytical working out of Example (G) shows in fact that when the motion is such as could be produced from rest by merely moving the boundary of the fluid without applying force to its individual particles otherwise than by the transmitted fluid pressure we have exactly the case of $C=0$, $C'=0$, etc.: and Lagrange's generalized equations with the kinetic energy expressed in terms of velocity-components completely specifying the motion of the boundary are available. Thus,

Kinetics of a perfect liquid.

320. Problems in fluid motion of remarkable interest and importance, not hitherto attacked, are very readily solved by the aid of Lagrange's generalized equations of motion. For brevity we shall designate a mass which is absolutely incompressible, and absolutely devoid of resistance to change of shape, by the simple appellation of a *liquid.* We need scarcely say that matter perfectly satisfying this definition does not exist in nature: but we shall see (under properties of matter) how nearly it is approached by water and other common real liquids. And we shall find that much practical and interesting information regarding their true motions is obtained by deductions from the principles of abstract dynamics applied to the ideal perfect liquid of our definition. It follows from Example

Kinetics of a perfect liquid.

(G) above (and several other proofs, some of them more synthetical in character, will be given in our Second Volume,) that the motion of a homogeneous liquid, whether of infinite extent, or contained in a finite closed vessel of any form, with any rigid or flexible bodies moving through it, if it has ever been at rest, is the same at each instant as that determinate motion (fulfilling, § 312, the condition of having the least possible kinetic energy) which would be impulsively produced from rest by giving instantaneously to every part of the bounding surface, and of the surface of each of the solids within it, its actual velocity at that instant. So that, for example, however long it may have been moving, if all these surfaces were suddenly or gradually brought to rest, the whole fluid mass would come to rest at the same time. Hence, if none of the surfaces is flexible, but we have one or more rigid bodies moving in any way through the liquid, under the influence of any forces, the kinetic energy of the whole motion at any instant will depend solely on the finite number of co-ordinates and component velocities, specifying the position and motion of those bodies, whatever may be the positions reached by particles of the fluid (expressible only by an infinite number of co-ordinates). And an expression for the whole kinetic energy in terms of such elements, finite in number, is precisely what is wanted, as we have seen, as the foundation of Lagrange's equations in any particular case.

It will clearly, in the hydrodynamical, as in all other cases, be a homogeneous quadratic function of the components of velocity, if referred to an invariable co-ordinate system; and the coefficients of the several terms will in general be functions of the co-ordinates, the determination of which follows immediately from the solution of the minimum problem of Example (3) § 317, in each particular case.

Example (1).—*A ball set in motion through a mass of incompressible fluid extending infinitely in all directions on one side of an infinite plane, and originally at rest.* Let x, y, z be the co-ordinates of the centre of the ball at time t, with reference to rectangular axes through a fixed point O of the bounding plane, with OX perpendicular to this plane. If at any instant either component $\dot{y}$ or $\dot{z}$ of the velocity be reversed, the kinetic energy will clearly be unchanged, and hence no terms $\dot{y}\dot{z}$, $\dot{z}\dot{x}$, or $\dot{x}\dot{y}$ can appear in the expression for the kinetic energy: which, on this account, and because of the symmetry of circumstances with reference to y and z, is

Kinetics of a perfect liquid.

$$T = \tfrac{1}{2}\{P\dot{x}^2 + Q(\dot{y}^2 + \dot{z}^2)\}.$$

Also, we see that P and Q are functions of x simply, since the circumstances are similar for all values of y and z. Hence, by differentiation,

$$\frac{dT}{d\dot{x}} = P\dot{x},\quad \frac{dT}{d\dot{y}} = Q\dot{y},\quad \frac{dT}{d\dot{z}} = Q\dot{z},$$

$$\frac{d}{dt}\left(\frac{dT}{d\dot{x}}\right) = P\ddot{x} + \frac{dP}{dx}\dot{x}^2,\quad \frac{d}{dt}\left(\frac{dT}{d\dot{y}}\right) = Q\ddot{y} + \frac{dQ}{dx}\dot{y}\dot{x},\ \text{etc.},$$

$$\frac{dT}{dx} = \tfrac{1}{2}\left\{\frac{dP}{dx}\dot{x}^2 + \frac{dQ}{dx}(\dot{y}^2 + \dot{z}^2)\right\},\quad \frac{dT}{dy} = 0,\ \text{etc.},$$

and the equations of motion are

$$P\ddot{x} + \tfrac{1}{2}\left\{\frac{dP}{dx}\dot{x}^2 - \frac{dQ}{dx}(\dot{y}^2 + \dot{z}^2)\right\} = X,$$

$$Q\ddot{y} + \frac{dQ}{dx}\dot{y}\dot{x} = Y,\quad Q\ddot{z} + \frac{dQ}{dx}\dot{z}\dot{x} = Z.$$

Principles sufficient for a practical solution of the problem of determining P and Q will be given later. In the meantime, it is obvious that each decreases as x increases. Hence the equations of motion show that

321. A ball projected through a liquid perpendicularly *from* an infinite plane boundary, and influenced by no other forces than those of fluid pressure, experiences a gradual acceleration, quickly approximating to a limiting velocity which it sensibly reaches when its distance from the plane is many times its diameter. But if projected *parallel* to the plane, it experiences, as the resultant of fluid pressure, a resultant attraction towards the plane. The former of these results is easily proved by first considering projection *towards* the plane (in which case the motion of the ball will obviously be retarded), and by taking into account the general principle of reversibility (§ 272) which has perfect application in the ideal case of a perfect liquid. The second result is less easily foreseen without

Effect of a rigid plane on the motion of a ball through a liquid.

the aid of Lagrange's analysis; but it is an obvious consequence of the Hamiltonian form of his equations, as stated in words in § 319 above. In the precisely equivalent case, of a liquid extending infinitely in all directions, and given at rest; and two equal balls projected through it with equal velocities perpendicular to the line joining their centres—the result that the two balls will seem to attract one another is most remarkable, and very suggestive.

Seeming attraction between two ships moving side by side in the same direction.

Hydrodynamical examples continued.

Example (2).—*A solid symmetrical round an axis, moving through a liquid so as to keep its axis always in one plane.* Let ω be the angular velocity of the body at any instant about any axis perpendicular to the fixed plane, and let u and q be the component velocities along and perpendicular to the axis of figure, of any chosen point, C, of the body in this line. By the general principle stated in § 320 (since changing the sign of u cannot alter the kinetic energy), we have

$$T = \tfrac{1}{2}(Au^2 + Bq^2 + \mu'\omega^2 + 2E\omega q) \qquad \ldots\ldots (a),$$

where A, B, μ', and E are constants depending on the figure of the body, its mass, and the density of the liquid. Now let v denote the velocity, perpendicular to the axis, of a point which we shall call the *centre of reaction*, being a point in the axis and at a distance $\frac{E}{B}$ from C, so that (§ 87) $q = v - \frac{E}{B}\omega$. Then, denoting $\mu' - \frac{E^2}{B}$ by μ, we have $T = \tfrac{1}{2}(Au^2 + Bv^2 + \mu\omega^2) \ldots\ldots (a')$.

"Centre of reaction" defined.

Let x and y be the co-ordinates of the centre of reaction relatively to any fixed rectangular axes in the plane of motion of the axis of figure, and let θ be the angle between this line and OX, at any instant, so that

$$\omega = \dot\theta,\ u = \dot x \cos\theta + \dot y \sin\theta,\ v = -\dot x \sin\theta + \dot y \cos\theta \qquad \ldots\ldots (b).$$

Substituting in T, differentiating, and retaining the notation u, v where convenient for brevity, we have

$$\left.\begin{aligned} &\frac{dT}{d\dot\theta} = \mu\dot\theta,\ \frac{dT}{d\dot x} = Au\cos\theta - Bv\sin\theta,\ \frac{dT}{d\dot y} = Au\sin\theta + Bv\cos\theta, \\ &\frac{dT}{d\theta} = (A - B)\,uv,\ \frac{dT}{dx} = 0,\ \frac{dT}{dy} = 0, \end{aligned}\right\} \quad (c).$$

Hydro-dynamical examples continued.

Hence the equations of motion are

$$\left.\begin{array}{l}\mu\ddot{\theta}-(A-B)uv=L,\\ \dfrac{d(Au\cos\theta-Bv\sin\theta)}{dt}=X,\quad \dfrac{d(Au\sin\theta+Bv\cos\theta)}{dt}=Y\end{array}\right\}\ (d),$$

where X, Y are the component forces in lines through C parallel to OX and OY, and L the couple, applied to the body.

Denoting by λ, ξ, η the impulsive couple, and the components of impulsive force through C, required to produce the motion at any instant, we have of course [§ 313 (c)],

$$\lambda=\frac{dT}{d\dot{\theta}}=\mu\dot{\theta},\ \xi=\frac{dT}{d\dot{x}},\ \eta=\frac{dT}{d\dot{y}}\ \ldots\ldots\ldots\ldots\ldots\ldots\ (e),$$

and therefore by (c), and (b),

$$u=\frac{1}{A}(\xi\cos\theta+\eta\sin\theta),\ v=\frac{1}{B}(-\xi\sin\theta+\eta\cos\theta),\ \ \dot{\theta}=\frac{\lambda}{\mu}\ldots\ldots(f),$$

$$\left.\begin{array}{l}\dot{x}=\left(\dfrac{\cos^2\theta}{A}+\dfrac{\sin^2\theta}{B}\right)\xi+\left(\dfrac{1}{A}-\dfrac{1}{B}\right)\sin\theta\cos\theta\eta,\\ \dot{y}=\left(\dfrac{1}{A}-\dfrac{1}{B}\right)\sin\theta\cos\theta\xi+\left(\dfrac{\sin^2\theta}{A}+\dfrac{\cos^2\theta}{B}\right)\eta\end{array}\right\}\ldots\ldots\ (g),$$

and the equations of motion become

$$\mu\frac{d^2\theta}{dt^2}-\frac{A-B}{2AB}\{(-\xi^2+\eta^2)\sin 2\theta+2\xi\eta\cos 2\theta\}=L,\ \frac{d\xi}{dt}=X,\ \frac{d\eta}{dt}=Y,\ \ (h).$$

The simple case of $X=0$, $Y=0$, $L=0$, is particularly interesting. In it ξ and η are each constant; and we may therefore choose the axes OX, OY, so that η shall vanish. Thus we have, in (g), two first integrals of the equations of motion; and they become

$$\dot{x}=\xi\left(\frac{\cos^2\theta}{A}+\frac{\sin^2\theta}{B}\right),\ \ \dot{y}=-\frac{A-B}{2AB}\xi\sin 2\theta\ \ldots\ldots\ldots\ldots\ (k):$$

and the first of equations (h) becomes

$$\mu\frac{d^2\theta}{dt^2}+\frac{A-B}{2AB}\xi^2\sin 2\theta=0\ldots\ldots\ldots\ldots\ldots\ldots\ldots\ldots\ (l).$$

In this let, for a moment, $2\theta=\phi$, and $\dfrac{A-B}{AB}\xi^2=ghW$. It becomes

$$\mu\frac{d^2\phi}{dt^2}+ghW\sin\phi=0,$$

which is the equation of motion of a common pendulum, of mass W, moment of inertia μ round its fixed axis, and length

Hydro-dynamical examples continued.

h from axis to centre of gravity; if ϕ be the angle from the position of equilibrium to the position at time t. As we shall see, under kinetics, the final integral of this equation expresses ϕ in terms of t by means of an elliptic function. By using the value thus found for θ or $\frac{1}{2}\phi$, in (k), we have equations giving x and y in terms of t by common integration; and thus the full solution of our present problem is reduced to quadratures. The detailed working out to exhibit both the actual curve described by the centre of reaction, and the position of the axis of the body at any instant, is highly interesting. It is very easily done approximately for the case of very small angular vibrations; that is to say, when either $A-B$ is positive, and ϕ always very small, or $A-B$ negative, and ϕ very nearly equal to $\frac{1}{2}\pi$. But without attending at present to the final integrals, rigorous or approximate, we see from (k) and (l) that

322. If a solid of revolution in an infinite liquid, be set in motion round any axis perpendicular to its axis of figure, or simply projected in any direction without rotation, it will move with its axis always in one plane, and every point of it moving only parallel to this plane; and the strange evolutions which it will, in general, perform, are perfectly defined by comparison with the common pendulum thus. First, for brevity, we shall call by the name of *quadrantal pendulum* (which will be further exemplified in various cases described later, under electricity and magnetism; for instance, an elongated mass of soft iron pivoted on a vertical axis, in a "uniform field of magnetic force"), a body moving about an axis, according to the same law with reference to a quadrant on each side of its position of equilibrium, as the common pendulum with reference to a half circle on each side.

Quadrantal pendulum defined.

Let now the body in question be set in motion by an impulse, ξ, in any line through the centre of reaction, and an impulsive couple λ in the plane of that line and the axis. This will (as will be proved later in the theory of statical couples) have the same effect as a simple impulse ξ (applied to a point, if not of the real body, connected with it by an imaginary infinitely light framework) in a certain fixed line, which we shall call the line of resultant impulse, or of resultant momentum, being parallel to the former line, and at a distance from it equal to $\frac{\lambda}{\xi}$. The whole momentum of the motion generated is of course (§ 295) equal to ξ. The body will move ever afterwards according to the following conditions:—(1.) The angular velocity follows the law of the quadrantal pendulum. (2.) The distance of the centre of reaction from the line of resultant impulse varies simply as the angular velocity. (3.) The velocity of the centre of reaction parallel to the line of impulse is found by dividing the excess of the whole constant energy of the motion above the part of it due to the angular velocity round the centre of reaction, by half the momentum. (4.) If A, B, and μ denote constants, depending on the mass of the solid and its distribution, the density of the liquid, and the form and dimensions of the solid, such that $\frac{\xi}{A}$, $\frac{\xi}{B}$, $\frac{\lambda}{\mu}$ are the linear velocities, and the angular velocity, respectively produced by an impulse ξ along the axis, an impulse ξ in a line through the centre of reaction perpendicular to the axis, and an impulsive couple λ in a plane through the axis; the length of the simple gravitation pendulum, whose motion would keep time with the periodic motion in question, is $\frac{g\mu AB}{\xi^2(A-B)}$, and, when the angular motion is vibratory, the vibrations will, according as $A > B$, or $A < B$, be of the axis, or of a line perpendicular to the axis, vibrating on each side of the line of impulse. The angular motion will in fact be vibratory if the distance of the line of resultant impulse from the centre of reaction is anything less than $\sqrt{\frac{(A \sim B)\mu \cos 2\alpha}{AB}}$ where α denotes the inclination of the impulse to the initial position of the axis. In this case the path of the centre of reaction will be a sinuous curve symmetrical on the two sides of the line of impulse; every time it cuts this line, the angular motion will reverse, and the maximum inclination will be attained; and every time the centre of reaction is at its greatest distance on either side, the angular velocity will be at its greatest, positive or negative, value, and the linear velocity of

Motion of a solid of revolution through a liquid.

Motion of a solid of revolution through a liquid.

the centre of reaction will be at its least. If, on the other hand, the line of the resultant impulse be at a greater distance than $\sqrt{\frac{(A \sim B)\mu\cos 2\alpha}{AB}}$ from the centre of reaction, the angular motion will be always in one direction, but will increase and diminish periodically, and the centre of reaction will describe a sinuous curve on one side of that line; being at its greatest and least deviations when the angular velocity is greatest and least. At the same points the curvature of the path will be greatest and least respectively, and the linear velocity of the describing point will be least and greatest.

323. At any instant the component linear velocities along and perpendicular to the axis of the solid will be $\frac{\xi\cos\theta}{A}$ and $-\frac{\xi\sin\theta}{B}$ respectively, if θ be its inclination to the line of resultant impulse; and the angular velocity will be $\frac{\xi y}{\mu}$ if y be the distance of the centre of reaction from that line. The whole kinetic energy of the motion will be

$$\frac{\xi^2\cos^2\theta}{2A} + \frac{\xi^2\sin^2\theta}{2B} + \frac{\xi^2 y^2}{2\mu},$$

and the last term is what we have referred to above as the part due to rotation round the centre of reaction (defined in § 321). To stop the whole motion at any instant, a simple impulse equal and opposite to ξ in the fixed "line of resultant impulse" will suffice (or an equal and parallel impulse in any line through the body, with the proper impulsive couple, according to the principle already referred to).

324. From Lagrange's equations applied as above to the case of a solid of revolution moving through a liquid, the couple which must be kept applied to it to prevent it from turning is immediately found to be

$$uv(A - B),$$

if u and v be the component velocities along and perpendicular to the axis, or [§ 321 (f)]

Motion of a solid of revolution through a liquid.

$$\xi^2 \frac{(A-B)\sin 2\theta}{2AB},$$

if, as before, ξ be the generating impulse, and θ the angle between its line and the axis. The direction of this couple must be such as to prevent θ from diminishing or from increasing, according as A or B is the greater. The former will clearly be the case of a flat disc, or oblate spheroid; the latter that of an elongated, or oval-shaped body. The actual values of A and B we shall learn how to calculate (hydrodynamics) for several cases, including a body bounded by two spherical surfaces cutting one another at any angle a submultiple of two right angles; two complete spheres rigidly connected; and an oblate or a prolate spheroid.

Observed phenomena.

325. The tendency of a body to turn its flat side, or its length (as the case may be), across the direction of its motion through a liquid, to which the accelerations and retardations of rotatory motion described in § 322 are due, and of which we have now obtained the statical measure, is a remarkable illustration of the statement of § 319; and is closely connected with the dynamical explanation of many curious observations well known in practical mechanics, among which may be mentioned:—

(1) That the course of a symmetrical square-rigged ship sailing in the direction of the wind with rudder amidships is unstable, and can only be *kept* by manipulating the rudder to check infinitesimal deviations;—and that a child's toy-boat, whether "square-rigged" or "fore-and-aft rigged*," cannot be

* "Fore-and-aft" rig is any rig in which (as in "cutters" and "schooners") the chief sails come into the plane of mast or masts and keel, by the action of the wind upon the sails when the vessel's head is to wind. This position of the sails is unstable when the wind is right astern. Accordingly, in "wearing" a fore-and-aft rigged vessel (that is to say turning her round stern to wind, from sailing with the wind on one side to sailing with the wind on the other side) the mainsail must be hauled in as closely as may be towards the middle position before the wind is allowed to get on the other side of the sail from that on which it had been pressing, so that when the wind

Applications to nautical dynamics

got to sail permanently before the wind by any permanent adjustment of rudder and sails, and that (without a wind vane, or a weighted tiller, acting on the rudder to do the part of steersman) it always, after running a few yards before the wind, turns round till nearly in a direction perpendicular to the wind (either "gibing" first, or "luffing" without gibing if it is a cutter or schooner) :—

(2) That the towing rope of a canal boat, when the rudder is left straight, takes a position in a vertical plane cutting the axis *before* its middle point :—

(3) That a boat sculled rapidly across the direction of the wind, always (unless it is extraordinarily unsymmetrical in its draught of water, and in the amounts of surface exposed to the wind, towards its two ends) requires the weather oar to be worked hardest to prevent it from running up on the wind, and that for the same reason a sailing vessel generally "carries a weather helm*" or "gripes;" and that still more does so a steamer with sail even if only in the forward half of her length—griping so badly with any after canvass† that it is often impossible to steer :—

(4) That in a heavy gale it is exceedingly difficult, and often found impossible, to get a ship out of "the trough of the sea," and that it cannot be done at all without rapid motion ahead, whether by steam or sails :—

(5) That in a smooth sea with moderate wind blowing parallel to the shore, a sailing vessel heading towards the shore with not enough of sail set can only be saved from creeping ashore by setting more sail, and sailing rapidly towards the shore, or the danger that is to be avoided, so as to allow her to be steered away from it. The risk of going ashore in fulfilment

does get on the other side, and when therefore the sail dashes across through the mid-ship position to the other side, carrying massive boom and gaff with it, the range of this sudden motion, which is called "gibing," shall be as small as may be.

* The weather side of any object is the side of it towards the wind. A ship is said to "carry a weather helm" when it is necessary to hold the "helm" or "tiller" permanently on the weather side of its middle position (by which the rudder is held towards the lee side) to keep the ship on her course.

† Hence mizen masts are altogether condemned in modern war-ships by many competent nautical authorities.

of Lagrange's equations is a frequent incident of "getting under way" while lifting anchor, or even after slipping from moorings :—

(6) That an elongated rifle-bullet requires rapid rotation about its axis to keep its point foremost.

and gunnery.

(7) The curious motions of a flat disc, oyster-shell, or the like, when dropped obliquely into water, resemble, no doubt, to some extent those described in § 322. But it must be remembered that the real circumstances differ greatly, because of fluid friction, from those of the abstract problem, of which we take leave for the present.

Least action.

326. Maupertuis' celebrated principle of *Least Action* has been, even up to the present time, regarded rather as a curious and somewhat perplexing property of motion, than as a useful guide in kinetic investigations. We are strongly impressed with the conviction that a much more profound importance will be attached to it, not only in abstract dynamics, but in the theory of the several branches of physical science now beginning to receive dynamic explanations. As an extension of it, Sir W. R. Hamilton* has evolved his method of *Varying Action*, which undoubtedly must become a most valuable aid in future generalizations.

Action.

What is meant by "Action" in these expressions is, unfortunately, something very different from the *Actio Agentis* defined by Newton†, and, it must be admitted, is a much less judiciously chosen word. Taking it, however, as we find it, now universally used by writers on dynamics, we define the *Action of a Moving System* as proportional to the average kinetic energy, which the system has possessed during the time from any convenient epoch of reckoning, multiplied by the time. According to tho unit generally adopted, the action of a system which has not varied in its kinetic energy, is twice the amount of the energy multiplied by the time from the epoch. Or if the energy has been sometimes greater and sometimes less,

Time average of energy.

* *Phil. Trans.* 1834—1835.

† Which, however (§ 263), we have translated "activity" to avoid confusion.

Time average of energy doubled.

the action at time t is the double of what we may call the *time-integral* of the energy, that is to say, it is what is denoted in the integral calculus by

$$2\int_0^t T d\tau,$$

where T denotes the kinetic energy at any time τ, between the epoch and t.

Let m be the mass, and v the velocity at time τ, of any one of the material points of which the system is composed. We have

$$T = \Sigma \tfrac{1}{2} mv^2 \text{ .... } (1),$$

and therefore, if A denote the action at time t,

$$A = \int_0^t \Sigma mv^2 d\tau \text{ .... } (2).$$

This may be put otherwise by taking ds to denote the space described by a particle in time $d\tau$, so that $vd\tau = ds$, and therefore

$$A = \int \Sigma mvds \text{ .... } (3),$$

or, if x, y, z be the rectangular co-ordinates of m at any time,

$$A = \int \Sigma m \,(\dot{x}dx + \dot{y}dy + \dot{z}dz) \text{ .... } (4).$$

Hence we might, as many writers in fact have virtually done, define action thus :—

Space average of momentums.

The action of a system is equal to the sum of the *average momentums for the spaces* described by the particles from any era each multiplied by the length of its path.

Least action.

327. The principle of Least Action is this :—Of all the different sets of paths along which a conservative system may be guided to move from one configuration to another, with the sum of its potential and kinetic energies equal to a given constant, that one for which the action is the least is such that the system will require only to be started with the proper velocities, to move along it unguided. Consider the Problem :— Given the whole initial kinetic energy; find the initial velocities through one given configuration, which shall send the system unguided to another specified configuration. This problem is essentially determinate, but generally has multiple solutions (§ 363 below); (or only imaginary solutions.)

General Problem of *aim*.

High aim and low aim for same target. Target too far off to be reached.

Least action.

If there are any real solutions, there is one of them for which the action is less than for any other real solution, and less than for any constrainedly guided motion with proper sum of potential and kinetic energies. Compare §§ 346—366 below.

Let x, y, z be the co-ordinates of a particle, m, of the system, at time τ, and V the potential energy of the system in its particular configuration at this instant; and let it be required to find the way to pass from one given configuration to another with velocities at each instant satisfying the condition

$$\Sigma \tfrac{1}{2} m (\dot{x}^2 + \dot{y}^2 + \dot{z}^2) + V = E, \text{ a constant} \ldots\ldots\ldots\ldots\ldots (5),$$

so that A, or

$$\int \Sigma m (\dot{x}dx + \dot{y}dy + \dot{z}dz)$$

may be the least possible.

By the method of variations we must have $\delta A = 0$, where

$$\delta A = \int \Sigma m (\dot{x}d\delta x + \dot{y}d\delta y + \dot{z}d\delta z + \delta\dot{x}dx + \delta\dot{y}dy + \delta\dot{z}dz) \ldots\ldots\ldots (6).$$

Taking in this $dx = \dot{x}d\tau$, $dy = \dot{y}d\tau$, $dz = \dot{z}d\tau$, and remarking that

$$\Sigma m (\dot{x}\delta\dot{x} + \dot{y}\delta\dot{y} + \dot{z}\delta\dot{z}) = \delta T \ldots\ldots\ldots\ldots\ldots\ldots\ldots\ldots (7),$$

we have

$$\int \Sigma m (\delta\dot{x}dx + \delta\dot{y}dy + \delta\dot{z}dz) = \int_0^t \delta T d\tau \ldots\ldots\ldots\ldots\ldots (8).$$

Also by integration by parts,

$$\int \Sigma m (\dot{x}d\delta x + \ldots) = \{\Sigma m (\dot{x}dx + \ldots)\} - [\Sigma m (\dot{x}dx + \ldots)] - \int \Sigma m (\ddot{x}\delta x + \ldots) d\tau,$$

where $[\ldots]$ and $\{\ldots\}$ denote the values of the quantities enclosed, at the beginning and end of the motion considered, and where, further, it must be remembered that $d\dot{x} = \ddot{x}d\tau$, etc. Hence, from above,

$$\delta A = \{\Sigma m (\dot{x}\delta x + \dot{y}\delta y + \dot{z}\delta z)\} - [\Sigma m (\dot{x}\delta x + \dot{y}\delta y + \dot{z}\delta z)]$$
$$+ \int_0^t d\tau \, [\delta T - \Sigma m (\ddot{x}\delta x + \ddot{y}\delta y + \ddot{z}\delta z)] \ldots\ldots\ldots\ldots (9).$$

This, it may be observed, is a perfectly general kinematical expression, unrestricted by any terminal or kinetic conditions. Now in the present problem we suppose the initial and final positions to be invariable. Hence the terminal variations, δx, etc., must all vanish, and therefore the integrated expressions $\{\ldots\}$, $[\ldots]$ disappear. Also, in the present problem $\delta T = -\delta V$, by the equation of energy (5). Hence, to make $\delta A = 0$, since the intermediate variations, δx, etc., are quite arbitrary, subject only to the con-

Least action.

ditions of the system, we must have

$$\Sigma m(\ddot{x}\delta x+\ddot{y}\delta y+\ddot{z}\delta z)+\delta V=0 \quad\text{.......................}\quad (10),$$

which [(4), § 293 above] is the general variational equation of motion of a conservative system. This proves the proposition.

Principle of Least Action applied to find Lagrange's generalized equations of motion.

It is interesting and instructive as an illustration of the principle of least action, to derive directly from it, without any use of Cartesian co-ordinates, Lagrange's equations in generalized co-ordinates, of the motion of a conservative system [§ 318 (24)]. We have

$$A=\int 2Tdt,$$

where T denotes the formula of § 313 (2). If now we put

$$T=\tfrac{1}{2}\frac{ds^2}{dt^2},$$

so that $\qquad ds^2=(\psi,\psi)\,d\psi^2+2\,(\psi,\phi)\,d\psi d\phi+\text{etc.},$

we have $\qquad A=\int\frac{ds}{dt}\,ds.$

Hence

$$\delta A=\int\left(\delta\frac{ds}{dt}\,ds+\frac{ds\delta ds}{dt}\right)=\int dt\frac{ds}{dt}\,\delta\,\frac{ds}{dt}+\int\frac{\frac{1}{2}\delta\,(ds^2)}{dt}$$

$$=\int dt\delta T+\int\frac{(\psi,\psi)\,d\psi+(\psi,\phi)\,d\phi+\text{etc.}}{dt}\,\delta d\psi$$

$$+\int\frac{(\psi,\phi)\,d\psi+(\phi,\phi)\,d\phi+\text{etc.}}{dt}\,\delta d\phi+\text{etc.}+\int dt\delta_{(\psi,\phi,\text{etc.})}\,T,$$

where $\delta_{(\psi,\phi,\text{etc.})}$ denotes variation dependent on the explicit appearance of ψ, ϕ, etc. in the coefficients of the quadratic function T. The second chief term in the formula for δA is clearly equal to $\int\frac{dT}{d\dot{\psi}}\,d\delta\psi$, and this, integrated by parts, becomes

$$\frac{dT}{d\dot{\psi}}\,\delta\psi-\int d\,\frac{dT}{d\dot{\psi}}\,\delta\psi,\quad\text{or}\quad\left[\frac{dT}{d\dot{\psi}}\,\delta\psi\right]-\int dt\,\frac{d}{dt}\frac{dT}{d\dot{\psi}}\,\delta\psi,$$

where [ ] denotes the difference of the values of the bracketed expression, at the beginning and end of the time $\int dt$. Thus we have finally

$$\delta A=\left[\frac{dT}{d\dot{\psi}}\,\delta\psi+\frac{dT}{d\dot{\phi}}\,\delta\phi+\text{etc.}\right]$$

$$+\int dt\left\{-\left(\frac{d}{dt}\frac{dT}{d\dot{\psi}}\,\delta\psi+\frac{d}{dt}\frac{dT}{d\dot{\phi}}\,\delta\phi+\text{etc.}\right)+\delta T+\delta_{(\psi,\phi,\text{etc.})}\,T\right\}\text{....}\,(10)'.$$

So far we have a purely kinematical formula. Now introduce the dynamical condition [§ 293 (7)]

Principle of Least Action applied to find Lagrange's generalized equations of motion.

$$T = C - V \dots\dots\dots\dots\dots\dots\dots (10)''.$$

From it we find

$$\delta T = -\left(\frac{dV}{d\psi}\delta\psi + \frac{dV}{d\phi}\delta\phi + \text{etc.}\right) \dots\dots\dots\dots (10)'''.$$

Again, we have

$$\delta_{(\psi, \phi, \text{etc.})} T = \frac{dT}{d\psi}\delta\psi + \frac{dT}{d\phi}\delta\phi + \text{etc.} \dots\dots\dots\dots (10)^{\text{iv}}.$$

Hence $(10)'$ becomes

$$\delta A = \left[\frac{dT}{d\dot\psi}\delta\psi + \frac{dT}{d\dot\phi}\delta\phi + \text{etc.}\right]$$

$$+ \int dt \left\{\left(-\frac{d}{dt}\frac{dT}{d\dot\psi} + \frac{dT}{d\psi} + \frac{dV}{d\psi}\right)\delta\psi + (\text{etc.})\,\delta\phi + \text{etc.}\right\} \dots (10)^{\text{v}}.$$

To make this a minimum we have

$$-\frac{d}{dt}\frac{dT}{d\dot\psi} + \frac{dT}{d\psi} + \frac{dV}{d\psi} = 0, \text{ etc.} \dots\dots\dots\dots (10)^{\text{vi}},$$

which are the required equations [§ 318 (24)].

From the proposition that $\delta A = 0$ implies the equations of motion, it follows that

Why called "stationary action" by Hamilton.

328. In any unguided motion whatever, of a conservative system, the Action from any one stated position to any other, though not necessarily a minimum, fulfils the *stationary condition*, that is to say, the condition that the variation vanishes, which secures either a minimum or maximum, or maximum-minimum.

Stationary action.

This can scarcely be made intelligible without mathematical language. Let (x_1, y_1, z_1), (x_2, y_2, z_2), etc., be the co-ordinates of particles, m_1, m_2, etc., composing the system; at any time τ of the actual motion. Let V be the potential energy of the system, in this configuration; and let E denote the given value of the sum of the potential and kinetic energies. The equation of energy is—

$$\tfrac{1}{2}\{m_1(\dot x_1^2 + \dot y_1^2 + \dot z_1^2) + m_2(\dot x_1^2 + \dot y_1^2 + \dot z_1^2) + \text{etc.}\} + V = E \dots (5) \text{ bis.}$$

Choosing any part of the motion, for instance that from time 0 to time t, we have, for the action during it,

$$A = \int_0^t (E - V)\,d\tau = Et - \int_0^t V d\tau \dots\dots\dots\dots\dots (11).$$

Stationary action.

Let now the system be guided to move in any other way possible for it, with any other velocities, from the same initial to the same final configuration as in the given motion, subject only to the condition, that the sum of the kinetic and potential energies shall still be E. Let (x_1', y_1', z_1'), etc., be the co-ordinates, and V' the corresponding potential energy; and let $(\dot{x}_1', \dot{y}_1', \dot{z}_1')$, etc., be the component velocities, at time τ in this arbitrary motion; equation (2) still holding, for the accented letters, with only E unchanged. For the action we shall have

$$A' = Et' - \int_0^{t'} V'd\tau \quad\quad\quad\quad (12),$$

where t' is the time occupied by this supposed motion. Let now θ denote a small numerical quantity, and let ξ_1, η_1, etc., be finite lines such that

$$\frac{x_1' - x_1}{\xi_1} = \frac{y_1' - y_1}{\eta_1} = \frac{z_1' - z_1}{\zeta_1} = \frac{x_2' - x_2}{\xi_2} = \text{etc.} = \theta.$$

The "principle of stationary action" is, that $\frac{V' - V}{\theta}$ vanishes when θ is made infinitely small, for every possible deviation $(\xi_1\theta, \eta_1\theta$, etc.) from the natural way and velocities, subject only to the equation of energy and to the condition of passing through the stated initial and final configurations: and conversely, that if $\frac{V' - V}{\theta}$ vanishes with θ for every possible such deviation from a certain way and velocities, specified by (x_1, y_1, z_1), etc., as the co-ordinates at t, *this* way and *these* velocities are such that the system unguided will move accordingly if only started with proper velocities from the initial configuration.

Varying action.

329. From this principle of stationary action, founded, as we have seen, on a comparison between a natural motion, and any other motion, arbitrarily guided and subject only to the law of energy, the initial and final configurations of the system being the same in each case, Hamilton passes to the consideration of the variation of the action in a natural or unguided motion of the system produced by varying the initial and final configurations, and the sum of the potential and kinetic energies. The result is, that

330. The rate of *decrease* of the action per unit of increase of any one of the free (generalized) co-ordinates (§ 204) specifying the initial configuration, is equal to the corresponding (generalized) component momentum [§ 313, (*c*)] of the actual motion from that configuration: the rate of *increase* of the action per unit increase of any one of the free co-ordinates specifying the final configuration, is equal to the corresponding component momentum of the actual motion towards this second configuration: and the rate of increase of the action per unit increase of the constant sum of the potential and kinetic energies, is equal to the time occupied by the motion of which the action is reckoned.

Varying action.

To prove this we must, in our previous expression (9) for δA, now suppose the terminal co-ordinates to vary; δT to become $\delta E - \delta V$, in which δE is a constant during the motion; and each set of paths and velocities to belong to an unguided motion of the system, which requires (10) to hold. Hence

Action expressed as a function of initial and final co-ordinates and the energy;

$$\delta A = \{\Sigma m(\dot{x}\delta x + \dot{y}\delta y + \dot{z}\delta z)\} - [\Sigma m(\dot{x}\delta x + \dot{y}\delta y + \dot{z}\delta z)] + t\delta E \ldots (13).$$

If, now, in the first place, we suppose the particles constituting the system to be all free from constraint, and therefore (x, y, z) for each to be three independent variables, and if, for distinctness, we denote by (x_1', y_1', z_1') and (x_1, y_1, z_1) the co-ordinates of m_1 in its initial and final positions, and by $(\dot{x}_1', \dot{y}_1', \dot{z}_1')$, $(\dot{x}_1, \dot{y}_1, \dot{z}_1)$ the components of the velocity it has at those points, we have, from the preceding, according to the ordinary notation of partial differential coefficients,

$$\left.\begin{array}{lll} \dfrac{dA}{dx_1'} = -m_1\dot{x}_1', & \dfrac{dA}{dy_1'} = -m_1\dot{y}_1', & \dfrac{dA}{dz_1'} = -m_1\dot{z}_1', \text{ etc.} \\ \dfrac{dA}{dx_1} = m_1\dot{x}_1, & \dfrac{dA}{dy_1} = m_1\dot{y}_1, & \dfrac{dA}{dz_1} = m_1\dot{z}_1, \text{ etc.} \\ \text{and} & \dfrac{dA}{dE} = t. & \end{array}\right\} \ldots (14).$$

its differential coefficients equal respectively to initial and final momentums, and to the time from beginning to end.

In these equations we must suppose A to be expressed as a function of the initial and final co-ordinates, in all six times as many independent variables as there are of particles; and E, one more variable, the sum of the potential and kinetic energies.

If the system consist not of free particles, but of particles connected in any way forming either one rigid body or any number

Varying action.

of rigid bodies connected with one another or not, we might, it is true, be contented to regard it still as a system of free particles, by taking into account among the impressed forces, the forces necessary to compel the satisfaction of the conditions of connexion. But although this method of dealing with a system of connected particles is very simple, so far as the law of energy merely is concerned, Lagrange's methods, whether that of "equations of condition," or, what for our present purposes is much more convenient, his "generalized co-ordinates," relieve us from very troublesome interpretations when we have to consider the displacements of particles due to arbitrary variations in the configuration of a system.

Let us suppose then, for any particular configuration (x_1, y_1, z_1) (x_2, y_2, z_2)..., the expression

$$m_1(\dot{x}_1\delta x_1 + \dot{y}_1\delta y_1 + \dot{z}_1\delta z_1) + \text{etc., to become } \xi\delta\psi + \eta\delta\phi + \zeta\delta\theta + \text{etc.} \quad (15),$$

Same propositions for generalized co-ordinates.

when transformed into terms of ψ, ϕ, θ..., generalized co-ordinates, as many in number as there are of degrees of freedom for the system to move [§ 313, (c)].

The same transformation applied to the kinetic energy of the system would obviously give

$$\tfrac{1}{2}m_1(\dot{x}_1^2 + \dot{y}_1^2 + \dot{z}_1^2) + \text{etc.} = \tfrac{1}{2}(\xi\dot{\psi} + \eta\dot{\phi} + \zeta\dot{\theta} + \text{etc.}) \ldots\ldots (16),$$

and hence ξ, η, ζ, etc., are those linear functions of the generalized velocities which, in § 313 (e), we have designated as "generalized components of momentum;" and which, when T, the kinetic energy, is expressed as a quadratic function of the velocities (of course with, in general, functions of the co-ordinates ψ, ϕ, θ, etc., for the coefficients) are derivable from it thus:

$$\xi = \frac{dT}{d\dot{\psi}}, \quad \eta = \frac{dT}{d\dot{\phi}}, \quad \zeta = \frac{dT}{d\dot{\theta}}, \text{ etc.} \ldots\ldots\ldots\ldots (17).$$

Hence, taking as before non-accented letters for the second, and accented letters for the initial, configurations of the system respectively, we have

$$\left.\begin{aligned} &\frac{dA}{d\psi'} = -\xi', \quad \frac{dA}{d\phi'} = -\eta', \quad \frac{dA}{d\theta'} = -\zeta', \text{ etc.} \\ &\frac{dA}{d\psi} = \xi, \quad \frac{dA}{d\phi} = \eta, \quad \frac{dA}{d\theta} = \zeta, \text{ etc.} \\ &\text{and, as before,} \quad \frac{dA}{dE} = t, \end{aligned}\right\} \ldots\ldots\ldots (18).$$

These equations (18), including of course (14) as a particular case, express in mathematical terms the proposition stated in words above, as the *Principle of Varying Action.*

Varying action.

The values of the momentums, thus, (14) and (18), expressed in terms of differential coefficients of A, must of course satisfy the equation of energy. Hence, for the case of free particles,

$$\Sigma \frac{1}{m}\left(\frac{dA^2}{dx^2}+\frac{dA^2}{dy^2}+\frac{dA^2}{dz^2}\right)=2(E-V) \quad \text{............(19)},$$

$$\Sigma \frac{1}{m}\left(\frac{dA^2}{dx'^2}+\frac{dA^2}{dy'^2}+\frac{dA^2}{dz'^2}\right)=2(E-V') \quad \text{............(20)}.$$

Hamilton's "characteristic equation" of motion in Cartesian co-ordinates.

Or, in general, for a system of particles or rigid bodies connected in any way, we have, (16) and (18),

$$\dot\psi\frac{dA}{d\psi}+\dot\phi\frac{dA}{d\phi}+\dot\theta\frac{dA}{d\theta}+\text{etc.}=2(E-V)\text{.........(21)},$$

$$-\left(\dot\psi'\frac{dA}{d\psi'}+\dot\phi'\frac{dA}{d\phi'}+\dot\theta'\frac{dA}{d\theta'}+\text{etc.}\right)=2(E-V')\text{.....(22)},$$

Hamilton's characteristic equation of motion in generalized co-ordinates.

where $\dot\psi$, $\dot\phi$, etc., are expressible as linear functions of $\frac{dA}{d\psi}$, $\frac{dA}{d\phi}$, etc., by the solution of the equations

$$\left.\begin{array}{l}(\psi,\psi)\dot\psi+(\psi,\phi)\dot\phi+(\psi,\theta)\dot\theta+\text{etc.}=\xi=\dfrac{dA}{d\psi}\\[2ex](\phi,\psi)\dot\psi+(\phi,\phi)\dot\phi+(\phi,\theta)\dot\theta+\text{etc.}=\eta=\dfrac{dA}{d\phi}\\[2ex]\qquad\qquad\text{etc.}\qquad\qquad\qquad\text{etc.}\end{array}\right\}\text{..... (23)},$$

and $\dot\psi'$, $\dot\phi'$, etc., as similar functions of $-\frac{dA}{d\psi'}$, $-\frac{dA}{d\phi'}$, etc., by

$$\left.\begin{array}{l}(\psi',\psi')\dot\psi'+(\psi',\phi')\dot\phi'+(\psi',\theta')\dot\theta'+\text{etc.}=\xi'=-\dfrac{dA}{d\psi'}\\[2ex](\phi',\psi')\dot\psi'+(\phi',\phi')\dot\phi'+(\phi',\theta')\dot\theta'+\text{etc.}=\eta'=-\dfrac{dA}{d\phi'}\\[2ex]\qquad\qquad\text{otc.}\qquad\qquad\qquad\text{etc.}\end{array}\right\}\text{...(24)},$$

where it must be remembered that (ψ, ψ), (ψ, ϕ), etc., are functions of the specifying elements, ψ, ϕ, θ, etc., depending on the kinematical nature of the co-ordinate system alone, and quite independent of the dynamical problem with which we are now concerned; being the coefficients of the half squares and the products of the generalized velocities in the expression for the

Varying action.

kinetic energy of any motion of the system; and that (ψ', ψ'), (ψ', ϕ'), etc., are the same functions with ψ', ϕ', etc., written for ψ, ϕ, θ, etc.; but, on the other hand, that A is a function of all the elements ψ, ϕ, etc., ψ', ϕ', etc. Thus the first member of (21) is a quadratic function of $\frac{dA}{d\psi}$, $\frac{dA}{d\phi}$, etc., with coefficients, known functions of ψ, ϕ, etc., depending merely on the kinematical relations of the system, and the masses of its parts, but not at all on the actual forces or motions; while the second member is a function of the co-ordinates ψ, ϕ, etc., depending on the forces in the dynamical problem, and a constant expressing the particular value given to the sum of the potential and kinetic energies in the actual motion; and so for (22), and ψ', ϕ', etc.

Proof that the characteristic equation defines the motion, for free particles.

It is remarkable that the single linear partial differential equation (19) of the first order and second degree, for the case of free particles, or its equivalent (21), is sufficient to determine a function A, such that the equations (14) or (18) express the momentums in an actual motion of the system, subject to the given forces. For, taking the case of free particles first, and differentiating (19) still on the Hamiltonian understanding that A is expressed merely as a function of initial and final co-ordinates, and of E, the sum of the potential and kinetic energies, we have

$$2\Sigma \frac{1}{m}\left(\frac{dA}{dx}\frac{d^2A}{dx_1 dx} + \frac{dA}{dy}\frac{d^2A}{dx_1 dy} + \frac{dA}{dz}\frac{d^2A}{dx_1 dz}\right) = -2\frac{dV}{dx_1}.$$

But, by (14),

$$\frac{1}{m_1}\frac{dA}{dx_1} = \dot{x}_1, \quad \frac{1}{m_1}\frac{dA}{dy_1} = \dot{y}_1, \text{ etc.},$$

and therefore

$$\frac{d^2A}{dx_1^2} = m_1\frac{d\dot{x}_1}{dx_1}, \quad \frac{d^2A}{dx_1 dy_1} = m_1\frac{d\dot{y}_1}{dx_1} = m_1\frac{d\dot{x}_1}{dy_1}, \quad \frac{d^2A}{dx_1 dz_1} = m_1\frac{d\dot{z}_1}{dx_1} = m_1\frac{d\dot{x}_1}{dz_1},$$

$$\frac{d^2A}{dx_1 dx_2} = m_2\frac{d\dot{x}_2}{dx_1} = m_1\frac{d\dot{x}_1}{dx_2}, \text{ etc.}$$

Using these properly in the preceding and taking half; and writing out for two particles to avoid confusion as to the meaning of Σ, we have

$$m_1\left(\dot{x}_1\frac{d\dot{x}_1}{dx_1} + \dot{y}_1\frac{d\dot{x}_1}{dy_1} + \dot{z}_1\frac{d\dot{x}_1}{dz_1} + \dot{x}_2\frac{d\dot{x}_1}{dx_2} + \dot{y}_2\frac{d\dot{x}_1}{dy_2} + \dot{z}_2\frac{d\dot{x}_1}{dz_2} + \text{etc.}\right) = -\frac{dV}{dx_1} \quad (25).$$

Now if we multiply the first member by dt, we have clearly the change of the value of $m_1\dot{x}_1$ due to varying, still on the Hamiltonian supposition, the co-ordinates of all the points, that is to say, the configuration of the system, from what it is at any moment to what it becomes at a time dt later; and it is therefore the actual change in the value of $m\dot{x}_1$, in the natural motion, from the time, t, when the configuration is $(x_1, y_1, z_1, x_2, \ldots, E)$, to the time $t+dt$. It is therefore equal to $m_1\ddot{x}_1 dt$, and hence (25) becomes simply $m_1\ddot{x}_1 = -\frac{dV}{dx_1}$. Similarly we find

Varying action. **Proof that the characteristic equation defines the motion, for free particles.**

$$m_1\ddot{y}_1 = -\frac{dV}{dy_1},\quad m_1\ddot{z}_1 = -\frac{dV}{dz_1},\quad m_2\ddot{x}_2 = -\frac{dV}{dx_2},\ \text{etc.}$$

But these are [§ 293, (4)] the elementary differential equations of the motions of a conservative system composed of free mutually influencing particles.

If next we regard x_1, y_1, z_1, x_2, etc., as constant, and go through precisely the same process with reference to x_1', y_1', z_1', x_2', etc., we have exactly the same equations among the accented letters, with only the difference that $-A$ appears in place of A; and end with $m_1\ddot{x}_1' = \frac{dV'}{dx_1'}$, from which we infer that, if (20) is satisfied, the motion represented by (14) is a natural motion through the configuration $(x_1', y_1', z_1', x_2', \text{etc.})$.

Hence if both (19) and (20) are satisfied, and if when $x_1 = x_1'$, $y_1 = y_1'$, $z_1 = z_1'$, $x_2 = x_2'$, etc., we have $\frac{dA}{dx_1} = -\frac{dA}{dx_1'}$, etc., the motion represented by (14) is a natural motion through the two configurations $(x_1', y_1', z_1', x_2', \text{etc.})$, and $(x_1, y_1, z_1, x_2, \text{etc.})$. Although the signs in the preceding expressions have been fixed on the supposition that the motion is *from* the former, to the latter configuration, it may clearly be from either towards the other, since whichever way it is, the reverse is also a natural motion (§ 271), according to the general property of a conservative system.

To prove the same thing for a conservative system of particles or rigid bodies connected in any way, we have, in the first place, from (18)

Same proposition for a connected system, and generalized co-ordinates.

$$\frac{d\eta}{d\psi} = \frac{d\xi}{d\phi},\quad \frac{d\zeta}{d\psi} = \frac{d\xi}{d\theta},\ \text{etc.} \quad\ldots\ldots\ldots\ldots\ldots\ldots(26),$$

where, on the Hamiltonian principle, we suppose ψ, ϕ, etc., and ξ, η, etc., to be expressed as functions of ψ, ϕ, etc., ψ', ϕ', etc.,

Varying action.

and the sum of the potential and kinetic energies. On the same supposition, differentiating (21), we have

$$\dot{\psi}\frac{d\xi}{d\psi}+\dot{\phi}\frac{d\eta}{d\psi}+\dot{\theta}\frac{d\zeta}{d\psi}+\text{etc.}+\xi\frac{d\dot{\psi}}{d\psi}+\eta\frac{d\dot{\phi}}{d\psi}+\zeta\frac{d\dot{\theta}}{d\psi}+\text{etc.}=-2\frac{dV}{d\psi}\text{ ....(27).}$$

But, by (26), and by the considerations above, we have

$$\dot{\psi}\frac{d\xi}{d\psi}+\dot{\phi}\frac{d\eta}{d\psi}+\dot{\theta}\frac{d\zeta}{d\psi}+\text{etc.}=\dot{\psi}\frac{d\xi}{d\psi}+\dot{\phi}\frac{d\xi}{d\phi}+\dot{\theta}\frac{d\xi}{d\theta}+\text{etc.}=\dot{\xi}\text{ ....(28),}$$

where $\dot{\xi}$ denotes the rate of variation of ξ per unit of time in the actual motion.

Again, we have

$$\left.\begin{aligned}\frac{d\dot{\psi}}{d\psi}&=\frac{\partial\dot{\psi}}{d\xi}\frac{d\xi}{d\psi}+\frac{\partial\dot{\psi}}{d\eta}\frac{d\eta}{d\psi}+\text{etc.}+\frac{\partial\dot{\psi}}{d\psi}\\ \frac{d\dot{\phi}}{d\psi}&=\frac{\partial\dot{\phi}}{d\xi}\frac{d\xi}{d\psi}+\frac{\partial\dot{\phi}}{d\eta}\frac{d\eta}{d\psi}+\text{etc.}+\frac{\partial\dot{\phi}}{d\psi}\\ &\quad\text{etc.}\qquad\qquad\text{etc.}\end{aligned}\right\}\text{ .... (29),}$$

if, as in Hamilton's system of canonical equations of motion, we suppose $\dot{\psi}$, $\dot{\phi}$, etc., to be expressed as linear functions of ξ, η, etc., with coefficients involving ψ, ϕ, θ, etc., and if we take ∂ to denote the partial differentiation of these functions with reference to the system ξ, η,...ψ, ϕ,..., regarded as independent variables. Let the coefficients be denoted by $[\psi, \psi]$, etc., according to the plan followed above; so that, if the formula for the kinetic energy be

$$T=\tfrac{1}{2}\{[\psi,\psi]\,\xi^2+[\phi,\phi]\,\eta^2+\ldots+2\,[\psi,\phi]\,\xi\eta+\text{etc.}\}\text{.....(30),}$$

we have

$$\left.\begin{aligned}\dot{\psi}&=\frac{dT}{d\xi}=[\psi,\psi]\,\xi+[\psi,\phi]\,\eta+[\psi,\theta]\,\zeta+\text{etc.}\\ \dot{\phi}&=\frac{dT}{d\eta}=[\phi,\psi]\,\xi+[\phi,\phi]\,\eta+[\phi,\theta]\,\zeta+\text{etc.}\\ &\qquad\quad\text{etc.}\qquad\qquad\text{etc.}\end{aligned}\right\}\text{......(31),}$$

where of course $[\psi, \phi]$, and $[\phi, \psi]$, mean the same.

Hence $$\frac{\partial\dot{\psi}}{d\xi}=[\psi,\psi],\ \frac{\partial\dot{\phi}}{d\xi}=[\phi,\psi],\ \ldots;$$

$$\frac{\partial\dot{\psi}}{d\psi}=\frac{d\,[\psi,\psi]}{d\psi}\xi+\frac{d\,[\psi,\phi]}{d\psi}\eta+\text{etc.};\ \frac{\partial\dot{\phi}}{d\psi}=\frac{d\,[\phi,\psi]}{d\psi}\xi+\text{etc. etc.,}$$

and therefore, by (29),

$$\xi\frac{d\dot{\psi}}{d\psi}+\eta\frac{d\dot{\phi}}{d\psi}+\zeta\frac{d\dot{\theta}}{d\psi}+\text{etc.}=\{[\psi,\psi]\xi+[\phi,\psi]\eta+\text{etc.}\}\frac{d\xi}{d\psi}+\{[\psi,\phi]\xi+[\phi,\phi]\eta+\text{etc.}\}\frac{d\eta}{d\psi}$$

$$+\text{etc.}+\frac{d[\psi,\psi]}{d\psi}\xi^2+\frac{d[\phi,\phi]}{d\psi}\eta^2+\text{etc.}+2\frac{d[\psi,\phi]}{d\psi}\xi\eta+\text{etc.}$$

$$=\dot{\psi}\frac{d\xi}{d\psi}+\dot{\phi}\frac{d\eta}{d\psi}+\text{etc.}+2\frac{\partial T}{d\psi};$$

whence, by (28), we see that

Hamiltonian form of Lagrange's generalized equations deduced from characteristic equation.

$$\xi\frac{d\dot\psi}{d\psi}+\eta\frac{d\dot\phi}{d\psi}+\zeta\frac{d\dot\theta}{d\psi}+\text{etc.}=\dot\xi+2\frac{\partial T}{d\psi}\text{............(32).}$$

This, and (28), reduce the first member of (27) to $2\dot\xi+2\dfrac{\partial T}{d\psi}$, and therefore, halving, we conclude

$$\dot\xi+\frac{\partial T}{d\psi}=-\frac{dV}{d\psi}\text{, and similarly, }\dot\eta+\frac{\partial T}{d\phi}=-\frac{dV}{d\phi}\text{, etc...(33).}$$

These, in all as many differential equations as there are of variables, ψ, ϕ, etc., suffice for determining them in terms of t and twice as many arbitrary constants. But every solution of the dynamical problem, as has been demonstrated above, satisfies (21) and (23); and therefore it must satisfy these (33), which we have derived from them. These (33) are therefore *the* equations of motion, of the system referred to generalized co-ordinates, as many in number as it has of degrees of freedom. They are the Hamiltonian explicit equations of motion, of which a direct demonstration was given in § 318 above. Just as above, it appears therefore, that if (21) and (22) are satisfied, (18) expresses a natural motion of the system from one to another of the two configurations $(\psi, \phi, \theta, ...)\,(\psi', \phi', \theta', ...)$. Hence

331. The determination of the motion of any conservative system from one to another of any two configurations, when the sum of its potential and kinetic energies is given, depends on the determination of a single function of the co-ordinates of those configurations by solution of two quadratic partial differential equations of the first order, with reference to those two sets of co-ordinates respectively, with the condition that the corresponding terms of the two differential equations become separately equal when the values of the two sets of co-ordinates agree. The function thus determined and employed to express the solution of the kinetic problem was called the *Characteristic Function* by Sir W. R. Hamilton, to whom the method is due. It is, as we have seen, the "action" from one of the configurations to the other; but its peculiarity in Hamilton's system is, that it is to be expressed as a function of the co-ordinates and a constant, the whole energy, as explained above. It is evi-

Hence proof concluded.

Characteristic function.

dently symmetrical with respect to the two configurations, changing only in sign if their co-ordinates are interchanged.

Characteristic equation of motion.

Since not only the complete solution of the problem of motion gives a solution, A, of the partial differential equation (19) or (21), but, as we have just seen [§ 330 (33), etc.], every solution of this equation corresponds to an actual problem relative to the motion, it becomes an object of mathematical analysis, which could not be satisfactorily avoided, to find what character of completeness a solution or integral of the differential equation must have in order that a complete integral of the dynamical equations may be derivable from it—a question which seems to have been first noticed by Jacobi. What is called a "complete integral" of the differential equation; that is to say, an expression,

Complete integral of characteristic equation.

$$A = A_0 + F(\psi, \phi, \theta, \ldots \alpha, \beta, \ldots) \quad \ldots\ldots\ldots\ldots\ldots\ldots (34),$$

for A satisfying it and involving the same number i, let us suppose, of independent arbitrary constants, A_0, α, β,...as there are of the independent variables, ψ, ϕ, etc.; leads, as he found, to a complete final integral of the equations of motion, expressed as follows:—

$$\frac{dF}{d\alpha} = \mathfrak{A}, \quad \frac{dF}{d\beta} = \mathfrak{B} \quad \ldots\ldots\ldots\ldots\ldots\ldots (35),$$

and, as above,

$$\frac{dF}{dE} = t + \epsilon \quad \ldots\ldots\ldots\ldots\ldots\ldots (36),$$

where ϵ is the constant depending on the epoch, or era of reckoning, chosen, and $\mathfrak{A}, \mathfrak{B}, \ldots$ are $i-1$ other arbitrary constants, constituting in all, with E, α, β,..., the proper number, $2i$, of arbitrary constants. This is proved by remarking that (35) are the equations of the "course" (or *paths* in the case of a system of free particles), which is obvious. For they give

$$\left.\begin{aligned} 0 &= \frac{d}{d\psi}\frac{dF}{d\alpha}d\psi + \frac{d}{d\phi}\frac{dF}{d\alpha}d\phi + \frac{d}{d\theta}\frac{dF}{d\alpha}d\theta + \ldots \\ 0 &= \frac{d}{d\psi}\frac{dF}{d\beta}d\psi + \frac{d}{d\phi}\frac{dF}{d\beta}d\phi + \frac{d}{d\theta}\frac{dF}{d\beta}d\theta + \ldots \\ &\quad \text{etc.} \qquad\qquad\qquad\qquad \text{etc.} \end{aligned}\right\} \ldots\ldots (37),$$

in all $i-1$ equations to determine the ratios $d\psi : d\phi : d\theta :\ldots$ From these, and (21), we find

$$\frac{d\psi}{\dot\psi} = \frac{d\phi}{\dot\phi} = \frac{d\theta}{\dot\theta} \quad \ldots\ldots\ldots\ldots\ldots\ldots (38)$$

[since (37) are the same as the equations which we obtain by differentiating (21) and (23) with reference to $\alpha, \beta, \ldots$ successively, only that they have $d\psi, d\phi, d\theta, \ldots$ in place of $\dot{\psi}, \dot{\phi}, \dot{\theta}, \ldots$].

Complete integral of characteristic equation.

A perfectly general solution of the partial differential equation, that is to say, an expression for A including every function of $\psi, \phi, \theta, \ldots$ which can satisfy (21), may of course be found, by the regular process, from the complete integral (34), by eliminating $A_0, \alpha, \beta, \ldots$ from it by means of an arbitrary equation

General solution derived from complete integral.

$$f(A_0, \alpha, \beta, \ldots) = 0,$$

and the $(i-1)$ equations

$$\frac{1}{\frac{df}{dA_0}} = \frac{\frac{dF}{d\alpha}}{\frac{df}{d\alpha}} = \frac{\frac{dF}{d\beta}}{\frac{df}{d\beta}} = \ldots$$

where f denotes an arbitrary function of the i elements $A_0, \alpha, \beta, \ldots$ now made to be variables depending on $\psi, \phi, \ldots$ But the full meaning of the general solution of (21) will be better understood in connexion with the physical problem if we first go back to the Hamiltonian solution, and then from it to the general. Thus, first, let the equations (35) of the course be assumed to be satisfied for each of two sets $\psi, \phi, \theta, \ldots$, and $\psi', \phi', \theta', \ldots$, of the co-ordinates. They will give $2(i-1)$ equations for determining the $2(i-1)$ constants $\alpha, \beta, \ldots, \mathfrak{A}, \mathfrak{B}, \ldots$, in terms of $\psi, \phi, \ldots, \psi', \phi', \ldots$, to fulfil these conditions. Using the values of $\alpha, \beta, \ldots$, so found, and assigning A_0 so that A shall vanish when $\psi = \psi'$, $\phi = \phi'$, etc., we have the Hamiltonian expression for A in terms of $\psi, \phi, \ldots, \psi', \phi', \ldots$, and E, which is therefore equivalent to a "complete integral" of the partial differential equation (21). Now let $\psi', \phi', \ldots$, be connected by any single arbitrary equation

$$\mathbf{f}(\psi', \phi', \ldots) = 0 \quad \text{.........................} \quad (39),$$

and by means of this equation and the following $(i-1)$ equations, let their values be determined in terms of $\psi, \phi, \ldots$, and E:—

$$\frac{\frac{dA}{d\psi'}}{\frac{d\mathbf{f}}{d\psi'}} = \frac{\frac{dA}{d\phi'}}{\frac{d\mathbf{f}}{d\phi'}} = \frac{\frac{dA}{d\theta'}}{\frac{d\mathbf{f}}{d\theta'}} = \text{etc.} \quad \text{.........................} \quad (40).$$

Substituting the values thus found for ψ', ϕ', θ', etc., in the Hamiltonian A, we have an expression for A, which is the general

General solution derived from complete integral.

solution of (21). For we see immediately that (40) expresses that the values of A are equal for all configurations satisfying (39), that is to say, we have

$$\frac{dA}{d\psi'}d\psi' + \frac{dA}{d\phi'}d\phi' + \ldots = 0$$

when ψ', ϕ', etc., satisfy (39) and (40). Hence when, by means of these equations, ψ', ϕ', ..., are eliminated from the Hamiltonian expression for A, the complete Hamiltonian differential

$$dA = \left(\frac{dA}{d\psi}\right)d\psi + \left(\frac{dA}{d\phi}\right)d\phi + \ldots + \frac{dA}{d\psi'}d\psi' + \frac{dA}{d\phi'}d\phi' + \ldots\ldots\ldots (41)$$

becomes merely

$$dA = \left(\frac{dA}{d\psi}\right)d\psi + \left(\frac{dA}{d\phi}\right)d\phi + \ldots\ldots\ldots\ldots\ldots\ldots(42),$$

where $\left(\frac{dA}{d\psi}\right)$, etc., denote the differential coefficients in the Hamiltonian expression. Hence, A being now a function of ψ, ϕ, etc., both as these appear in the Hamiltonian expression and as they are introduced by the elimination of ψ', ϕ', etc., we have

$$\frac{dA}{d\psi} = \left(\frac{dA}{d\psi}\right),\ \frac{dA}{d\phi} = \left(\frac{dA}{d\phi}\right),\ \text{etc}\ldots\ldots\ldots\ldots\ldots(43):$$

and therefore the new expression satisfies the partial differential equation (21). That it is a completely general solution we see, because it satisfies the condition that the action is equal for all configurations fulfilling an absolutely arbitrary equation (39).

For the case of a single free particle, the interpretation of (39) is that the point (x', y', z') is on an arbitrary surface, and of (40) that each line of motion cuts this surface at right angles. Hence

Practical interpretation of the complete solution of the characteristic equation.

332. The most general possible solution of the quadratic, partial, differential equation of the first order, which Hamilton showed to be satisfied by his Characteristic Function (either terminal configuration alone varying), when interpreted for the case of a single free particle, expresses the action up to any point (x, y, z), from some point of a certain arbitrarily given surface, from which the particle has been projected, in the direction of the normal, and with the proper velocity to make the sum of the potential and actual energies have a given value. In other words, the physical problem solved by the most general solution of that partial differential equation, is this :—

Properties of surfaces of equal action.

Let free particles, not mutually influencing one another, be projected normally from all points of a certain arbitrarily given surface, each with the proper velocity to make the sum of its potential and kinetic energies have a given value. To find, for the particle which passes through a given point (x, y, z), the "action" in its course from the surface of projection to this point. The Hamiltonian principles stated above, show that the surfaces of equal action cut the paths of the particles at right angles; and give also the following remarkable properties of the motion :—

If, from all points of an arbitrary surface, particles not mutually influencing one another be projected with the proper velocities in the directions of the normals; points which they reach with equal actions lie on a surface cutting the paths at right angles. The infinitely small thickness of the space between any two such surfaces corresponding to amounts of action differing by any infinitely small quantity, is inversely proportional to the velocity of the particle traversing it; being equal to the infinitely small difference of action divided by the whole momentum of the particle.

Let λ, μ, ν be the direction cosines of the normal to the surface of equal action through (x, y, z). We have

$$\lambda = \frac{\frac{dA}{dx}}{\left(\frac{dA^2}{dx^2}+\frac{dA^2}{dy^2}+\frac{dA^2}{dz^2}\right)^{\frac{1}{2}}}, \text{ etc. } \ldots\ldots\ldots\ldots\ldots(1).$$

But $\frac{dA}{dx} = m\dot{x}$, etc., and, if q denote the resultant velocity,

$$mq = \left(\frac{dA^2}{dx^2}+\frac{dA^2}{dy^2}+\frac{dA^2}{dz^2}\right)^{\frac{1}{2}} \quad \ldots\ldots\ldots\ldots\ldots\ldots(2).$$

Hence
$$\lambda = \frac{\dot{x}}{q},\ \mu = \frac{\dot{y}}{q},\ \nu = \frac{\dot{z}}{q},$$

which proves the first proposition. Again, if δA denote the in-

Properties of surfaces of equal action.

finitely small difference of action from (x, y, z) to any other point $(x+\delta x, y+\delta y, z+\delta z)$, we have

$$\delta A = \frac{dA}{dx}\delta x + \frac{dA}{dy}\delta y + \frac{dA}{dz}\delta z.$$

Let the second point be at an infinitely small distance, e, from the first, in the direction of the normal to the surface of equal action; that is to say, let

$$\delta x = e\lambda, \quad \delta y = e\mu, \quad \delta z = e\nu.$$

Hence, by (1), $$\delta A = e\left(\frac{dA^2}{dx^2} + \frac{dA^2}{dy^2} + \frac{dA^2}{dz^2}\right)^{\frac{1}{2}} \quad \text{.............}(3);$$

whence, by (2), $$e = \frac{\delta A}{mq} \quad \text{........................}(4),$$

which is the second proposition.

Examples of varying action.

333. Irrespectively of methods for finding the "characteristic function" in kinetic problems, the fact that any case of motion whatever can be represented by means of a single function in the manner explained in § 331, is most remarkable, and, when geometrically interpreted, leads to highly important and interesting properties of motion, which have valuable applications in various branches of Natural Philosophy. One of the many applications of the general principle made by Hamilton* led to a general theory of optical instruments, comprehending the whole in one expression.

Some of its most direct applications; to the motions of planets, comets, etc., considered as free points, and to the celebrated problem of perturbations, known as the Problem of Three Bodies, are worked out in considerable detail by Hamilton (*Phil. Trans.*, 1834-35), and in various memoirs by Jacobi, Liouville, Bour, Donkin, Cayley, Boole, etc. The now abandoned, but still interesting, corpuscular theory of light furnishes a good and exceedingly simple illustration. In this theory light is supposed to consist of material particles not mutually influencing one another, but subject to molecular forces from the particles of bodies—not sensible at sensible distances, and therefore not causing any deviation from uniform rectilinear motion in a homogeneous medium, except within an indefinitely small dis-

* *On the Theory of Systems of Rays.* Trans. R. I. A., 1824, 1830, 1832.

tance from its boundary. The laws of reflection and of single refraction follow correctly from this hypothesis, which therefore suffices for what is called geometrical optics.

Examples of varying action.

We hope to return to this subject, with sufficient detail, in treating of Optics. At present we limit ourselves to state a theorem comprehending the known rule for measuring the magnifying power of a telescope or microscope (by comparing the diameter of the object-glass with the diameter of pencil of parallel rays emerging from the eye-piece, when a point of light is placed at a great distance in front of the object-glass), as a particular case.

Application to common optics,

334. Let any number of attracting or repelling masses, or perfectly smooth elastic objects, be fixed in space. Let two stations, O and O', be chosen. Let a shot be fired with a stated velocity, V, from O, in such a direction as to pass through O'. There may clearly be more than one natural path by which this may be done; but, generally speaking, when one such path is chosen, no other, not considerably diverging from it, can be found; and any infinitely small deviation in the line of fire from O, will cause the bullet to pass infinitely near to, but not through, O'. Now let a circle, with infinitely small radius r, be described round O as centre, in a plane perpendicular to the line of fire from this point, and let—all with infinitely nearly the same velocity, but fulfilling the condition that the sum of the potential and kinetic energies is the same as that of the shot from O—bullets be fired from all points of this circle, all directed infinitely nearly parallel to the line of fire from O, but each precisely so as to pass through O'. Let a target be held at an infinitely small distance, a', beyond O', in a plane perpendicular to the line of the shot reaching it from O. The bullets fired from the circumference of the circle round O, will, after passing through O', strike this target in the circumference of an exceedingly small ellipse, each with a velocity (corresponding of course to its position, under the law of energy) differing infinitely little from V', the common velocity with which they pass through O'. Let now a circle, equal to the former, be described round O', in the plane perpendicular to the central path through O', and let bullets be fired from points in its circumference, each

or kinetics of a single particle.

Application to common optics, or kinetics of a single particle.

with the proper velocity, and in such a direction infinitely nearly parallel to the central path as to make it pass through O. These bullets, if a target is held to receive them perpendicularly at a distance $a = a'\frac{V}{V'}$, beyond O, will strike it along the circumference of an ellipse equal to the former and placed in a "corresponding" position; and the points struck by the individual bullets will correspond; according to the following law of "correspondence":—Let P and P' be points of the first and second circles, and Q and Q' the points on the first and second targets which bullets from them strike; then if P' be in a plane containing the central path through O' and the position which Q would take if its ellipse were made circular by a pure strain (§ 183); Q and Q' are similarly situated on the two ellipses.

For, let XOY be a plane perpendicular to the central path through O; and $X'O'Y'$ the corresponding plane through O'. Let A be the "action" from O to O', and ϕ the action from a point $P(x, y, z)$, in the neighbourhood of O, specified with reference to the former axes of co-ordinates, to a point $P'(x', y', z')$, in the neighbourhood of O', specified with reference to the latter.

The function $\phi - A$ vanishes, of course, when $x = 0$, $y = 0$, $z = 0$, $x' = 0$, $y' = 0$, $z' = 0$. Also, for the same values of the co-ordinates, its differential coefficients $\frac{d\phi}{dx}$, $\frac{d\phi}{dy}$, and $\frac{d\phi}{dx'}$, $\frac{d\phi}{dy'}$, must vanish, and $\frac{d\phi}{dz}$, $-\frac{d\phi}{dz'}$ must be respectively equal to V and V', since, for any values whatever of the co-ordinates, $\frac{d\phi}{dx}$ and $\frac{d\phi}{dy}$ are the component velocities parallel to the two lines OX, OY, of the particle passing through P, when it comes from P', and $-\frac{d\phi}{dx'}$ and $-\frac{d\phi}{dy'}$ are the components parallel to OX', OY', of the velocity through P' directed so as to reach P. Hence by Taylor's (or Maclaurin's) theorem we have

$$\begin{aligned}\phi - A = &- V'z' + Vz \\ &+ \tfrac{1}{2}\{(X, X)x^2 + (Y, Y)y^2 + \ldots + (X', X')x'^2 + \ldots \\ &+ 2(Y, Z)yz + \ldots + 2(Y', Z')y'z' + \ldots \\ &+ 2(X, X')xx' + 2(Y, Y')yy' + 2(Z, Z')zz' \\ &+ 2(X, Y')xy' + 2(X, Z')xz' + \ldots + 2(Z, Y')zy'\} + R \ldots(1),\end{aligned}$$

where (X, X), (X, Y), etc., denote constants, viz., the values of the differential coefficients $\frac{d^2\phi}{dx^2}$, $\frac{d^2\phi}{dxdy}$, etc., when each of the six co-ordinates x, y, z, x', y', z' vanishes; and R denotes the remainder after the terms of the second degree. According to Cauchy's principles regarding the convergence of Taylor's theorem, we have a rigorous expression for $\phi - A$ in the same form, without R, if the coefficients (X, X), etc., denote the values of the differential coefficients with some variable values intermediate between 0 and the actual values of x, y, etc., substituted for these elements. Hence, provided the values of the differential coefficients are infinitely nearly the same for any infinitely small values of the co-ordinates as for the vanishing values, R becomes infinitely smaller than the terms preceding it, when x, y, etc., are each infinitely small. Hence when each of the variables x, y, z, x', y', z' is infinitely small, we may omit R in the expression (1) for $\phi - A$. Now, as in the proposition to be proved, let us suppose z and z' each to be rigorously zero: and we have

Application to common optics, or kinetics of a single particle.

$$\frac{d\phi}{dx} = (X, X)\,x + (X, Y)\,y + (X, X')\,x' + (X, Y')\,y';$$

$$\frac{d\phi}{dy} = (Y, Y)\,y + (X, Y)\,x + (Y, X')\,x' + (Y, Y')\,y'.$$

These expressions, if in them we make $x = 0$, and $y = 0$, become the component velocities parallel to OX, OY, of a particle passing through O having been projected from P'. Hence, if ξ, η, ζ denote its co-ordinates, an infinitely small time, $\frac{a}{V}$, after it passes through O, we have $\zeta = a$, and

$$\xi = \{(X, X')\,x' + (X, Y')\,y'\}\frac{a}{V},\quad \eta = \{(Y, X')\,x' + (Y, Y')\,y'\}\frac{a}{V} \ldots (2).$$

Here ξ and η are the rectangular co-ordinates of the point Q' in which, in the second case, the supposed target is struck. And by hypothesis

$$x'^2 + y'^2 = r^2 \ldots\ldots\ldots\ldots\ldots\ldots\ldots\ldots (3).$$

If we eliminate x', y' between these three equations, we have clearly an ellipse; and the former two express the relation of the "corresponding" points. Corresponding equations with x and y for x' and y'; with ξ', η' for ξ, η; and with $-(X, X')$, $-(Y, X')$, $-(X, Y')$, $-(Y, Y')$, in place of (X, X'), (X, Y'),

Application to common optics, or kinetics of a single particle.

(Y, X'), (Y, Y'), express the first case. Hence the proposition, as is most easily seen by choosing OX and $O'X'$ so that (X, Y') and (Y, X') may each be zero.

Application to common optics.

335. The most obvious optical application of this remarkable result is, that in the use of any optical apparatus whatever, if the eye and the object be interchanged without altering the position of the instrument, the magnifying power is unaltered. This is easily understood when, as in an ordinary telescope, microscope, or opera-glass (Galilean telescope), the instrument is symmetrical about an axis, and is curiously contradictory of the common idea that a telescope "diminishes" when looked through the wrong way, which no doubt is true if the telescope is simply reversed about the middle of its length, eye and object remaining fixed. But if the telescope be removed from the eye till its eye-piece is close to the object, the part of the object seen will be seen enlarged to the same extent as when viewed with the telescope held in the usual manner. This is easily verified by looking from a distance of a few yards, in through the object-glass of an opera-glass, at the eye of another person holding it to his eye in the usual way.

The more general application may be illustrated thus:—Let the points, O, O' (the centres of the two circles described in the preceding enunciation), be the optic centres of the eyes of two persons looking at one another through any set of lenses, prisms, or transparent media arranged in any way between them. If their pupils are of equal sizes in reality, they will be seen as similar ellipses of equal apparent dimensions by the two observers. Here the imagined particles of light, projected from the circumference of the pupil of either eye, are substituted for the projectiles from the circumference of either circle, and the retina of the other eye takes the place of the target receiving them, in the general kinetic statement.

Application to system of free mutually influencing particles.

336. If instead of one free particle we have a conservative system of any number of mutually influencing free particles, the same statement may be applied with reference to the initial position of one of the particles and the final position of another, or with reference to the initial positions or to the final positions of two of the particles. It serves to show how the influence of an infinitely small change in one of those positions, on the direction of the other particle passing through the other position, is related to the influence on the direction of the former particle passing through the former position produced by an infinitely small change in the latter position. A corresponding statement, in terms of generalized co-ordinates, may of course be adapted to a system of rigid bodies or particles connected in any way. All such statements are included in the following very general proposition:—

Application to system of free mutually influencing particles,

and to generalized system.

The rate of increase of any one component momentum, corresponding to any one of the co-ordinates, per unit of increase of any other co-ordinate, is equal to the rate of increase of the component momentum corresponding to the latter per unit increase or diminution of the former co-ordinate, according as the two co-ordinates chosen belong to one configuration of the system, or one of them belongs to the initial configuration and the other to the final.

Let ψ and χ be two out of the whole number of co-ordinates constituting the argument of the Hamiltonian characteristic function A; and ξ, η the corresponding momentums. We have [§ 330 (18)]

$$\frac{dA}{d\psi} = \pm\xi, \; \frac{dA}{d\chi} = \pm\eta,$$

the upper or lower sign being used according as it is a final or an initial co-ordinate that is concerned. Hence

$$\frac{d^2A}{d\psi d\chi} = \pm\frac{d\xi}{d\chi} = \pm\frac{d\eta}{d\psi},$$

and therefore

$$\frac{d\xi}{d\chi} = \frac{d\eta}{d\psi},$$

if both co-ordinates belong to one configuration, or

$$\frac{d\xi}{d\chi} = -\frac{d\eta}{d\psi},$$

if one belongs to the initial configuration, and the other to the final, which is the second proposition. The geometrical interpretation of this statement for the case of a free particle, and two co-ordinates both belonging to one position, its final position, for

Application to system of free mutually influencing particles, and to generalized system.

instance, gives merely the proposition of § 332 above, for the case of particles projected from one point, with equal velocities in all directions; or, in other words, the case of the arbitrary surface of that enunciation, being reduced to a point. To complete the set of variational equations derived from § 330 we have $\frac{dt}{d\chi} = \pm \frac{d\eta}{dE}$ which expresses another remarkable property of conservative motion.

Slightly disturbed equilibrium.

337. By the help of Lagrange's form of the equations of motion, § 318, we may now, as a preliminary to the consideration of stability of motion, investigate the motion of a system infinitely little disturbed from a position of equilibrium, and left free to move, the velocities of its parts being initially infinitely small. The resulting equations give the values of the independent co-ordinates at any future time, provided the displacements *continue* infinitely small; and the mathematical expressions for their values must of course show the nature of the equilibrium, giving at the same time an interesting example of the *coexistence of small motions*, § 89. The method consists simply in finding what the equations of motion, and their integrals, become for co-ordinates which differ infinitely little from values corresponding to a configuration of equilibrium—and for an infinitely small initial kinetic energy. The solution of these differential equations is always easy, as they are linear and have constant coefficients. If the solution indicates that these differences *remain infinitely small*, the position is one of stable equilibrium; if it shows that one or more of them may *increase indefinitely*, the result of an infinitely small displacement from or infinitely small velocity through the position of equilibrium may be a finite departure from it—and thus the equilibrium is unstable.

Since there is a position of equilibrium, the kinematic relations must be invariable. As before,

$$T = \tfrac{1}{2}\{(\psi, \psi)\dot{\psi}^2 + (\phi, \phi)\dot{\phi}^2 + 2(\psi, \phi)\dot{\psi}\dot{\phi} + \text{etc.}\ldots\}\ldots(1),$$

which cannot be negative for any values of the co-ordinates. Now, though the values of the coefficients in this expression are not generally constant, they are to be taken as constant in the approximate investigation, since their variations, depending on the infinitely small variations of ψ, ϕ, etc., can only give rise to terms of the third or higher orders of small quantities. Hence Lagrange's equations become simply

Slightly disturbed equilibrium.

$$\frac{d}{dt}\left(\frac{dT}{d\dot{\psi}}\right) = \Psi,\ \frac{d}{dt}\left(\frac{dT}{d\dot{\phi}}\right) = \Phi,\ \text{etc.} \quad\ldots\ldots\ldots\ldots(2),$$

and the first member of each of these equations is a linear function of $\ddot{\psi}$, $\ddot{\phi}$, etc., with constant coefficients.

Now, since we may take what origin we please for the generalized co-ordinates, it will be convenient to assume that ψ, ϕ, θ, etc., are measured from the position of equilibrium considered; and that their values are therefore always infinitely small.

Hence, infinitely small quantities of higher orders being neglected, and the forces being supposed to be independent of the velocities, we shall have linear expressions for Ψ, Φ, etc., in terms of ψ, ϕ, etc., which we may write as follows:—

$$\left.\begin{array}{l}\Psi = a\psi + b\phi + c\theta + \ldots \\ \Phi = a'\psi + b'\phi + c'\theta + \ldots \\ \quad\text{etc.} \qquad \text{etc.}\end{array}\right\} \quad\ldots\ldots\ldots\ldots(3).$$

Equations (2) consequently become linear differential equations of the second order, with constant coefficients; as many in number as there are variables ψ, ϕ, etc., to be determined.

The regular processes explained in elementary treatises on differential equations, lead of course, independently of any particular relation between the coefficients, to a general form of solution (§ 343 below). But this form has very remarkable characteristics in the case of a conservative system; which we therefore examine particularly in the first place. In this case we have

$$\Psi = -\frac{dV}{d\psi},\quad \Phi = -\frac{dV}{d\phi},\ \text{etc.}$$

where V is, in our approximation, a homogeneous quadratic function of ψ, ϕ, ... if we take the origin, or configuration of equilibrium, as the configuration from which (§ 273) the potential energy is reckoned. Now, it is obvious*, from the theory

* For in the first place any such assumption as

Simultaneous transformation of two quadratic functions to sums of squares.

$$\begin{array}{c}\psi = A\psi_{,} + B\phi_{,} + \ldots \\ \phi = A'\psi_{,} + B'\phi_{,} + \ldots \\ \text{etc., etc.}\end{array}$$

gives equations for $\dot{\psi}$, $\dot{\phi}$, etc., in terms of $\dot{\psi}_{,}$, $\dot{\phi}_{,}$, etc., with the same coefficients, A, B, etc., if these are independent of t. Hence (the co-ordinates being i in

Slightly disturbed equilibrium.

of the transformation of quadratic functions, that we may, by a determinate linear transformation of the co-ordinates, reduce the

Simultaneous transformation of two quadratic functions to sums of squares.

number) we have i^2 quantities $A, A', A'', \ldots B, B', B'', \ldots$ etc., to be determined by i^2 equations expressing that in $2T$ the coefficients of $\dot\psi_,^2$, $\dot\phi_,^2$, etc. are each equal to unity, and of $\dot\psi_,\dot\phi_,$ etc. each vanish, and that in V the coefficients of $\psi_,\phi_,$, etc. each vanish. But, particularly in respect to our dynamical problem, the following process in two steps is instructive:—

(1) Let the quadratic expression for T in terms of $\dot\psi^2$, $\dot\phi^2$, $\dot\psi\dot\phi$, etc., be reduced to the form $\dot\psi_,^2+\dot\phi_,^2+\ldots$ by proper assignment of values to A, B, etc. This may be done arbitrarily, in an infinite number of ways, without the solution of any algebraic equation of degree higher than the first; as we may easily see by working out a synthetical process algebraically according to the analogy of finding first the conjugate diametral plane to any chosen diameter of an ellipsoid, and then the diameter of its elliptic section, conjugate to any chosen diameter of this ellipse. Thus, of the $\frac{i(i-1)}{2}$ equations expressing that the coefficients of the products $\dot\psi_,\dot\phi_,$, $\dot\psi_,\dot\theta_,$, $\dot\phi_,\dot\theta_,$ etc. vanish in T, take first the one expressing that the coefficient of $\dot\psi_,\dot\phi_,$ vanishes, and by it find the value of one of the B's, supposing all the A's and all the B's but one to be known. Then take the two equations expressing that the coefficients of $\dot\psi_,\dot\theta_,$ and $\dot\phi_,\dot\theta_,$ vanish, and by them find two of the C's supposing all the C's but two to be known, as are now all the A's and all the B's: and so on. Thus, in terms of all the A's, all the B's but one, all the C's but two, all the D's but three, and so on, supposed known, we find by the solution of linear equations the remaining B's, C's, D's, etc. Lastly, using the values thus found for the unassumed quantities, B, C, D, etc., and equating to unity the coefficients of $\dot\psi_,^2$, $\dot\phi_,^2$, $\dot\theta_,^2$, etc. in the transformed expression for $2T$, we have i equations among the squares and products of the $\frac{(i+1)i}{2}$ assumed quantities, (i) A's, $(i-1)$ B's, $(i-2)$ C's, etc., by which any one of the A's, any one of the B's, any one of the C's, and so on, are given immediately in terms of the $\frac{i(i-1)}{2}$ ratios of the others to them. Thus the thing is done, and $\frac{i(i-1)}{2}$ disposable ratios are left undetermined.

(2) These quantities may be determined by the $\frac{i(i-1)}{2}$ equations expressing that also in the transformed quadratic V the coefficients of $\psi_,\phi_,$, $\psi_,\theta_,$, $\phi_,\theta_,$, etc. vanish.

Or, having made the first transformation as in (1) above, with assumed values for $\frac{i(i-1)}{2}$ disposable ratios, make a second transformation determinately thus:

Generalized orthogonal transformation of co-ordinates.

—Let

$$\psi_,=l\psi_{,,}+m\phi_{,,}+\ldots$$
$$\phi_,=l'\psi_{,,}+m'\phi_{,,}+\ldots$$
etc., etc.,

where the i^2 quantities $l, m, \ldots, l', m', \ldots$ satisfy the $\frac{1}{2}i(i+1)$ equations

$$ll'+mm'+\ldots=0,\ l'l''+m'm''+\ldots=0, \text{ etc.},$$

expression for $2T$, which is essentially positive, to a sum of squares of generalized component velocities, and at the same time V to a sum of the squares of the corresponding co-ordinates, each multiplied by a constant, which may be either positive or negative, but is essentially real. [In the case of an equality or of any number of equalities among the values of these constants (α, β, etc. in the notation below), roots as they are of a determinantal equation, the linear transformation ceases to be wholly determinate; but the degree or degrees of indeterminacy which supervene is the reverse of embarrassing in respect to either the process of obtaining the solution, or the interpretation and use of it when obtained.] Hence ψ, ϕ, ... may be so chosen that

Simplified expressions for the kinetic and potential energies.

$$T = \tfrac{1}{2}(\dot{\psi}^2 + \dot{\phi}^2 + \text{etc.}) \quad \ldots\ldots\ldots\ldots\ldots\ldots(4),$$

and

$$V = \tfrac{1}{2}(\alpha\psi^2 + \beta\phi^2 + \text{etc.}) \ldots\ldots\ldots\ldots\ldots\ldots(5),$$

α, β, etc., being real positive or negative constants. Hence Lagrange's equations become

$$\ddot{\psi} = -\alpha\psi, \quad \ddot{\phi} = -\beta\phi, \quad \text{etc.} \ldots\ldots\ldots\ldots\ldots\ldots(6).$$

The solutions of these equations are

$$\psi = A\cos(t\sqrt{\alpha} - e), \quad \phi = A'\cos(t\sqrt{\beta} - e'), \quad \text{etc.} \ldots\ldots(7),$$

Integrated equations of motion, expressing the fundamental modes of vibration.

A, e, A', e', etc., being the arbitrary constants of integration. Hence we conclude the motion consists of a simple harmonic variation of each co-ordinate, provided that α, β, etc., are all positive. This condition is satisfied when V is a true minimum at the configuration of equilibrium; which, as we have seen (§ 292), is necessarily the case when the equilibrium is stable. If any one or more of α, β, ... vanishes, the equilibrium might

and $$l^2 + m^2 + \ldots = 1, \quad l'^2 + m'^2 + \ldots = 1, \text{ etc.},$$

Simultaneous transformation of two quadratic functions to sums of squares.

leaving $\frac{1}{2}i(i-1)$ disposables.

We shall still have, obviously, the same form for $2T$, that is:—

$$2T = \dot{\psi}_{\prime\prime}^2 + \dot{\phi}_{\prime\prime}^2 + \ldots$$

And, according to the known theory of the transformation of quadratic functions, we may determine the $\frac{1}{2}i(i-1)$ disposables of l, m, ..., l', m', ... so as to make the $\frac{1}{2}i(i-1)$ products of the co-ordinates $\psi_{\prime\prime}$, $\phi_{\prime\prime}$, etc. disappear from the expression for V, and give

$$2V = \alpha\psi_{\prime\prime}^2 + \beta\phi_{\prime\prime}^2 + \ldots,$$

where α, β, γ, etc., are the roots, necessarily real, of an equation of the ith degree of which the coefficients depend on the coefficients of the squares and products in the expression for V in terms of $\psi_\prime$, $\phi_\prime$, etc. Later [(7′), (8) and (9) of § 343 *f*], a *single process* for carrying out this investigation will be worked out.

Integrated equations of motion, expressing the fundamental modes of vibration; or of falling away from configuration of unstable equilibrium.

be either stable or unstable, or neutral; but terms of higher orders in the expansion of V in ascending powers and products of the co-ordinates would have to be examined to test it; and if it were stable, the period of an infinitely small oscillation in the value of the corresponding co-ordinate or co-ordinates would be infinitely great. If any or all of $\alpha, \beta, \gamma, \ldots$ are negative, V is not a minimum, and the equilibrium is (§ 292) essentially unstable. The form (7) for the solution, for each co-ordinate for which this is the case, becomes imaginary, and is to be changed into the exponential form, thus; for instance, let $-\alpha = p$, a positive quantity. Thus

Infinitely small disturbance from unstable equilibrium.

$$\psi = C\epsilon^{+t\sqrt{p}} + K\epsilon^{-t\sqrt{p}} \quad \ldots\ldots\ldots\ldots(8),$$

which (unless the disturbance is so adjusted as to make the arbitrary constant C vanish) indicates an unlimited increase in the deviation. This form of solution expresses the approximate law of falling away from a configuration of unstable equilibrium. In general, of course, the approximation becomes less and less accurate as the deviation increases.

Potential and Kinetic energies expressed as functions of time.

We have, by (5), (4), (7) and (8),

$$\left.\begin{aligned} V &= \tfrac{1}{4}\alpha A^2 \left[1 + \cos 2\left(t\sqrt{\alpha} - e\right)\right] + \text{etc.} \\ \text{or} \quad V &= -\tfrac{1}{2}p\left[2CK + C^2\epsilon^{2t\sqrt{p}} + K^2\epsilon^{-2t\sqrt{p}}\right] - \text{etc.} \end{aligned}\right\} \ldots\ldots(9),$$

$$\left.\begin{aligned} \text{and} \quad T &= \tfrac{1}{4}\alpha A^2 \left[1 - \cos 2\left(t\sqrt{\alpha} - e\right)\right] + \text{etc.} \\ \text{or} \quad T &= \tfrac{1}{2}p\left[-2CK + C^2\epsilon^{2t\sqrt{p}} + K^2\epsilon^{-2t\sqrt{p}}\right] + \text{etc.} \end{aligned}\right\} \ldots\ldots(10);$$

and, verifying the constancy of the sum of potential and kinetic energies,

$$\left.\begin{aligned} T + V &= \tfrac{1}{2}\left(\alpha A^2 + \beta A'^2 + \text{etc.}\right) \\ \text{or} \quad T + V &= -2\left(pCK + qC'K' + \text{etc.}\right) \end{aligned}\right\} \ldots\ldots\ldots\ldots(11).$$

Example of fundamental modes.

One example for the present will suffice. Let a solid, immersed in an infinite *liquid* (§ 320), be prevented from any motion of rotation, and left only freedom to move parallel to a certain fixed plane, and let it be influenced by forces subject to the conservative law, which vanish in a particular position of equilibrium. Taking any point of reference in the body, choosing its position when the body is in equilibrium, as origin of rectangular co-ordinates OX, OY, and reckoning the potential energy from it, we shall have, as in general,

$$2T = A\dot{x}^2 + B\dot{y}^2 + 2C\dot{x}\dot{y}; \quad 2V = ax^2 + by^2 + 2cxy,$$

the principles stated in § 320 above, allowing us to regard the co-ordinates x and y as fully specifying the system, provided always, that if the body is given at rest, or is brought to rest, the whole liquid is at rest (§ 320) at the same time. By solving the obviously determinate problem of finding that pair of conjugate diameters which are in the same directions for the ellipse

Example of fundamental modes.

$$Ax^2 + By^2 + 2Cxy = \text{const.},$$

and the ellipse or hyperbola,

$$ax^2 + by^2 + 2cxy = \text{const.},$$

and choosing these as oblique axes of co-ordinates (x_1, y_1), we shall have

$$2T = A_1\dot{x}_1^2 + B_1\dot{y}_1^2, \text{ and } 2V = a_1x_1^2 + b_1y_1^2.$$

And, as A_1, B_1 are essentially positive, we may, to shorten our expressions, take $x_1\sqrt{A_1} = \psi$, $y_1\sqrt{B_1} = \phi$; so that we shall have

$$2T = \dot{\psi}^2 + \dot{\phi}^2, \quad 2V = \alpha\psi^2 + \beta\phi^2,$$

the normal expressions, according to the general forms shown above in (4) and (5).

The interpretation of the general solution is as follows :—

338. If a conservative system is infinitely little displaced from a configuration of stable equilibrium, it will ever after vibrate about this configuration, remaining infinitely near it; each particle of the system performing a motion which is composed of simple harmonic vibrations. If there are i degrees of freedom to move, and we consider any system (§ 202) of generalized co-ordinates specifying its position at any time, the deviation of any one of these co-ordinates from its value for the configuration of equilibrium will vary according to a complex harmonic function (§ 68), composed of i simple harmonics generally of incommensurable periods, and therefore (§ 67) the whole motion of the system will not in general recur periodically through the same series of configurations. There are, however, i distinct displacements, generally quite determinate, which we shall call *the normal displacements*, fulfilling the condition, that if any one of them be produced alone, and the system then left to itself for an instant at rest, this displacement will diminish and increase periodically according to a simple harmonic func-

General theorem of fundamental modes of infinitely small motion about a configuration of equilibrium.

Normal displacements from equilibrium.

Fundamental modes of vibration.

tion of the time, and consequently every particle of the system will execute a simple harmonic movement in the same period. This result, we shall see later (Vol. II.), includes cases in which there are an infinite number of degrees of freedom; as for instance a stretched cord; a mass of air in a closed vessel; waves in water, or oscillations of water in a vessel of limited extent, or of an elastic solid; and in these applications it gives the theory of the so-called "fundamental vibration," and successive "harmonics" of a cord or organ-pipe, and of all the different possible simple modes of vibration in the other cases. In all these cases it is convenient to give the name "fundamental mode" to any one of the possible simple harmonic vibrations, and not to restrict it to the gravest simple harmonic mode, as has been hitherto usual in respect to vibrating cords and organ-pipes.

Theorem of kinetic energy; of potential energy.

The whole kinetic energy of any complex motion of the system is [§ 337 (4)] equal to the sum of the kinetic energies of the fundamental constituents; and [§ 337 (5)] the potential energy of any displacement is equal to the sum of the potential energies of its normal components.

Infinitesimal motions in neighbourhood of configuration of unstable equilibrium.

Corresponding theorems of normal constituents and fundamental modes of motion, and the summation of their kinetic and potential energies in complex motions and displacements, hold for motion in the neighbourhood of a configuration of *unstable* equilibrium. In this case, some or all of the constituent motions are fallings away from the position of equilibrium (according as the potential energies of the constituent normal vibrations are negative).

Case of equality among periods.

339. If, as may be in particular cases, the periods of the vibrations for two or more of the normal displacements are equal, any displacement compounded of them will also fulfil the condition of being a normal displacement. And if the system be displaced according to any one such normal displacement, and projected with velocity corresponding to another, it will execute a movement, the resultant of two simple harmonic movements in equal periods. The graphic representation of the variation of the corresponding co-ordinates of the system, laid down as two rectangular co-ordinates in a plane diagram, will consequently (§ 65) be a circle or an ellipse; which will therefore,

Graphic representation.

of course, be the form of the orbit of any particle of the system which has a distinct direction of motion, for two of the displacements in question. But it must be remembered that some of the principal parts [as for instance the body supported on the fixed axis, in the illustration of § 319, *Example* (C)] may have only one degree of freedom; or even that each part of the system may have only one degree of freedom, as for instance if the system is composed of a set of particles each constrained to remain on a given line, or of rigid bodies on fixed axes, mutually influencing one another by elastic cords or otherwise. In such a case as the last, no particle of the system can move otherwise than in one line; and the ellipse, circle, or other graphical representation of the composition of the harmonic motions of the system, is merely an aid to comprehension, and is not the orbit of a motion actually taking place in any part of the system.

Graphic representation.

340. In nature, as has been said above (§ 278), every system uninfluenced by matter external to it is conservative, when the ultimate molecular motions constituting heat, light, and magnetism, and the potential energy of chemical affinities, are taken into account along with the palpable motions and measurable forces. But (§ 275) practically we are obliged to admit forces of friction, and resistances of the other classes there enumerated, as causing losses of energy, to be reckoned, in abstract dynamics, without regard to the equivalents of heat or other molecular actions which they generate. Hence when such resistances are to be taken into account, forces opposed to the motions of various parts of a system must be introduced into the equations. According to the approximate knowledge which we have from experiment, these forces are independent of the velocities when due to the friction of solids; but are simply proportional to the velocities when due to fluid viscosity directly, or to electric or magnetic influences; with corrections depending on varying temperature, and on the varying configuration of the system. In consequence of the last-mentioned cause, the resistance of a *real liquid* (which is always more or less viscous) against a body moving rapidly enough through it, to leave a great deal of irregular motion, in the shape of

Dissipative systems.

Views of Stokes on resistance to a solid moving through a liquid.

"eddies," in its wake, seems, when the motion of the solid has been kept long enough uniform, to be nearly in proportion to the square of the velocity; although, as Stokes has shown, at the lowest speeds the resistance is probably in simple proportion to the velocity, and for all speeds, after long enough time of one speed, may, it is probable, be approximately expressed as the sum of two terms, one simply as the velocity, and the other as the square of the velocity. If a solid is started from rest in an incompressible fluid, the initial law of resistance is no doubt simple proportionality to velocity, (however great, if suddenly enough given;) until by the gradual growth of eddies the resistance is increased gradually till it comes to fulfil Stokes' law.

Stokes' probable law.

Friction of solids.

341. The effect of friction of solids rubbing against one another is simply to render impossible the *infinitely* small vibrations with which we are now particularly concerned; and to allow any system in which it is present, to rest balanced when displaced, within certain finite limits, from a configuration of frictionless equilibrium. In mechanics it is easy to estimate its effects with sufficient accuracy when any practical case of finite oscillations is in question. But the other classes of dissipative agencies give rise to resistances simply as the velocities, without the corrections referred to, when the motions are infinitely small; and can never balance the system in a configuration deviating to any extent, however small, from a configuration of equilibrium. In the theory of infinitely small vibrations, they are to be taken into account by adding to the expressions for the generalized components of force, proper (§ 343 *a*, below) linear functions of the generalized velocities, which gives us equations still remarkably amenable to rigorous mathematical treatment.

Resistances varying as velocities.

The result of the integration for the case of a single degree of freedom is very simple; and it is of extreme importance, both for the explanation of many natural phenomena, and for use in a large variety of experimental investigations in Natural Philosophy. Partial conclusions from it are as follows:—

If the resistance per unit velocity is less than a certain critical value, in any particular case, the motion is a simple harmonic oscillation, with amplitude decreasing in the same ratio in equal successive intervals of time. But if the resistance equals or exceeds the critical value, the system when displaced from its position of equilibrium, and left to itself, returns gradually towards its position of equilibrium, never oscillating through it to the other side, and only reaching it after an infinite time.

Resistances varying as velocities.

In the unresisted motion, let n^2 be the rate of acceleration, when the displacement is unity; so that (§ 57) we have $T=\frac{2\pi}{n}$: and let the rate of retardation due to the resistance corresponding to unit velocity be k. Then the motion is of the oscillatory or non-oscillatory class according as $k^2<(2n)^2$ or $k^2>(2n)^2$. In the first case, the period of the oscillation is increased by the resistance from T to $T\frac{n}{(n^2-\frac{1}{4}k^2)^{\frac{1}{2}}}$, and the rate at which the Napierian logarithm of the amplitude diminishes per unit of time is $\frac{1}{2}k$. If a negative value be given to k, the case represented will be one in which the motion is assisted, instead of resisted, by force proportional to the velocity: but this case is purely ideal.

Effect of resistance varying as velocity in a simple motion.

The differential equation of motion for the case of one degree of motion is

$$\ddot{\psi}+k\dot{\psi}+n^2\psi=0\,;$$

of which the complete integral is

$$\psi=\{A\sin n't+B\cos n't\}\epsilon^{-\frac{1}{2}kt},\ \text{where } n'=\surd(n^2-\tfrac{1}{4}k^2),$$

or, which is the same,

$$\psi=(C\epsilon^{-n_{,}t}+C'\epsilon^{n_{,}t})\epsilon^{-\frac{1}{2}kt},\ \text{where } n_{,}=\surd(\tfrac{1}{4}k^2-n^2),$$

A and B in one case, or C and C' in the other, being the arbitrary constants of integration. Hence the propositions above. In the case of $k^2=(2n)^2$ the general solution is $\psi=(C+C't)\,\epsilon^{-\frac{1}{2}kt}$.

Case of equal roots.

342. The general solution [§ 343 *a* (2) and § 345ⁱ] of the problem, to find the motion of a system having any number, i, of degrees of freedom, when infinitely little disturbed from a position of stable equilibrium, and left to move subject to resistances proportional to velocities, shows that the whole motion may be resolved, in general determinately, into $2i$ different motions each

Infinitely small motion of a dissipative system.

Infinitely small motion of a dissipative system.

either simple harmonic with amplitude diminishing according to the law stated above, or non-oscillatory and consisting of equi-proportionate diminutions of the components of displacement in equal successive intervals of time.

Cycloidal system defined.

Easy and instructive lecture illustration.

343. It is now convenient to cease limiting our ideas to infinitely small motions of an absolutely general system through configurations infinitely little different from a configuration of equilibrium, and to consider any motions large or small of a system so constituted that the positional* forces are proportional to displacements and the motional* to velocities, and that the kinetic energy is a quadratic function of the velocities with constant coefficients. Such a system we shall call a cycloidal† system; and we shall call its motions cycloidal motions. A good and instructive illustration is presented in the motion of one two or more weights in a vertical line, hung one from another, and the highest from a fixed point, by spiral springs.

343 *a*. If now instead of $\psi, \phi, \ldots$ we denote by $\psi_1, \psi_2, \ldots$ the generalized co-ordinates, and if we take 11, 12, 21, 22..., ɪɪ, ɪ2, 2ɪ, 22,... to signify constant coefficients (not numbers as in the ordinary notation of arithmetic), the most general equations of motions of a cycloidal system may be written thus:

Positional and Motional Forces.

* Much trouble and verbiage is to be avoided by the introduction of these adjectives, which will henceforth be in frequent use. They tell their own meanings as clearly as any definition could.

† A single adjective is needed to avoid a sea of troubles here. The adjective 'cycloidal' is already classical in respect to any motion with one degree of freedom, curvilineal or rectilineal, lineal or angular (Coulomb-torsional, for example), following the same law as the cycloidal pendulum, that is to say:—*the displacement a simple harmonic function of the time*. The motion of a particle on a cycloid with vertex up may as properly be called cycloidal; and in it the displacement is an imaginary simple harmonic, or a real exponential, or the sum of two real exponentials of the time

$$\left(C\epsilon^{t\sqrt{\frac{g}{l}}} + C'\epsilon^{-t\sqrt{\frac{g}{l}}}\right).$$

In cycloidal motion as defined in the text, each component of displacement is proved to be a sum of exponentials $(C\epsilon^{\lambda t} + C'\epsilon^{\lambda' t} + \text{etc.})$ real or imaginary, reducible to a sum of products of real exponentials and real simple harmonics $[C\epsilon^{mt}\cos(nt-e) + C'\epsilon^{m't}\cos(n't-e') + \text{etc.}]$.

Differential equations of complex cycloidal motion.

$$\left.\begin{array}{l}\frac{d}{dt}\left(\frac{dT}{d\dot{\psi}_1}\right)+11\dot{\psi}_1+12\dot{\psi}_2+\ldots+\mathrm{11}\psi_1+\mathrm{12}\psi_2+\ldots=0\\ \frac{d}{dt}\left(\frac{dT}{d\dot{\psi}_2}\right)+21\dot{\psi}_1+22\dot{\psi}_2+\ldots+\mathrm{21}\psi_1+\mathrm{22}\psi_2+\ldots=0\\ \quad\text{etc.}\qquad\qquad\text{etc.}\end{array}\right\}\ldots\ldots(1).$$

Positional forces of the non-conservative class are included by *not* assuming $\mathrm{12}=\mathrm{21}$, $\mathrm{13}=\mathrm{31}$, $\mathrm{23}=\mathrm{32}$, etc.

The theory of simultaneous linear differential equations with constant coefficients shows that the general solution for each co-ordinate is the sum of particular solutions, and that every particular solution is of the form

$$\psi_1=a_1\epsilon^{\lambda t},\ \ \psi_2=a_2\epsilon^{\lambda t}\ldots\ldots\ldots\ldots\ldots\ldots(2).$$

Their solution.

Assuming, then, this to be a solution, and substituting in the differential equations, we have

$$\left.\begin{array}{l}\lambda^2\frac{d\mathfrak{T}}{da_1}+\lambda(11a_1+12a_2+\ldots)+\mathrm{11}a_1+\mathrm{12}a_2+\ldots=0\\ \lambda^2\frac{d\mathfrak{T}}{da_2}+\lambda(21a_1+22a_2+\ldots)+\mathrm{21}a_1+\mathrm{22}a_2+\ldots=0\\ \quad\text{etc.}\qquad\qquad\text{etc.}\end{array}\right\}\ldots\ldots(3),$$

where $\mathfrak{T}$ denotes the same homogeneous quadratic function of $a_1, a_2\ldots$, that T is of $\dot{\psi}_1, \dot{\psi}_2,\ldots$. These equations, i in number, determine λ by the determinantal equation

$$\left|\begin{array}{l}(\mathrm{11})\lambda^2+11\lambda+\mathrm{11},\ (\mathrm{12})\lambda^2+12\lambda+\mathrm{12},\ldots\\ (\mathrm{21})\lambda^2+21\lambda+\mathrm{21},\ (\mathrm{22})\lambda^2+22\lambda+\mathrm{22},\ldots\\ \ldots\ldots\ldots\ldots\ldots\ldots\ldots\ldots\\ \ldots\ldots\ldots\ldots\ldots\ldots\ldots\ldots\end{array}\right|=0\ldots..(4),$$

where $(\mathrm{11})$, $(\mathrm{22})$, $(\mathrm{12})$, $(\mathrm{21})$, etc. denote the coefficients of squares and doubled products in the quadratic, $2T$; with identities

$$(\mathrm{12})=(\mathrm{21}),\ \ (\mathrm{13})=(\mathrm{31}),\ \text{etc}\ldots\ldots\ldots\ldots\ldots(5).$$

The equation (4) is of the degree $2i$, in λ; and if any one of its roots be used for λ in the i linear equations (3), these become harmonized and give the $i-1$ ratios a_2/a_1, a_3/a_1, etc.; and we have then, in (2), a particular solution with one arbitrary constant, a_1. Thus, from the $2i$ roots, when unequal, we have $2i$ distinct particular solutions, each with an arbitrary constant; and the addition of these solutions, as explained above, gives the general solution.

Solution of differential equations of complex cycloidal motion.

343 *b*. To show explicitly the determination of the ratios a_2/a_1, a_3/a_1, etc. put for brevity

$$(11)\lambda^2 + 11\lambda + 11 = 1{\cdot}1, \quad (12)\lambda^2 + 12\lambda + 12 = 1{\cdot}2, \text{ etc.,}$$

$$(32)\lambda^2 + 32\lambda + 32 = 3{\cdot}2, \text{ etc. } \ldots\ldots(5)';$$

and generally let $j{\cdot}k$ denote the coefficient of a_k in the jth equation of (3), or the kth term of the jth line of the determinant (to be called D for brevity) constituting the first member of (4).

Algebra of linear equations.

Let $M(j{\cdot}k)$ denote the factor of $j{\cdot}k$ in D so that $j{\cdot}k\,.\,M(j{\cdot}k)$ is the sum of all the terms of D which contain $j{\cdot}k$, and we have

$$D = \frac{1}{i}\sum_{j=1}^{j=i}\sum_{k=1}^{k=i} j{\cdot}k\,.\,M(j{\cdot}k) \ldots\ldots\ldots\ldots\ldots\ldots(5)'',$$

because in the sum $\Sigma\Sigma$ each term of D clearly occurs i times: and taking different groupings of terms, but each one only once, we have

$$\left.\begin{aligned}
D &= 1{\cdot}1\,M(1{\cdot}1) + 1{\cdot}2\,M(1{\cdot}2) + 1{\cdot}3\,M(1{\cdot}3) + \text{etc.}\\
&= 2{\cdot}1\,M(2{\cdot}1) + 2{\cdot}2\,M(2{\cdot}2) + 2{\cdot}3\,M(2{\cdot}3) + \text{etc.}\\
&= 3{\cdot}1\,M(3{\cdot}1) + 3{\cdot}2\,M(3{\cdot}2) + 3{\cdot}3\,M(3{\cdot}3) + \text{etc.}\\
&\ldots\ldots\ldots\ldots\ldots\ldots\ldots\ldots\ldots\ldots\ldots\ldots\\
&= 1{\cdot}1\,M(1{\cdot}1) + 2{\cdot}1\,M(2{\cdot}1) + 3{\cdot}1\,M(3{\cdot}1) + \text{etc.}\\
&= 1{\cdot}2\,M(1{\cdot}2) + 2{\cdot}2\,M(2{\cdot}2) + 3{\cdot}2\,M(3{\cdot}2) + \text{etc.}\\
&= 1{\cdot}3\,M(1{\cdot}3) + 2{\cdot}3\,M(2{\cdot}3) + 3{\cdot}3\,M(3{\cdot}3) + \text{etc.}\\
&\ldots\ldots\ldots\ldots\ldots\ldots\ldots\ldots\ldots\ldots\ldots\ldots
\end{aligned}\right\}\ldots.(5)'''$$

in all $2i$ different expressions for D.

Minor determinants.

Farther, by the elementary law of formation of determinants we see that

$$M(j-1{\cdot}k-1) = (-1)^{(i-1)(j+k)}\begin{vmatrix}
j{\cdot}k, & j{\cdot}(k+1),\ j{\cdot}(k+2), \ldots j{\cdot}i,\ j{\cdot}1,\ j{\cdot}2, \ldots, \ j{\cdot}(k-2)\\
(j+1){\cdot}k, & \ldots\ldots\ldots\ldots\ldots\ldots\ldots\ldots\\
(j+2){\cdot}k, & \ldots\ldots\ldots\ldots\ldots\ldots\ldots\ldots\\
\ldots\ldots & \ldots\ldots\ldots\ldots\ldots\ldots\ldots\ldots\\
\ldots\ldots & \ldots\ldots\ldots\ldots\ldots\ldots\ldots\ldots\\
i{\cdot}k, & \ldots\ldots\ldots\ldots\ldots\ldots\ldots\ldots\\
1{\cdot}k, & \ldots\ldots\ldots\ldots\ldots\ldots\ldots\ldots\\
2{\cdot}k, & \ldots\ldots\ldots\ldots\ldots\ldots\ldots\ldots\\
\ldots\ldots & \ldots\ldots\ldots\ldots\ldots\ldots\ldots\ldots\\
\ldots\ldots & \ldots\ldots\ldots\ldots\ldots\ldots\ldots\ldots\\
\ldots\ldots & \ldots\ldots\ldots\ldots\ldots\ldots\ldots\ldots\\
(j-2){\cdot}k, & \ldots\ldots\ldots\ldots\ldots\ldots, \quad (j-2){\cdot}(k-2)
\end{vmatrix}$$

$$\ldots\ldots\ldots\ldots\ldots\ldots\ldots\ldots(5)^{\text{iv}}.$$

The quantities $M(1\cdot 1)$, $M(1\cdot 2)$, ... $M(j\cdot k)$, thus defined are what are commonly called the first minors of the determinant D, with just this variation from ordinary usage that the proper signs are given to them by the factor

Minors of a determinant.

$$(-1)^{(i-1)(j+k)}$$

in 5^{iv} so that in the formation of D the ordinary complication of alternate positive and negative signs when i is even and all signs positive when i is odd is avoided. In terms of the notation (5)′ the linear equations (3) become

$$\left.\begin{array}{l} 1\cdot 1a_1 + 1\cdot 2a_2 + \ldots\ldots + 1\cdot ia_i = 0 \\ 2\cdot 1a_1 + 2\cdot 2a_2 + \ldots\ldots + 2\cdot ia_i = 0 \\ \ldots\ldots\ldots\ldots\ldots\ldots\ldots\ldots\ldots\ldots \\ i\cdot 1a_1 + i\cdot 2a_2 + \ldots\ldots + i\cdot ia_i = 0 \end{array}\right\} \ldots\ldots\ldots\ldots(5)^{v},$$

and when $D=0$, which is required to harmonize them, they may be put under any of the following i different but equivalent forms,

$$\left.\begin{array}{ll} & \dfrac{a_1}{M(1\cdot 1)} = \dfrac{a_2}{M(1\cdot 2)} = \dfrac{a_3}{M(1\cdot 3)} = \text{etc.} \\ \text{or} & \dfrac{a_1}{M(2\cdot 1)} = \dfrac{a_2}{M(2\cdot 2)} = \dfrac{a_3}{M(2\cdot 3)} = \text{etc.} \\ \text{or} & \dfrac{a_1}{M(3\cdot 1)} = \dfrac{a_2}{M(3\cdot 2)} = \dfrac{a_3}{M(3\cdot 3)} = \text{etc.} \end{array}\right\} \ldots\ldots\ldots(5)^{vi},$$

from which we find

$$\left.\begin{array}{l} \dfrac{a_2}{a_1} = \dfrac{M(1\cdot 2)}{M(1\cdot 1)} = \dfrac{M(2\cdot 2)}{M(2\cdot 1)} = \dfrac{M(3\cdot 2)}{M(3\cdot 1)} = \text{etc.} \\ \dfrac{a_3}{a_1} = \dfrac{M(1\cdot 3)}{M(1\cdot 1)} = \dfrac{M(2\cdot 3)}{M(2\cdot 1)} = \dfrac{M(3\cdot 3)}{M(3\cdot 1)} = \text{etc.} \\ \ldots\ldots\ldots\ldots\ldots\ldots\ldots\ldots\ldots\ldots\ldots\ldots \end{array}\right\} \ldots\ldots\ldots(5)^{vii}.$$

The remarkable relations here shown among the minors, due to the evanescence of the major determinant D, are well known in algebra. They are all included in the following formula,

Relations among the minors of an evanescent determinant.

$$M(j\cdot k)\,.\,M(l\cdot n) - M(j\cdot n)\,.\,M(l\cdot k) = 0\ldots\ldots\ldots(5)^{viii},$$

which is given in Salmon's *Higher Algebra* (§ 33 Ex. 1), as a consequence of the formula

$$M(j\cdot k)\,.\,M(l\cdot n) - M(j\cdot n)\,.\,M(l\cdot k) = D\,.\,M(j,\, l\cdot k,\, n)\ldots(5)^{ix},$$

where $M(j,\, l\cdot k,\, n)$ denotes the second minor formed by suppressing the jth and l^{th} columns and the kth and n^{th} lines.

Relations among the minors of an evanescent determinant.

343 c. When there are equalities among the roots the problem has generally solutions of the form

$$\psi_1 = (c_1 t + b_1)\,\epsilon^{\lambda t}, \quad \psi_2 = (c_2 t + b_2)\,\epsilon^{\lambda t}, \quad \text{etc.} \quad \ldots\ldots\ldots\ldots(6).$$

To prove this let λ, λ' be two unequal roots which become equal with some slight change of the values of some or all of the given constants (11), 11, 11, (12), 12, 12, etc.; and let

$$\psi_1 = A_1'\epsilon^{\lambda' t} - A_1\epsilon^{\lambda t}, \quad \psi_2 = A_2'\epsilon^{\lambda' t} - A_2\epsilon^{\lambda t}, \quad \text{etc.} \ldots\ldots\ldots\ldots(6)'$$

be a particular solution of (1) corresponding to these roots.

Now let

$$\left.\begin{array}{lll} c_1 = A_1'(\lambda' - \lambda), & c_2 = A_2'(\lambda' - \lambda), & \text{etc.} \\ \text{and} \quad b_1 = A_1' - A_1, & b_2 = A_2' - A_2, & \text{etc.} \end{array}\right\} \ldots\ldots\ldots\ldots(6)''.$$

Using these in $(6)'$ we find

$$\psi_1 = c_1 \frac{\epsilon^{\lambda' t} - \epsilon^{\lambda t}}{\lambda' - \lambda} + b_1\epsilon^{\lambda t}, \quad \psi_2 = c_2 \frac{\epsilon^{\lambda' t} - \epsilon^{\lambda t}}{\lambda' - \lambda} + b_2\epsilon^{\lambda t}, \quad \text{etc.} \ldots(6)'''.$$

To find proper equations for the relations among $b_1, b_2, \ldots c_1, c_2, \ldots$ in order that $(6)'''$ may be a solution of (1), proceed thus:—first write down equations (3) for the λ' solution, with constants A_1', A_2', etc.: then subtract from these the corresponding equations for the λ solution: thus, and introducing the notation $(6)''$, we find

$$\left.\begin{array}{l} \{(11)\lambda'^2 + 11\lambda' + 11\}\,c_1 + \{(12)\lambda'^2 + 12\lambda' + 12\}\,c_2 + \text{etc.} = 0 \\ \{(21)\lambda'^2 + 21\lambda' + 21\}\,c_1 + \{(22)\lambda'^2 + 22\lambda' + 22\}\,c_2 + \text{etc.} = 0 \\ \qquad \text{etc.} \qquad\qquad \text{etc.} \end{array}\right\} \ldots(6)^{\text{iv}},$$

and

$$\left.\begin{array}{l} \{(11)\lambda'^2 + 11\lambda' + 11\}\,b_1 + \{(12)\lambda'^2 + 12\lambda' + 12\}\,b_2 + \text{etc.} \\ \quad = -[c_1 - b_1(\lambda' - \lambda)]\{(11)(\lambda + \lambda') + 11\} \\ \qquad\qquad - [c_2 - b_2(\lambda' - \lambda)]\{(12)(\lambda + \lambda') + 12\} - \text{etc.} \\ \{(21)\lambda'^2 + 21\lambda' + 21\}\,b_1 + \{(22)\lambda'^2 + 22\lambda' + 22\}\,b_2 + \text{etc.} \\ \quad = -[c_1 - b_1(\lambda' - \lambda)]\{(21)(\lambda + \lambda') + 21\} \\ \qquad\qquad - [c_2 - b_2(\lambda' - \lambda)]\{(22)(\lambda + \lambda') + 22\} + \text{etc.} \\ \qquad \text{etc.} \qquad\qquad \text{etc.} \end{array}\right\} \ldots(6)^{\text{v}}.$$

Case of equal roots.

Equations $(6)^{\text{iv}}$ require that λ' be a root of the determinant, and $i - 1$ of them determine $i - 1$ of the quantities c_1, c_2, etc. in terms of one of them assumed arbitrarily. Supposing now c_1, c_2, etc. to be thus all known, the i equations $(6)^{\text{v}}$ fail to determine the i quantities b_1, b_2, etc. in terms of the right-hand members because λ' is a root of the determinant. The two sets of equations $(6)^{\text{iv}}$ and $(6)^{\text{v}}$ require that λ be also a root of the determinant: and $i - 1$ of the equations $(6)^{\text{v}}$ determine $i - 1$ of the quantities b_1, b_2, etc. in terms of c_1, c_2, etc. (supposed already known as above) and a properly assumed value of one of the b's.

Case of equal roots.

343 *d*. When λ' is infinitely nearly equal to λ, $(6)'''$ becomes infinitely nearly the same as (6), and $(6)^{iv}$ and $(6)^{v}$ become in terms of the notation $(5)'$

$$\left.\begin{array}{l} 1{\cdot}1\,c + 1{\cdot}2\,c_2 + \text{etc.} = 0 \\ 2{\cdot}1\,c_1 + 2{\cdot}2\,c_2 + \text{etc.} = 0 \\ \quad\text{etc.} \qquad \text{etc.} \end{array}\right\}\ldots\ldots\ldots\ldots(6)^{vi},$$

$$\left.\begin{array}{l} 1{\cdot}1\,b_1 + 1{\cdot}2\,b_2 + \text{etc.} = -c_1\dfrac{d1{\cdot}1}{d\lambda} - c_2\dfrac{d1{\cdot}2}{d\lambda} - \text{etc.} \\ 2{\cdot}1\,b_1 + 2{\cdot}2\,b_2 + \text{etc.} = -c_1\dfrac{d2{\cdot}1}{d\lambda} - c_2\dfrac{d2{\cdot}2}{d\lambda} - \text{etc.} \\ \quad\text{etc.} \qquad\qquad\qquad \text{etc.} \end{array}\right\}\ldots(6)^{vii}.$$

These, $(6)^{vi}$, $(6)^{vii}$, are clearly the equations which we find simply by trying if (6) is a solution of (1). $(6)^{vi}$ requires that λ be a root of the determinant D; and they give by $(5)^{vi}$ with c substituted for a the values of $i-1$ of the quantities c_1, c_2, etc. in terms of one of them assumed arbitrarily. And by the way we have found them we know that $(6)^{vii}$ superadded to $(6)^{vi}$ shows that λ must be a dual root of the determinant. To verify this multiply the first of them by $M(1{\cdot}1)$, the second by $M(2{\cdot}1)$, etc., and add. The coefficients of b_2, b_3, etc. in the sum are each identically zero in virtue of the elementary constitution of determinants, and the coefficient of b_1 is the major determinant D. Thus irrespectively of the value of λ we find in the first place,

$$Db_1 = -c_1\left\{M(1{\cdot}1)\frac{d1{\cdot}1}{d\lambda} + M(2{\cdot}1)\frac{d2{\cdot}1}{d\lambda} + \text{etc.}\right\}$$
$$-c_2\left\{M(1{\cdot}1)\frac{d1{\cdot}2}{d\lambda} + M(2{\cdot}1)\frac{d2{\cdot}2}{d\lambda} + \text{etc.}\right\} - \text{etc.}\ldots(6)^{viii}.$$

Now in virtue of $(6)^{vi}$ and $(5)^{v}$ we have

$$\frac{c_2}{c_1} = \frac{a_2}{a_1},\quad \frac{c_3}{c_1} = \frac{a_3}{a_1},\ \text{etc.}$$

Using successively the several expressions given by $(5)^{vii}$ for these ratios, in $(6)^{viii}$, and putting $D = 0$, we find

$$0 = \Sigma_1^i\,\Sigma_1^i\,M(j{\cdot}k)\frac{dj{\cdot}k}{d\lambda},\quad \text{or } 0 = \frac{dD}{d\lambda},$$

which with $D = 0$ shows that λ is a double root.

Suppose now that one of the c's has been assumed, and the others found by $(6)^{vi}$: let one of the b's be assumed: the other

Case of equal roots.

$i-1$ b's are to be calculated by $i-1$ of the equations $(6)^{vii}$. Thus for example take $b_1=0$. In the first place use all except the first of equations $(6)^{vii}$ to determine b_2, b_3, etc.: we thus find

$$\left.\begin{aligned} M(1\cdot 1)b_2 = &-\left\{M(1,2\cdot 1,2)\frac{d2\cdot 1}{d\lambda}+M(1,2\cdot 1,3)\frac{d3\cdot 1}{d\lambda}+\text{etc.}\right\}c_1 \\ &-\left\{M(1,2\cdot 1,2)\frac{d2\cdot 2}{d\lambda}+M(1,2\cdot 1,3)\frac{d3\cdot 2}{d\lambda}+\text{etc.}\right\}c_2-\text{etc.} \\ M(1\cdot 1)b_3 = \text{etc.} \quad & M(1\cdot 1)b_4=\text{etc.} \qquad \text{etc.} \qquad \text{etc.} \end{aligned}\right\} \ldots\ldots(6)^{ix}.$$

Secondly, use all except the second of $(6)^{vii}$ to find b_2, b_3, etc.: we thus find

$$M(2\cdot 1)\,b_2=\text{etc.},\quad M(2\cdot 1)\,b_3=\text{etc.},\quad M(2\cdot 1)\,b_4=\text{etc.}\ldots\ldots(6)^{x}.$$

Thirdly, by using all of $(6)^{vii}$ except the third, fourthly, all except the fourth, and so on, we find

$$M(3\cdot 1)\,b_2=\text{etc.},\quad M(3\cdot 1)\,b_3=\text{etc.},\quad M(3\cdot 1)\,b_4=\text{etc.}\ldots\ldots(6)^{xi}.$$

343 *e*. In certain cases of equality among the roots (343 *m*) it is found that values of the coefficients (11), 11, 11, etc. differing infinitely little from particular values which give the equality give values of a_1 and a_1', a_2 and a_2', etc., which are not infinitely nearly equal. In such cases we see by $(6)''$ that b_1, b_2, etc. are finite, and c_1, c_2, etc. vanish: and so the solution does not contain terms of the form $t\epsilon^{\lambda t}$: but the requisite number of arbitrary constants is made up by a proper degree of indeterminateness in the residuary equations for the ratios b_2/b_1, b_3/b_1, etc.

Now when $c_1=0$, $c_2=0$, etc. the second members of equations $(6)^{ix}$, $(6)^{x}$, $(6)^{xi}$, etc. all vanish, and as b_2, b_3, b_4, etc. do not all vanish, it follows that we have

$$M(1\cdot 1)=0,\quad M(2\cdot 1)=0,\quad M(3\cdot 1)=0,\quad \text{etc.}\ldots\ldots(6)^{xii}.$$

Case of equal roots and evanescent minors:

Hence by $(5)^{vii}$ or $(5)^{viii}$ we infer that all the first minors are zero for any value of λ which is doubly a root, and which yet does not give terms of the form $t\epsilon^{\lambda t}$ in the solution. This important proposition is due to Routh*, who, escaping the errors of previous writers (§ 343 *m* below), first gave the complete theory of equal roots of the determinant in cycloidal motion.

* *Stability of Motion* (Adams Prize Essay for 1877), chap. I. § 5.

He also remarked that the factor t does not necessarily imply instability, as terms of the form $t\epsilon^{-pt}$, or $t\epsilon^{-pt}\cos(nt-e)$, when p is positive, do not give instability, but on the contrary correspond to non-oscillatory or oscillatory subsidence to equilibrium.

Routh's theorem.

343 *f*. We fall back on the case of no motional forces by taking $11=0$, $12=0$, etc., which reduces the equations (3) for determining the ratios a_2/a_1, a_3/a_1, etc. to

Case of no motional forces.

$$\lambda^2\frac{d\mathfrak{T}}{da_1}+11a_1+12a_2+\text{etc.}=0,\quad \lambda^2\frac{d\mathfrak{T}}{da_2}+21a_1+22a_2+\text{etc.}=0,\ \text{etc.}\ (7),$$

or, expanded,

$$\left.\begin{array}{l}[(11)\lambda^2+11]a_1+[(12)\lambda^2+12]a_2+\text{etc.}=0\\ [(21)\lambda^2+21]a_1+[(22)\lambda^2+22]a_2+\text{etc.}=0\end{array}\right\}\ldots(7').$$

The determinantal equation (4) to harmonize these simplified equations (7) or (7′) becomes

$$\begin{vmatrix}(11)\lambda^2+11, & (12)\lambda^2+12, & \ldots\\ (21)\lambda^2+21, & (22)\lambda^2+22, & \ldots\\ \ldots & \ldots & \ldots\end{vmatrix}=0\ldots\ldots\ldots\ldots(8).$$

This is of degree i, in λ^2: therefore λ has i pairs of oppositely signed equal values, which we may now denote by

$$\pm\lambda,\ \pm\lambda',\ \pm\lambda'',\ \ldots;$$

and for each of these pairs the series of ratio-equations (7′) are the same. Hence the complete solution of the differential equations of motion may be written as follows, to show its arbitraries explicitly :—

$$\left.\begin{array}{l}\psi_1=(A\epsilon^{\lambda t}+B\epsilon^{-\lambda t})+(A'\epsilon^{\lambda' t}+B'\epsilon^{-\lambda' t})+(A''\epsilon^{\lambda'' t}+B''\epsilon^{-\lambda'' t})+\text{etc.}\\ \psi_2=\frac{a_2}{a_1}(A\epsilon^{\lambda t}+B\epsilon^{-\lambda t})+\frac{a_2'}{a_1'}(A'\epsilon^{\lambda' t}+B'\epsilon^{-\lambda' t})+\frac{a_2''}{a_1''}(A''\epsilon^{\lambda'' t}+B''\epsilon^{-\lambda'' t})+\text{etc.}\\ \psi_3=\frac{a_3}{a_1}(A\epsilon^{\lambda t}+B\epsilon^{-\lambda t})+\frac{a_3'}{a_1'}(A'\epsilon^{\lambda' t}+B'\epsilon^{-\lambda' t})+\frac{a_3''}{a_1''}(A''\epsilon^{\lambda'' t}+B''\epsilon^{-\lambda'' t})+\text{etc.}\\ \qquad\text{etc.}\qquad\qquad\text{etc.}\qquad\qquad\text{etc.}\end{array}\right\}\ldots\ldots(9),$$

where A, B; A', B'; A'', B''; etc. denote $2i$ arbitrary constants, and

$$\frac{a_2}{a_1},\ \frac{a_3}{a_1},\ \ldots,\ \frac{a_2'}{a_1'},\ \frac{a_3'}{a_1'},\ \ldots,\ \frac{a_2''}{a_1''},\ \frac{a_3''}{a_1''},\ \ldots,\ \text{etc.},$$

are i sets of $i-1$ ratios each, the values of which, when all the i roots of the determinantal equation in λ^2 have different values, are fully determined by giving successively these i values to λ^2 in (7′).

Case of no motional forces.

343 *g*. When there are equal roots, the solution is to be completed according to § 343 *d* or *e*, as the case may be. The case of a conservative system (343 *h*) necessarily falls under § 343 *e*, as is proved in § 343 *m*. The same form, (9), still represents the complete solutions when there are equalities among the roots, but with changed conditions as to arbitrariness of the elements appearing in it. Suppose $\lambda^2 = \lambda'^2$ for example. In this case any value may be chosen arbitrarily for a_2/a_1, and the remainder of the set a_3/a_1, a_4/a_1 ... are then fully determined by (7′); again another value may be chosen for a_2'/a_1', and with it a_3'/a_1', a_4'/a_1', ... are determined by a fresh application of (7′) with the same value for λ^2: and the arbitraries now are $A + A'$, $B + B'$,

$$\frac{a_2}{a_1}A + \frac{a_2'}{a_1'}A', \quad \frac{a_2}{a_1}B + \frac{a_2'}{a_1'}B', \quad A'', \quad B'', \quad A''', \quad B''', \quad \ldots A^{(i-1)}, \text{ and } B^{(i-1)},$$

numbering still $2i$ in all. Similarly we see how, beginning with the form (9), convenient for the general case of i different roots, we have in it also the complete solution when λ^2 is triply, or quadruply, or any number of times a root, and when any other root or roots also are double or multiple.

Cycloidal motion. Conservative positional, and no motional, forces.

343 *h*. For the case of a conservative system, that is to say, the case in which

$$12 = 21, \quad 13 = 31, \quad 23 = 32, \text{ etc., etc.} \qquad \ldots\ldots\ldots\ldots(10),$$

the differential equations of motion, (1), become

$$\frac{d}{dt}\left(\frac{dT}{d\dot{\psi}}\right) + \frac{dV}{d\psi} = 0, \quad \frac{d}{dt}\left(\frac{dT}{d\dot{\phi}}\right) + \frac{dV}{d\phi} = 0, \text{ etc.} \qquad \ldots\ldots\ldots\ldots(10'),$$

and the solving linear algebraic equations, (3), become

$$\frac{d\mathfrak{T}}{da_1} + \frac{d\mathfrak{V}}{da_1} = 0, \quad \frac{d\mathfrak{T}}{da_2} + \frac{d\mathfrak{V}}{da_2} = 0 \qquad \ldots\ldots\ldots\ldots(10''),$$

where

$$V = \tfrac{1}{2}(11\psi_1^2 + 2.12\psi_1\psi_2 + \text{etc.}), \text{ and } \mathfrak{V} = \tfrac{1}{2}(11a_1^2 + 2.12a_1a_2 + \text{etc.}) \ldots (10''').$$

In this case the i roots, λ^2, of the determinantal equation are the negatives of the values of α, β, ... of our first investigation; and thus in (10″), (8), and (9) we have the promised solution by one completely expressed process. From § 337 and its footnote we infer that in the present case the roots λ^2 are all real, whether negative or positive.

Cycloidal motion. Conservative positional, and no motional, forces.

In § 337 it was expressly assumed that T (as it must be in the dynamical problem) is essentially positive; but the investigation was equally valid for any case in which either of the quadratics T or V is incapable of changing sign for real values of the variables ($\dot\psi_1$, $\dot\psi_2$, etc. for T, or ψ_1, ψ_2, etc. for V). Thus we see that the roots λ^2 are all real when the relations (5) and (9) are satisfied, and when the magnitudes of the residual independent coefficients (11), (22), (12), ... and 11, 22, 12, ... are such that of the resulting quadratics, $\mathfrak{T}$, $\mathfrak{V}$, one or other is essentially positive or essentially negative. This property of the determinantal equation (7′) is very remarkable. A more direct algebraic proof is to be desired. Here is one:—

343 *k*. Writing out (7′) for λ^2, and for λ'^2, multiplying the first for λ^2 by $\frac{1}{2}a_1'$, the second by $\frac{1}{2}a_2'$, and so on, and adding; and again multiplying the first for λ'^2 by $\frac{1}{2}a_1$, the second by $\frac{1}{2}a_2$, and so on, and adding, we find

$$\left.\begin{array}{l}\lambda^2\mathfrak{T}(a,\ a')+\mathfrak{V}(a,\ a')=0\\ \text{and}\qquad \lambda'^2\mathfrak{T}(a,\ a')+\mathfrak{V}(a,\ a')=0\end{array}\right\}\ldots\ldots\ldots\ldots\ldots(11),$$

where

$$\left.\begin{array}{l}\mathfrak{T}(a,\ a')=\tfrac{1}{2}\{(11)\,a_1a_1'+(12)(a_1a_2'+a_2a_1')+\text{etc.}\}\\ \text{and}\ \ \mathfrak{V}(a,\ a')=\tfrac{1}{2}\{\ 11\ \ a_1a_1'+12\ \ (a_1a_2'+a_2a_1')+\text{etc.}\}\end{array}\right\}\ldots(12).$$

Remark that according to this (12) notation $\mathfrak{T}$ $(a,\ a)$ means the same thing as $\mathfrak{T}$ simply, according to the notation of (3) etc. above, and $\mathfrak{T}$ $(\dot\psi,\ \dot\psi)$ the same thing as T. Remark farther that $\mathfrak{T}$ $(a,\ a')$ is a linear function of a_1, a_2, ... with coefficients each involving a_1', a_2', ... linearly; and that it is symmetrical with reference to a_1, a_1', and a_2, a_2', etc.; and that we therefore have

$$\left.\begin{array}{l}\mathfrak{T}(mp,\ p')=m\mathfrak{T}(p,\ p')=\mathfrak{T}(p,\ mp')\ \text{and}\\ \mathfrak{T}(mp+nq,\ m'p+n'q)=mm'\,\mathfrak{T}(p,p)+(mn'+m'n)\,\mathfrak{T}(p,\ q)+nn'\,\mathfrak{T}(q,q)\end{array}\right\}\ldots(13).$$

Precisely similar statements and formulas hold for $\mathfrak{V}$ $(a,\ a')$.

From (11) we infer that if λ^2 and λ'^2 be unequal we must have

$$\mathfrak{T}(a,\ a')=0,\ \text{and}\ \mathfrak{V}(a,\ a')=0\ldots\ldots\ldots\ldots\ldots(14).$$

Now if there can be imaginary roots, λ^2, let $\lambda^2=\rho+\sigma\sqrt{-1}$ and $\lambda'^2=\rho-\sigma\sqrt{-1}$ be a pair of them, ρ and σ being real. And, p_1, q_1, p_2, q_2, etc. being all real, let $p_1+q_1\sqrt{-1}$, $p_1-q_1\sqrt{-1}$, be arbitrarily chosen values of a_1, a_1', and let

$$p_2+q_2\sqrt{-1},\ p_3+q_3\sqrt{-1},\ \ldots,\ p_2-q_2\sqrt{-1},\ p_3-q_3\sqrt{-1},\ \ldots$$

Cycloidal motion. Conservative positional, and no motional, forces.

be the *determinately deduced* values * of $a_2, a_3, \ldots, a_2', a_3', \ldots$ according to (7′); we have, by (13), with

$$m = m' = 1,\ n = \sqrt{-1},\ n' = -\sqrt{-1},$$

$$\left.\begin{array}{rl} & \mathfrak{T}(a, a') = \mathfrak{T}(p, p) + \mathfrak{T}(q, q) \\ \text{and} & \mathfrak{V}(a, a') = \mathfrak{V}(p, p) + \mathfrak{V}(q, q) \end{array}\right\} \ldots\ldots\ldots\ldots\ldots\ldots (14').$$

Now by hypothesis either $\mathfrak{T}(x, x)$, or $\mathfrak{V}(x, x)$ is essentially of one sign for all real values of x_1, x_2, etc. Hence the second member of one or other of equations (14′) cannot be zero, because $p_1, p_2, \ldots$, and $q_1, q_2, \ldots$ are all real. But by (14) the first member of each of the equations (14′) is zero if λ^2 and λ'^2 are unequal: hence they are equal: hence either $p_1 = 0$, $p_2 = 0$, etc., or $q_1 = 0$, $q_2 = 0$, etc., that is to say the roots λ^2 are all necessarily real, whether negative or positive.

343 *l*. Farther we now see by going back to (11):—

(*a*) if for all real values of $x_1, x_2, \ldots$ the values of $\mathfrak{T}(x, x)$ and $\mathfrak{V}(x, x)$ have the same unchanging sign, the roots λ^2 are all negative;

(*b*) if for different real values of x_1, x_2, etc., one of the two $\mathfrak{T}(x, x)$, $\mathfrak{V}(x, x)$ has different signs (the other by hypothesis having always one sign), some of the roots λ^2 are negative and some positive;

(*c*) if the values of $\mathfrak{T}$ and $\mathfrak{V}$ have essentially opposite signs (and each therefore according to hypothesis unchangeable in sign), the roots λ^2 are all positive.

The (*a*) and (*c*) of this tripartite conclusion we see by taking $\lambda'^2 = \lambda^2$ in (11), which reduces them to

$$\lambda^2 \mathfrak{T}(a, a) + \mathfrak{V}(a, a) = 0 \ldots\ldots\ldots\ldots\ldots\ldots (15),$$

and remarking that a_2, a_3, etc. are now all real if we please to give a real value to a_1. The (*b*) is proved in § 343 *o* below.

Case of equal roots.

343 *m*. From (14) we see that when two roots λ^2, λ'^2, are infinitely nearly equal there is no approach to equality between a_1 and a_1', a_2 and a_2', and therefore, when there are no motional forces, and when the positional forces are conservative, equality of roots essentially falls under the case of § 343 *e* above. This may be proved explicitly as follows:—let

$$\psi_1 = (a_1 t + b_1)\,\epsilon^{\lambda t}, \quad \psi_2 = (a_2 t + b_2)\,\epsilon^{\lambda t}, \text{ etc.} \ldots\ldots\ldots (15)'$$

* Cases of equalities among the roots are disregarded for the moment merely to avoid circumlocutions, but they obviously form no exception to the reasoning and conclusion.

be the complete solution corresponding to the root λ supposed to be a dual root. Using this in equations (1) and equating to zero in each equation so found the coefficients of $t\epsilon^{\lambda t}$ and of $\epsilon^{\lambda t}$, with the notation of (12) we find

Cycloidal motion. Conservative positional, and no motional, forces.

$$\lambda^2 \frac{d\mathfrak{T}(a,a)}{da_1} + \frac{d\mathfrak{V}(a,a)}{da_1} = 0, \quad \lambda^2 \frac{d\mathfrak{T}(a,a)}{da_2} + \frac{d\mathfrak{V}(a,a)}{da_2} = 0, \text{ etc.} \ldots\ldots (15)'',$$

$$\left.\begin{array}{l} \lambda^2 \dfrac{d\mathfrak{T}(b,b)}{db_1} + 2\lambda \dfrac{d\mathfrak{T}(a,a)}{da_1} + \dfrac{d\mathfrak{V}(b,b)}{db_1} = 0, \\ \lambda^2 \dfrac{d\mathfrak{T}(b,b)}{db_2} + 2\lambda \dfrac{d\mathfrak{T}(a,a)}{da_2} + \dfrac{d\mathfrak{V}(b,b)}{db_2} = 0, \qquad \text{etc.} \end{array}\right\} \ldots (15)'''.$$

Multiplying the first, second, third, etc. of (15)″ by b_1, b_2, b_3, etc. and adding we find

$$\lambda^2\mathfrak{T}(a,b) + \mathfrak{V}(a,b) = 0 \ldots\ldots\ldots\ldots\ldots\ldots\ldots (15)^{\text{iv}};$$

and similarly from (15)‴ with multipliers a_1, a_2, etc.

$$\lambda^2\mathfrak{T}(a,b) + \mathfrak{V}(a,b) + 2\lambda\mathfrak{T}(a,a) = 0 \ldots\ldots\ldots\ldots (15)^{\text{v}}.$$

Subtracting $(15)^{\text{iv}}$ from $(15)^{\text{v}}$ we see that $\mathfrak{T}(a,a) = 0$. Hence we must have $a_1 = 0$, $a_2 = 0$, etc., that is to say there are no terms of the form $t\epsilon^{\lambda t}$ in the solution. It is to be remarked that the inference of $a_1 = 0$, $a_2 = 0$, etc. from $\mathfrak{T}(a,a) = 0$, is not limited to real roots λ because λ^2 in the present case is essentially real, and whether it be positive or negative the ratios a_2/a_1, a_3/a_1, etc., are essentially real.

It is remarkable that both Lagrange and Laplace fell into the error of supposing that equality among roots necessarily implies terms in the solution of the form $t\epsilon^{\lambda t}$ (or $t \cos \rho t$), and therefore that for stability the roots must be all unequal. This we find in the *Mécanique Analytique*, Seconde Partie, section VI. Art. 7 of the second edition of 1811 published three years before Lagrange's death, and repeated without change in the posthumous edition of 1853. It occurs in the course of a general solution of the problem of the infinitely small oscillations of a system of bodies about their positions of equilibrium, with conservative forces of position and no motional forces, which from the "Avertissement" (p. vi.) prefixed to the 1811 edition seems to have been first published in the 1811 edition, and not to have appeared in the original edition of 1788*. It would be

* Since this statement was put in type, the first edition of the *Mécanique Analytique* (which had been inquired for in vain in the University libraries of Cambridge and Glasgow) has been found in the University library of Edinburgh,

Cycloidal motion. Conservative positional, and no motional, forces.

curious if such an error had remained for twenty-three years in Lagrange's mind. It could scarcely have existed even during the writing and printing of the Article for his last edition if he had been in the habit of considering particular applications of his splendid analytical work: if he had he would have seen that a proposition which asserted that the equilibrium of a particle in the bottom of a frictionless bowl is unstable if the bowl be a figure of revolution with its axis vertical, cannot be true. No such obvious illustration presents itself to suggest or prove the error as Laplace has it in the *Mécanique Céleste* (Première Partie, Livre II. Art. 57) in the course of an investigation of the secular inequalities of the planetary system. But as [by a peculiarly simple case of the process of § 345^{xlv} (54)] he has reduced his analysis of this problem virtually to the same as that of conservative oscillations about a configuration of equilibrium, the physical illustrations which abound for this case suffice to prove the error in Laplace's statement, different and comparatively recondite as its dynamical subject is. An error the converse of that of Laplace and Lagrange occurred in page 278 of our First Edition where it was said that "Cases in which "there are equal roots leave a corresponding number of degrees "of indeterminateness in the ratios $l : m$, $l : n$, etc., and so allow "the requisite number of arbitrary constants to be made up," without limiting this statement to the case of conservative positional and no motional forces, for which its truth is obvious from the nature of the problem, and for which alone it is obvious at first sight; although for the cases of adynamic oscillations, and of stable precessions, § 345^{xxiv}, it is also essentially true. The correct theory of equal roots in the generalized problem of cycloidal motion has been so far as we know first given by Routh in his investigation referred to above (§ 343 *e*).

343 *n*. Returning to § 343 *l*, to make more of (*b*), and to understand the efficiency of the oppositely signed roots, λ^2, asserted in it, let $\sigma^2 = -\lambda^2$ in any case in which λ^2 is negative, and let

$$\psi_1 = r_1 \cos(\sigma t - e),\quad \psi_2 = r_2 \cos(\sigma t - e),\ \text{etc.}\ldots\ldots(16),$$

be the corresponding particular solution in fully *realized* terms,

and it does contain the problem of infinitely small oscillations, with the remarkable error referred to in the text.

as in § 337 (6) above but with somewhat different notation. By substituting in (1) and multiplying the first of the resulting equations by r_1, the second by r_2, and so on and adding, virtually as we found (15), we now find

Cycloidal motion. Conservative positional, and no motional, forces.

$$-\sigma^2 \mathfrak{T}(r, r) + \mathfrak{V}(r, r) = 0 \quad \text{.................(17).}$$

Adopting now the notation of (9) for the real positive ones of the roots λ^2, but taking, for brevity, $a_1 = 1$, $a_1' = 1$, $a_1'' = 1$, etc., we have for the complete solution when there are both negative and positive roots of the determinantal equation (7′);

$$\left.\begin{array}{l} \psi_1 = (A\epsilon^{\lambda t}+B\epsilon^{-\lambda t})+ (A'\epsilon^{\lambda' t}+B'\epsilon^{-\lambda' t})+\text{etc.}+r_1\cos(\sigma t-e)+r_1'\cos(\sigma' t-e')+\text{etc.} \\ \psi_2 = a_2(A\epsilon^{\lambda t}+B\epsilon^{-\lambda t})+a_2'(A'\epsilon^{\lambda' t}+B'\epsilon^{-\lambda' t})+\text{etc.}+r_2\cos(\sigma t-e)+r_2'\cos(\sigma' t-e')+\text{etc.} \\ \psi_3 = \text{etc.,} \qquad \psi_4 = \text{etc.} \qquad \text{etc.} \qquad \text{etc.} \end{array}\right\} \ldots(18).$$

343 *o*. Using this in the general expressions for T and V, with the notation (12), and remarking that the products $\epsilon^{\lambda t} \times \epsilon^{\lambda' t}$, etc. and $\epsilon^{\lambda t} \times \sin(\sigma t - e)$, etc., and $\sin(\sigma t - e) \times \sin(\sigma' t - e')$, etc., disappear from the terms in virtue of (11), we find

$$\left.\begin{array}{l} T = \lambda^2 \mathfrak{T}(a, a)(A\epsilon^{\lambda t} - B\epsilon^{-\lambda t})^2 + \lambda'^2 \mathfrak{T}(a', a')(A'\epsilon^{\lambda' t} - B'\epsilon^{-\lambda' t})^2 + \text{etc.} \\ \quad + \sigma^2 \mathfrak{T}(r, r)\sin^2(\sigma t - e) + \sigma'^2 \mathfrak{T}(r', r')\sin^2(\sigma' t - e') + \text{etc.} \end{array}\right\} (19),$$

and

$$\left.\begin{array}{l} V = \mathfrak{V}(a, a)(A\epsilon^{\lambda t} + B\epsilon^{-\lambda t})^2 + \mathfrak{V}(a', a')(A'\epsilon^{\lambda' t} + B'\epsilon^{-\lambda' t})^2 + \text{etc.} \\ \quad + \mathfrak{V}(r, r)\cos^2(\sigma t - e) + \mathfrak{V}(r', r')\cos^2(\sigma' t - e') + \text{etc.} \end{array}\right\} (20).$$

The factors which appear with

$$\mathfrak{T}(a, a),\ \mathfrak{T}(a', a'), \ldots \mathfrak{T}(r, r),\ \mathfrak{T}(r', r')$$

in this expression (19) for T are all essentially positive; and the same is true of $\mathfrak{V}$ in (20) for V. Now for every set of real co-ordinates and velocity-components the potential and kinetic energies are expressible by the formulas (20) and (19) because (18) is the complete solution with $2i$ arbitraries. Hence if the value of V can change sign with real values of the co-ordinates, the quantities $\mathfrak{V}(a, a)$, $\mathfrak{V}(a', a')$, etc., and $\mathfrak{V}(r, r)$, $\mathfrak{V}(r', r')$, etc., for the several roots must be some of them positive and some of them negative; and if the value of T could change sign with real values of the velocity-components, some of the quantities $\mathfrak{T}(a, a)$, $\mathfrak{T}(a', a')$, etc., and $\mathfrak{T}(r, r)$, $\mathfrak{T}(r', r')$, etc. would need to be positive and some negative. So much being learned from (20) and (19) we must now recal to mind that according to hypothesis one only of the two quadratics T and V can change sign, to conclude from (15) and (17) that there are both positive and negative roots λ^2 when either T or V can change sign. Thus (*b*) of the tripartite conclusion above is rigorously proved.

Cycloidal motion. Conservative positional, and no motional, forces.

343 *p*. A short algebraic proof of (*b*) could no doubt be easily given; but our somewhat elaborate discussion of the subject is important as showing in (15)...(20) the whole relation between the previous short algebraic investigation, conducted in terms involving quantities which are essentially imaginary for the case of oscillations about a configuration of stable equilibrium, and the fully realized solution, with formulas for the potential and kinetic energies realized both for oscillations and for fallings away from unstable equilibrium.

Equation of energy in realized general solution.

We now see definitively by (15) and (17) that, in *real* dynamics (that is to say T essentially positive) the factors $\mathfrak{V}(a, a)$, $\mathfrak{V}(a', a')$, etc., are all negative, and $\mathfrak{V}(r, r)$, $\mathfrak{V}(r', r')$, etc., all positive in the expression (20) for the potential energy. Adding (20) to (19) and using (15) and (17) in the sum, we find

$$\left.\begin{array}{l} T+V=-4AB\lambda^2\mathfrak{T}(a, a)-4A'B'\lambda'^2\mathfrak{T}(a', a'), \text{ etc.} \\ \quad +\sigma^2\mathfrak{T}(r, r)+\sigma'^2\mathfrak{T}(r', r')+\text{etc.} \end{array}\right\}\ldots(21).$$

It is interesting to see in this formula how the constancy of the sum of the potential and kinetic energies is attained in any solution of the form $A\epsilon^{\lambda t}+B\epsilon^{-\lambda t}$ [which, with $\lambda=\sigma\sqrt{-1}$, includes the form $r\cos(\sigma t-e)$], and to remark that for any single solution $a\epsilon^{\lambda t}$, or solution compounded of single solutions depending on unequal values of λ^2 (whether real or imaginary), the sum of the potential and kinetic energies is essentially zero.

Artificial or ideal accumulative system.

344. When the positional forces of a system violate the law of conservatism, we have seen (§ 272) that energy without limit may be drawn from it by guiding it perpetually through a returning cycle of configurations, and we have inferred that in every real system, not supplied with energy from without, the positional forces fulfil the conservative law. But it is easy to arrange a system artificially, in connexion with a source of energy, so that its positional forces shall be non-conservative; and the consideration of the kinetic effects of such an arrangement, especially of its oscillations about or motions round a configuration of equilibrium, is most instructive, by the contrasts which it presents to the phenomena of a natural system. The preceding formulas, (7)...(9) of § 343 *f* and § 343 *g*, express the general solution of the problem—to find the infinitely small motion of a cycloidal system, when, without motional forces, there is deviation from conservatism by the character of the positional forces.

In this case [(10) not fulfilled,] just as in the case of motional forces fulfilling the conservative law (10), the character of the equilibrium as to stability or instability is discriminated according to the character of the roots of an algebraic equation of degree equal to the number of degrees of freedom of the system.

Artificial or ideal accumulative system. Criterion of stability.

If the roots (λ^2) of the determinantal equation § 343 (8) are all real and negative, the equilibrium is stable: in every other case it is unstable.

345. But although, when the equilibrium is stable, no possible infinitely small displacement and velocity given to the system can cause it, when left to itself, to go on moving farther and farther away till either a finite displacement is reached, or a finite velocity acquired; it is very remarkable that stability should be possible, considering that even in the case of stability an endless increase of velocity may, as is easily seen from § 272, be obtained merely by *constraining* the system to a particular closed course, or circuit of configurations, nowhere deviating by more than an infinitely small amount from the configuration of equilibrium, and leaving it at rest anywhere in a certain part of this circuit. This result, and the distinct peculiarities of the cases of stability and instability, will be sufficiently illustrated by the simplest possible example, that of a material particle moving in a plane.

Let the mass be unity, and the components of force parallel to two rectangular axes be $ax+by$, and $a'x+b'y$, when the position of the particle is (x, y). The equations of motion will be

$$\ddot{x}=ax+by,\quad \ddot{y}=a'x+b'y \text{.....................(1)}.$$

Let $\frac{1}{2}(a'+b)=c$, and $\frac{1}{2}(a'-b)=e$:

the components of the force become

$$ax+cy-ey, \text{ and } cx+b'y+ex,$$

or

$$-\frac{dV}{dx}-ey, \text{ and } -\frac{dV}{dy}+ex,$$

where $V=-\frac{1}{2}(ax^2+b'y^2+2cxy)$.

The terms $-ey$ and $+ex$ are clearly the components of a force $e(x^2+y^2)^{\frac{1}{2}}$, perpendicular to the radius-vector of the particle. Hence if we turn the axes of co-ordinates through any angle, the

Artificial or ideal accumulative system.

corresponding terms in the transformed components are still $-ey$ and $+ex$. If, therefore, we choose the axes so that

$$V = \tfrac{1}{2}(\alpha x^2 + \beta y^2) \quad\ldots\ldots\ldots\ldots(2),$$

the equations of motion become, without loss of generality,

$$\ddot{x} = -\alpha x - ey, \quad \ddot{y} = -\beta y + ex.$$

To integrate these, assume, as in general [§ 343 (2)],

$$x = l\epsilon^{\lambda t}, \quad y = m\epsilon^{\lambda t}.$$

Then, as before [§ 343 (7)],

$$(\lambda^2 + \alpha)\, l + em = 0, \text{ and } -el + (\lambda^2 + \beta)\, m = 0.$$

Whence
$$(\lambda^2 + \alpha)(\lambda^2 + \beta) = -e^2 \quad\ldots\ldots\ldots\ldots(3),$$

which gives

$$\lambda^2 = -\tfrac{1}{2}(\alpha + \beta) \pm \{\tfrac{1}{4}(\alpha - \beta)^2 - e^2\}^{\frac{1}{2}}.$$

This shows that the equilibrium is stable if both $\alpha\beta + e^2$ and $\alpha + \beta$ are positive and $e^2 < \frac{1}{4}(\alpha - \beta)^2$ but unstable in every other case.

But let the particle be constrained to remain on a circle, of radius r. Denoting by θ its angle-vector from OX, and transforming (§ 27) the equations of motion, we have

$$\ddot{\theta} = -(\beta - \alpha)\sin\theta\cos\theta + e = -\tfrac{1}{2}(\beta - \alpha)\sin 2\theta + e \quad\ldots\ldots(4).$$

If we had $e = 0$ (a conservative system of force) the positions of equilibrium would be at $\theta = 0$, $\theta = \frac{1}{2}\pi$, $\theta = \pi$, and $\theta = \frac{3}{2}\pi$; and the motion would be that of the quadrantal pendulum. But when e has any finite value less than $\frac{1}{2}(\beta - \alpha)$ which, for convenience, we may suppose positive, there are positions of equilibrium at

$$\theta = \vartheta, \quad \theta = \frac{\pi}{2} - \vartheta, \quad \theta = \pi + \vartheta, \text{ and } \theta = \frac{3\pi}{2} - \vartheta,$$

where ϑ is half the acute angle whose sine is $\dfrac{2e}{\beta - \alpha}$: the first and third being positions of stable, and the second and fourth of unstable, equilibrium. Thus it appears that the effect of the constant tangential force is to displace the positions of stable and unstable equilibrium forwards and backwards on the circle through angles each equal to ϑ. And, by multiplying (4) by $2\dot{\theta}dt$ and integrating, we have as the integral equation of energy

$$\dot{\theta}^2 = C + \tfrac{1}{2}(\beta - \alpha)\cos 2\theta + 2e\theta \quad\ldots\ldots\ldots\ldots(5).$$

From this we see that the value of C, to make the particle just reach the position of unstable equilibrium, is

Artificial or ideal accumulative system.

$$C = -\tfrac{1}{2}(\beta - \alpha)\cos(\pi - 2\vartheta) - e(\pi - 2\vartheta),$$

$$= \sqrt{\frac{(\beta-\alpha)^2}{4} - e^2} - e\left(\pi - \sin^{-1}\frac{2e}{\beta - \alpha}\right),$$

and by equating to zero the expression (5) for $\dot{\theta}^2$, with this value of C substituted, we have a transcendental equation in θ, of which the least negative root, $\theta_{,}$, gives the limit of vibrations on the side reckoned backwards from a position of stable equilibrium. If the particle be placed at rest on the circle at any distance less than $\frac{\pi}{2} - 2\vartheta$ *before* a position of stable equilibrium, or less than $\vartheta - \theta_{,}$ *behind* it, it will vibrate. But if placed anywhere beyond those limits and left either at rest or moving with any velocity in either direction, it will end by flying round and round forwards with a periodically increasing and diminishing velocity, but increasing every half turn by equal additions to its squares.

If on the other hand $e > \frac{1}{2}(\beta - \alpha)$, the positions both of stable and unstable equilibrium are imaginary; the tangential force predominating in every position. If the particle be left at rest in any part of the circle it will fly round with continually increasing velocity, but periodically increasing and diminishing acceleration.

345'. Leaving now the ideal case of positional forces violating the law of conservatism, interestingly curious as it is, and instructive in respect to the contrast it presents with the positional forces of nature which are essentially conservative, let us henceforth suppose the positional forces of our system to be conservative and let us admit infringement of conservatism only as in nature through motional forces. We shall soon see (§ 345^{vii} and ^{viii}) that we may have motional forces which do not violate the law of conservatism. At present we make no restriction upon the motional forces and no other restriction on the positional forces than that they are conservative.

Cycloidal system with conservative positional forces and unrestricted motional forces.

The differential equations of motion, taken from (1) of 343a above, with the relations (10), and with V to denote the potential energy, are,

Cycloidal system with conservative positional forces and unrestricted motional forces.

$$\left.\begin{array}{l}\dfrac{d}{dt}\left(\dfrac{dT}{d\dot\psi_1}\right)+11\dot\psi_1+12\dot\psi_2+\ldots+\dfrac{dV}{d\psi_1}=0\\[2ex] \dfrac{d}{dt}\left(\dfrac{dT}{d\dot\psi_2}\right)+21\dot\psi_1+22\dot\psi_2+\ldots+\dfrac{dV}{d\psi_2}=0\\ \qquad\text{etc.}\qquad\qquad\text{etc.}\end{array}\right\}\ldots\ldots\ldots\ldots(1).$$

Multiplying the first of these by $\dot\psi_1$, the second by $\dot\psi_2$, adding and transposing, we find

$$\frac{d(T+V)}{dt}=-Q\ldots\ldots\ldots\ldots\ldots\ldots\ldots\ldots(2),$$

where

$$Q=11\dot\psi_1^{\,2}+(12+21)\,\dot\psi_1\dot\psi_2+22\dot\psi_2^{\,2}+(13+31)\,\dot\psi_1\dot\psi_3+\text{etc.}\ldots\ldots(3).$$

345[ii]. The quadratic function of the velocities here denoted by Q has been called by Lord Rayleigh* the Dissipation Function. We prefer to call it *Dissipativity.* It expresses the rate at which the *palpable energy* of our supposed cycloidal system is lost, not, as we now know, annihilated but (§§ 278, 340, 341, 342) dissipated away into other forms of energy. *It is essentially positive when the assumed motional forces are such as can exist in nature.* That it is equal to a quadratic function of the velocities is an interesting and important theorem.

Dissipativity defined.

Lord Rayleigh's theorem of Dissipativity.

Integral equation of energy.

Multiplying (2) by dt, and integrating, we find

$$T+V=E_0-\int_0^t Qdt\ldots\ldots\ldots\ldots\ldots\ldots\ldots\ldots(4),$$

where E_0 is a constant denoting the sum of the kinetic and potential energies at the instant $t=0$. Now T and Q are each of them essentially positive except when the system is at rest, and then each of them is zero. Therefore $\int_0^t Qdt$ must increase to infinity unless the system comes more and more nearly to rest as time advances. Hence either this must be the case, or V must diminish to $-\infty$. It follows that when V is positive for all real values of the co-ordinates the system must as time advances come more and more nearly to rest in its zero-configuration, whatever may have been the initial values of the co-ordinates and velocities. Even if V is negative for some or for all values of the co-ordinates, the system may be projected from *some given*

* *Proceedings of the London Mathematical Society*, May, 1873; *Theory of Sound*, Vol. I. § 81.

configurations with such velocities that when $t = \infty$ it shall be at rest in its zero configuration: this we see by taking, as a particular solution, the terms of (9) § 345^{iv} below, for which m is negative. But this equilibrium is essentially unstable, unless V is positive for all real values of the co-ordinates. To prove this imagine the system placed in any configuration in which V is negative, and left there either at rest or with any motion of kinetic energy less than or at the most equal to $-V$: thus E_0 will be negative or zero; $T+V$ will therefore have increasing negative value as time advances; therefore V must always remain negative; and therefore the system can never reach its zero configuration. It is clear that $-V$ and T must each on the whole increase though there may be fluctuations, of T diminishing for a time, during which $-V$ must also diminish so as to make the excess $(-V)-T$ increase at the rate equal to Q per unit of time according to formula (2).

Cycloidal system with conservative positional forces and unrestricted motional forces.

345'''. To illustrate the circumstances of the several cases let $\lambda = m + n\sqrt{-1}$ be a root of the determinantal equation, m and n being both real. The corresponding realized solution of the dynamical problem is

$$\psi_1 = r_1\epsilon^{mt}\cos(nt - e_1),\ \psi_2 = r_2\epsilon^{mt}\cos(nt - e_2),\ \text{etc.}\ldots\ldots\ldots(5),$$

where the differences of epochs $e_2 - e_1$, $e_3 - e_1$, etc. and the ratios r_2/r_1, etc., in all $2i-2$ numerics*, are determined by the $2i$ simultaneous linear equations (3) of § 343 harmonized by taking for $\lambda = m + n\sqrt{-1}$, and again $\lambda = m - n\sqrt{-1}$. Using these expressions for ψ_1, ψ_2, etc. in the expressions for V, Q, T, we find,—

$$\left.\begin{aligned} V &= \epsilon^{2mt}(C + A\cos 2nt + B\sin 2nt) \\ Q &= \epsilon^{2mt}(C' + A'\cos 2nt + B'\sin 2nt) \\ T &= \epsilon^{2mt}(C'' + A''\cos 2nt + B''\sin 2nt) \end{aligned}\right\}\ldots\ldots\ldots\ldots\ldots(6),$$

* The term numeric has been recently introduced by Professor James Thomson to denote a number, or a proper fraction, or an improper fraction, or an incommensurable ratio (such as π or ϵ). It must also to be useful in mathematical analysis include imaginary expressions such as $m+n\sqrt{-1}$, where m and n are real numerics. "Numeric" may be regarded as an abbreviation for "numerical expression." It lets us avoid the intolerable verbiage of integer or proper or improper fraction which mathematical writers hitherto are so often compelled to use; and is more appropriate for mere number or ratio than the designation "quantity," which rather implies quantity of something than the mere numerical expression by which quantities of any measurable things are reckoned in terms of the unit of quantity.

Cycloidal system with conservative positional forces and unrestricted motional forces.

where $C, A, B, C', A', B', C'', A'', B''$, are determinate constants: and in order that Q and T may be positive we have

$$C' > +\sqrt{(A'^2 + B'^2)}, \text{ and } C'' > +\sqrt{(A''^2 + B''^2)} \ldots\ldots\ldots\ldots(7).$$

Substituting these in (2), and equating coefficients of corresponding terms, we find

$$\left.\begin{aligned} 2m(C + C'') &= -C' \\ 2\{m(A + A'') + n(B + B'')\} &= -A' \\ 2\{m(B + B'') - n(A + A'')\} &= -B' \end{aligned}\right\} \ldots\ldots\ldots\ldots(8).$$

Real part of every root proved negative when V positive for all real co-ordinates;

The first of these shows that $C + C''$ and m must be of contrary signs. Hence if V be essentially positive [which requires that C be greater than $+\sqrt{(A^2+B^2)}$], every value of m must be negative.

positive for some roots when V has negative values;

345^iv. If V have negative values for some or all real values of the co-ordinates, m must clearly be positive for some roots, but there must still, and always, be roots for which m is negative. To prove this last clause let us instead of (5) take sums of particular solutions corresponding to different roots

$$\lambda = m \pm n\sqrt{-1},\ \lambda' = m' \pm n'\sqrt{-1}, \text{ etc.},$$

m and n denoting real numerics. Thus we have

$$\left.\begin{aligned} \psi_1 &= r_1\epsilon^{mt}\cos(nt - e_1) + r'_1\epsilon^{m't}\cos(n't - e'_1) + \text{etc.} \\ \psi_2 &= r_2\epsilon^{mt}\cos(nt - e_2) + r'_2\epsilon^{m't}\cos(n't - e'_2) + \text{etc.} \\ &\text{etc.} \end{aligned}\right\} \quad (9).$$

but always negative for some roots.

Suppose now m, m', etc. to be all positive: then for $t = -\infty$, we should have $\psi_1 = 0, \psi_2 = 0, \dot{\psi}_1 = 0, \dot{\psi}_2 = 0$, etc., and therefore $V = 0, T = 0$. Hence, for finite values of t, T would in virtue of (4) be less than $-V$ (which in this case is essentially positive): but we may place the system in any configuration and project it with any velocity we please, and therefore the amount of kinetic energy we may give it is unlimited. Hence, if (9) be the complete solution, it must include some negative value or values of m, and therefore of all the roots λ, λ', etc. there must be some of which the real part is negative. This conclusion is also obvious on purely algebraic grounds, because the coefficient of λ^{2i-1} in the determinant is obviously $11 + 22 + 33 + \ldots$, which is essentially positive when Q is positive for all real values of the co-ordinates.

345^v. It is an important subject for investigation, interesting both in mere Algebra and in Dynamics, to find how many roots there are with m positive, or how many with m negative in any particular case or class of cases; also to find under what conditions n disappears [or the motion non-oscillatory (compare § 341)]. We hope to return to it in our second volume, and should be very glad to find it taken up and worked out fully by mathematicians in the mean time. At present it is obvious that if V be negative for all real values of ψ_1, ψ_2, etc., the motion must be non-oscillatory for every mode (or every value of λ must be real) if Q be but large enough: but as we shall see immediately with Q *not too large*, n may appear in some or in all the roots, even though V be negative for all real co-ordinates, when there are forces of the gyroscopic class [§ 319, Examp. (G) above and § 345^{x} below). When the motional forces are wholly of the viscous class it is easily seen that n can only appear if V is positive for some or all real values of the co-ordinates: n must disappear if V is negative for all real values of the co-ordinates (again compare § 341).

Non-oscillatory subsidence to stable equilibrium, or falling away from unstable.

Oscillatory subsidence to stable equilibrium, or falling away from unstable.

Falling away from wholly unstable equilibrium is essentially non-oscillatory if motional forces wholly viscous.

345^{vi}. A chief part of the substance of §§ 345^{ii} ... 345^{v} above may be expressed shortly without symbols thus:—When there is any dissipativity the equilibrium in the zero position is stable or unstable according as the same system with no motional forces, but with the same positional forces, is stable or unstable. The gyroscopic forces which we now proceed to consider may convert instability into stability, as in the gyrostat § 345^{x} below, *when there is no dissipativity*:—but when there is any dissipativity gyroscopic forces may convert rapid falling away from an unstable configuration into falling by (as it were) exceedingly gradual spirals, but they cannot convert instability into stability if there be any dissipativity.

Stability of Dissipative system.

The theorem of Dissipativity [§ 345^{i}, (2) and (3)] suggests the following notation,—

$$\left.\begin{array}{l} \tfrac{1}{2}(12+21)=[12] \text{ or } [21],\ \tfrac{1}{2}(13+31)=[13] \text{ or } [31], \text{ etc.} \\ \text{and } \tfrac{1}{2}(12-21)=12] \text{ or } -21],\ \tfrac{1}{2}(13-31)=13] \text{ or } -31], \text{ etc.} \end{array}\right\} \quad (10),$$

so that the symbols [12], [21], [13], etc., and 12], 21], 13], etc. denote quantities which respectively fulfil the following mutual relations,

$$\left.\begin{array}{l} [12]=[21],\ [13]=[31],\ [23]=[32], \text{ etc.} \\ 12]=-21],\ 13]=-31],\ 23]=-32], \text{ etc.} \end{array}\right\} \ldots\ldots\ldots \quad (11).$$

Thus (3) of § 345^{i} becomes

$$Q = 11\dot{\psi}_1^2 + 2\,[12]\,\dot{\psi}_1\dot{\psi}_2 + 22\dot{\psi}_2^2 + 2\,[13]\,\dot{\psi}_1\dot{\psi}_3 + \text{etc.} \ldots\ldots (12),$$

and going back to (1), with (10) and (12) we have

$$\left.\begin{array}{l}\frac{d}{dt}\frac{dT}{d\psi_1}+\frac{dQ}{d\psi_1}+12]\dot{\psi}_2+13]\dot{\psi}_3+\text{etc.}+\frac{dV}{d\psi_1}=0\\ \frac{d}{dt}\frac{dT}{d\psi_2}+\frac{dQ}{d\psi_2}+21]\dot{\psi}_1+23]\dot{\psi}_3+\text{etc.}+\frac{dV}{d\psi_2}=0\\ \frac{d}{dt}\frac{dT}{d\psi_3}+\frac{dQ}{d\psi_3}+31]\dot{\psi}_1+32]\dot{\psi}_2+\text{etc.}+\frac{dV}{d\psi_3}=0\\ \dots\dots\dots\dots\dots\dots\dots\dots\end{array}\right\}\dots\dots(13).$$

Various origins of gyroscopic terms.

In these equations the terms $12]\dot{\psi}_2$, $21]\dot{\psi}_1$, $13]\dot{\psi}_3$, $31]\dot{\psi}_1$, etc. represent what we may call gyroscopic forces, because, as we have seen in § 319, Ex. G, they occur when fly-wheels each given in a state of rapid rotation form part of the system by being mounted on frictionless bearings connected through framework with other parts of the system; and because, as we have seen in § 319, Ex. F, they occur when the motion considered is motion of the given system relatively to a rigid body revolving with a constrainedly constant angular velocity round a fixed axis This last reason is especially interesting on account of Laplace's dynamical theory of the tides at the foundation of which it lies, and in which it is answerable for some of the most curious and instructive results, such as the beautiful vortex problem presented by what Laplace calls "Oscillations of the First Species*."

Equation of energy.

345vii. The gyrostatic terms disappear from the equation of energy as we see by § 345i, (2) and (3), and as we saw previously by § 319, Example G (19), and in § 319, Ex. F′ (f). Comparing § 319 (f) and (g), we see that in the case of motion relatively to a body revolving uniformly round a fixed axis it is not the equation of total absolute energy but the equation of energy of the *relative motion* that the gyroscopic terms disappear from, as (f) of § 319; and (2) and (3) of § 345i when the subject of their application is to such relative motion.

* The integrated equation for this species of tidal motions, in an ideal ocean equally deep over the whole solid rotating spheroid, is given in a form ready for numerical computation in "Note on the 'Oscillations of the First Species' in Laplace's Theory of the Tides" (W. Thomson), *Phil. Mag.* Oct. 1875.

345^{viii}. To discover something of the character of the gyroscopic influence on the motion of a system, suppose there to be no resistances (or viscous influences), that is to say let the dissipativity, Q, be zero. The determinantal equation (4) becomes

Gyrostatic conservative system:

$$\begin{vmatrix} (11)\lambda^2 \qquad +11, & (12)\lambda^2 + 12]\lambda + 12, \ldots \\ (21)\lambda^2 + 21]\lambda + 21, & (22)\lambda^2 \qquad +22, \ldots \\ \ldots\ldots\ldots\ldots\ldots\ldots & \\ \ldots\ldots\ldots\ldots\ldots\ldots & \end{vmatrix} = 0 \ldots\ldots (14).$$

Now by the relations $(12)=(21)$, etc., $12=21$, etc., and $12]=-21]$, we see that if λ be changed into $-\lambda$ the determinant becomes altered merely by interchange of terms between columns and rows, and hence the value of the determinant remains unchanged. Hence the first member of (14) cannot contain odd powers of λ, and therefore its roots must be in pairs of oppositely signed equals. The condition for stability of equilibrium in the zero configuration is therefore that the roots λ^2 of the determinantal equation be each real and negative.

345^{ix}. The equations are simplified by transforming the co-ordinates (§ 337) so as to reduce T to a sum of squares with positive coefficients and V to a sum of squares with positive or negative coefficients as the case may be, or which is the same thing to adopt for co-ordinates those displacements which would correspond to "fundamental modes" (§ 338), if the positional forces were as they are and there were no motional forces. Suppose farther the unit values of the co-ordinates to be so chosen that the coefficients of the squares of the velocities in $2T$ shall be each unity; and let us put ϖ_1, ϖ_2, ϖ_3, etc. instead of the coefficients 11, 22, 33, etc., remaining in $2V$. Thus we have

Simplification of its equations.

$$T = \tfrac{1}{2}(\dot{\psi}_1^2 + \dot{\psi}_2^2 + \text{etc.}), \text{ and } V = \tfrac{1}{2}(\varpi_1\psi_1^2 + \varpi_2\psi_2^2 + \text{etc.}) \ldots\ldots (15).$$

If now we omit the half brackets] as no longer needed to avoid ambiguity, and understand that $12=-21$, $13=-31$, $23=-32$, etc., the equations of motion are

$$\left.\begin{array}{l} \ddot{\psi}_1 + 12\dot{\psi}_2 + 13\dot{\psi}_3 + \ldots\ldots + \varpi_1\psi_1 = 0 \\ \ddot{\psi}_2 + 21\dot{\psi}_1 + 23\dot{\psi}_3 + \ldots\ldots + \varpi_2\psi_2 = 0 \\ \ddot{\psi}_3 + 31\dot{\psi}_1 + 32\dot{\psi}_2 + \ldots\ldots + \varpi_3\psi_3 = 0 \\ \ldots\ldots\ldots\ldots\ldots\ldots\ldots\ldots \end{array}\right\} \ldots\ldots\ldots (16),$$

and the determinantal equation becomes

$$\begin{vmatrix} \lambda^2+\varpi_1, & 12\lambda, & 13\lambda,\ldots \\ 21\lambda, & \lambda^2+\varpi_2, & 23\lambda,\ldots \\ 31\lambda, & 32\lambda, & \lambda^2+\varpi_3,\ldots \\ \ldots\ldots & \ldots\ldots & \ldots\ldots \end{vmatrix} = 0 \ldots\ldots\ldots (17).$$

Determinant of gyrostatic conservative system.

The determinant (which for brevity we shall denote by D) in this case is what has been called by Cayley a skew determinant. What it would become if zero were substituted for $\lambda^2+\varpi_1$, $\lambda^2+\varpi_2$, etc. in its principal diagonal is what is called a skew symmetric determinant. The known algebra of skew and skew symmetric determinants gives

$$\left.\begin{aligned} D = &(\lambda^2+\varpi_1)(\lambda^2+\varpi_2)\ldots(\lambda^2+\varpi_i) \\ &+\lambda^2\Sigma(\lambda^2+\varpi_3)(\lambda^2+\varpi_4)\ldots(\lambda^2+\varpi_i)\,12^2 \\ &+\lambda^4\Sigma(\lambda^2+\varpi_5)(\lambda^2+\varpi_6)\ldots(\lambda^2+\varpi_i)(12\,.\,34+31\,.\,24+23\,.\,14)^2 \\ &+\lambda^6\Sigma(\lambda^2+\varpi_7)(\lambda^2+\varpi_8)\ldots(\lambda^2+\varpi_i)(\mathfrak{S}\,12\,.\,34\,.\,56)^2+\text{etc.} \\ &+\lambda^i(\mathfrak{S}\,12\,.\,34\,.\,56\,.\,\ldots\,.\,i-1,\;i)^2 \end{aligned}\right\} (18),$$

when i is even. For example see (30) below. When λ is odd the last term is

$$\lambda^{i-1}\Sigma(\lambda^2+\varpi_i)(\mathfrak{S}\,12\,.\,34\,.\,56\ldots\,i-2,\;i-1)^2 \ldots\ldots\ldots\ldots (18'),$$

and no other change in the formula is necessary. In each case the small $\mathfrak{S}$ denotes the sum of the products obtained by making every possible permutation of the numbers in the line of factors following it, with orders chosen accccording to a proper rule to render the sign of each product positive (Salmon's *Higher Algebra*, Lesson v. Art. 40). This sum is in each case the square root of a certain corresponding skew symmetric determinant.

Square roots of skew symmetrics.

An easy rule to find other products from any one given to begin with is this:—Invert the order in any one factor, and make a simple interchange of any two numbers in different factors. Thus, in the last $\mathfrak{S}$ of (18) alter $i-1$, i to i, $i-1$, and interchange $i-1$ with 3: so we find $12\,.\,i-1,\,4\,.\,56\ldots\ldots i,\,3$ for a term: similarly $12\,.\,64\,.\,53\ldots i,\,i-1$, and $62\,.\,14\,.\,53\ldots i-1,\,i$, for two others. The same number must not occur more than once in any one product. Two products differing only in the orders of the two numbers in factors are not admitted. If n be the number of factors in each term, the whole number of factors is clearly $1\,.\,3\,.\,5\ldots(2n-1)$, and they may be found in regular progression thus: Begin with a single factor and single term 12. Then apply to it the factor 34, and permute to suit 24 instead of 34, and permute the result to suit 14 instead of 24. Thirdly, apply to the sum thus found the factor 56, and permute successively from 56 to 46, from 46 to 36, from 36 to 26, and from 26 to 16. Fourthly, introduce the factor 78; and so on. Thus we find

Square roots of skew symmetrics.

$$\left.\begin{aligned}
&\sqrt{\begin{vmatrix} 0, & 12 \\ 21, & 0 \end{vmatrix}} = 12 \\
&\sqrt{\begin{vmatrix} 0, & 12, & 13, & 14 \\ 21, & 0, & 23, & 24 \\ 31, & 32, & 0, & 34 \\ 41, & 42, & 43, & 0 \end{vmatrix}} = 12.34 + 31.24 + 23.14 \\
&\sqrt{\begin{vmatrix} 0, & 12, & 13, & 14, & 15, & 16 \\ 21, & 0, & 23, & 24, & 25, & 26 \\ 31, & 32, & 0, & 34, & 35, & 36 \\ 41, & 42, & 43, & 0, & 45, & 46 \\ 51, & 52, & 53, & 54, & 0, & 56 \\ 61, & 62, & 63, & 64, & 65, & 0 \end{vmatrix}} = \begin{aligned} &(12.34+31.24+23.14)56 \\ &+(12.53+13.52+23.51)46 \\ &+(12.45+41.52+42.51)36 \\ &+(31.45+41.35+34.51)26 \\ &+(23.45+24.35+34.25)16 \end{aligned}
\end{aligned}\right\} \quad (19).$$

The second member of the last of these equations is what is denoted by $\Sigma 12.34.56$ in (18).

345^x. Each term of the determinant D except

$$(\lambda^2 + \varpi_1)(\lambda^2 + \varpi_2)\ldots(\lambda^2 + \varpi_i)$$

contains λ^2 as a factor. Hence, when all are expanded in powers of λ^2, the term independent of λ is $\varpi_1 \varpi_2 \ldots \varpi_i$. If this be negative there must be at least one real positive and one real negative root λ^2. Hence for stability either must all of ϖ_1, ϖ_2, ..., ϖ_i be positive or an even number of them negative. Ex.:—Two modes of motion, x and y the co-ordinates. Let the equations of motion be

Gyrostatic system with two freedoms.

$$\left.\begin{aligned} I\ddot{x} + g\dot{y} + Ex = 0 \\ J\ddot{y} - g\dot{x} + Fy = 0 \end{aligned}\right\} \quad \ldots\ldots\ldots\ldots\ldots\ldots\ldots\ldots (20),$$

and the determinantal equation is

$$(I\lambda^2 + E)(J\lambda^2 + F) + g^2\lambda^2 = 0.$$

If we put

$$x = \xi / \sqrt{I}, \quad y = \eta / \sqrt{J} \quad \ldots\ldots\ldots\ldots\ldots\ldots (21),$$

and

$$E = \varpi I, \; F = \zeta J, \text{ and } g = \gamma \sqrt{(IJ)} \ldots\ldots\ldots (22),$$

Gyrostatic system with two freedoms.

equations (20) and the determinantal equation become

$$\left.\begin{array}{l}\ddot{\xi}+\gamma\dot{\eta}+\varpi\xi=0\\ \ddot{\eta}-\gamma\dot{\xi}+\zeta\eta=0\end{array}\right\}\ldots\ldots\ldots\ldots\ldots\ldots\ldots(23),$$

and $$(\lambda^2+\varpi)(\lambda^2+\zeta)+\gamma^2\lambda^2=0\ldots\ldots\ldots\ldots\ldots\ldots(24).$$

The solution of this quadratic in λ^2 may be put under the following forms,—

$$\left.\begin{array}{l}-\lambda^2=\frac{1}{2}(\gamma^2+\varpi+\zeta)\pm\frac{1}{2}\{[\gamma^2+(\sqrt{\varpi}+\sqrt{\zeta})^2]\,[\gamma^2+(\sqrt{\varpi}-\sqrt{\zeta})^2]\}^{\frac{1}{2}}\\ -\lambda^2=\frac{1}{2}(\gamma^2+\varpi+\zeta)\pm\frac{1}{2}\{[\gamma^2-(\sqrt{-\varpi}+\sqrt{-\zeta})^2][\gamma^2-(\sqrt{-\varpi}-\sqrt{-\zeta})^2]\}^{\frac{1}{2}}\end{array}\right\}\ldots(25).$$

To make both values of $-\lambda^2$ real and positive ϖ and ζ must be of the same sign. If they are both positive no farther condition is necessary. If they are both negative we must have

$$\gamma>\sqrt{-\varpi}+\sqrt{-\zeta}\ldots\ldots\ldots\ldots\ldots\ldots\ldots\ldots(26).$$

These are the conditions that the zero configuration may be stable. Remark that when (as practically in all the gyrostatic illustrations) γ^2 is very great in comparison with $\sqrt{(\varpi\zeta)}$, the greater value of $-\lambda^2$ is approximately equal to γ^2, and therefore (as the product of the two roots is exactly $\varpi\zeta$), the less is approximately equal to $\varpi\zeta/\gamma^2$. Remark also that $2\pi/\sqrt{\varpi}$ and $2\pi/\sqrt{\zeta}$ are the periods of the two fundamental vibrations of a system otherwise the same as the given system, but with $\gamma=0$. Hence, using the word irrotational to refer to the system with $g=0$, and gyroscopic, or gyrostatic, or gyrostat, to refer to the actual system;

Gyrostatic influence dominant.

From the preceding analysis we have the curious and interesting result that, in a system with two freedoms, two irrotational instabilities are converted into complete gyrostatic stability (each freedom stable) by sufficiently rapid rotation; but that with one irrotational stability the gyrostat is essentially unstable, with one of its freedoms unstable and the other stable, if there be one irrotational instability. Various good illustrations of gyrostatic systems with two, three, and four freedoms (§§ 345^x, ^xi and ^xii) are afforded by the several different modes of mounting shown in the accompanying sketches, applied to the ordinary gyrostat* (a rapidly rotating fly-wheel pivoted as finely as possible within a rigid case, having a convex curvilinear polygonal border, in the plane perpendicular to the axis through the centre of gravity of the whole).

Gyrostatic stability.

* *Nature*, No. 379, Vol. 15 (February 1, 1877), page 297.

Ordinary gyrostats.

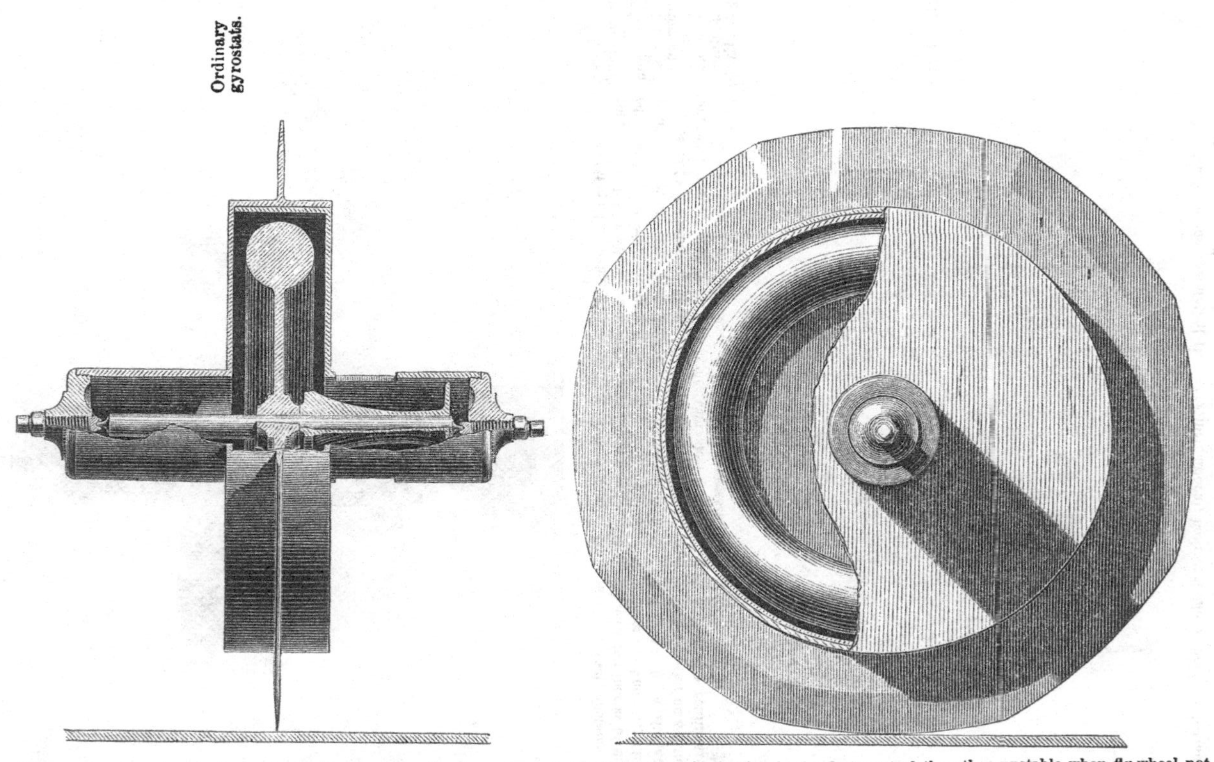

(Translational motions not considered) there are two freedoms, one azimuthal the other inclinational; the first neutral the other unstable when fly-wheel not rotating; the first still neutral the second stable when fly-wheel in rapid rotation. Equations (23) with $\zeta = 0$ express the problem, and (24) and (25) its solution.

Gyrostats,

on gimbals;

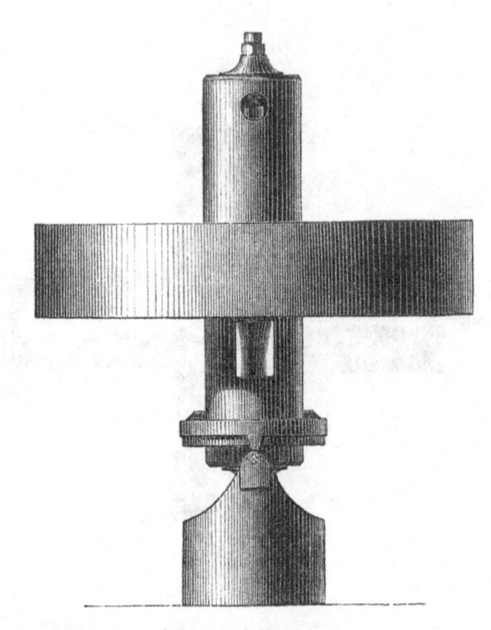

on universal-flexure-joint (§ 109) in place of gimbals; constituting an inverted gyroscopic pendulum (§ 319, Ex. D).

Gyrostat on knife-edge gimbals with its axis vertical. Two freedoms; each unstable without rotation of the fly-wheel; each stable when it is rotating rapidly. Neglecting inertia of the knife-edges and gimbal-ring we have $I = J$ in (20), and supposing the levels of the knife-edges to be the same, we have $E = F$. Thus its determinantal equation is $(I\lambda^2 + E)^2 + \mathfrak{g}^2\lambda^2 = 0$. A similar result, expressed by the same equations of motion, is obtained by supporting the gyrostat on a little elastic universal flexure-joint of, for example, thin steel pianoforte-wire one or two centimetres long between end clamps or solderings. A drawing is unnecessary.

on stilts;

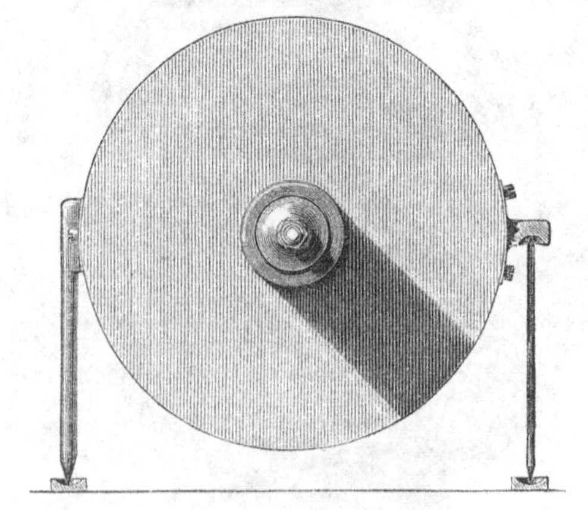

Two freedoms, one azimuthal the other inclinational, both unstable without, both stable with rapid rotation of the fly-wheel.

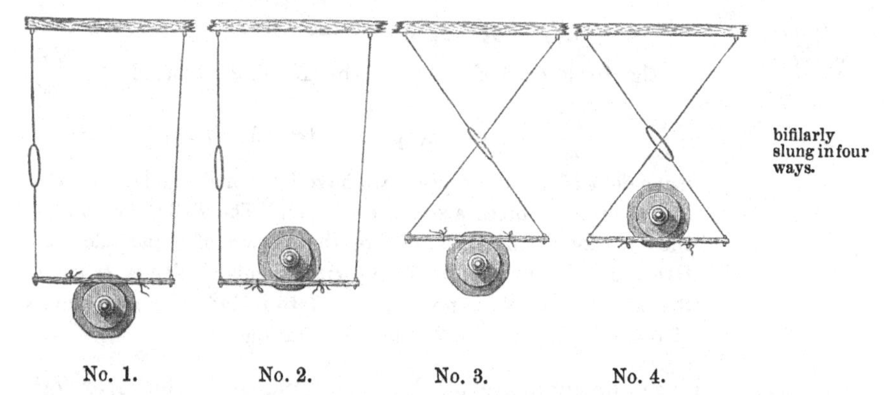

bifilarly slung in four ways.

Four freedoms, reducible to three if desired by a third thread in each case, diagonal in the first and second, lateral in the third and fourth, the freedom thus annulled being in each case stable and independent of the rotation of the fly-wheel. Three modes essentially involved in the gyrostatic system in each case, two inclinational and one azimuthal.

No. 1.—Azimuthally stable without rotation; with rotation all three modes stable.

No. 2.—Azimuthally stable, one inclinational mode unstable the other stable without rotation; with rotation two unstable, one stable.

No. 3.—The azimuthal mode unstable, two inclinational modes stable the other unstable, without rotation; with rotation one azimuthal mode and one inclinational mode unstable, and one inclinational mode stable.

No. 4.—Azimuthally and one inclinational mode unstable, one inclinational mode stable, without rotation; with rotation all three stable.

Gyrostatic system with three freedoms.

345xi. Take for another example a system having three freedoms (that is to say, three independent co-ordinates ψ_1, ψ_2, ψ_3), (16) become

$$\left.\begin{array}{l}\ddot{\psi}_1 + g_3\dot{\psi}_2 - g_2\dot{\psi}_3 + \varpi_1\psi_1 = 0\\ \ddot{\psi}_2 + g_1\dot{\psi}_3 - g_3\dot{\psi}_1 + \varpi_2\psi_2 = 0\\ \ddot{\psi}_3 + g_2\dot{\psi}_1 - g_1\dot{\psi}_2 + \varpi_3\psi_3 = 0\end{array}\right\} \qquad (27),$$

where g_1, g_2, g_3 denote the values of the three pairs of equals 23 or -32, 31 or -13, 12 or -21. Imagine ψ_1, ψ_2, ψ_3 to be rectangular co-ordinates of a material point, and let the co-ordinates be transformed to other axes OX, OY, OZ, so chosen that OZ coincides with the line whose direction cosines relatively to the ψ_1-, ψ_2-, ψ_3- axes are proportional to g_1, g_2, g_3. The equations become

$$\left.\begin{array}{l}\ddot{x} - 2\omega\dot{y} = X\\ \ddot{y} + 2\omega\dot{x} = Y\\ \ddot{z} \qquad\quad = Z\end{array}\right\} \qquad (28),$$

Gyrostatic system with three freedoms:

where $\omega = \sqrt{(g_1^2 + g_2^2 + g_3^2)}$, and the force-components parallel to the fresh axes are denoted by X, Y, Z (instead of $-\frac{dV}{dx}$, $-\frac{dV}{dy}$, $-\frac{dV}{dz}$, because the present transformation is clearly independent of the assumption we have been making latterly that the positional forces are conservative). These (28) are simply the equations [§ 319, Ex. (E)] of the motion of a particle relatively to co-ordinates revolving with angular velocity ω round the axis OZ, if we suppose X, Y, Z to include the components of the centrifugal force due to this rotation.

reduced to a mere rotating system.

Hence the influence of the gyroscopic terms however originating in any system with three freedoms (and therefore also in any system with only two freedoms) may be represented by the motion of a material particle supported by massless springs attached to a rigid body revolving uniformly round a fixed axis. It is an interesting and instructive exercise to imagine or to actually construct mechanical arrangements for the motion of a material particle to illustrate the experiments described in § 345^{x}.

345^{xii}. Consider next the case of a system with four freedoms. The equations are

$$\left.\begin{array}{l} \ddot{\psi}_1 + 12\dot{\psi}_2 + 13\dot{\psi}_3 + 14\dot{\psi}_4 + \varpi_1\psi_1 = 0 \\ \ddot{\psi}_2 + 21\dot{\psi}_1 + 23\dot{\psi}_3 + 24\dot{\psi}_4 + \varpi_2\psi_2 = 0 \\ \ddot{\psi}_3 + 31\dot{\psi}_1 + 32\dot{\psi}_2 + 34\dot{\psi}_4 + \varpi_3\psi_3 = 0 \\ \psi_4 + 41\dot{\psi}_1 + 42\dot{\psi}_2 + 43\dot{\psi}_3 + \varpi_4\psi_4 = 0 \end{array}\right\} \quad \ldots\ldots\ldots (29).$$

Denoting by D the determinant we have, by (18),

$$\left.\begin{array}{l} D = (\lambda^2 + \varpi_1)(\lambda^2 + \varpi_2)(\lambda^2 + \varpi_3)(\lambda^2 + \varpi_4) \\ \quad + \lambda^2 \{34^2(\lambda^2 + \varpi_1)(\lambda^2 + \varpi_2) + 12^2(\lambda^2 + \varpi_3)(\lambda^2 + \varpi_4) + 42^2(\lambda^2 + \varpi_1)(\lambda^2 + \varpi_3) \\ \quad\quad + 13^2(\lambda^2 + \varpi_4)(\lambda^2 + \varpi_2) + 23^2(\lambda^2 + \varpi_1)(\lambda^2 + \varpi_4) + 14^2(\lambda^2 + \varpi_2)(\lambda^2 + \varpi_3)\} \\ \quad + \lambda^4 (12\ 34 + 13\ 42 + 14\ 23)^2 \end{array}\right\} \ldots (30).$$

If ϖ_1, ϖ_2, ϖ_3, ϖ_4 be each zero, D becomes

$$\lambda^8 + (12^2 + 13^2 + 14^2 + 23^2 + 42^2 + 34^2)\lambda^6 + (12\ 34 + 13\ 42 + 14\ 23)^2\lambda^4.$$

This equated to zero and viewed as an equation for λ^2 has two roots each equal to 0, and two others given by the residual quadratic

Quadruply free gyrostatic system without force.

$$\lambda^4 + (12^2+13^2+14^2+23^2+24^2+34^2)\lambda^2 + (12\,34+13\,42+14\,23)^2=0 \ldots (31).$$

Now remarking that the solution of $z^2+pz+q^2=0$ may be written

$$-z = \tfrac{1}{2}\{p \pm \sqrt{(p+2q)(p-2q)}\} = \tfrac{1}{4}\{\sqrt{(p+2q)} \pm \sqrt{(p-2q)}\}^2,$$

we have from (31)

$$\left.\begin{aligned} -\lambda^2 &= \tfrac{1}{2}(12^2+13^2+14^2+23^2+24^2+34^2 \pm \sqrt{s}) \\ &= \tfrac{1}{4}(r \pm s)^2 \end{aligned}\right\} \ldots\ldots (32),$$

$$\left.\begin{aligned} \text{where}\quad r &= \sqrt{\{(12+34)^2+(13+42)^2+(14+23)^2\}} \\ \text{and}\quad s &= \sqrt{\{(12-34)^2+(13-42)^2+(14-23)^2\}} \end{aligned}\right\} \ldots\ldots (33).$$

As 12, 34, 13, etc. are essentially real, r and s are real, and (unless $12\,43 + 13\,42 + 14\,23 = 0$, when one of the values of λ^2 is zero, a case which must be considered specially, but is excluded for the present,) they are unequal. Hence the two values of $-\lambda^2$ given by (32) are real and positive. Hence two of the four freedoms are stable. The other two (corresponding to $-\lambda^2=0$) are neutral.

Excepted case of failing gyrostatic predominance.

345[xiii]. Now suppose ϖ_1, ϖ_2, ϖ_3, ϖ_4 to be not zero, but each very small. The determinantal equation will be a biquadratic in λ^2, of which two roots (the two which vanish when ϖ_1, etc. vanish) are approximately equal to the roots of the quadratic

Quadruply free cycloidal system, gyrostatically dominated.

$$(12\,34+13\,42+14\,23)^2\lambda^4 + (12^2\varpi_3\varpi_4 + 13^2\varpi_2\varpi_4 + 14^2\varpi_2\varpi_3$$
$$+ 23^2\varpi_1\varpi_4 + 24^2\varpi_1\varpi_3 + 34^2\varpi_1\varpi_2)\lambda^2 + \varpi_1\varpi_2\varpi_3\varpi_4 = 0 \ldots (34),$$

and the other two roots are approximately equal to those of the previous residual quadratic (31).

To solve equation (34), first write it thus:—

$$\left(\frac{1}{\lambda^2}\right)^2 + (12'^2+13'^2+14'^2+23'^2+24'^2+34'^2)\frac{1}{\lambda^2} + (12'34'+13'42'+14'23')^2 = 0$$
$$\ldots\ldots\ldots (35),$$

where

$$12' = \frac{12}{\sqrt{(\varpi_1\varpi_2)}},\quad 13' = \frac{13}{\sqrt{(\varpi_1\varpi_3)}},\quad 23' = \frac{23}{\sqrt{(\varpi_2\varpi_3)}},\ \text{etc.} \ldots\ldots (36).$$

Quadruply free cycloidal system, gyrostatically dominated.

Thus, taken as a quadratic for λ^{-2}, it has the same form as (31) for λ^2, and so, as before in (32) and (33), we find

$$\frac{-1}{\lambda^2} = \tfrac{1}{4}(r' \pm s')^2 \text{.....................} \quad (37),$$

$$\left.\begin{array}{ll}\text{where} & r' = \sqrt{\{(12'+34')^2+(13'+42')^2+(14'+23')^2\}} \\ \text{and} & s' = \sqrt{\{(12'-34')^2+(13'-42')^2+(14'-23')^2\}}\end{array}\right\} \text{......} (38).$$

Four irrotational stabilities confirmed, four irrotational instabilities rendered stable, by gyrostatic links.

Now if $\varpi_1, \varpi_2, \varpi_3, \varpi_4$ be all four positive or all four negative, $12', 34', 13'$, etc. are all real, and therefore both the values of $-\frac{1}{\lambda^2}$ given by (37) are real and positive (the excluded case referred to at the end of § 345^{xii}, which makes

$$12'34' + 13'42' + 14'23' = 0,$$

and therefore the smaller value of $-\frac{1}{\lambda^2} = 0$, being still excluded). Hence the corresponding freedoms are stable. But it is not *necessary* for stability that $\varpi_1, \varpi_2, \varpi_3, \varpi_4$ be all four of one sign: it *is* necessary that their product be positive: since if it were negative the values of λ^2 given by (34) would both be real, but one only negative and the other positive. Suppose two of them, ϖ_3, ϖ_4 for example, be negative, and the other two, ϖ_1, ϖ_2, positive: this makes $\varpi_1\varpi_3, \varpi_1\varpi_4, \varpi_2\varpi_3$, and $\varpi_2\varpi_4$ negative, and therefore $13', 14', 23'$, and $24'$ imaginary. Instead of four of the six equations (36), put therefore

$$13'' = \frac{13}{\sqrt{(-\varpi_1\varpi_3)}},\ 14'' = \frac{14}{\sqrt{(-\varpi_1\varpi_4)}},\ 23'' = \frac{23}{\sqrt{(-\varpi_2\varpi_3)}},\ 24'' = \frac{24}{\sqrt{(-\varpi_2\varpi_4)}} \quad (39).$$

Thus $13''$ etc. are real, and $13' = 13''\sqrt{-1}$ etc., and (38) become

$$\left.\begin{array}{l} r' = \sqrt{\{(12'+34')^2-(13''+42'')^2-(14''+23'')^2\}} \\ s' = \sqrt{\{(12'-34')^2-(13''-42'')^2-(14''-23'')^2\}}\end{array}\right\} \text{.......} (40).$$

Hence for stability it is necessary and sufficient that

$$\left.\begin{array}{ll} & (12'+34')^2 > (13''+42'')^2+(14''+23'')^2 \\ \text{and} & (12'-34')^2 > (13''-42'')^2+(14''-23'')^2\end{array}\right\} \text{..........} (41).$$

Combined dynamic and gyrostatic sta-

If these inequalities are reversed, the stabilities due to ϖ_1, ϖ_2 and $34'$ are undone by the gyrostatic connexions $13'', 42'', 14''$ and $23''$.

345^xiv. Going back to (29) we see that for the particular solution $\psi_1 = a_1\epsilon^{\lambda t}$, $\psi_2 = a_2\epsilon^{\lambda t}$, etc., given by the first pair of roots of (32), they become approximately

bility gyrostatically counteracted.

$$\left.\begin{array}{l} \lambda a_1 + 12\, a_2 + 13\, a_3 + 14\, a_4 = 0 \\ \lambda a_2 + 21\, a_1 + 23\, a_3 + 24\, a_4 = 0 \\ \lambda a_3 + 31\, a_1 + 32\, a_2 + 34\, a_4 = 0 \\ \lambda a_4 + 41\, a_1 + 42\, a_2 + 43\, a_3 = 0 \end{array}\right\} \ldots\ldots\ldots\ldots\ldots(42);$$

being in fact the linear algebraic equations for the solution in the form $\epsilon^{\lambda t}$ of the simple simultaneous differential equations (53) below. And if we take

Completed solution.

$$\psi_1 = \frac{b_1}{\sqrt{\varpi_1}}\epsilon^{\lambda t}, \quad \psi_2 = \frac{b_2}{\sqrt{\varpi_2}}\epsilon^{\lambda t}, \text{ etc. } \ldots\ldots\ldots\ldots(43),$$

for either particular approximate solution of (29) corresponding to (37), we find from (29) approximately

$$\left.\begin{array}{l} \lambda^{-1} b_1 + 12' b_2 + 13' b_3 + 14' b_4 = 0 \\ \lambda^{-1} b_2 + 21' b_1 + 23' b_3 + 24' b_4 = 0 \\ \lambda^{-1} b_3 + 31' b_1 + 32' b_2 + 34' b_4 = 0 \\ \lambda^{-1} b_4 + 41' b_1 + 42' b_2 + 43' b_3 = 0 \end{array}\right\} \ldots\ldots\ldots\ldots\ldots(44).$$

Remark that in (42) the coefficients of the first terms are imaginary and those of all the others real. Hence the ratios a_1/a_2, a_1/a_3, etc., are imaginary. To realize the equations put

$$\lambda = n\sqrt{-1}, \quad a_1 = p_1 + q_1\sqrt{-1}, \quad a_2 = p_2 + q_2\sqrt{-1}, \text{ etc.} \ldots(45),$$

and let p_1, q_1, p_2, etc. be real; we find, as equivalent to (42),

Realization of completed solution.

$$\left.\begin{array}{l} \begin{cases} -nq_1 + 12\, p_2 + 13\, p_3 + 14\, p_4 = 0 \\ \;\; np_1 + 12\, q_2 + 13\, q_3 + 14\, q_4 = 0 \end{cases} \\ \begin{cases} -nq_2 + 21\, p_1 + 23\, p_3 + 24\, p_4 = 0 \\ \;\; np_2 + 21\, q_1 + 23\, q_3 + 24\, q_4 = 0 \end{cases} \\ \qquad \text{etc.} \qquad\qquad \text{etc.} \end{array}\right\} \ldots\ldots\ldots\ldots(46).$$

Eliminating q_1, q_2, etc. from the seconds by the firsts of these pairs, we find

$$\left.\begin{array}{llll} (n^2 + 11)p_1 + & 12\, p_2 + & 13\, p_3 + & 14\, p_4 = 0 \\ 21\, p_1 + (n^2 + 22)p_2 + & & 23\, p_3 + & 24\, p_4 = 0 \\ 31\, p_1 + & 32\, p_2 + (n^2 + 33)\, p_3 + & & 34\, p_4 = 0 \\ 41\, p_1 + & 42\, p_2 + & 43\, p_3 + (n^2 + 44)p_4 = 0 & \end{array}\right\} \; (47);$$

and by eliminating p_1, p_2, etc. similarly we find similar equations

Realization of completed solution.

for the q's; with the same coefficients 11, 12, etc., given by the following formulas:—

$$\left.\begin{array}{r} 11 = 12\ 21 + 13\ 31 + 14\ 41 \\ 12 = 13\ 32 + 14\ 42 \\ 13 = 12\ 23 + 14\ 43 \\ 21 = 23\ 31 + 24\ 41 \\ \text{etc.} \qquad \text{etc.} \end{array}\right\} \quad \ldots\ldots\ldots\ldots (48).$$

Remember now that

$$12 = -21,\ 13 = -31,\ 32 = -23, \text{ etc.} \ldots\ldots\ldots\ldots (49),$$

and we see in (48) that

$$12 = 21,\ 13 = 31,\ 23 = 31, \text{ etc.} \ldots\ldots\ldots\ldots (50);$$

and farther, that 11, 12, etc. are the negatives of the coefficients of $\frac{1}{2} a_1^2$, $a_1 a_2$, etc. in the quadratic

$$\tfrac{1}{2}\{(12\, a_2 + 13\, a_3 + 14\, a_4)^2 + (21\, a_1 + 23\, a_3 + 24\, a_4)^2 + \text{etc.}\} \ldots (51)$$

Resultant motion reduced to motion of a conservative system with four fundamental periods equal two and two.

expanded. Hence if $G(aa)$ denote this quadratic, and $G(pp)$, $G(qq)$ the same of the p's and the q's, we may write (47) and the corresponding equations for the q's as follows:

$$\left.\begin{array}{l} -n^2 p_1 + \dfrac{dG(pp)}{dp_1} = 0, \quad -n^2 p_2 + \dfrac{dG(pp)}{dp_2} = 0, \text{ etc.} \\ -n^2 q_1 + \dfrac{dG(qq)}{dq_1} = 0, \quad -n^2 q_2 + \dfrac{dG(qq)}{dq_2} = 0, \text{ etc.} \end{array}\right\} \quad (52).$$

These equations are harmonized by, and as is easily seen, only by, assigning to n^2 one or other of the two values of $-\lambda^2$ given in (32), above. Hence their determinantal equation, a biquadratic in n^2, has two pairs of equal real positive roots. We readily verify this by verifying that the square of the determinant of (42), with λ^2 replaced by $-n^2$, is equal to the determinant of (47) with 11, 12, etc. replaced by their values (48). Hence (§ 343g) there is for each root an indeterminacy in the ratios p_1/p_2, p_1/p_3, p_1/p_4, according to which one of them may be assumed arbitrarily and the two others then determined by two of the equations (47); so that with two of the p's assumed arbitrarily the four are known: then the corresponding set of four q's is determined explicitly by the firsts of the pairs (46). Similarly the other root, n'^2, of the determinantal equation gives another solution with two fresh arbitraries. Thus we have the complete solution of the four equations

Algebraic theorem.

Details of realized solution.

$$\left.\begin{array}{l}\frac{d\psi_1}{dt} + 12\,\psi_2 + 13\,\psi_3 + 14\,\psi_4 = 0 \\ \frac{d\psi_2}{dt} + 21\,\psi_1 + 23\,\psi_3 + 24\,\psi_4 = 0 \\ \text{etc.} \qquad\qquad \text{etc.}\end{array}\right\} \dots\dots\dots\dots (53),$$

Details of realized solution.

with its four arbitraries. The formulas (46)...(52) are clearly the same as we should have found if we had commenced with assuming

$$\psi_1 = p_1 \sin nt + q_1 \cos nt, \quad \psi_2 = p_2 \sin nt + q_2 \cos nt, \text{ etc.} \dots (54),$$

as a particular solution of (53).

345[xv]. Important properties of the solution of (53) are found thus:—

Orthogonalities proved between two components of one fundamental oscillation:

(*a*) Multiply the firsts of (46) by p_1, p_2, p_3, p_4 and add: or the seconds by q_1, q_2, q_3, q_4 and add: either way we find

$$p_1q_1 + p_2q_2 + p_3q_3 + p_4q_4 = 0 \dots\dots\dots\dots\dots\dots (55).$$

(*b*) Multiply the firsts of (46) by q_1, q_2, q_3, q_4 and add: multiply the seconds by p_1, p_2, p_3, p_4 and add: and compare the results: we find

$$n\Sigma p^2 = n\Sigma q^2 = \Sigma 12\,(p_2q_1 - p_1q_2) \dots\dots\dots\dots (56),$$

and equality of their energies.

where Σ of the last member denotes a sum of such double terms as the sample without repetition of their equals, such as $21\,(p_1q_2 - p_2q_1)$.

(*c*) Let n^2, n'^2 denote the two values of $-\lambda^2$ given in (32), and let (54) and

Orthogonalities proved between different fundamental oscillations.

$$\psi_1 = p'_1 \sin n't + q'_1 \cos n't, \quad \psi_2 = p'_2 \sin n't + q'_2 \cos n't, \text{ etc.} \dots (57)$$

be the two corresponding solutions of (53). Imagine (46) to be written out for n'^2 and call them (46′): multiply the firsts of (46) by p'_1, p'_2, p'_3, p'_4 and add: multiply the firsts of (46′) by p_1, p_2, p_3, p_4 and add. Proceed correspondingly with the seconds. Proceed similarly with multipliers q for the firsts and p for the seconds. By comparisons of the sums we find that when n' is not equal to n we must have

$$\left.\begin{array}{ll}\Sigma p'q = 0, & \Sigma 12\,(p'_1p_2 - p'_2p_1) = 0 \\ \Sigma q'p = 0, & \Sigma 12\,(q'_1q_2 - q'_2q_1) = 0 \\ \left.\begin{array}{l}\Sigma q'q = 0 \\ \Sigma p'p = 0\end{array}\right\} & \Sigma 12\,(q'_1p_2 - q'_2p_1) = 0, \quad \Sigma 12\,(p'_1q_2 - p'_2q_1) = 0\end{array}\right\} (58).$$

Case of equal periods.

345[xvi]. The case of $n=n'$ is interesting. The equations $\Sigma q'q=0$, $\Sigma p'p=0$, $\Sigma p'q=0$, $\Sigma q'p=0$, when n differs however little from n', show (as we saw in a corresponding case in § 343 *m*) that equality of n to n' does *not* bring into the solution terms of the form $Ct\cos nt$, and it must therefore come under § 343 *e*. The condition to be fulfilled for the equality of the roots is seen from (32) and (33) to be

$$12=34,\ 13=42,\ \text{and}\ 14=23 \quad\ldots\ldots\ldots\ldots\ldots(59):$$

and to give

$$n^2=12^2+13^2+14^2 \quad\ldots\ldots\ldots\ldots\ldots\ldots\ldots\ldots(60)$$

for the common value of the roots. It is easy to verify that these relations reduce to zero each of the first minors of (42), as they must according to Routh's theorem (§ 343 *e*), because each root, λ, of (42) is a double root. According to the same theorem all the first, second and third minors of (47) must vanish for each root, n^2, of (47) is a quadruple root: for this, as there are just four equations, it is necessary and sufficient that

$$11=22=33=44\ \text{and}\ 12=0,\ 13=0,\ 14=0,\ 23=0,\ \text{etc.}\ldots(60'),$$

which we see at once by (48) is the case when (59) are fulfilled. In fact, these relations immediately reduce (51) to

$$G(aa)=\tfrac{1}{2}(12^2+13^2+14^2)(a_1^2+a_2^2+a_3^2+a_4^2)\ldots\ldots\ldots(61).$$

In this case one particular solution is readily seen from (52) and (46) to be

$$\left.\begin{array}{llll} p_1=1, & p_2=0, & p_3=0, & p_4=0 \\ q_1=0, & q_2=-\dfrac{12}{n}, & q_3=-\dfrac{13}{n}, & q_4=-\dfrac{14}{n} \\ \psi_1=\sin nt, & \psi_2=-\dfrac{12}{n}\cos nt, & \psi_3=-\dfrac{13}{n}\cos nt, & \psi_4=-\dfrac{14}{n}\cos nt \end{array}\right\}\ (62).$$

Completed solution for case of equal periods.

Hence the general solution, with four arbitraries p_1, p_2, p_3, p_4, is

$$\left.\begin{array}{l} \psi_1=p_1\sin nt+\dfrac{1}{n}(12p_2+13p_3+14p_4)\cos nt \\ \psi_2=p_2\sin nt+\dfrac{1}{n}(-12p_1+14p_3-13p_4)\cos nt \\ \psi_3=p_3\sin nt+\dfrac{1}{n}(-13p_1-14p_2+12p_4)\cos nt \\ \psi_4=p_4\sin nt+\dfrac{1}{n}(-14p_1+13p_2-12p_3)\cos nt \end{array}\right\}\ldots\ldots(63).$$

It is easy to verify that this satisfies the four differential equations (53).

345^xvii. Quite as we have dealt with (42), (45), (53), (54) in § 345^xiv, we may deal with (44) and the simple simultaneous equations for the solution of which they serve, which are

Two higher, and two lower, of the four fundamental oscillations,

$$\left.\begin{array}{l} 12\dfrac{d\psi_2}{dt} + 13\dfrac{d\psi_3}{dt} + 14\dfrac{d\psi_4}{dt} + \varpi_1\psi_1 = 0, \\ 21\dfrac{d\psi_1}{dt} + 23\dfrac{d\psi_3}{dt} + 24\dfrac{d\psi_4}{dt} + \varpi_2\psi_2 = 0, \\ \qquad \text{etc.} \qquad \text{etc.} \end{array}\right\} \text{............(64)};$$

and all the formulas which we meet in so doing are real when $\varpi_1, \varpi_2, \varpi_3, \varpi_4$ are all of one sign, and therefore $12'$, $13'$, etc., all real. In the case of some of the ϖ's negative and some positive there is no difficulty in realizing the formulas, but the consideration of the simultaneous reduction of the two quadratics,

similarly dealt with by solution of two similar quadratics,

$$\left.\begin{array}{l} \tfrac{1}{2}\left\{\dfrac{(12\,a_2 + 13\,a_3 + 14\,a_4)^2}{\varpi_1} + \dfrac{(21\,a_2 + 23\,a_3 + 24\,a_4)^2}{\varpi_2} + \text{etc.}\right\} \\ \text{and} \qquad \tfrac{1}{2}(\varpi_1 a_1^2 + \varpi_2 a_2^2 + \varpi_3 a_3^2 + \varpi_4 a_4^2) \end{array}\right\} (65),$$

to which we are led when we go back from the notation $12'$, etc. of (36), is not completely instructive in respect to stability, as was our previous explicit working out of the two roots of the determinantal equation in (37), (38), and (40).

345^xviii. The conditions to be fulfilled that the system may be dominated by gyrostatic influence are that the smaller value of $-\lambda^2$ found from (31) and the greater found from (34) be respectively very great in comparison with the greatest and very small in comparison with the smallest, of the four quantities $\varpi_1, \varpi_2, \varpi_3, \varpi_4$ irrespectively of their signs. Supposing ϖ_1 to be the greatest and ϖ_4 the smallest, these conditions are easily proved to be fulfilled when, and only when,

provided that gyrostatic influence be fully dominant.

$$\frac{(12\,.\,34 + 13\,.\,42 + 14\,.\,23)^2}{12^2 + 13^2 + 14^2 + 34^2 + 42^2 + 23^2} >> \pm\varpi_1 \text{.........(66)},$$

and

$$\frac{(12\,.\,34 + 13\,.\,42 + 14\,.\,23)^2}{12^2\varpi_3\varpi_4 + 13^2\varpi_4\varpi_2 + 14^2\varpi_2\varpi_3 + 34^2\varpi_1\varpi_2 + 42^2\varpi_1\varpi_3 + 23^2\varpi_1\varpi_4} >> \pm\varpi_4^{-1} \text{ (67)},$$

where >> denotes "*very great in comparison with.*" When these conditions are fulfilled, let 12, 13, 23, etc., be each increased in the ratio of N to 1. The two greater values of n (or $\lambda\sqrt{-1}$) will be increased in the same ratio, N to 1; and the two smaller will be diminished each in the inverse ratio, 1 to N. Again, let $\sqrt{\pm\varpi_1}$, $\sqrt{\pm\varpi_2}$, $\sqrt{\pm\varpi_3}$, $\sqrt{\pm\varpi_4}$ be each diminished in the ratio M to 1; the two larger values of n will be sensibly unaltered; and the two smaller will be diminished in the ratio M^2 to 1.

Limits of smallest and second smallest of the four periods.

345^xix. Remark that

(*a*) When (66) is satisfied the two greater values of n are each

$$\left.\begin{array}{l} <\sqrt{\{(12^2+13^2+14^2+34^2+42^2+23^2)\}} \\ \text{and} \quad > \dfrac{12\,.\,34+13\,.\,42+14\,.\,23}{\sqrt{(12^2+13^2+14^2+34^2+42^2+23^2)}} \end{array}\right\}\text{........(68);}$$

and that when they are very unequal the greater is approximately equal to the former limit and the less to the latter.

(*b*) When (67) is satisfied, and when the equilibrium is stable, the two smaller values of n are each

$$\left.\begin{array}{l} < \dfrac{\sqrt{(12^2\varpi_3\varpi_4+13^2\varpi_4\varpi_2+14^2\varpi_2\varpi_3+34^2\varpi_1\varpi_2+42^2\varpi_1\varpi_3+23^2\varpi_1\varpi_4)}}{12\,.\,34+13\,.\,42+14\,.\,23} \\ \text{and} \\ > \dfrac{\sqrt{(\varpi_1\varpi_2\varpi_3\varpi_4)}}{\sqrt{\{(12^2\varpi_3\varpi_4+13^2\varpi_4\varpi_2+14^2\varpi_2\varpi_3+34^2\varpi_1\varpi_2+42^2\varpi_1\varpi_3+23^2\varpi_1\varpi_4)\}}} \end{array}\right\}(69),$$

Limits of the next greatest and greatest of the four periods.

and that when they are very unequal the greater of the two is approximately equal to the former limit, and the less to the latter.

345^xx. Both (66) and (67) must be satisfied in order that the four periods may be found approximately by the solution of the two quadratics (31), (34). If (66) is satisfied but not (67), the biquadratic determinant still splits into two quadratics, of which one is approximately (31) but the other is not approximately (34). Similarly, if (67) is satisfied but not (66), the biquadratic splits into two quadratics of which one is approximately (34) but the other not approximately (31).

Quadruply free cycloidal system with non-dominant gyrostatic influences.

345^xxi. When neither (66) nor (67) is fulfilled there is not generally any splitting of the biquadratic into two rational quadratics; and the conditions of stability, the determination of the fundamental periods, and the working out of the complete solution depend essentially on the roots of a biquadratic equation. When ϖ_1, ϖ_2, ϖ_3, ϖ_4 are all positive it is clear from the equation of energy [345ii, (4), with $Q=0$] that the motion is stable whatever be the values of the gyrostatic coefficients 12, 34, 13, etc. and therefore in this case each of the four roots λ^2 of the biquadratic is real and negative, a proposition included in the general theorem of § 345xxvi below. To illustrate the interesting questions which occur when the ϖ's are not all positive put

Quadruply free cycloidal system with non-dominant gyrostatic influences.

$$12 = {}_{12}g,\ 34 = {}_{34}g,\ 13 = {}_{13}g, \text{ etc.} \dots\dots\dots\dots\dots\dots(70),$$

where ${}_{12}$, ${}_{34}$, ${}_{13}$, etc. denote any numerics whatever subject only to the condition that they do not make zero of

$${}_{12} \cdot {}_{34} + {}_{13} \cdot {}_{42} + {}_{14} \cdot {}_{23}.$$

When ϖ_1, ϖ_2, ϖ_3, ϖ_4, are all negative each root λ^2 of the biquadratic is as we have seen in § 345xiii real and negative when the gyrostatic influences dominate. It becomes an interesting question to be answered by treatment of the biquadratic, how small may g be to keep all the roots λ^2 real and negative, and how large may g be to render them other than real and positive as they are when $g=0$? Similar questions occur in connexion with the case of two of the ϖ's negative and two positive, when the gyrostatic influences are so proportioned as to fulfil 345xiii (41), so that when g is infinitely great there is complete gyrostatic stability, though when $g=0$ there are two instabilities and two stabilities.

345xxii. Returning now to 345x and 345vii, 345viii, and 345ix, for a gyrostatic system with any number of freedoms, we see by 345vi that the roots λ^2 of the determinantal equation (14) or (17) are necessarily real and negative when ϖ_1, ϖ_2, ϖ_3, ϖ_4, etc. are all positive. This conclusion is founded on the reasoning of § 345ii regarding the equation of energy (4) applied to the case $Q=0$, for which it becomes $T+V=E_0$, or the same as for the case of no motional forces. It is easy of course to eliminate dynamical considerations from the reasoning and to give a purely algebraic proof that the roots λ^2 of the determinantal equation (14) of 345viii are necessarily real and negative, provided both of the two quadratic functions $(11)\,a_1^2 + 2\,(12)\,a_1a_2 + \text{etc.}$, and $11a_1^2 + 2\;12a_1a_2 + \text{etc.}$ are positive for all real values of a_1, a_2, etc. But the equations (14) of § 343 (*k*), which we obtained and used in the course of the corresponding demonstration for the case of no motional forces, do not hold in our present case of gyrostatic motional forces. Still for this present case we have the con-

Gyrostatic system with any number of freedoms.

Case of equal roots with stability.

clusion of § 343 (m) that equality among the roots falls essentially under the case of § 343 (e) above. For we know from the consideration of energy, as in § 345[ii], that no particular solution can be of the form $t\epsilon^{\lambda t}$ or $t \sin \sigma t$, when the potential energy is positive for all displacements: yet [though there cannot be equal roots for the gyrostatic system of two freedoms (§ 345[x]) as we see from the solution (25) of the determinantal equation for this case] there obviously may be equality of roots* in a quadruply free gyrostatic system, or in one with more than four freedoms. Hence, if both the quadratic functions have the same sign for all real values of a_1, a_2, etc., all the first minors

Application of Routh's theorem.

* Examples of this may be invented *ad libitum* by commencing with pairs of equations such as (23) and altering the variables by (generalized) orthogonal transformations. For one very simple example put $\zeta = \varpi$ and take (23) as one pair of equations of motion, and as a second pair take

$$\ddot{\xi}' + \gamma\dot{\eta}' + \varpi\xi' = 0,$$
$$\ddot{\eta}' - \gamma\dot{\xi}' + \varpi\eta' = 0.$$

The second of (23) and the first of these multiplied respectively by $\cos a$ and $\sin a$, and again by $\sin a$ and $\cos a$, and added and subtracted, give

$$\ddot{\psi}_2 - \gamma \cos a\dot{\xi} + \gamma \sin a\dot{\eta}' + \varpi\psi_2 = 0,$$

and

$$\ddot{\psi}_3 + \gamma \sin a\dot{\xi} + \gamma \cos a\dot{\eta}' + \varpi\psi_3 = 0,$$

where

$$\psi_2 = \xi' \sin a + \eta \cos a,$$

and

$$\psi_3 = \xi' \cos a - \eta \sin a.$$

Eliminating ξ' and η by these last equations, from the first and fourth of the equations of motion, and for symmetry putting ψ_1 instead of ξ, and ψ_4 instead of η', and for simplicity putting $\gamma \cos a = g$, and $\gamma \sin a = h$, and collecting the equations of motion in order, we have the following,—

$$\ddot{\psi}_1 + g\dot{\psi}_2 - h\dot{\psi}_3 + \varpi\psi_1 = 0,$$
$$\ddot{\psi}_2 - g\dot{\psi}_1 + h\dot{\psi}_4 + \varpi\psi_2 = 0,$$
$$\ddot{\psi}_3 + h\dot{\psi}_1 + g\dot{\psi}_4 + \varpi\psi_3 = 0,$$
$$\ddot{\psi}_4 - h\dot{\psi}_2 - g\dot{\psi}_3 + \varpi\psi_4 = 0,$$

for the equations of motion of a quadruply free gyrostatic system having two equalities among its four fundamental periods. The two different periods are the two values of the expression

$$2\pi/\{\sqrt{(\tfrac{1}{4}g^2 + \tfrac{1}{4}h^2)} \pm \sqrt{(\tfrac{1}{4}g^2 + \tfrac{1}{4}h^2 + \varpi)}\}.$$

When these two values are unequal the equalities among the roots *do not* give rise to terms of the form $t\epsilon^{\lambda t}$ or $t \cos \sigma t$ in the solution. But if $\varpi = -(\tfrac{1}{4}g^2 + \tfrac{1}{4}h^2)$, which makes these two values equal, and therefore all four roots equal, terms of the form $t \cos \sigma t$ *do appear* in the solution, and the equilibrium is unstable in the transitional case though it is stable if $-\varpi$ be less than $\tfrac{1}{4}g^2 + \tfrac{1}{4}h^2$ by ever so small a difference.

of the determinantal equation (14), § 345^viii, must vanish for each double, triple, or multiple root of the equation, if it has any such roots.

Application of Routh's theorem.

It will be interesting to find a purely algebraic proof of this theorem, and we leave it as an exercise to the student; remarking only that, when the quadratic functions have contrary signs for some real values of a_1, a_2, etc., there may be equality among the roots without the evanescence of all the first minors; or, in dynamical language, there may be terms of the form $te^{\lambda t}$, or $t \sin \sigma t$, in the solution expressing the motion of a gyrostatic system, in transitional cases between stability and instability. It is easy to invent examples of such cases, taking for instance the quadruply free gyrostatic system, whether gyrostatically dominated as in § 345^xiii, but in this case with some of the four quantities negative, and some positive; or, as in § 345^xxi, not gyrostatically dominated, with either some or all of the quantities $\varpi_1, \varpi_2, \ldots, \varpi_i$ negative. All this we recommend to the student as interesting and instructive exercise.

Equal roots with instability in transitional cases between stability and instability.

345^xxiii. When all the quantities $\varpi_1, \varpi_2, \ldots, \varpi_i$ are of the same sign it is easy to find the conditions that must be fulfilled in order that the system may be gyrostatically dominated. For if $\rho_1, \rho_2, \ldots, \rho_n$ are the roots of the equation

Conditions of gyrostatic domination.

$$c_0 z^n + c_1 z^{n-1} + \ldots + c_{n-1} z + c_n = 0,$$

we have

$$-(\rho_1 + \rho_2 + \ldots + \rho_n) = \frac{c_1}{c_0}, \text{ and } -\left(\frac{1}{\rho_1} + \frac{1}{\rho_2} + \ldots + \frac{1}{\rho_n}\right) = \frac{c_{n-1}}{c_n}.$$

Hence if $-\rho_1, -\rho_2, \ldots -\rho_n$ be each positive, c_1/nc_0 is their arithmetic mean, and nc_n/c_{n-1} is their harmonic mean. Hence c_1/nc_0 is greater than nc_n/c_{n-1}, and the greatest of $-\rho_1, -\rho_2, \ldots, -\rho_n$ is greater than c_1/nc_0, and the least of them is less than nc_n/c_{n-1}. Take now the two following equations:

$$\lambda^i + \lambda^{i-2} \Sigma\, 12^2 + \lambda^{i-4} \Sigma (\Sigma 12 \,.\, 34)^2 + \lambda^{i-6} \Sigma (\Sigma 12 \,.\, 34 \,.\, 56)^2 + \text{etc.} = 0 \quad \ldots\ldots\ldots (71),$$

$$\left(\frac{1}{\lambda}\right)^i + \left(\frac{1}{\lambda}\right)^{i-2} \Sigma\, 12'^2 + \left(\frac{1}{\lambda}\right)^{i-4} \Sigma (\Sigma 12' \,.\, 34')^2 + \left(\frac{1}{\lambda}\right)^{i-6} \Sigma (\Sigma 12' \,.\, 34' \,.\, 56')^2 + \text{etc.} = 0 \quad (72),$$

where

$$12' = \frac{12}{\sqrt{(\varpi_1 \varpi_2)}},\ 13' = \frac{13}{\sqrt{(\varpi_1 \varpi_3)}},\ 34' = \frac{34}{\sqrt{(\varpi_3 \varpi_4)}}, \ldots,\ i-1, i' = \frac{i-1, i}{\sqrt{(\varpi_{i-1} \varpi_i)}} \quad (73).$$

Cycloidal motion. Conditions of gyrostatic domination.

Suppose for simplicity i to be even. All the roots λ^2 of (71) are (§ 345^{xxvi} below) essentially real and negative. So are those of (72) provided $\varpi_1, \varpi_2, \ldots, \varpi_i$ are all of one sign as we now suppose them to be. Hence the smallest root $-\lambda^2$ of (71) is less than

$$\frac{\frac{1}{2}i\Sigma(12 \cdot 34 \cdot 56, \ldots, i-1, i)^2}{\Sigma(\Sigma 12 \cdot 34 \cdot 56, \ldots, i-3, i-2)^2} \quad \ldots\ldots\ldots\ldots\ldots\ldots(74),$$

and the greatest root $-\lambda^2$ of (72) is greater than

$$\frac{\Sigma(\Sigma 12' \cdot 34' \cdot 56', \ldots, i-3, i-2')^2}{\frac{1}{2}i\Sigma(12' \cdot 34' \cdot 56', \ldots, i-1, i')} \quad \ldots\ldots\ldots\ldots(75).$$

Hence the conditions for gyrostatic domination are that (74) must be much greater than the greatest of the positive quantities $\pm\varpi_1$, $\pm\varpi_2, \ldots, \pm\varpi_i$, and that (75) must be very much less than the least of these positive quantities. When these conditions are fulfilled the i roots of (18) § 345^{ix} equated to zero are separable into two groups of $\frac{1}{2}i$ roots which are infinitely nearly equal to the roots of equations (71) and (72) respectively, conditions of reality of which are investigated in § 345^{xxvi} below. The interpretation leads to the following interesting conclusions:—

Gyrostatic links explained.

345^{xxiv}. Consider a cycloidal system provided with non-rotating flywheels mounted on frames so connected with the moving parts as to give infinitesimal angular motions to the axes of the flywheels proportional to the motions of the system. Let the number of freedoms of the system exclusive of the ignored co-ordinates [§ 319, Ex. (G)] of the flywheels relatively to their frames be even. Let the forces of the system be such that when the flywheels are given at rest, when the system is at rest, the equilibrium is either stable for all the freedoms, or unstable for all the freedoms. Let the number and connexions of the gyrostatic links be such as to permit gyrostatic domination (§ 345^{xxi}) when each of the flywheels is set into sufficiently rapid rotation. Now let the flywheels be set each into sufficiently rapid rotation to fulfil the conditions of gyrostatic domination (§ 345^{xxi}): the equilibrium of the system becomes stable: with half the whole number i of its modes of vibration exceedingly rapid, with frequencies equal to the roots of a certain algebraic equation of the degree $\frac{1}{2}i$; and the other half of

Gyrostatically dominated system:

its modes of vibration very slow, with frequencies given by the roots of another algebraic equation of degree $\frac{1}{2}i$. The first class of fundamental modes may be called adynamic because they are the same as if no forces were applied to the system, or acted between its moving parts, except actions and reactions in the normals between mutually pressing parts (depending on the inertias of the moving parts). The second class of fundamental modes may be called precessional because the precession of the equinoxes, and the slow precession of a rapidly spinning top supported on a very fine point, are familiar instances of it. Remark however that the obliquity of the ecliptic should be infinitely small to bring the precession of the equinoxes precisely within the scope of the equations of our "cycloidal system."

Gyrostatically dominated system: its adynamic oscillations (very rapid); and precessional oscillations (very slow).

345xxv. If the angular velocities of all the flywheels be altered in the same proportion the frequencies of the adynamic oscillations will be altered in the same proportion directly, and those of the precessional modes in the same proportion inversely. Now suppose there to be either no inertia in the system except that of the flywheels round their pivoted axes and round their equatorial diameters, or suppose the effective inertia of the connecting parts to be comparable with that of the flywheels when given without rotation. The period of each of the adynamic modes is comparable with the periods of the flywheels. And the periods of the precessional modes are comparable with a third proportional to a mean of the periods of the flywheels and a mean of the irrotational periods of the system, if the system be stable when the flywheels are deprived of rotation. For the last mentioned term of the proportion we may, in the case of irrotational *instability*, substitute the time of increasing a displacement a thousandfold, supposing the system to be falling away from its configuration of equilibrium according to one of its fundamental modes of motion ($\epsilon^{\lambda t}$). The reciprocal of this time we shall call, for brevity, the rapidity of the system, for convenience of comparison with the frequency of a vibrator or of a rotator, which is the name commonly given to the reciprocal of its period.

Comparison between adynamic frequencies, rotational frequencies of the flywheels, precessional frequencies of the system, and frequencies or rapidities of the system, with flywheels deprived of rotation.

Proof of reality of adynamic and of precessional periods when system's irrotational periods are either all real or all imaginary.

345^{xxvi}. It remains to prove that the roots λ^2 of (71), and of (72) also when $\varpi_1, \varpi_2, \ldots, \varpi_i$ are all of one sign, are essentially real and negative. (71) is the determinantal equation of § 345^{xiv} (42) with any even number of equations instead of only four. The treatment of §§ 345^{xiv} and 345^{xv} is all directly applicable without change to this extension; and it proves that the roots λ^2 are real and negative by bringing the problem to that of the orthogonal reduction of the essentially positive quadratic function

$$G(aa)=\tfrac{1}{2}\{(12a_2+13a_3+\text{etc.})^2+(21a_1+23a_3+\text{etc.})^2+(31a_1+32a_2+\text{etc.})^2+\text{etc.}\} \quad (76):$$

it proves also the equalities of energies of (56), § 345^{xv}, and the orthogonalities of (55), (58) § 345^{xv}: also the curious algebraic theorem that the determinantal roots of the quadratic function consist of $\frac{1}{2}i$ pairs of equals.

Algebraic theorem.

Inasmuch as (72) is the same as (71) with λ^{-1} put for λ and $12'$, $13'$, $23'$, etc. for 12, 13, 23, etc., all the formulas and propositions which we have proved for (71) hold correspondingly for (72) when $12'$, $13'$, $23'$, etc. are all real, as they are when $\varpi_1, \varpi_2, \ldots \varpi_i$ are all of one sign.

345^{xxvii}. Going back now to § 345^{viii}, and taking advantage of what we have learned in § 345^{ix} and the consequent treatment of the problem, particularly that in § 345^{xiv}, we see now how to simplify equations (14) of § 345^{viii} otherwise than was done in § 345^{ix}, by a new method which has the advantage of being applicable also to materially simplify the general equations (13) of § 345^{vi}. Apply orthogonal transformation of the co-ordinates to reduce to a sum of squares of simple co-ordinates, the quadratic function (76). Thus denoting by $G(\psi\psi)$ what $G(aa)$ becomes when ψ_1, ψ_2, etc. are substituted for a_1, a_2, etc.; and denoting by $n_1^2, n_2^2, \ldots, n_{\frac{1}{2}i}^2$ the values of the pairs of roots of the determinantal equation of degree i, which are simply the negative of the roots λ^2 of equation (71) of degree $\frac{1}{2}i$ in λ^2; and denoting by $\xi_1, \eta_1, \xi_2, \eta_2, \ldots \xi_{\frac{1}{2}i}\eta_{\frac{1}{2}i}$, the fresh co-ordinates, we have

$$G(\psi\psi)=\tfrac{1}{2}\{n_1^2(\xi_1^2+\eta_1^2)+n_2^2(\xi_2^2+\eta_2^2)+\ldots+n_{\frac{1}{2}i}^2(\xi_{\frac{1}{2}i}^2+\eta_{\frac{1}{2}i}^2)\}\ldots(77).$$

It is easy to see that the general equations of cycloidal motion (13) of § 345^{vi} transformed to the ξ-co-ordinates come out in $\frac{1}{2}i$ pairs as follows:

Cycloidal motion.

$$\left.\begin{array}{l}\left\{\begin{array}{l}\frac{d}{dt}\frac{dT}{d\dot{\xi}_1}+\frac{dQ}{d\dot{\xi}_1}+n_1\dot{\eta}_1+\frac{dV}{d\xi_1}=0\\ \frac{d}{dt}\frac{dT}{d\dot{\eta}_1}+\frac{dQ}{d\dot{\eta}_1}-n_1\dot{\xi}_1+\frac{dV}{d\eta_1}=0\end{array}\right.\\ \left\{\begin{array}{l}\frac{d}{dt}\frac{dT}{d\dot{\xi}_2}+\frac{dQ}{d\dot{\xi}_2}+n_2\dot{\eta}_2+\frac{dV}{d\xi_2}=0\\ \frac{d}{dt}\frac{dT}{d\dot{\eta}_2}+\frac{dQ}{d\dot{\eta}_2}-n_2\dot{\xi}_2+\frac{dV}{d\eta_2}=0\end{array}\right.\\ \cdots\cdots\cdots\cdots\cdots\cdots\\ \left\{\begin{array}{l}\frac{d}{dt}\frac{dT}{d\dot{\xi}_{\frac{1}{2}i}}+\frac{dQ}{d\dot{\xi}_{\frac{1}{2}i}}+n_{\frac{1}{2}i}\dot{\eta}_{\frac{1}{2}i}+\frac{dV}{d\xi_{\frac{1}{2}i}}=0\\ \frac{d}{dt}\frac{dT}{d\dot{\eta}_{\frac{1}{2}i}}+\frac{dQ}{d\dot{\eta}_{\frac{1}{2}i}}-n_{\frac{1}{2}i}\dot{\xi}_{\frac{1}{2}i}+\frac{dV}{d\eta_{\frac{1}{2}i}}=0\end{array}\right.\end{array}\right\}\quad\cdots\cdots\cdots\cdots(78).$$

345xxviii. Considerations of space and time prevent us from detailed treatment at present of gyrostatic systems with odd numbers of degrees of freedom, but it is obvious from § 345xxvii and 345xi that the general equations (13) of § 345vi may, when i the number of freedoms is odd, by proper transformation from co-ordinates ψ_1, ψ_2, etc. to a set of co-ordinates ζ, ξ_1, η_2, ... $\xi_{\frac{1}{2}(i-1)}$, $\eta_{\frac{1}{2}(i-i)}$ be reduced to the following form:

$$\left.\begin{array}{l}\left\{\begin{array}{l}\frac{d}{dt}\frac{dT}{d\dot{\xi}_1}+\frac{dQ}{d\dot{\xi}_1}+n_1\dot{\eta}_1+\frac{dV}{d\xi_1}=0\\ \frac{d}{dt}\frac{dT}{d\dot{\eta}_1}+\frac{dQ}{d\dot{\eta}_1}-n_1\dot{\xi}_1+\frac{dV}{d\eta_1}=0\end{array}\right.\\ \left\{\begin{array}{l}\frac{d}{dt}\frac{dT}{d\dot{\xi}_2}+\frac{dQ}{d\dot{\xi}_2}+n_2\dot{\eta}_2+\frac{dV}{d\xi_2}=0\\ \frac{d}{dt}\frac{dT}{d\dot{\eta}_2}+\frac{dQ}{d\dot{\eta}_2}-n_2\dot{\xi}_2+\frac{dV}{d\eta_2}=0\end{array}\right.\\ \cdots\cdots\cdots\cdots\cdots\cdots\\ \left\{\begin{array}{l}\frac{d}{dt}\frac{dT}{d\dot{\xi}_{\frac{1}{2}(i-1)}}+\frac{dQ}{d\dot{\xi}_{\frac{1}{2}(i-1)}}+n_{\frac{1}{2}(i-1)}\dot{\eta}_{\frac{1}{2}(i-1)}+\frac{dV}{d\xi_{\frac{1}{2}(i-1)}}=0\\ \frac{d}{dt}\frac{dT}{d\dot{\eta}_{\frac{1}{2}(i-1)}}+\frac{dQ}{d\dot{\eta}_{\frac{1}{2}(i-1)}}-n_{\frac{1}{2}(i-1)}\dot{\xi}_{\frac{1}{2}(i-1)}+\frac{dV}{d\eta_{\frac{1}{2}(i-1)}}=0\end{array}\right.\\ \frac{d}{dt}\frac{dT}{d\dot{\zeta}}+\frac{dQ}{d\dot{\zeta}}+\frac{dV}{d\zeta}=0\end{array}\right\}\quad\cdots\cdots(79).$$

Kinetic stability.

346. There is scarcely any question in dynamics more important for Natural Philosophy than the stability or instability of motion. We therefore, before concluding this chapter, propose to give some general explanations and leading principles regarding it.

Conservative disturbance.

A "conservative disturbance of motion" is a disturbance in the motion or configuration of a conservative system, not altering the sum of the potential and kinetic energies. A conservative disturbance of the motion through any particular configuration is a change in velocities, or component velocities, not altering the whole kinetic energy. Thus, for example, a conservative disturbance of the motion of a particle through any point, is a change in the direction of its motion, unaccompanied by change of speed.

Kinetic stability and instability discriminated.

347. The actual motion of a system, from any particular configuration, is said to be *stable* if every possible infinitely small conservative disturbance of its motion through that configuration may be compounded of conservative disturbances, any one of which would give rise to an alteration of motion which would bring the system again to some configuration belonging to the undisturbed path, in a finite time, and without more than an infinitely small digression. If this condition is not fulfilled, the motion is said to be *unstable*.

Examples.

348. For example, if a body, A, be supported on a fixed vertical axis; if a second, B, be supported on a parallel axis belonging to the first; a third, C, similarly supported on B, and so on; and if B, C, etc., be so placed as to have each its centre of inertia as far as possible from the fixed axis, and the whole set in motion with a common angular velocity about this axis, the motion will be stable, from every configuration, as is evident from the principles regarding the resultant centrifugal force on a rigid body, to be proved later. If, for instance, each of the bodies is a flat rectangular board hinged on one edge, it is obvious that the whole system will be kept stable by centrifugal force, when all are in one plane and as far out from the axis as possible. But if A consist partly of a shaft and crank, as a common spinning-wheel, or the fly-wheel and crank of a steam-engine, and if B be supported on the crank-pin as axis, and turned inwards (towards the fixed axis, or across the fixed axis), then, even although the centres of inertia of C, D, etc., are placed as far from the fixed axis as possible, consistent with this position of B, the motion of the system will be unstable.

Examples.

349. The rectilinear motion of an elongated body lengthwise, or of a flat disc edgewise, through a fluid is unstable. But the motion of either body, with its length or its broadside perpendicular to the direction of motion, is stable. This is demonstrated for the ideal case of a perfect liquid (§ 320), in § 321, Example (2); and the results explained in § 322 show, for a solid of revolution, the precise character of the motion consequent upon an infinitely small disturbance in the direction of the motion from being exactly along or exactly perpendicular to the axis of figure; whether the infinitely small oscillation, in a definite period of time, when the rectilineal motion is stable, or the swing round to an infinitely nearly inverted position when the rectilineal motion is unstable. Observation proves the assertion we have just made, for real fluids, air and water, and for a great variety of circumstances affecting the motion. Several illustrations have been referred to in § 325; and it is probable we shall return to the subject later, as being not only of great practical importance, but profoundly interesting although very difficult in theory.

Kinetic stability. Hydrodynamic example.

350. The motion of a single particle affords simpler and not less instructive illustrations of stability and instability. Thus if a weight, hung from a fixed point by a light inextensible cord, be set in motion so as to describe a circle about a vertical line through its position of equilibrium, its motion is stable. For, as we shall see later, if disturbed infinitely little in direction without gain or loss of energy, it will describe a sinuous path, cutting the undisturbed circle at points successively distant from one another by definite fractions of the circumference, depending upon the angle of inclination of the string to the vertical. When this angle is very small, the motion is sensibly the same as that of a particle confined to one plane and moving under the influence of an attractive

Circular simple pendulum.

force towards a fixed point, simply proportional to the distance; and the disturbed path cuts the undisturbed circle four times in a revolution. Or if a particle confined to one plane, move under the influence of a centre in this plane, attracting with a force inversely as the square of the distance, a path infinitely little disturbed from a circle will cut the circle twice in a revolution. Or if the law of central force be the nth power of the distance, and if $n+3$ be positive, the disturbed path will cut the undisturbed circular orbit at successive angular intervals, each equal to $\pi/\sqrt{n+3}$. But the motion will be unstable if n be negative, and $-n>3$.

Circular orbit.

Kinetic stability in circular orbit.

The criterion of stability is easily investigated for circular motion round a centre of force from the differential equation of the general orbit (§ 36),

$$\frac{d^2u}{d\theta^2}+u=\frac{P}{h^2u^2}.$$

Let the value of h be such that motion in a circle of radius a^{-1} satisfies this equation. That is to say, let $P/h^2u^2=u$, when $u=a$. Let now $u=a+\rho$, ρ being infinitely small. We shall have

$$u-\frac{P}{h^2u^2}=\alpha\rho,$$

if α denotes the value of $\frac{d}{du}\left(u-\frac{P}{h^2u^2}\right)$ when $u=a$: and therefore the differential equation for motion infinitely nearly circular is

$$\frac{d^2\rho}{d\theta^2}+\alpha\rho=0.$$

The integral of this is most conveniently written

$$\rho=A\sin(\theta\sqrt{\alpha}+\beta)$$

when α is positive, and

$$\rho=C\epsilon^{\theta\sqrt{-\alpha}}+C'\epsilon^{-\theta\sqrt{-\alpha}}$$

when α is negative.

Hence we see that the circular motion is stable in the former case, and unstable in the latter.

For instance, if $P = \mu r^n = \mu u^{-n}$, we have

Kinetic stability in circular orbit.

$$\frac{d}{du}\left(u - \frac{P}{h^2u^2}\right) = 1 + (n+2)\frac{P}{h^2u^3};$$

and putting $\frac{P}{h^2u^2} = u = a$, in this we find $\alpha = n + 3$; whence the result stated above.

Or, taking Example (B) of § 319, and putting mP for P, and mh for h,

$$\frac{P}{h^2u^2} = \frac{m'}{m+m'}\left(\frac{g}{h^2}u^{-2} + u\right),$$

$$\frac{d}{du}\left(u - \frac{P}{h^2u^2}\right) = \frac{m + \frac{2m'g}{h^2u^3}}{m+m'}.$$

Hence, putting $u = a$, and making $h^2 = gm'/ma^3$ so that motion in a circle of radius a^{-1} may be possible, we find

$$\alpha = \frac{3m}{m+m'}.$$

Hence the circular motion is always stable; and the period of the variation produced by an infinitely small disturbance from it is

$$2\pi\sqrt{\frac{m+m'}{3m}}.$$

351. The case of a particle moving on a smooth fixed surface under the influence of no other force than that of the constraint, and therefore always moving along a geodetic line of the surface, affords extremely simple illustrations of stability and instability. For instance, a particle placed on the inner circle of the surface of an anchor-ring, and projected in the plane of the ring, would move perpetually in that circle, but unstably, as the smallest disturbance would clearly send it away from this path, never to return until after a digression round the outer edge. (We suppose of course that the particle is held to the surface, as if it were placed in the infinitely narrow space between a solid ring and a hollow one enclosing it.) But if a particle is placed on the outermost, or greatest,

Kinetic stability of a particle moving on a smooth surface.

Kinetic stability of a particle moving on a smooth surface.

circle of the ring, and projected in its plane, an infinitely small disturbance will cause it to describe a sinuous path cutting the circle at points round it successively distant by angles each equal to $\pi\sqrt{b/a}$, or intervals of time, $\pi\sqrt{b}/\omega\sqrt{a}$, where a denotes the radius of that circle, ω the angular velocity in it, and b the radius of the circular cross section of the ring. This is proved by remarking that an infinitely narrow band from the outermost part of the ring has, at each point, a and b for its principal radii of curvature, and therefore (§ 150) has for its geodetic lines the great circles of a sphere of radius $\sqrt{ab}$, upon which (§ 152) it may be bent.

352. In all these cases the undisturbed motion has been circular or rectilineal, and, when the motion has been stable, the effect of a disturbance has been *periodic*, or recurring with the same phases in equal successive intervals of time. An illustration of thoroughly stable motion in which the effect of a disturbance is not "periodic," is presented by a particle sliding down an inclined groove under the action of gravity. To take the simplest case, we may consider a particle sliding down along the lowest straight line of an inclined hollow cylinder. If slightly disturbed from this straight line, it will oscillate on each side of it perpetually in its descent, but not with a uniform periodic motion, though the durations of its excursions to each side of the straight line are all equal.

Kinetic stability. Incommensurable oscillations.

353. A very curious case of stable motion is presented by a particle constrained to remain on the surface of an anchor-ring fixed in a vertical plane, and projected along the great circle from any point of it, with any velocity. An infinitely small disturbance will give rise to a disturbed motion of which the path will cut the vertical circle over and over again for ever, at unequal intervals of time, and unequal angles of the circle; and obviously not recurring periodically in any cycle, except with definite particular values for the whole energy, some of which are less and an infinite number are greater than that which just suffices to bring the particle to the highest point of the ring. The full mathematical investigation of these circumstances would afford an excellent exercise in the theory of differential equations, but it is not necessary for our present illustrations.

354. In this case, as in all of stable motion with only two degrees of freedom, which we have just considered, there has been stability throughout the motion; and an infinitely small disturbance from any point of the motion has given a disturbed path which intersects the undisturbed path over and over again at finite intervals of time. But, for the sake of simplicity at present confining our attention to two degrees of freedom, we have a *limited* stability in the motion of an unresisted projectile, which satisfies the criterion of stability only at points of its upward, not of its downward, path. Thus if $MOPQ$ be

Oscillatory kinetic stability.

Limited kinetic stability.

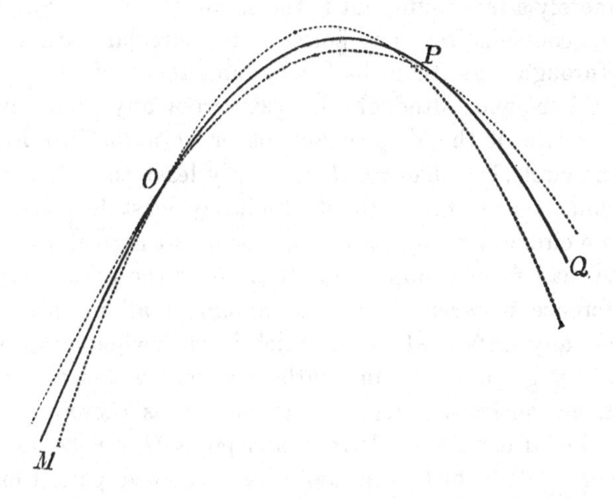

the path of a projectile, and if at O it be disturbed by an infinitely small force either way perpendicular to its instantaneous direction of motion, the disturbed path will cut the undisturbed infinitely near the point P where the direction of motion is perpendicular to that at O: as we easily see by considering that the line joining two particles projected from one point at the same instant with equal velocities in the directions of any two lines, will always remain perpendicular to the line bisecting the angle between these two lines.

Kinetic stability of a projectile.

General criterion.

355. The principle of varying action gives a mathematical criterion for stability or instability in every case of motion. Thus in the first place it is obvious, and it will be proved below (§§ 358, 361), that if the action is a true minimum in the motion of a system from any one configuration to the configuration reached at any other time, however much later, the motion is thoroughly unstable. For instance, in the motion of a particle constrained to remain on a smooth fixed surface, and uninfluenced by gravity, the action is simply the length of the path, multiplied by the constant velocity. Hence in the particular case of a particle uninfluenced by gravity, moving round the inner circle in the plane of an anchor-ring considered above, the action, or length of path, is clearly a minimum from any one point to the point reached at any subsequent time. (The action is not merely a minimum, but is the smaller of two minimums, when the course is from any point of the circular path to any other, through less than half a circumference of the circle.) On the other hand, although the path from any point in the greatest circle of the ring to any other at a distance from it along the circle, less than $\pi\sqrt{ab}$, is clearly least possible if along the circumference; the path of absolutely least length is not along the circumference between two points at a greater circular distance than $\pi\sqrt{ab}$ from one another, nor is the path along the circumference between them a minimum at all in this latter case. On any surface whatever which is everywhere anticlastic, or along a geodetic of any surface which passes altogether through an anticlastic region, the motion is thoroughly unstable. For if it were stable from any point O, we should have the given undisturbed path, and the disturbed path from O cutting it at some point Q;—two different geodetic lines joining two points; which is impossible on an anticlastic surface, inasmuch as the sum of the exterior angles of any closed figure of geodetic lines exceeds four right angles (§ 136) when the integral curvature of the enclosed area is negative, which (§§ 138, 128) is the case for every portion of surface thoroughly anticlastic. But, on the other hand, it is easily proved that if we have an endless rigid band of curved surface everywhere synclastic, with a geodetic line running through its

Examples.

Motion on an anticlastic surface proved unstable.

Motion of a particle on an anticlastic surface, unstable;

middle, the motion of a particle projected along this line will be stable throughout, and an infinitely slight disturbance will give a disturbed path cutting the given undisturbed path again and again for ever at successive distances differing according to the different specific curvatures of the intermediate portions of the surface. If from any point, N, of the undisturbed path, a perpendicular be drawn to cut the infinitely near disturbed path in E, the angles OEN and NOE must (§ 138) be together greater than a right angle by an amount equal to the integral curvature of the area EON. From this the differential equation of the disturbed path may be obtained immediately.

On a synclastic surface, stable.

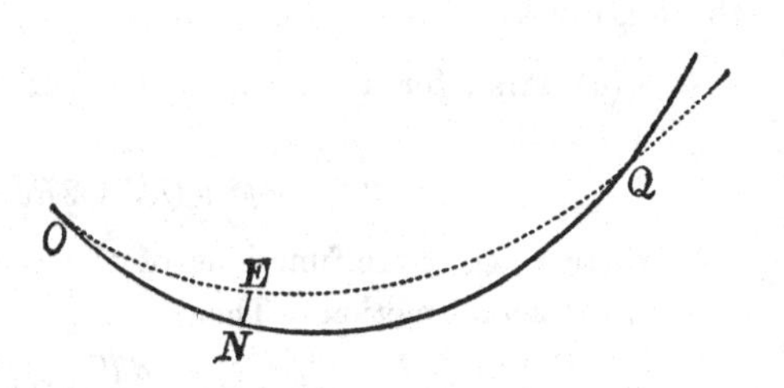

Differential equation of disturbed path.

Let $\angle EON = a$, $ON = s$, and $NE = u$; and let ϑ, a known function of s, be the specific curvature (§ 136) of the surface in the neighbourhood of N. Let also, for a moment, ϕ denote the complement of the angle OEN. We have

$$a - \phi = \int_0^s \vartheta u ds.$$

Hence

$$\frac{d\phi}{ds} = -\vartheta u.$$

But, obviously,

$$\phi = \frac{du}{ds};$$

hence

$$\frac{d^2u}{ds^2} + \vartheta u = 0.$$

When ϑ is constant (as in the case of the equator of a surface of revolution considered above, § 351), this gives

$$u = A \cos (s \sqrt{\vartheta} + E),$$

agreeing with the result (§ 351) which we obtained by development into a spherical surface.

The case of two or more bodies supported on parallel axes in the manner explained above in § 348, and rotating with the centre of inertia of the whole at the least possible distance from the fixed axis, affords a very good illustration also of this proposition which may be safely left as an exercise to the student.

General investigation of disturbed path.

356. To investigate the effect of an infinitely small conservative disturbance produced at any instant in the motion of any conservative system, may be reduced to a practicable problem (however complicated the required work may be) of mathematical analysis, provided the undisturbed motion is thoroughly known.

General equation of motion free in two degrees.

(*a*) First, for a system having but two degrees of freedom to move, let

$$2T = P\dot{\psi}^2 + Q\dot{\phi}^2 + 2R\dot{\psi}\dot{\phi} \quad \ldots\ldots\ldots\ldots\ldots\ldots(1),$$

where P, Q, R are functions of the co-ordinates not depending on the actual motion. Then

$$\left.\begin{aligned} &\frac{dT}{d\dot{\psi}} = P\dot{\psi} + R\dot{\phi}, \quad \frac{dT}{d\dot{\phi}} = Q\dot{\phi} + R\dot{\psi} \\ &\frac{d}{dt}\frac{dT}{d\dot{\psi}} = P\ddot{\psi} + R\ddot{\phi} + \frac{dP}{d\psi}\dot{\psi}^2 + \left(\frac{dP}{d\phi} + \frac{dR}{d\psi}\right)\dot{\psi}\dot{\phi} + \frac{dR}{d\phi}\dot{\phi}^2 \end{aligned}\right\}\ldots(2);$$

and the Lagrangian equations of motion [§ 318 (24)] are

$$\left.\begin{aligned} &P\ddot{\psi} + R\ddot{\phi} + \tfrac{1}{2}\left\{\frac{dP}{d\psi}\dot{\psi}^2 + 2\frac{dP}{d\phi}\dot{\psi}\dot{\phi} + \left(2\frac{dR}{d\phi} - \frac{dQ}{d\psi}\right)\dot{\phi}^2\right\} = \Psi \\ &R\ddot{\psi} + Q\ddot{\phi} + \tfrac{1}{2}\left\{\left(2\frac{dR}{d\psi} - \frac{dP}{d\phi}\right)\dot{\psi}^2 + 2\frac{dQ}{d\psi}\dot{\psi}\dot{\phi} + \frac{dQ}{d\phi}\dot{\phi}^2\right\} = \Phi \end{aligned}\right\}\ldots(3).$$

We shall suppose the system of co-ordinates so chosen that none of the functions P, Q, R, nor their differential coefficients $\frac{dP}{d\phi}$, etc., can ever become infinite.

(*b*) To investigate the effects of an infinitely small disturbance, we may consider a motion in which, at any time t, the co-ordinates are $\psi + p$ and $\phi + q$, p and q being infinitely small; and, by simply taking the variations of equations (3) in the usual manner, we arrive at two simultaneous differential equations of the second degree, linear with respect to

$$p,\ q,\ \dot{p},\ \dot{q},\ \ddot{p},\ \ddot{q},$$

but having variable coefficients which, when the undisturbed motion ψ, ϕ is fully known, may be supposed to be known functions of t. In these equations obviously none of the coefficients can at any time become infinite if the data correspond to a real dynamical problem, provided the system of co-ordinates is properly chosen (*a*); and the coefficients of $\ddot{p}$ and $\ddot{q}$ are the values, at the time t, of P, R, and R, Q, respectively, in the order in which they appear in (3), P, Q, R being the coefficients of a homogeneous quadratic function (1) which is essentially positive. These properties being taken into account, it may be shown that in no case can an infinitely small interval of time be the solution of the problem presented (§ 347) by the question of kinetic stability or instability, which is as follows :—

General investigation of disturbed path.

(c) The component velocities $\dot{\psi}$, $\dot{\phi}$ are at any instant changed to $\dot{\psi}+\alpha$, $\dot{\phi}+\beta$, subject to the condition of not changing the value of T. Then, α and β being infinitely small, it is required to find the interval of time until q/p first becomes equal to $\dot{\phi}/\dot{\psi}$.

(d) The differential equations in p and q reduce this problem, and in fact the full problem of finding the disturbance in the motion when the undisturbed motion is given, to a practicable form. But, merely to prove the proposition that the disturbed course cannot meet the undisturbed course until after some finite time, and to estimate a limit which this time must exceed in any particular case, it may be simpler to proceed thus :—

(e) To eliminate t from the general equations (3), let them first be transformed so as not to have t independent variable. We must put

$$\ddot{\psi}=\frac{dtd^2\psi-d\psi d^2t}{dt^3},\quad \ddot{\phi}=\frac{dtd^2\phi-d\phi d^2t}{dt^3}\dots\dots\dots\dots\dots(4).$$

And by the equation of energy we have

$$dt=\frac{(Pd\psi^2+Qd\phi^2+2Rd\psi d\phi)^{\frac{1}{2}}}{\{2(E-V)\}^{\frac{1}{2}}}\dots\dots\dots\dots\dots\dots(5),$$

it being assumed that the system is conservative. Eliminating dt and d^2t between this and the two equations (3), we find a differential equation of the second degree between ψ and ϕ, which is the differential equation of the course. For simplicity, let us suppose one of the co-ordinates, ϕ for instance, to be independent variable; that is, let $d^2\phi=0$. We have, by (4),

$$d^2t=-\ddot{\phi}\frac{dt^3}{d\phi},$$

and therefore

$$\ddot{\psi}dt^2=d^2\psi+\frac{d\psi}{d\phi}\ddot{\phi}dt^2,$$

General investigation of disturbed path.

and the result of the elimination becomes

$$(PQ-R^2)\frac{d^2\psi}{d\phi^2}+F\left(\frac{d\psi}{d\phi}\right)=\frac{\left(P\frac{d\psi^2}{d\phi^2}+2R\frac{d\psi}{d\phi}+Q\right)\left[\left(Q+R\frac{d\psi}{d\phi}\right)\Psi-\left(R+P\frac{d\psi}{d\phi}\right)\Phi\right]}{2(E-V)} \quad \text{..........................(6)},$$

$F\left(\frac{d\psi}{d\phi}\right)$ denoting a function of $\frac{d\psi}{d\phi}$ of the third degree, with variable coefficients, none of which can become infinite as long as $E-V$, the kinetic energy, is finite.

(*f*) Taking the variation of this equation on the supposition that ψ becomes $\psi+p$, where p is infinitely small, we have

$$(PQ-R^2)\frac{d^2p}{d\phi^2}+L\frac{dp}{d\phi}+Mp=0 \quad \text{..................(7)},$$

where L and M denote known functions of ϕ, neither of which has any infinitely great value. This determines the deviation, p, of the course. Inasmuch as the quadratic (1) is essentially always positive, $PQ-R^2$ must be always positive. Hence, if for a particular value of ϕ, p vanishes, and $\frac{dp}{d\phi}$ has a given value which defines the disturbance we suppose made at any instant, ϕ must increase by a finite amount (and therefore a finite time must elapse) before the value of p can be again zero; that is to say, before the disturbed course can again cut the undisturbed course.

(*g*) The same proposition consequently holds for a system having any number of degrees of freedom. For the preceding proof shows it to hold for the system subjected to any frictionless constraint, leaving it only two degrees of freedom; including that particular frictionless constraint which would not alter either the undisturbed or the disturbed course. The full general investigation of the disturbed motion, with more than two degrees of freedom, takes a necessarily complicated form, but the principles on which it is to be carried out are sufficiently indicated by what we have done.

(*h*) If for $L/PQ-R^2$ we substitute a constant $2a$, less than its least value, irrespectively of sign, and for $M/PQ-R^2$, a constant β greater algebraically than its greatest value, we have an equation

General investigation of disturbed path.

$$\frac{d^2p}{d\phi^2} + 2\alpha\frac{dp}{d\phi} + \beta = 0 \text{..........................(8).}$$

Here the value of p vanishes for values of ϕ successively exceeding one another by $\pi/\sqrt{\beta - \alpha^2}$, which is clearly less than the increase that ϕ must have in the actual problem before p vanishes a second time. Also, we see from this that if $\alpha^2 > \beta$ the actual motion is unstable. It might of course be unstable even if $\alpha^2 < \beta$; and the proper analytical methods for finding either the rigorous solution of (7), or a sufficiently near practical solution, would have to be used to close the criterion of stability or instability, and to thoroughly determine the disturbance of the course.

(*i*) When the system is only a single particle, confined to a plane, the differential equation of the deviation may be put under a remarkably simple form, useful for many practical problems. Let N be the normal component of the force, per unit of the mass, at any instant, v the velocity, and ρ the radius of curvature of the path. We have (§ 259)

Differential equation of disturbed path of single particle in a plane.

$$N = \frac{v^2}{\rho}.$$

Let, in the diagram, ON be the undisturbed, and QE the disturbed path. Let EN, cutting ON at right angles, be denoted by u, and ON by s. If further we denote by ρ' the radius of curvature in the disturbed path,

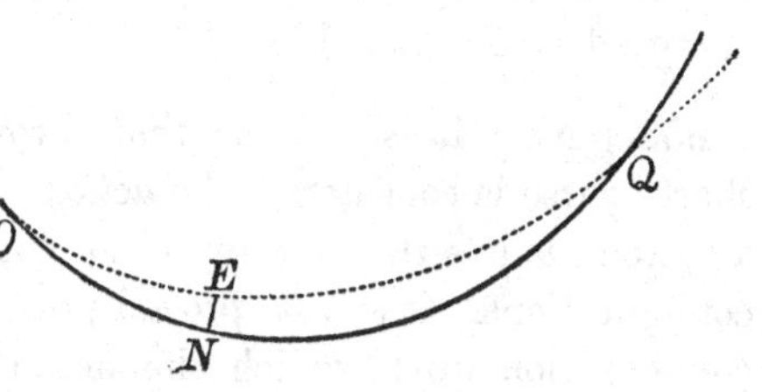

remembering that u is infinitely small, we easily find

$$\frac{1}{\rho'} = \frac{1}{\rho} + \frac{d^2u}{ds^2} + \frac{u}{\rho^2} \text{..........................(9).}$$

Hence, using δ to denote variations from N to E, we have

$$\delta N = \delta\frac{v^2}{\rho} = \frac{\delta(v^2)}{\rho} + v^2\left(\frac{d^2u}{ds^2} + \frac{u}{\rho^2}\right) \text{............(10).}$$

Differential equation of disturbed, path of single particle in a plane.

But, by the equation of energy,

$$v^2 = 2(E - V),$$

and therefore

$$\delta(v^2) = -2\delta V = 2Nu = \frac{2v^2}{\rho} u.$$

Hence (10) becomes

$$\frac{d^2u}{ds^2} + \frac{3u}{\rho^2} - \frac{\delta N}{v^2} = 0 \ldots\ldots\ldots\ldots\ldots\ldots\ldots(11),$$

or, if we denote by ζ the rate of variation of N, per unit of distance from the point N in the normal direction, so that $\delta N = \zeta u$,

$$\frac{du^2}{ds^2} + \left(\frac{3}{\rho^2} - \frac{\zeta}{v^2}\right) u = 0 \ldots\ldots\ldots\ldots\ldots\ldots\ldots(12).$$

This includes, as a particular case, the equation of deviation from a circular orbit, investigated above (§ 350).

Kinetic foci.

357. If, from any one configuration, two courses differing infinitely little from one another have again a configuration in common, this second configuration will be called a kinetic focus relatively to the first: or (because of the reversibility of the motion) these two configurations will be called conjugate kinetic foci. Optic foci, if for a moment we adopt the corpuscular theory of light, are included as a particular case of kinetic foci in general. By § 356 (*g*) we see that there must be finite intervals of space and time between two conjugate foci in every motion of every kind of system, only provided the kinetic energy does not vanish.

Theorem of minimum action.

Action never a minimum in a course including kinetic foci.

358. Now it is obvious that, provided only a sufficiently short course is considered, the *action*, in any natural motion of a system, is less than for any other course between its terminal configurations. It will be proved presently (§ 361) that the first configuration up to which the action, reckoned from a given initial configuration, ceases to be a minimum, is the first kinetic focus; and conversely, that when the first kinetic focus is passed, the action, reckoned from the initial configuration, ceases to be a minimum; and therefore of course can never again be a minimum, because a course of shorter action, deviating infinitely little from it, can be found for a part, without altering the remainder of the whole, natural course.

359. In such statements as this it will frequently be convenient to indicate particular configurations of the system by single letters, as O, P, Q, R; and any particular course, in which it moves through configurations thus indicated, will be called the course $O...P...Q...R$. The *action* in any natural course will be denoted simply by the terminal letters, taken in the order of the motion. Thus OR will denote the action from O to R; and therefore $OR = -RO$. When there are more real natural courses from O to R than one, the analytical expression for OR will have more than one real value; and it may be necessary to specify for which of these courses the action is reckoned. Thus we may have

Notation for configurations, courses, and action.

$$OR \text{ for } O...E...R,$$
$$OR \text{ for } O...E'...R,$$
$$OR \text{ for } O...E''...R,$$

three different values of one algebraic irrational expression.

360. In terms of this notation the preceding statement (§ 358) may be expressed thus:—If, for a conservative system, moving on a certain course $O...P...O'...P'$, the first kinetic focus conjugate to O be O', the action OP', in this course, will be less than the action along any other course deviating infinitely little from it: but, on the other hand, OP' is greater than the actions in some courses from O to P' deviating infinitely little from the specified natural course $O...P...O'...P'$.

Theorem of minimum action.

361. It must not be supposed that the action along OP is necessarily *the least possible* from O to P. There are, in fact, cases in which the action ceases to be *least of all possible*, before

Two or more courses of minimum action possible.

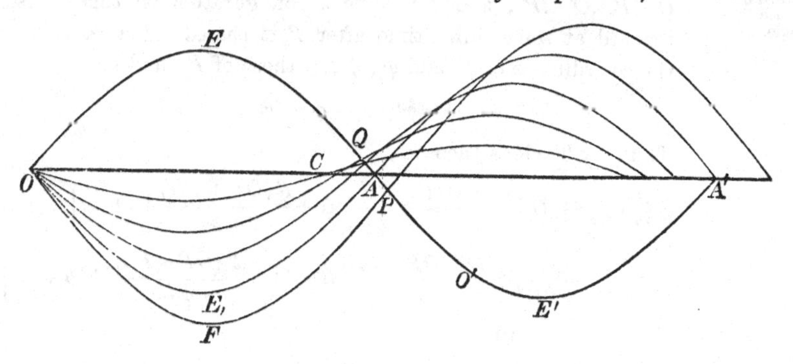

Two or more courses of minimum action possible.

a kinetic focus is reached. Thus if $OEAPO'E'A'$ be a sinuous geodetic line cutting the outer circle of an anchor-ring, or the equator of an oblate spheroid, in successive points O, A, A', it is easily seen that O', the first kinetic focus conjugate to O, must lie somewhat beyond A. But the length $OEAP$, although a *minimum* (a stable position for a stretched string), is not the shortest distance on the surface from O to P, as *this* must obviously be a line lying entirely on one side of the great circle. From O, to any point, Q, short of A, the distance along the geodetic $OEQA$ is clearly the least possible: but if Q be near enough to A (that is to say, between A and the point in which the envelope of the geodetics drawn from O, cuts OEA), there will also be two other geodetics from O to Q. The length of one of these will be a minimum, and that of the other not a minimum. If Q is moved forward to A, the former becomes $OE_{,}A$, equal and similar to OEA, but on the other side of the great circle: and the latter becomes the great circle from O to A. If now Q be moved on, to P, beyond A, the minimum geodetic $OEAP$ ceases to be the less of the two minimums, and the geodetic OFP lying altogether on the other side of the great circle becomes the least possible line from O to P. But until P is advanced beyond the point, O', in which it is cut by another geodetic from O lying infinitely nearly along it, the length $OEAP$ remains a minimum, according to the general proposition of § 358, which we now proceed to prove.

Case of two minimum, and one not minimum, geodetic lines between two points.

Difference between two sides and the third of a kinetic triangle.

(*a*) Referring to the notation of § 360, let $P_{,}$ be any configuration differing infinitely little from P, but not on the course $O \ldots P \ldots O' \ldots P'$; and let S be a configuration on this course, reached at some finite time after P is passed. Let ψ, ϕ,... be the co-ordinates of P, and $\psi_{,}$, $\phi_{,}$,... those of $P_{,}$, and let

$$\psi_{,} - \psi = \delta\psi, \quad \phi_{,} - \phi = \delta\phi, \ldots$$

Thus, by Taylor's theorem,

$$OP_{,} + P_{,}S = OS + \left\{ \frac{d(OP+PS)}{d\psi}\delta\psi + \frac{d(OP+PS)}{d\phi}\delta\phi + \ldots \right\}$$

$$+ \tfrac{1}{2}\left\{ \frac{d^2(OP+PS)}{d\psi^2}(\delta\psi)^2 + 2\frac{d^2(OP+PS)}{d\psi d\phi}\delta\psi\delta\phi + \ldots \right\}$$

$$+ \text{etc.}$$

Difference between two sides and the third of a kinetic triangle.

But if $\xi, \eta, \ldots$ denote the components of momentum at P in the course $O \ldots P$, which are the same as those at P in the continuation, $P \ldots S$, of this course, we have [§ 330 (18)]

$$\xi = \frac{dOP}{d\psi} = -\frac{dPS}{d\psi}, \quad \eta = \frac{dOP}{d\phi} = -\frac{dPS}{d\phi}, \ldots$$

Hence the coefficients of the terms of the first degree of $\delta\psi$, $\delta\phi$, in the preceding expression vanish, and we have

$$\left.\begin{array}{l} OP_{,} + P_{,}S - OS = \frac{1}{2}\left\{\frac{d^2(OP+PS)}{d\psi^2}\delta\psi^2 + 2\frac{d^2(OP+PS)}{d\psi d\phi}\delta\psi\delta\phi + \ldots\right\} \\ \qquad + \text{etc.} \end{array}\right\} \text{(1)}.$$

(*b*) Now, assuming

$$\left.\begin{array}{l} x_1 = \alpha_1 \delta\psi + \beta_1 \delta\phi + \ldots \\ x_2 = \alpha_2 \delta\psi + \beta_2 \delta\phi + \ldots \\ \quad \text{etc.} \qquad \text{etc.} \end{array}\right\} \ldots\ldots\ldots\ldots\ldots(2),$$

according to the known method of linear transformations, let $\alpha_1, \beta_1, \ldots \alpha_2, \beta_2, \ldots$ be so chosen that the preceding quadratic function be reduced to the form

$$A_1 x_1^2 + A_2 x_2^2 + \ldots + A_i x_i^2,$$

the whole number of degrees of freedom being i.

This may be done in an infinite variety of ways; and, towards fixing upon one particular way, we may take $\alpha_i = \dot{\psi}$, $\beta_i = \dot{\phi}$, etc.; and subject the others to the conditions

$$\dot{\psi}\alpha_1 + \dot{\phi}\beta_1 + \ldots = 0, \quad \dot{\psi}\alpha_2 + \dot{\phi}\beta_2 + \ldots = 0, \text{ etc.}$$

This will make $A_i = 0$: for if for a moment we suppose $P_{,}$ to be on the course $O \ldots P \ldots O'$, we have

$$\frac{\delta\psi}{\dot{\psi}} = \frac{\delta\phi}{\dot{\phi}} = \ldots,$$

and therefore

$$x_i = \frac{\dot{\psi}}{\delta\psi}(\delta\psi^2 + \delta\phi^2 + \ldots), \quad x_{i-1} = 0, \quad \ldots x_2 = 0, \quad x_1 = 0.$$

But in this case $OP_{,} + P_{,}S = OS$; and therefore the value of the quadratic must be zero; that is to say, we must have $A_i = 0$. Hence we have

$$\left.\begin{array}{l} OP_{,} + P_{,}S - OS = \frac{1}{2}(A_1 x_1^2 + A_2 x_2^2 + \ldots + A_{i-1} x_{i-1}^2) \\ \qquad + R \end{array}\right\} \ldots\ldots\ldots(3)$$

where R denotes a remainder consisting of terms of the third and higher degrees in $\delta\psi$, $\delta\phi$, etc., or in x_1, x_2, etc.

Difference between two sides and the third of a kinetic triangle.

(c) Another form, which will be used below, may be given to the same expression thus:—Let $(\xi_{,}, \eta_{,}, \zeta_{,},\ldots)$ and $(\xi_{,}', \eta_{,}', \zeta_{,}',\ldots)$ be the components of momentum at $P_{,}$, in the courses $OP_{,}$ and $P_{,}S$ respectively. By § 330 (18) we have

$$\xi_{,} = \frac{dOP'}{d\psi_{,}},$$

and therefore by Taylor's theorem

$$\xi_{,} = \frac{dOP}{d\psi} + \frac{d^2OP}{d\psi^2}\delta\psi + \frac{d^2OP}{d\psi d\phi}\delta\phi + \ldots + \text{etc.}$$

Similarly,

$$-\xi_{,}' = \frac{dPS}{d\psi} + \frac{d^2PS}{d\psi^2}\delta\psi + \frac{d^2PS}{d\psi d\phi}\delta\phi + \ldots + \text{etc.};$$

and therefore, as $\dfrac{dOP}{d\psi} = -\dfrac{dPS}{d\psi}$,

$$\xi_{,}' - \xi_{,} = -\left\{\frac{d^2(OP+PS)}{d\psi^2}\delta\psi + \frac{d^2(OP+PS)}{d\psi d\phi}\delta\phi + \ldots\right\} + \text{etc.} \ldots (4),$$

and so for $\eta_{,}' - \eta_{,}$, etc. Hence (1) is the same as

$$\left.\begin{array}{l} OP_{,} + P_{,}S - OS = -\tfrac{1}{2}\{(\xi_{,}' - \xi')\delta\psi + (\eta_{,}' - \eta_{,})\delta\phi + \ldots\} \\ \qquad + R \end{array}\right\} \ldots\ldots (5),$$

where R denotes a remainder consisting of terms of the third and higher degrees. Also the transformation from $\delta\psi$, $\delta\phi$, ... to x_1, x_2, ..., gives clearly

$$\left.\begin{array}{l} \xi_{,}' - \xi_{,} = -(A_1\alpha_1x_1 + A_2\alpha_2x_2 + \ldots + A_{i-1}\alpha_{i-1}x_{i-1}) \\ \eta_{,}' - \eta_{,} = -(A_1\beta_1x_1 + A_2\beta_2x_2 + \ldots + A_{i-1}\beta_{i-1}x_{i-1}) \\ \qquad \text{etc.} \qquad\qquad \text{etc.} \end{array}\right\} \ldots\ldots\ldots\ldots (6).$$

(d) Now for any infinitely small time the velocities remain sensibly constant; as also do the coefficients (ψ, ψ), (ψ, ϕ), etc., in the expression [§ 313 (2)] for T: and therefore for the action we have

$$\begin{aligned} \int 2Tdt &= \sqrt{2T}\int\sqrt{2T}dt \\ &= \sqrt{2T}\{(\psi, \psi)(\psi-\psi_0)^2 + 2(\psi, \phi)(\psi-\psi_0)(\phi-\phi_0) + \text{etc.}\}^{\frac{1}{2}} \end{aligned}$$

where $(\psi_0, \phi_0, \ldots)$ are the co-ordinates of the configuration from which the action is reckoned. Hence, if P, P', P'' be any three configurations infinitely near one another, and if Q, with the proper differences of co-ordinates written after it, be used to denote square roots of quadratic functions such as that in the preceding expression, we have

$$\left.\begin{aligned} PP' &= \sqrt{2T}\,.\,Q\{(\psi-\psi'),\ (\phi-\phi'),\ \ldots\} \\ P'P'' &= \sqrt{2T}\,.\,Q\{(\psi'-\psi''),\ (\phi'-\phi''),\ \ldots\} \\ P''P &= \sqrt{2T}\,.\,Q\{(\psi''-\psi'),\ (\phi''-\phi'),\ \ldots\} \end{aligned}\right\}\ldots\ldots\ldots\ldots(7).$$

Difference between two sides and the third of a kinetic triangle.

In the particular case of a single free particle, these expressions become simply proportional to the distances PP', $P'P''$, $P''P$; and by Euclid we have

$$P'P + PP'' < P'P''$$

unless P is in the straight line $P'P''$.

The verification of this proposition by the preceding expressions (7) is merely its proof by co-ordinate geometry with an oblique rectilineal system of co-ordinates, and is necessarily somewhat complicated. If $(\psi, \phi) = (\phi, \theta) = (\theta, \psi) = 0$, the co-ordinates become rectangular and the algebraic proof is easy. There is no difficulty, by following the analogies of these known processes, to prove that, for any number of co-ordinates, ψ, ϕ, etc., we have

$$P'P + PP'' > P'P'',$$

unless

$$\frac{\psi-\psi'}{\psi''-\psi'} = \frac{\phi-\phi'}{\phi''-\phi'} = \frac{\theta-\theta'}{\theta''-\theta'} = \ldots$$

(expressing that P is on the course from P to P''), in which case

$$P'P + PP'' = P'P'',$$

$P'P$, etc., being given by (7). And further, by the aid of (1), it is easy to find the proper expression for $P'P + PP'' - P'P''$, when P is infinitely little off the course from P' to P'': but it is quite unnecessary for us here to enter on such purely algebraic investigations.

(*e*) It is obvious indeed, as has been already said (§ 358), that the action along any natural course is *the least possible between its terminal configurations* if only a sufficiently short course is included. Hence for all cases in which the time from O to S is less than some particular amount, the quadratic term in the expression (3) for $OP_{,} + P_{,}S - OS$ is necessarily positive, for all values of x_1, x_2, etc.; and therefore $A_1, A_2, \ldots A_{i-1}$ must each be positive.

(*f*) Let now S be removed further and further from O, along the definite course $O \ldots P \ldots O'$, until it becomes O'. When it is O', let $P_{,}$ be taken on a natural course through O and O', de-

Actions on different courses infinitely near one another

between two conjugate kinetic foci, proved ultimately equal.

viating infinitely little from the course OPO'. Then, as $OP_,O'$ is a natural course,

$$\xi_,' - \xi_, = \eta_,' - \eta_,, \ldots = 0;$$

and therefore (5) becomes

$$OP_, + P_,O' - OO' = R,$$

which proves that the chief, or quadratic, term in the other expression (3) for the same, vanishes. Hence one at least of the coefficients $A_1, A_2, \ldots$ must vanish, and if one only, $A_{i-1} = 0$ for instance, we must have

$$x_1 = 0,\ x_2 = 0, \ldots x_{i-2} = 0.$$

These equations express the condition that $P_,$ lies on a natural course from O to O'.

If two sides, deviating infinitely little from the third, are together equal to it, they constitute an unbroken natural course.

(g) Conversely if one or more of the coefficients A_1, A_2, etc., vanishes, if for instance $A_{i-1} = 0$, S must be a kinetic focus. For if we take $P_,$ so that

$$x_1 = 0,\ x_2 = 0, \ldots x_{i-2} = 0,$$

we have, by (6),

$$\xi_,' - \xi_, = \eta_,' - \eta' = \ldots = 0.$$

(h) Thus we have proved that at a kinetic focus conjugate to O the action from O is not a minimum of the first order*, and that the last configuration, up to which the action from O *is* a minimum of the first order, is a kinetic focus conjugate to O.

(i) It remains to be proved that the action from O ceases to be a minimum when the first kinetic focus conjugate to O is passed. Let, as above (§ 360), $O \ldots P \ldots O' \ldots P'$ be a natural course extending beyond O', the first kinetic focus conjugate to O. Let P and P' be so near one another that there is no focus conjugate to either, between them; and let $O \ldots P_, \ldots O'$ be a natural course from O to O' deviating infinitely little from $O \ldots P \ldots O'$. By what we have just proved (e), the action OO' along $O \ldots P_, \ldots O'$ differs only by R, an infinitely small quantity of the third order, from the action OO' along $O \ldots P \ldots O'$, and therefore

Natural course proved not a course of minimum action, beyond a kinetic focus.

$$\begin{aligned} Ac.\,(O \ldots P \ldots O' \ldots P') &= Ac.\,(O \ldots P_, \ldots O') + O'P' + R \\ &= OP_, + P_,O' + O'P' + R. \end{aligned}$$

* A maximum or minimum "of the first order" of any function of one or more variables, is one in which the differential of the first degree vanishes, but not that of the second degree.

But, by a proper application of (e) we see that

Natural course proved not a course of minimum action, beyond a kinetic focus.

$$P_{,}O' + O'P' = P_{,}P' + Q$$

where Q denotes an infinitely small quantity of the second order, which is essentially positive. Hence

$$Ac\,(O...P...O'...P') = OP_{,} + P_{,}P' + Q + R,$$

and therefore, as R is infinitely small in comparison with Q,

$$Ac\,(O...P...O'...P') > OP_{,} + P_{,}P'.$$

Hence the broken course $O...P_{,}$, $P_{,}...P'$ has less action than the natural course $O...P...O'...P'$, and therefore, as the two are infinitely near one another, the latter is not a minimum.

A course which includes no focus conjugate to either extremity includes no pair of conjugate foci.

362. As it has been proved that the action from any configuration ceases to be a minimum at the first conjugate kinetic focus, we see immediately that if O' be the first kinetic focus conjugate to O, reached after passing O, no two configurations on this course from O to O' can be kinetic foci to one another. For, the action from O just ceasing to be a minimum when O' is reached, the action between any two intermediate configurations of the same course is necessarily a minimum.

How many kinetic foci in any case.

363. When there are i degrees of freedom to move there are in general, on any natural course from any particular configuration, O, at least $i-1$ kinetic foci conjugate to O. Thus, for example, on the course of a ray of light emanating from a luminous point O, and passing through the centre of a convex lens held obliquely to its path, there are two kinetic foci conjugate to O, as defined above, being the points in which the line of the central ray is cut by the so-called "focal lines"* of a pencil of rays diverging from O and made convergent after passing through the lens. But some or all of these kinetic foci may be on the course previous to O; as for instance in the case of a common projectile when its course passes obliquely downwards through O. Or some or all may be lost; as when, in the optical illustration just referred to, the lens is only strong enough to produce convergence in one of the principal planes, or too weak to produce convergence in either. Thus

* In our second volume we hope to give all necessary elementary explanations on this subject.

How many kinetic foci in any case.

also in the case of the undisturbed rectilineal motion of a point, or in the motion of a point uninfluenced by force, on an anticlastic surface (§ 355), there are no real kinetic foci. In the motion of a projectile (not confined to one vertical plane) there can only be one kinetic focus on each path, conjugate to one given point; though there are three degrees of freedom. Again, there may be any number more than $i-1$, of foci in one course, all conjugate to one configuration, as for instance on the course of a particle uninfluenced by force, moving round the surface of an anchor-ring, along either the outer great circle, or along a sinuous geodetic such as we have considered in § 361, in which clearly there are an infinite number of foci each conjugate to any one point of the path, at equal successive distances from one another.

Referring to the notation of § 361 (f), let S be gradually moved on until first one of the coefficients, A_{i-1} for instance, vanishes; then another, A_{i-2}', etc.; and so on. We have seen that each of these positions of S is a kinetic focus: and thus by the successive vanishing of the $i-1$ coefficients we have $i-1$ foci. If none of the coefficients can ever vanish, there are no kinetic foci. If one or more of them, after vanishing, comes to a minimum, and again vanishes, as S is moved on, there may be any number more than $i-1$ of foci each conjugate to the same configuration, O.

Theorem of maximum action.

364. If $i-1$ distinct* courses from a configuration O, each differing infinitely little from a certain natural course

$$O\ldots E\ldots O_1\ldots O_2\ldots O_{i-1}\ldots Q,$$

cut it in configurations O_1, O_2, O_3,...O_{i-1}, and if, besides these, there are not on it any other kinetic foci conjugate to O, between O and Q, and no focus at all, conjugate to E, between E and Q, the action in this natural course from O to Q is the maximum for all courses $O\ldots P_{,}$, $P_{,}\ldots Q$; $P_{,}$ being a configuration infinitely nearly agreeing with some configuration between E and O_1 of the standard course $O\ldots E\ldots O_1\ldots O_2\ldots O_{i-1}\ldots Q$, and $O\ldots P_{,}$, $P_{,}\ldots Q$

* Two courses are not called distinct if they differ from one another only in the absolute magnitude, not in the proportions of the components, of the deviations by which they differ from the standard course.

denoting the natural courses between O and $P_{\prime\prime}$, and $P_{\prime}$ and Q, which deviate infinitely little from this standard course.

Theorem of maximum action.

In § 361 (i), let O' be any one, O_1, of the foci $O_1, O_2, \ldots O_{i-1}$, and let $P_{\prime}$ be called P_1 in this case. The demonstration there given shows that

$$OQ > OP_1 + P_1Q.$$

Hence there are $i-1$ different broken courses

$$O \ldots P_1,\ P_1 \ldots Q;\ O \ldots P_2,\ P_2 \ldots Q;\ \text{etc.},$$

in each of which the action is less than in the standard course from O to Q. But whatever be the deviation of $P_{\prime\prime}$, it may clearly be compounded of deviations P to P_1, P to P_2, P to P_3, ..., P to P_{i-1}, corresponding to these $i-1$ cases respectively; and it is easily seen from the analysis that

$$OP_{\prime} + P_{\prime}Q - OQ = (OP_1 + P_1Q - OQ) + (OP_2 + P_2Q - OQ) + \ldots$$

Hence $OP_{\prime} + P_{\prime}Q < OQ$, which was to be proved.

Applications to two degrees of freedom.

365. Considering now, for simplicity, only cases in which there are but two degrees (§§ 195, 204) of freedom to move, we see that after any infinitely small conservative disturbance of a system in passing through a certain configuration, the system will first again pass through a configuration of the undisturbed course, at the first configuration of the latter at which the action in the undisturbed motion ceases to be a minimum. For instance, in the case of a particle, confined to a surface, and subject to any conservative system of force, an infinitely small conservative disturbance of its motion through any point, O, produces a disturbed path, which cuts the undisturbed path at the first point, O', at which the action in the undisturbed path from O ceases to be a minimum. Or, if projectiles, under the influence of gravity alone, be thrown from one point, O, in all directions with equal velocities, in one vertical plane, their paths, as is easily proved, intersect one another consecutively in a parabola, of which the focus is O, and the vertex the point reached by the particle projected directly upwards. The actual course of each particle from O is the course of least possible action to any point, P, reached before the enveloping parabola, but is not a course of minimum action to any point, Q, in its path after the envelope is passed.

Applications to two degrees of freedom.

366. Or again, if a particle slides round along the greatest circle of the smooth inner surface of a hollow anchor-ring, the "action," or simply the length of path, from point to point, will be least possible for lengths (§ 351) less than $\pi\sqrt{ab}$. Thus, if a string be tied round outside on the greatest circle of a perfectly smooth anchor-ring, it will slip off unless held in position by staples, or checks of some kind, at distances of not less than $\pi\sqrt{ab}$ from one another in succession round the circle. With reference to this example, see also § 361, above.

Or, of a particle sliding down an inclined cylindrical groove, the action from any point will be the least possible along the straight path to any other point reached in a time less than that of the vibration one way of a simple pendulum of length equal to the radius of the groove, and influenced by a force equal $g \cos i$, instead of g the whole force of gravity. But the action will not be a minimum from any point, along the straight path, to any other point reached in a longer time than this. The case in which the groove is horizontal ($i=0$) and the particle is projected along it, is particularly simple and instructive, and may be worked out in detail with great ease, without assuming any of the general theorems regarding action.

Hamilton's second form.

367. In the preceding account of the Hamiltonian principle, and of developments and applications which it has received, we have adhered to the system (§§ 328, 330) in which the initial and final co-ordinates and the constant sum of potential and kinetic energies are the elements of which the action is supposed to be a function. Another system was also given by Hamilton, according to which the action is expressed in terms of the initial and final co-ordinates and the *time prescribed for the motion;* and a set of expressions quite analogous to those with which we have worked, are established. For practical applications this method is generally less convenient than the other; and the analytical relations between the two are so obvious that we need not devote any space to them here.

Liouville's kinetic theorem.

368. We conclude by calling attention to a very novel analytical investigation of the motion of a conservative system, by Liouville (*Comptes Rendus,* June 16, 1856), which leads immediately to the principle of least action, and the Hamiltonian principle with the developments by Jacobi and others; but which also establishes a very remarkable and absolutely new theorem regarding the amount of the action along any constrained course. For brevity we shall content ourselves with giving it for a single free particle, referring the reader to the original article for Liouville's complete investigation in terms of generalized co-ordinates, applicable to any conservative system whatever.

Liouville's kinetic theorem.

Let (x, y, z) be the co-ordinates of any point through which the particle may move: V its potential energy in this position: E the sum of the potential and kinetic energies of the motion in question: A the action, from any position (x_0, y_0, z_0) to (x, y, z) along any course arbitrarily chosen (supposing, for instance, the particle to be guided along it by a frictionless guiding tube). Then (§ 326), the mass of the particle being taken as unity,

$$A = \int v ds = \int \sqrt{2(E-V)}\sqrt{(dx^2+dy^2+dz^2)}.$$

Now let ϑ be a function of x, y, z, which satisfies the partial differential equation

$$\frac{d\vartheta^2}{dx^2}+\frac{d\vartheta^2}{dy^2}+\frac{d\vartheta^2}{dz^2}=2(E-V).$$

Then

$$A = \int\sqrt{\left(\frac{d\vartheta^2}{dx^2}+\frac{d\vartheta^2}{dy^2}+\frac{d\vartheta^2}{dz^2}\right)(dx^2+dy^2+dz^2)}$$

$$=\int\sqrt{\left[\left(\frac{d\vartheta}{dx}dx+\frac{d\vartheta}{dy}dy+\frac{d\vartheta}{dz}dz\right)^2+\left(\frac{d\vartheta}{dz}dy-\frac{d\vartheta}{dy}dz\right)^2+\left(\frac{d\vartheta}{dx}dz-\frac{d\vartheta}{dz}dx\right)^2+\left(\frac{d\vartheta}{dy}dx-\frac{d\vartheta}{dx}dy\right)^2\right]}.$$

But

$$\frac{d\vartheta}{dx}dx+\frac{d\vartheta}{dy}dy+\frac{d\vartheta}{dz}dz=d\vartheta,$$

and, if $\dot{x}, \dot{y}, \dot{z}$ denote the actual component velocities along the arbitrary path, and $\dot{\vartheta}$ the rate at which ϑ increases per unit of time in this motion,

$$dx=\dot{x}dt,\quad dy=\dot{y}dt,\quad dz=\dot{z}dt,\quad d\vartheta=\dot{\vartheta}dt.$$

Hence the preceding becomes

$$A=\int d\vartheta\sqrt{\left\{1+\frac{\left(\dot{y}\frac{d\vartheta}{dz}-\dot{z}\frac{d\vartheta}{dy}\right)^2+\left(\dot{z}\frac{d\vartheta}{dx}-\dot{x}\frac{d\vartheta}{dz}\right)^2+\left(\dot{x}\frac{d\vartheta}{dy}-\dot{y}\frac{d\vartheta}{dx}\right)^2}{\dot{\vartheta}^2}\right\}}.$$

CHAPTER III.

EXPERIENCE.

Observation and experiment.

369. By the term Experience, in physical science, we designate, according to a suggestion of Herschel's, our means of becoming acquainted with the material universe and the laws which regulate it. In general the actions which we see ever taking place around us are *complex*, or due to the simultaneous action of many causes. When, as in astronomy, we endeavour to ascertain these causes by simply watching their effects, we *observe;* when, as in our laboratories, we interfere arbitrarily with the causes or circumstances of a phenomenon, we are said to *experiment*.

Observation.

370. For instance, supposing that we are possessed of instrumental means of measuring time and angles, we may trace out by successive observations the relative position of the sun and earth at different instants; and (the method is not susceptible of any accuracy, but is alluded to here only for the sake of illustration) from the variations in the apparent diameter of the former we may calculate the ratios of our distances from it at those instants. We have thus a set of observations involving time, angular position with reference to the sun, and ratios of distances from it: sufficient (if numerous enough) to enable us to discover the laws which connect the variations of these co-ordinates.

Similar methods may be imagined as applicable to the motion of any planet about the sun, of a satellite about its primary, or of one star about another in a binary group.

371. In general all the data of Astronomy are determined in this way, and the same may be said of such subjects as

Tides and Meteorology. Isothermal Lines, Lines of Equal Dip, Lines of Equal Intensity, Lines of Equal "Variation" (or "Declination" as it has still less happily been sometimes called), the Connexion of Solar Spots with Terrestrial Magnetism, and a host of other data and phénomena, to be explained under the proper heads in the course of the work, are thus deducible from *Observation* merely. In these cases the apparatus for the gigantic experiments is found ready arranged in Nature, and all that the philosopher has to do is to watch and measure their progress to its last details.

Observation.

372. Even in the instance we have chosen above, that of the planetary motions, the observed effects are complex; because, unless possibly in the case of a double star, we have no instance of the *undisturbed* action of one heavenly body on another; but to a first approximation the motion of a planet about the sun is found to be the same as if no other bodies than these two existed; and the approximation is sufficient to indicate the probable law of mutual action, whose full confirmation is obtained when, *its* truth being assumed, the disturbing effects thus calculated are allowed for, and found to account completely for the observed deviations from the consequences of the first supposition. This may serve to give an idea of the mode of obtaining the laws of phenomena, which can only be observed in a complex form—and the method can always be directly applied when one cause is known to be pre-eminent.

Experiment.

373. Let us take cases of the other kind—in which the effects are so complex that we cannot deduce the causes from the observation of combinations arranged in Nature, but must endeavour to form for ourselves other combinations which may enable us to study the effects of each cause separately, or at least with only slight modification from the interference of other causes.

374. A stone, when dropped, falls to the ground; a brick and a boulder, if dropped from the top of a cliff at the same moment, fall side by side, and reach the ground together. But a brick and a slate do not; and while the former falls in a nearly vertical direction, the latter describes a most complex

Experiment.

path. A sheet of paper or a fragment of gold leaf presents even greater irregularities than the slate. But by a slight modification of the circumstances, we gain a considerable insight into the nature of the question. The paper and gold leaf, if rolled into balls, fall nearly in a vertical line. Here, then, there are evidently at least two causes at work, one which tends to make all bodies fall, and fall vertically; and another which depends on the form and substance of the body, and tends to retard its fall and alter its course from the vertical direction. How can we study the effects of the former on all bodies without sensible complication from the latter? The effects of Wind, etc., at once point out *what* the latter cause is, the air (whose existence we may indeed suppose to have been discovered by such effects); and to study the nature of the action of the former it is necessary to get rid of the complications arising from the presence of air. Hence the necessity for *Experiment*. By means of an apparatus to be afterwards described, we remove the greater part of the air from the interior of a vessel, and in *that* we try again our experiments on the fall of bodies; and now a general law, simple in the extreme, though most important in its consequences, is at once apparent—viz., that *all* bodies, of whatever size, shape, or material, if dropped side by side at the same instant, fall side by side in a space void of air. Before experiment had thus separated the phenomena, hasty philosophers had rushed to the conclusion that some bodies possess the quality of *heaviness*, others that of *lightness*, etc. Had this state of confusion remained, the law of gravitation, vigorous though its action be throughout the universe, could never have been recognised as a general principle by the human mind.

Mere observation of lightning and its effects could never have led to the discovery of their relation to the phenomena presented by rubbed amber. A modification of the course of nature, such as the collecting of atmospheric electricity in our laboratories, was necessary. Without experiment we could never even have learned the existence of terrestrial magnetism.

Rules for the conduct of experiments.

375. When a particular agent or cause is to be studied, experiments should be arranged in such a way as to lead if possible to results depending on it alone; or, if this cannot be done, they should be arranged so as to show differences produced by varying it.

Rules for the conduct of experiments.

376. Thus to determine the resistance of a wire against the conduction of electricity through it, we may measure the whole strength of current produced in it by electromotive force between its ends when the amount of this electromotive force *is given*, or can be ascertained. But when the wire is that of a submarine telegraph cable there is always an *unknown* and ever varying electromotive force between its ends, due to the earth (producing what is commonly called the "earth-current"), and to determine its resistance, the difference in the strength of the current produced by suddenly adding to or subtracting from the terrestrial electromotive force the electromotive force of a given voltaic battery, is to be *very quickly* measured; and this is to be done over and over again, to eliminate the effect of variation of the earth-current during the few seconds of time which must elapse before the electrostatic induction permits the current due to the battery to reach nearly enough its full strength to practically annul error on this score.

377. Endless patience and perseverance in designing and trying different methods for investigation are necessary for the advancement of science: and indeed, in discovery, he is the most likely to succeed who, not allowing himself to be disheartened by the non-success of one form of experiment, judiciously varies his methods, and thus interrogates in every conceivably useful manner the subject of his investigations.

Residual phenomena.

378. A most important remark, due to Herschel, regards what are called *residual* phenomena. When, in an experiment, all known causes being allowed for, there remain certain unexplained effects (excessively slight it may be), these must be carefully investigated, and every conceivable variation of arrangement of apparatus, etc., tried; until, if possible, we manage so to isolate the residual phenomenon as to be able to detect its cause. It is here, perhaps, that in the present state of science we may most reasonably look for extensions of our knowledge; at all events we are warranted by the recent history of Natural Philosophy in so doing. Thus, to take only

Residual phenomena.

a very few instances, and to say nothing of the discovery of electricity and magnetism by the ancients, the peculiar smell observed in a room in which an electrical machine is kept in action, was long ago observed, but called the "smell of electricity," and thus left unexplained. The sagacity of Schönbein led to the discovery that this is due to the formation of Ozone, a most extraordinary body, of great chemical activity; whose nature is still uncertain, though the attention of chemists has for years been directed to it.

379. Slight anomalies in the motion of Uranus led Adams and Le Verrier to the discovery of a new planet; and the fact that the oscillations of a magnetized needle about its position of equilibrium are "damped" by placing a plate of copper below it, led Arago to his beautiful experiment showing a resistance to relative motion between a magnet and a piece of copper; which was first supposed to be due to magnetism in motion, but which soon received its correct explanation from Faraday, and has since been immensely extended, and applied to most important purposes. In fact, from this accidental remark about the oscillation of a needle was evolved the grand discovery of the Induction of Electrical Currents by magnets or by other currents.

We need not enlarge upon this point, as in the following pages the proofs of the truth and usefulness of the principle will continually recur. Our object has been not so much to give applications as principles, and to show how to attack a new combination, with the view of separating and studying in detail the various causes which generally conspire to produce observed phenomena, even those which are apparently the simplest.

Unexpected agreement or discordance of results of different trials.

380. If on repetition several times, an experiment continually gives different results, it must either have been very carelessly performed, or there must be some disturbing cause not taken account of. And, on the other hand, in cases where no very great coincidence is likely on repeated trials, an unexpected degree of agreement between the results of various trials should be regarded with the utmost suspicion, as probably due to some unnoticed peculiarity of the apparatus employed. In either of these cases, however, careful observation cannot fail to detect the cause of the discrepancies or of the unexpected agreement, and may possibly lead to discoveries in a totally unthought-of quarter. Instances of this kind may be given without limit; one or two must suffice.

Unexpected agreement or discordance of results of different trials.

381. Thus, with a *very* good achromatic telescope a star appears to have a sensible disc. But, as it is observed that the discs of all stars appear to be of equal angular diameter, we of course suspect some common error. Limiting the aperture of the object-glass *increases* the appearance in question, which, on full investigation, is found to have nothing to do with discs at all. It is, in fact, a diffraction phenomenon, and will be explained in our chapters on Light.

382. Again, in measuring the velocity of Sound by experiments conducted at night with cannon, the results at one station were never found to agree exactly with those at the other; sometimes, indeed, the differences were very considerable. But a little consideration led to the remark, that on those nights in which the discordance was greatest a strong wind was blowing nearly from one station to the other. Allowing for the obvious effect of this, or rather eliminating it altogether, the mean velocities on different evenings were found to agree very closely.

Hypotheses.

383. It may perhaps be advisable to say a few words here about the use of hypotheses, and especially those of very different gradations of value which are promulgated in the form of Mathematical Theories of different branches of Natural Philosophy.

384. Where, as in the case of the planetary motions and disturbances, the forces concerned are thoroughly known, the mathematical theory is absolutely true, and requires only analysis to work out its remotest details. It is thus, in general, far ahead of observation, and is competent to predict effects not yet even observed—as, for instance, Lunar Inequalities due to the action of Venus upon the Earth, etc. etc., to which no amount of observation, unaided by theory, could ever have enabled us to assign the true cause. It may also, in such subjects as Geometrical Optics, be carried to developments far beyond the reach

Hypotheses.

of experiment; but in this science the assumed bases of the theory are only approximate; and it fails to explain in all their peculiarities even such comparatively simple phenomena as Halos and Rainbows—though it is perfectly successful for the practical purposes of the maker of microscopes and telescopes, and has enabled really scientific instrument-makers to carry the construction of optical apparatus to a degree of perfection which merely tentative processes never could have reached.

385. Another class of mathematical theories, based to some extent on experiment, is at present useful, and has even in certain cases pointed to new and important results, which experiment has subsequently verified. Such are the Dynamical Theory of Heat, the Undulatory Theory of Light, etc. etc. In the former, which is based upon the conclusion from experiment that *heat is a form of energy*, many formulæ are at present obscure and uninterpretable, because we do not know the mechanism of the motions or distortions of the particles of bodies. Results of the theory in which these are not involved, are of course experimentally verified. The same difficulties exist in the Theory of Light. But before this obscurity can be perfectly cleared up, we must know something of the ultimate, or *molecular*, constitution of the bodies, or groups of molecules, at present known to us only in the aggregate.

Deduction of most probable result from a number of observations.

386. A third class is well represented by the Mathematical Theories of Heat (Conduction), Electricity (Statical), and Magnetism (Permanent). Although we do not know *how* Heat is propagated in bodies, nor *what* Statical Electricity or Permanent Magnetism are—the laws of their fluxes and forces are as certainly known as that of Gravitation, and can therefore like it be developed to their consequences, by the application of Mathematical Analysis. The works of Fourier*, Green†, and Poisson‡ areremarkable instances of such development. Another good example is Ampère's Theory of Electro-dynamics.

* *Théorie analytique de la Chaleur*, Paris, 1822.

† *Essay on the Application of Mathematical Analysis to the Theories of Electricity and Magnetism.* Nottingham, 1828. Reprinted in Crelle's Journal.

‡ *Mémoires sur le Magnétisme.* Mém. de l'Acad. des Sciences, 1811.

387. When the most probable result is required from a number of observations of the same quantity which do not exactly agree, we must appeal to the mathematical theory of probabilities to guide us to a method of combining the results of experience, so as to eliminate from them, as far as possible, the inaccuracies of observation. Of course it is to be understood that we do not here class as *inaccuracies of observation* any errors which may affect alike every one of a series of observations, such as the inexact determination of a zero point, or of the essential units of time and space, the personal equation of the observer, etc. The process, whatever it may be, which is to be employed in the elimination of errors, is applicable even to these, but only when *several distinct series* of observations have been made, with a change of instrument, or of observer, or of both.

Deduction of most probable result from a number of observations.

388. We understand as inaccuracies of observation the whole class of errors which are as likely to lie in one direction as in another in successive trials, and which we may fairly presume would, on the average of an infinite number of repetitions, exactly balance each other in excess and defect. Moreover, we consider only errors of such a kind that their probability is the less the greater they are; so that such errors as an accidental reading of a wrong number of whole degrees on a divided circle (which, by the way, can in general be "probably" corrected by comparison with other observations) are not to be included.

389. Mathematically considered, the subject is by no means an easy one, and many high authorities have asserted that the reasoning employed by Laplace, Gauss, and others, is not well founded; although the results of their analysis have been generally accepted. As an excellent treatise on the subject has recently been published by Airy, it is not necessary for us to do more than to sketch in the most cursory manner a simple and apparently satisfactory method of arriving at what is called the *Method of Least Squares.*

390. Supposing the zero-point and the graduation of an instrument (micrometer, mural circle, thermometer, electrometer,

Deduction of most probable result from a number of observations.

galvanometer, etc.) to be *absolutely* accurate, successive readings of the value of a quantity (linear distance, altitude of a star, temperature, potential, strength of an electric current, etc.) may, and in general do, continually differ. What is most probably the true value of the observed quantity?

The most probable value, in all such cases, if the observations are all equally trustworthy, will evidently be the simple mean; or if they are not equally trustworthy, the mean found by attributing *weights* to the several observations in proportion to their presumed exactness. But if several such means have been taken, or several single observations, and if these several means or observations have been differently qualified for the determination of the sought quantity (some of them being likely to give a more exact value than others), we must assign *theoretically* the best practical method of combining them.

391. Inaccuracies of observation are, in general, as likely to be in excess as in defect. They are also (as before observed) more likely to be small than great; and (practically) large errors are not to be expected at all, as such would come under the class of *avoidable mistakes.* It follows that in any one of a series of observations of the same quantity the probability of an error of magnitude x must depend upon x^2, and must be expressed by some function whose value diminishes very rapidly as x increases. The probability that the error lies between x and $x + \delta x$, where δx is very small, must also be proportional to δx.

Hence we may assume the probability of an error of any magnitude included in the range of x to $x + \delta x$ to be

$$\phi(x^2)\,\delta x.$$

Now the error must be included between $+\infty$ and $-\infty$. Hence, as a first condition,

$$\int_{-\infty}^{+\infty} \phi(x^2)\,dx = 1 \quad\ldots\ldots\ldots\ldots\ldots\ldots\ldots(1).$$

The consideration of a very simple case gives us the means of determining the form of the function ϕ involved in the preceding expression*.

* Compare Boole, *Trans. R.S.E.*, 1857. See also Tait, *Trans. R.S.E.*, 1864.

Suppose a stone to be let fall with the object of hitting a mark on the ground. Let two horizontal lines be drawn through the mark at right angles to one another, and take them as axes of x and y respectively. The chance of the stone falling at a distance between x and $x + \delta x$ from the axis of y is $\phi(x^2)\,\delta x$.

Deduction of most probable result from a number of observations.

Of its falling between y and $y + \delta y$ from the axis of x the chance is

$$\phi(y^2)\,\delta y.$$

The chance of its falling on the elementary area $\delta x \delta y$, whose co-ordinates are x, y, is therefore (since these are independent events, and it is to be observed that this is the assumption on which the whole investigation depends)

$$\phi(x^2)\,\phi(y^2)\,\delta x \delta y, \text{ or } a\phi(x^2)\,\phi(y^2),$$

if a denote the indefinitely small area about the point xy.

Had we taken any other set of rectangular axes with the same origin, we should have found for the same probability the expression

$$a\phi(x'^2)\,\phi(y'^2),$$

x', y' being the new co-ordinates of a. Hence we must have

$$\phi(x^2)\,\phi(y^2) = \phi(x'^2)\,\phi(y'^2), \text{ if } x^2 + y^2 = x'^2 + y'^2.$$

From this functional equation we have at once

$$\phi(x^2) = A\epsilon^{mx^2},$$

where A and m are constants. We see at once that m must be negative (as the chance of a large error is very small), and we may write for it $-\dfrac{1}{h^2}$, so that h will indicate the degree of delicacy or coarseness of the system of measurement employed.

Substituting in (1) we have

$$A\int_{-\infty}^{+\infty} \epsilon^{-\frac{x^2}{h^2}}\,dx = 1,$$

whence $A = \dfrac{1}{h\sqrt{\pi}}$, and the law of error is

$$\frac{1}{\sqrt{\pi}}\,\epsilon^{-\frac{x^2}{h^2}}\,\frac{\delta x}{h}.$$

Law of error.

The law of error, as regards *distance from the mark, without reference to the direction* of error, is evidently

$$\iint \phi(x^2)\,\phi(y^2)\,dxdy,$$

taken through the space between concentric circles whose radii are r and $r + \delta r$, and is therefore

$$\frac{2}{h^2}\,\epsilon^{-\frac{r^2}{h^2}}\,r\delta r,$$

Law of error.

which is of the same form as the law of error to the right or left of a line, with the additional factor r for the greater space for error at greater distances from the centre. As a verification, we see at once that

$$\frac{2}{h^2}\int_0^\infty \epsilon^{-\frac{r^2}{h^2}} r dr = 1$$

as was to be expected.

Probable error.

392. The *Probable Error* of an observation is a numerical quantity such that the error of the observation is as likely to exceed as to fall short of it in magnitude.

If we assume the law of error just found, and call P the probable error in one trial,

$$\int_0^P \epsilon^{-\frac{x^2}{h^2}} dx = \int_P^\infty \epsilon^{-\frac{x^2}{h^2}} dx.$$

The solution of this equation by trial and error leads to the approximate result

$$P = 0{\cdot}477\, h.$$

Probable error of a sum, difference, or multiple.

393. The probable error of any given multiple of the value of an observed quantity is evidently the same multiple of the probable error of the quantity itself.

The probable error of the sum or difference of two quantities, affected by *independent* errors, is the square root of the sum of the squares of their separate probable errors.

To prove this, let us investigate the *law* of error of

$$X \pm Y = Z$$

where the laws of error of X and Y are

$$\frac{1}{\sqrt{\pi}}\epsilon^{-\frac{x^2}{a^2}}\frac{dx}{a}, \text{ and } \frac{1}{\sqrt{\pi}}\epsilon^{-\frac{y^2}{b^2}}\frac{dy}{b},$$

respectively. The chance of an error in Z, of a magnitude included between the limits z, $z + \delta z$, is evidently

$$\frac{1}{\pi ab}\int_{-\infty}^{+\infty} \epsilon^{-\frac{x^2}{a^2}} dx \int_{z-x}^{z+\delta z - x} \epsilon^{-\frac{y^2}{b^2}} dy.$$

For, whatever value is assigned to x, the value of y is given by the limits $z - x$ and $z + \delta z - x$ [or $z + x$, $z + \delta z + x$; but the chances of $\pm x$ are the same, and both are included in the limits ($\pm\infty$) of integration with respect to x].

The value of the above integral becomes, by effecting the integration with respect to y,

Probable error of a sum, difference, or multiple.

$$\frac{\delta z}{\pi ab}\int_{-\infty}^{+\infty} \epsilon^{-\frac{x^2}{a^2}} \epsilon^{-\frac{(z-x)^2}{b^2}} dx,$$

and this is easily reduced to

$$\frac{1}{\sqrt{\pi}} \epsilon^{-\frac{z^2}{a^2+b^2}} \frac{\delta z}{\sqrt{a^2+b^2}}.$$

Thus the probable error is $0{\cdot}477\sqrt{a^2+b^2}$, whence the proposition. And the same theorem is evidently true for *any* number of quantities.

394. As above remarked, the principal use of this theory is in the deduction, from a large series of observations, of the values of the quantities sought in such a form as to be liable to the smallest probable error. As an instance—by the principles of physical astronomy, the place of a planet is calculated from assumed values of the elements of its orbit, and tabulated in the *Nautical Almanac*. The *observed* places do not exactly agree with the predicted places, for two reasons—first, the data for calculation are not exact (and in fact the main object of the observation is to correct their assumed values); second, each observation is in error to some unknown amount. Now the difference between the observed, and the calculated, places depends on the errors of assumed elements and of observation. The methods are applied to eliminate as far as possible the second of these, and the resulting equations give the required corrections of the elements.

Practical application.

Thus if θ be the calculated R.A. of a planet: δa, δe, $\delta\varpi$, etc., the corrections required for the assumed elements—the true R.A. is
$$\theta + A\delta a + E\delta e + \Pi\delta\varpi + \text{etc.},$$
where A, E, Π, etc., are approximately known. Suppose the observed R.A. to be Θ, then

Method of least squares.

$$\theta + A\delta a + E\delta e + \Pi\delta\varpi + \ldots = \Theta$$

or
$$A\delta a + E\delta e + \Pi\delta\varpi + \ldots = \Theta - \theta,$$

a known quantity, subject to error of observation. Every observation made gives us an equation of the same *form* as this, and in general the number of observations greatly exceeds that of the quantities δa, δe, $\delta\varpi$, etc., to be found. But it will be sufficient to consider the simple case where only *one* quantity is to be found.

Method of least squares.

Suppose a number of observations, of the same quantity x, lead to the following equations:—

$$x = B_1, \quad x = B_2, \quad \text{etc.},$$

and let the probable errors be E_1, E_2, ... Multiply the terms of each equation by numbers inversely proportional to E_1, E_2, ... This will make the probable errors of the second members of all the equations the same, e suppose. The equations have now the general form $\qquad ax = b,$

and it is required to find a system of linear factors, by which these equations, being multiplied in order and added, shall lead to a final equation giving the value of x with the probable error a minimum. Let them be f_1, f_2, etc. Then the final equation is

$$(\Sigma af)\, x = \Sigma\, (bf)$$

and therefore $\qquad P^2 (\Sigma af)^2 = e^2 \Sigma\, (f^2)$

by the theorems of § 393, if P denote the probable error of x.

Hence $\dfrac{\Sigma\,(f^2)}{(\Sigma af)^2}$ is a minimum, and its differential coefficients with respect to each separate factor f must vanish.

This gives a series of equations, whose general form is

$$f\Sigma\,(af) - a\Sigma\,(f^2) = 0,$$

which give evidently $f_1 = a_1$, $f_2 = a_2$, etc.

Hence the following rule, which may easily be seen to hold for any number of linear equations containing a smaller number of unknown quantities,

Make the probable error of the second member the same in each equation, by the employment of a proper factor; multiply each equation by the coefficient of x in it and add all, for one of the final equations; and so, with reference to y, z, etc., for the others. The probable errors of the values of x, y, etc., found from these final equations will be less than those of the values derived from any other *linear* method of combining the equations.

This process has been called the method of *Least Squares*, because the values of the unknown quantities found by it are such as to render the sum of the squares of the errors of the original equations a minimum.

That is, in the simple case taken above,

$$\Sigma\,(ax - b)^2 = \text{minimum}.$$

For it is evident that this gives, on differentiating with respect to x,

Method of least squares.

$$\Sigma a(ax-b)=0,$$

which is the law above laid down for the formation of the single equation.

Methods of representing experimental results.

395. When a series of observations of the same quantity has been made at different times, or under different circumstances, the law connecting the value of the quantity with the time, or some other variable, may be derived from the results in several ways—all more or less approximate. Two of these methods, however, are so much more extensively used than the others, that we shall devote a page or two here to a preliminary notice of them, leaving detailed instances of their application till we come to Heat, Electricity, etc. They consist in (1) a *Curve*, giving a graphic representation of the relation between the ordinate and abscissa, and (2) an *Empirical Formula* connecting the variables.

Curves.

396. Thus if the abscissæ represent intervals of time, and the ordinates the corresponding height of the barometer, we may construct curves which show at a glance the dependence of barometric pressure upon the time of day; and so on. Such curves may be accurately drawn by photographic processes on a sheet of sensitive paper placed behind the mercurial column, and made to move past it with a uniform horizontal velocity by clockwork. A similar process is applied to the Temperature and Electrification of the atmosphere, and to the components of terrestrial magnetism.

397. When the observations are not, as in the last section, continuous, they give us only a series of points in the curve, from which, however, we may in general approximate very closely to the result of continuous observation by drawing, *liberâ manu*, a curve passing through these points. This process, however, must be employed with great caution; because, unless the observations are sufficiently close to each other, most important fluctuations in the curve may escape notice. It is applicable, with abundant accuracy, to all cases where the quantity observed changes very slowly. Thus, for instance, weekly observations of the temperature at depths of from 6 to

Curves.

24 feet underground were found by Forbes sufficient for a very accurate approximation to the law of the phenomenon.

Interpolation and empirical formulæ.

398. As an instance of the processes employed for obtaining an empirical formula, we may mention methods of *Interpolation*, to which the problem can always be reduced. Thus from sextant observations, at known intervals, of the altitude of the sun, it is a common problem of astronomy to determine at what instant the altitude is greatest, and what is that greatest altitude. The first enables us to find the true solar time at the place; and the second, by the help of the *Nautical Almanac*, gives the latitude. The differential calculus, and the calculus of finite differences, give us formulæ for any required data; and Lagrange has shown how to obtain a very useful one by elementary algebra.

By Taylor's Theorem, if $y = f(x)$, we have

$$y = f(x_0 + \overline{x - x_0}) = f(x_0) + (x - x_0) f'(x_0) + \frac{(x - x_0)^2}{1 \, . \, 2} f''(x_0) + \ldots$$
$$+ \frac{(x - x_0)^n}{1 \, . \, 2 \ldots n} f^{(n)} [x_0 + \theta (x - x_0)] \ldots\ldots\ldots (1),$$

where θ is a proper fraction, and x_0 is *any* quantity whatever. This formula is useful only when the successive derived values of $f(x_0)$ diminish very rapidly.

In finite differences we have

$$f(x + h) = D^h f(x) = (1 + \Delta)^h f(x)$$
$$= f(x) + h \Delta f(x) + \frac{h(h-1)}{1 \, . \, 2} \Delta^2 f(x) + \ldots\ldots\ldots\ldots\ldots (2);$$

a very useful formula when the higher differences are small.

(1) suggests the proper form for the required expression, but it is only in rare cases that $f'(x_0)$, $f''(x_0)$, etc., are derivable directly from observation. But (2) is useful, inasmuch as the successive differences, $\Delta f(x)$, $\Delta^2 f(x)$, etc., are easily calculated from the tabulated results of observation, provided these have been taken for equal successive increments of x.

If for values $x_1, x_2, \ldots x_n$ a function takes the values $y_1, y_2, y_3, \ldots y_n$, Lagrange gives for it the obvious expression

$$\left[\frac{y_1}{x - x_1} \frac{1}{(x_1 - x_2)(x_1 - x_3) \ldots (x_1 - x_n)} + \frac{y_2}{x - x_2} \frac{1}{(x_2 - x_1)(x_2 - x_3) \ldots (x_2 - x_n)} + \ldots\right] (x - x_1)(x - x_2) \ldots (x - x_n).$$

Here it is of course assumed that the function required is a rational and integral one in x of the $n-1^{\text{th}}$ degree; and, in general, a similar limitation is in practice applied to the other formulæ above; for in order to find the complete expression for $f(x)$ in either, it is necessary to determine the values of $f'(x_0)$, $f''(x_0)$, ... in the first, or of $\Delta f(x)$, $\Delta^2 f(x)$, ... in the second. If n of the coefficients be required, so as to give the n chief terms of the general value of $f(x)$, we must have n observed simultaneous values of x and $f(x)$, and the expressions become determinate and of the $n-1^{\text{th}}$ degree in $x-x_0$ and h respectively.

Interpolation and empirical formulæ.

In practice it is usually sufficient to employ at most three terms of either of the first two series. Thus to express the length l of a rod of metal as depending on its temperature t, we may assume from (1)

$$l = l_0 + A(t-t_0) + B(t-t_0)^2,$$

l_0 being the measured length at any temperature t_0.

398′. These formulæ are practically useful for calculating the probable values of any observed element, for values of the independent variable lying within the range for which observation has given values of the element. But except for values of the independent variable either actually within this range, or not far beyond it in either direction, these formulæ express functions which, in general, will differ more and more widely from the truth the further their application is pushed beyond the range of observation.

Periodic functions.

In a large class of investigations the observed element is in its nature a periodic function of the independent variable. The harmonic analysis (§ 77) is suitable for all such. When the values of the independent variable for which the element has been observed are not equidifferent the coefficients, determined according to the method of least squares, are found by a process which is necessarily very laborious; but when they are equidifferent, and especially when the difference is a submultiple of the period, the equation derived from the method of least squares becomes greatly simplified. Thus, if θ denote an angle increasing in proportion to t, the time, through four right angles in the period, T, of the phenomenon; so that

$$\theta = \frac{2\pi t}{T};$$

Periodic functions.

let
$$f(\theta) = A_0 + A_1 \cos\theta + A_2 \cos 2\theta + \ldots$$
$$+ B_1 \sin\theta + B_2 \sin 2\theta + \ldots$$
where $A_0, A_1, A_2, \ldots B_1, B_2, \ldots$ are unknown coefficients, to be determined so that $f(\theta)$ may express the most probable value of the element, not merely at times between observations, but through all time as long as the phenomenon is strictly periodic. By taking as many of these coefficients as there are of distinct data by observation, the formula is made to agree precisely with these data. But in most applications of the method, the periodically recurring part of the phenomenon is expressible by a small number of terms of the harmonic series, and the higher terms, calculated from a great number of data, express either irregularities of the phenomenon not likely to recur, or errors of observation. Thus a comparatively small number of terms may give values of the element even for the very times of observation, more probable than the values actually recorded as having been observed, if the observations are numerous but not minutely accurate.

The student may exercise himself in writing out the equations to determine five, or seven, or more of the coefficients according to the method of least squares; and reducing them by proper formulæ of analytical trigonometry to their simplest and most easily calculated forms where the values of θ for which $f(\theta)$ is given are equidifferent. He will thus see that when the difference is $\frac{2\pi}{i}$, i being any integer, and when the number of the data is i or any multiple of it, the equations contain each of them only one of the unknown quantities: so that the method of least squares affords the most probable values of the coefficients, by the easiest and most direct elimination.

CHAPTER IV.

MEASURES AND INSTRUMENTS.

399. HAVING seen in the preceding chapter that for the investigation of the laws of nature we must carefully watch experiments, either those gigantic ones which the universe furnishes, or others devised and executed by man for special objects—and having seen that in all such observations accurate measurements of Time, Space, Force, etc., are absolutely necessary, we may now appropriately describe a few of the more useful of the instruments employed for these purposes, and the various standards or units which are employed in them.

Necessity of accurate measurements.

400. Before going into detail we may give a rapid *résumé* of the principal Standards and Instruments to be described in this chapter. As most, if not all, of them depend on physical principles to be detailed in the course of this work—we shall assume in anticipation the establishment of such principles, giving references to the future division or chapter in which the experimental demonstrations are more particularly explained. This course will entail a slight, but unavoidable, confusion—slight, because Clocks, Balances, Screws, etc., are familiar even to those who know nothing of Natural Philosophy; unavoidable, because it is in the very nature of our subject that no one part can grow alone, each requiring for its full development the utmost resources of all the others. But if one of our departments thus borrows from others, it is satisfactory to find that it more than repays by the power which its improvement affords them.

Classes of instruments.

401. We may divide our more important and fundamental instruments into four classes—

Those for measuring Time;
" " Space, linear or angular;
" " Force;
" " Mass.

Other instruments, adapted for special purposes such as the measurement of Temperature, Light, Electric Currents, etc., will come more naturally under the head of the particular physical energies to whose measurement they are applicable. Descriptions of self-recording instruments such as tide-gauges, and barometers, thermometers, electrometers, recording photographically or otherwise the continuously varying pressure, temperature, moisture, electric potential of the atmosphere, and magnetometers recording photographically the continuously varying direction and magnitude of the terrestrial magnetic force, must likewise be kept for their proper places in our work.

Calculating Machines have also important uses in assisting physical research in a great variety of ways. They belong to two classes:—

I. Purely Arithmetical, dealing with integral numbers of units. All of this class are evolved from the primitive use of the calculuses or little stones for counters (from which are derived the very names *calculation* and "The Calculus"), through such mechanism as that of the Chinese Abacus, still serving its original purpose well in infant schools, up to the Arithmometer of Thomas of Colmar and the grand but partially realized conceptions of calculating machines by Babbage.

II. Continuous Calculating Machines. As these are not only useful as auxiliaries for physical research but also involve dynamical and kinematical principles belonging properly to our subject, some of them have been described in the Appendix to this Chapter, from which dynamical illustrations will be taken in our chapters on Statics and Kinetics.

402. We shall consider in order the more prominent fundamental instruments of the four classes, and some of their most important applications:— **Classes of instruments.**

Clock, Chronometer, Chronoscope, Applications to Observation and to self-registering Instruments.

Vernier and Screw-Micrometer, Cathetometer, Spherometer, Dividing Engine, Theodolite, Sextant or Circle.

Common Balance, Bifilar Balance, Torsion Balance, Pendulum, Ergometer.

Among Standards we may mention—

1. *Time.*—Day, Hour, Minute, Second, sidereal and solar.
2. *Space.*—Yard and Mètre: Radian, Degree, Minute, Second.
3. *Force.*—Weight of a Pound or Kilogramme, etc., in any particular locality (gravitation unit); poundal, or dyne (kinetic unit).
4. *Mass.* Pound, Kilogramme, etc.

403. Although without instruments it is impossible to procure or apply any standard, yet, as without the standards no instrument could give us *absolute* measure, we may consider the standards first—referring to the instruments as if we already knew their principles and applications.

404. First we may notice the standards or units of angular measure: **Angular measure.**

Radian, or angle whose arc is equal to radius;

Degree, or ninetieth part of a right angle, and its successive subdivisions into sixtieths called *Minutes*, *Seconds*, *Thirds*, etc. The division of the right angle into 90 degrees is convenient because it makes the half-angle of an equilateral triangle ($\sin^{-1} \frac{1}{2}$) an integral number (30) of degrees. It has long been universally adopted by all Europe. The decimal division of the right angle, decreed by the French Republic when it successfully introduced other more sweeping changes, utterly and deservedly failed.

The division of the degree into 60 minutes and of the minute into 60 seconds is not convenient; and tables of the

Angular measure.

circular functions for degrees and hundredths of the degree are much to be desired. Meantime, when reckoning to tenths of a degree suffices for the accuracy desired, in any case the ordinary tables suffice, as 6′ is $\frac{1}{10}$ of a degree.

The decimal system is exclusively followed in reckoning by radians. The value of two right angles in this reckoning is 3·14159..., or π. Thus π radians is equal to 180°. Hence $180° \div \pi$ is 57°·29578..., or 57° 17′ 44″·8 is equal to one radian. In mathematical analysis, angles are uniformly reckoned in terms of the radian.

Measure of time.

405. The practical standard of time is the *Sidereal Day*, being the period, nearly constant*, of the earth's rotation about its axis (§ 247). From it is easily derived the *Mean Solar Day*, or the mean interval which elapses between successive passages of the sun across the meridian of any place. This is not so nearly as the Sidereal Day, an absolute or invariable unit:

* In our first edition it was stated in this section that Laplace had calculated from ancient observations of eclipses that the period of the earth's rotation about its axis had not altered by $\frac{1}{10000000}$ of itself since 720 B.C. In § 830 it was pointed out that this conclusion is overthrown by farther information from Physical Astronomy acquired in the interval between the printing of the two sections, in virtue of a correction which Adams had made as early as 1863 upon Laplace's dynamical investigation of an acceleration of the moon's mean motion, produced by the sun's attraction, showing that only about half of the observed acceleration of the moon's mean motion relatively to the angular velocity of the earth's rotation was accounted for by this cause. [Quoting from the first edition, § 830] "In 1859 Adams communicated to Delaunay his final result:—that at "the end of a century the moon is 5″·7 before the position she would have, "relatively to a meridian of the earth, according to the angular velocities of the "two motions, at the beginning of the century, and the acceleration of the "moon's motion truly calculated from the various disturbing causes then recog- "nized. Delaunay soon after verified this result: and about the beginning of "1866 suggested that the true explanation may be a retardation of the earth's "rotation by tidal friction. Using this hypothesis, and allowing for the conse- "quent retardation of the moon's mean motion by tidal reaction (§ 276), Adams, "in an estimate which he has communicated to us, founded on the rough as- "sumption that the parts of the earth's retardation due to solar and lunar tides "are as the squares of the respective tide-generating forces, finds 22^s as the "error by which the earth would in a century get behind a perfect clock rated "at the beginning of the century. If the retardation of rate giving this integral "effect were uniform (§ 35, *b*), the earth, as a timekeeper, would be going slower "by ·22 of a second per year in the middle, or ·44 of a second per year at the "end, than at the beginning of a century."

secular changes in the period of the earth's rotation about the sun affect it, though very slightly. It is divided into 24 hours, and the hour, like the degree, is subdivided into successive sixtieths, called minutes and seconds. The usual subdivision of seconds is decimal.

Measure of time.

It is well to observe that seconds and minutes of time are distinguished from those of angular measure by notation. Thus we have for time $13^h\ 43^m\ 27^s{\cdot}58$, but for angular measure $13^0\ 43'\ 27''{\cdot}58$.

When long periods of time are to be measured, the mean solar year, consisting of 366·242203 sidereal days, or 365·242242 mean solar days, or the century consisting of 100 such years, may be conveniently employed as the unit.

406. The ultimate standard of accurate chronometry must (if the human race live on the earth for a few million years) be founded on the physical properties of some body of more constant character than the earth: for instance, a carefully arranged metallic spring, hermetically sealed in an exhausted glass vessel. The time of vibration of such a spring would be necessarily more constant from day to day than that of the balance-spring of the best possible chronometer, disturbed as this is by the train of mechanism with which it is connected: and it would almost certainly be more constant from age to age than the time of rotation of the earth (cooling and shrinking, as it certainly is, to an extent that must be very considerable in fifty million years).

Necessity for a perennial standard. A spring suggested.

407. The British standard of length is the *Imperial Yard*, defined as the distance between two marks on a certain metallic bar, preserved in the Tower of London, when the whole has a temperature of 60° Fahrenheit. It was not directly derived from any fixed quantity in nature, although some important relations with such have been measured with great accuracy. It has been carefully compared with the length of a seconds pendulum vibrating at a certain station in the neighbourhood of London, so that if it should again be destroyed, as it was at the burning of the Houses of Parliament in 1834, and should all exact copies of it, of which several are preserved in various

Measure of length, founded on artificial metallic standards.

Earth's dimensions not constant,

places, be also lost, it can be restored by pendulum observations. A less accurate, but still (except in the event of earthquake disturbance) a very good, means of reproducing it exists in the measured base-lines of the Ordnance Survey, and the thence calculated distances between definite stations in the British Islands, which have been ascertained in terms of it with a degree of accuracy sometimes within an inch per mile, that is to say, within about $\frac{1}{60000}$.

408. In scientific investigations, we endeavour as much as possible to keep to one unit at a time, and the foot, which is defined to be one-third part of the yard, is, for British measurement, generally the most convenient. Unfortunately the inch, or one-twelfth of a foot, must sometimes be used. The statute mile, or 1760 yards, is most unhappily often used when great lengths are considered. The British measurements of area and volume are infinitely inconvenient and wasteful of brain-energy, and of plodding labour. Their contrast with the simple, uniform, metrical system of France, Germany, and Italy, is but little creditable to English intelligence.

nor easily measured with great accuracy.

409. In the French metrical system the decimal division is exclusively employed. The standard, (unhappily) called the *Mètre*, was defined originally as the ten-millionth part of the length of the quadrant of the earth's meridian from the pole to the equator; but it is now defined practically by the accurate standard metres laid up in various national repositories in Europe. It is somewhat longer than the yard, as the following Table shows:

Measure of length.

Inch = 25·39977 millimètres.	Centimètre = ·3937043 inch.
Foot = 3·047972 decimètres.	Mètre = 3·280869 feet.
British statute mile = 1609·329 mètres.	Kilomètre = ·6213767 British statute mile.

Measure of surface.

410. The unit of superficial measure is in Britain the square yard, in France the mètre carré. Of course we may use square inches, feet, or miles, as also square millimètres, kilomètres, etc., or the *Hectare* = 10,000 square mètres.

Measure of surface.

Square inch = 6·451483 square centimètres.
„ foot = 9·290135 „ decimètres.
„ yard = 83·61121 „ decimètres.
Acre = ·4046792 of a hectare.
Square British statute mile = 258·9946 hectares.
Hectare = 2·471093 acres.

411. Similar remarks apply to the cubic measure in the two countries, and we have the following Table:—

Measure of volume.

Cubic inch = 16·38661 cubic centimètres.
„ foot = 28·31606 „ decimètres or *Litres.*
Gallon = 4·543808 litres.
„ = 277·274 cubic inches, by Act of Parliament now repealed.
Litre = ·035315 cubic feet.

412. The British unit of mass is the Pound (defined by standards only); the French is the *Kilogramme*, defined originally as a litre of water at its temperature of maximum density; but now practically defined by existing standards.

Measure of mass.

Grain = 64·79896 milligrammes.	Gramme = 15·43235 grains.
Pound = 453·5927 grammes.	Kilogramme = 2·20462125 lbs.

Professor W. H. Miller finds (*Phil. Trans.* 1857) that the "*kilogramme des Archives*" is equal in mass to 15432·34874 grains; and the "*kilogramme type laiton*," deposited in the Ministère de l'Intérieure in Paris, as standard for French commerce, is 15432·344 grains.

413. The measurement of force, whether in terms of the weight of a stated mass in a stated locality, or in terms of the *absolute* or *kinetic* unit, has been explained in Chap. II. (See §§ 220—226). From the measures of force and length, we derive at once the measure of work or mechanical effect. That practically employed by engineers is founded on the gravitation measure of force. Neglecting the difference of gravity at London and Paris, we see from the above tables that the following relations exist between the London and the Parisian reckoning of work:—

Measure of force.

Foot-pound = 0·13825 kilogramme-mètre.
Kilogramme-mètre = 7·2331 foot-pounds.

Clock.

414. A *Clock* is primarily an instrument which, by means of a train of wheels, records the number of vibrations executed by a pendulum; a *Chronometer* or *Watch* performs the same duty for the oscillations of a flat spiral spring—just as the train of wheel-work in a gas-metre counts the number of revolutions of the main shaft caused by the passage of the gas through the machine. As, however, it is impossible to avoid friction, resistance of air, etc., a pendulum or spring, left to itself, would not long continue its oscillations, and, while its motion continued, would perform each oscillation in less and less time as the arc of vibration diminished: a continuous supply of energy is furnished by the descent of a weight, or the uncoiling of a powerful spring. This is so applied, through the train of wheels, to the pendulum or balance-wheel by means of a mechanical contrivance called an *Escapement,* that the oscillations are maintained of nearly uniform extent, and therefore of nearly uniform duration. The construction of escapements, as well as of trains of clock-wheels, is a matter of *Mechanics*, with the details of which we are not concerned, although it may easily be made the subject of mathematical investigation. The means of avoiding errors introduced by changes of temperature, which have been carried out in *Compensation* pendulums and balances, will be more properly described in our chapters on Heat. It is to be observed that there is little inconvenience if a clock lose or gain *regularly;* that can be easily and accurately allowed for: irregular rate is fatal.

Electrically controlled clocks.

415. By means of a recent application of electricity to be afterwards described, one good clock, carefully regulated from time to time to agree with astronomical observations, may be made (without injury to its own performance) to control any number of other less-perfectly constructed clocks, so as to compel their pendulums to vibrate, beat for beat, with its own.

Chronoscope.

416. In astronomical observations, time is estimated to tenths of a second by a practised observer, who, while watching the phenomena, counts the beats of the clock. But for the *very* accurate measurement of short intervals, many instruments have been devised. Thus if a small orifice be opened in a large and deep vessel full of mercury, and if we know by trial the weight of metal that escapes say in five minutes, a simple proportion gives the interval which elapses during the escape of any given weight. It is easy to contrive an adjustment by which a vessel may be placed under, and withdrawn from, the issuing stream at the time of occurrence of any two successive phenomena.

Chronoscope.

417. Other contrivances, called Stop-watches, Chronoscopes, etc., which can be read off at rest, started on the occurrence of any phenomenon, and stopped at the occurrence of a second, then again read off; or which allow of the making (by pressing a stud) a slight mark, on a dial revolving at a given rate, at the instant of the occurrence of each phenomenon to be noted, are common enough. But, of late, these have almost entirely given place to the Electric Chronoscope, an instrument which will be fully described later, when we shall have occasion to refer to experiments in which it has been usefully employed.

418. We now come to the measurement of space, and of angles, and for these purposes the most important instruments are the *Vernier* and the *Screw*.

Diagonal scale.

419. Elementary geometry, indeed, gives us the means of dividing any straight line into any assignable number of equal parts; but in practice this is by no means an accurate or reliable method. It was formerly used in the so-called Diagonal Scale, of which the construction is evident from the diagram. The reading is effected by a sliding-piece whose edge is perpendicular to the length of the scale. Suppose that it is PQ whose position on the scale is required. This can evidently cut only *one* of the transverse lines. *Its* number gives the number of tenths of an inch [4 in the figure], and the horizontal line next above the point of intersection gives evidently the number of hundredths [in the present case 4]. Hence the reading is 7·44. As an idea of the comparative uselessness of this

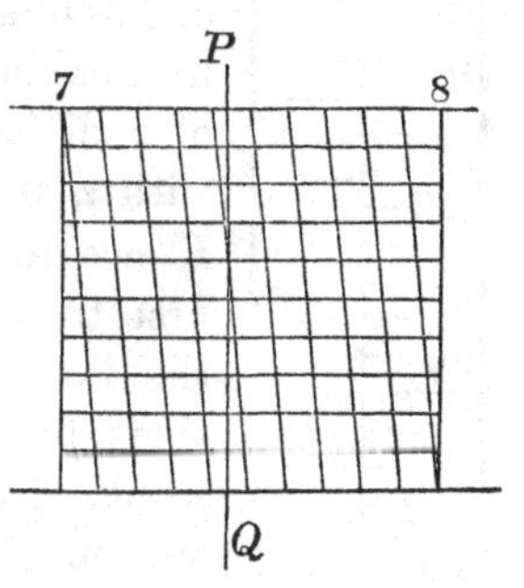

Diagonal scale.

method, we may mention that a quadrant of 3 feet radius, which belonged to Napier of Merchiston, and is divided on the limb by this method, reads to minutes of a degree; no higher accuracy than is now attainable by the pocket sextants made by Troughton and Simms, the radius of whose arc is virtually little more than an inch. The latter instrument is read by the help of a Vernier.

Vernier.

420. The Vernier is commonly employed for such instruments as the Barometer, Sextant, and Cathetometer, while the Screw is micrometrically applied to the more delicate instruments, such as Astronomical Circles, and Micrometers, and the Spherometer.

421. The vernier consists of a slip of metal which slides along a divided scale, the edges of the two being coincident. Hence, when it is applied to a divided circle, its edge is circular, and it moves about an axis passing through the centre of the divided limb.

In the sketch let 0, 1, 2,...10 be the divisions on the vernier, 0, 1, 2, etc., any set of consecutive divisions on the limb or scale along whose edge it slides. If, when 0 and 0 coincide, 10 and 11 coincide also, then 10 divisions of the vernier are equal in length to 11 on the limb; and therefore each division on the vernier is $\frac{11}{10}$ths or $1\frac{1}{10}$ of a division on the limb. If, then, the vernier be moved till 1 coincides with 1, 0 will be $\frac{1}{10}$th of a division of the limb beyond 0; if 2 coincide with 2, 0 will be $\frac{2}{10}$ths beyond 0; and so on. Hence to read the vernier in any position, note first the division next to 0, and behind it on the limb. This is the *integral* number of divisions to be read. For the fractional part, see which division of the vernier is in a line with one on the limb; if it be the 4th (as in the figure), that indicates an addition to the reading of $\frac{4}{10}$ths of a division of the limb; and so on. Thus, if the figure represent a barometer scale divided into inches and tenths, the reading is $30^{\text{in}}{\cdot}34$, the zero line of the vernier being adjusted to the level of the mercury.

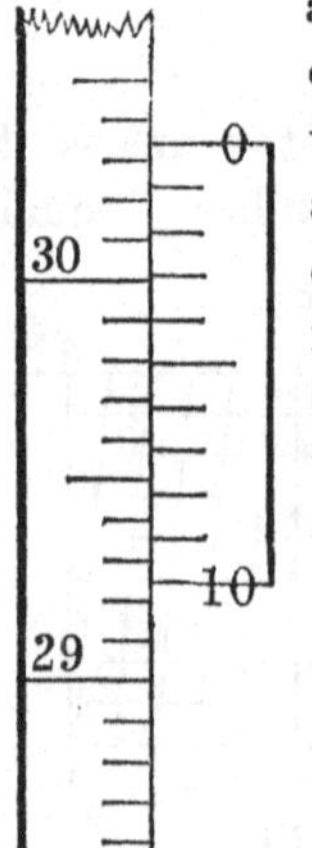

422. If the limb of a sextant be divided, as it usually is, to third parts of a degree, and the vernier be formed by dividing 21 of these into 20 equal parts, the instrument can be read to twentieths of divisions on the limb, that is, to minutes of arc.

Vernier.

If no line on the vernier coincide with one on the limb, then since the divisions of the former are the longer there will be one of the latter included between the two lines of the vernier, and it is usual in practice to take the mean of the readings which would be given by a coincidence of either pair of bounding lines.

423. In the above sketch and description, the numbers on the scale and vernier have been supposed to run *opposite* ways. This is generally the case with British instruments. In some foreign ones the divisions run in the same direction on vernier and limb, and in that case it is easy to see that to read to tenths of a scale division we must have ten divisions of the vernier equal to *nine* of the scale.

In general, to read to the nth part of a scale division, n divisions of the vernier must equal $n+1$ or $n-1$ divisions on the limb, according as these run in opposite or similar directions.

424. The principle of the *Screw* has been already noticed (§ 102). It may be used in either of two ways, *i.e.*, the nut may be fixed, and the screw advance through it, or the screw may be prevented from moving longitudinally by a fixed collar, in which case the nut, if prevented by fixed guides from rotating, will move in the direction of the common axis. The advance in either case is evidently proportional to the angle through which the screw has turned about its axis, and this may be measured by means of a divided head fixed perpendicularly to the screw at one end, the divisions being read off by a pointer or vernier attached to the frame of the instrument. The nut carries with it either a tracing point (as in the dividing engine) or a wire, thread, or half the object-glass of a telescope (as in micrometers), the thread or wire, or the play of the tracing point, being at right angles to the axis of the screw.

Screw.

425. Suppose it be required to divide a line into any number of equal parts. The line is placed parallel to the axis

Screw.

of the screw with one end exactly under the tracing point, or under the fixed wire of a microscope carried by the nut, and the screw-head is read off. By turning the head, the tracing point or microscope wire is brought to the other extremity of the line; and the number of turns and fractions of a turn required for the whole line is thus ascertained. Dividing this by the number of equal parts required, we find at once the number of turns and fractional parts corresponding to *one* of the required divisions, and by giving that amount of rotation to the screw over and over again, drawing a line after each rotation, the required division is effected.

Screw-Micrometer.

426. In the Micrometer, the movable wire carried by the nut is parallel to a fixed wire. By bringing them into optical contact the zero reading of the head is known; hence when another reading has been obtained, we have by subtraction the number of turns corresponding to the length of the object to be measured. The *absolute* value of a turn of the screw is determined by calculation from the number of threads in an inch, or by actually applying the micrometer to an object of known dimensions.

Spherometer.

427. For the measurement of the thickness of a plate, or the curvature of a lens, the *Spherometer* is used. It consists of a screw nut rigidly fixed in the middle of a very rigid three-legged table, with its axis perpendicular to the plane of the three feet (or finely rounded ends of the legs), and an accurately cut screw working in this nut. The lower extremity of the screw is also finely rounded. The number of turns, whole or fractional, of the screw, is read off by a divided head and a pointer fixed to the stem. Suppose it be required to measure the thickness of a plate of glass. The three feet of the instrument are placed upon a nearly enough flat surface of a hard body, and the screw is gradually turned until its point touches and presses the surface. The muscular sense of touch perceives resistance to the turning of the screw when, after touching the hard body, it presses on it with a force somewhat exceeding the weight of the screw. The first effect of the contact is a diminution of resistance to the turning, due to the weight of the screw coming to be borne on its fine pointed end instead of on the thread of the nut. The *sudden* increase of resistance at the instant when the screw commences to bear part of the weight of the nut finds the sense prepared to perceive it with remarkable delicacy on account of its contrast with the immediately preceding diminution of resistance. The screw-head is now read off, and the screw turned backwards until room is left for the insertion, beneath its point, of the plate whose thickness is to be measured. The screw is again turned until increase of resistance is again perceived; and the screw-head is again read off. The difference of the readings of the head is equal to the thickness of the plate, reckoned in the proper unit of the screw and the division of its head.

Sphero-meter.

428. If the curvature of a lens is to be measured, the instrument is first placed, as before, on a plane surface, and the reading for the contact is taken. The same operation is repeated on the spherical surface. The difference of the screw readings is evidently the greatest thickness of the glass which would be cut off by a plane passing through the three feet. This enables us to calculate the radius of the spherical surface (the distance from foot to foot of the instrument being known).

Let a be the distance from foot to foot, l the length of screw corresponding to the difference of the two readings, R the radius of the spherical surface; we have at once $2R = \frac{a^2}{3l} + l$, or, as l is generally very small compared with a, the diameter is, very approximately, $\frac{a^2}{3l}$.

429. The *Cathetometer* is used for the accurate determination of differences of level—for instance, in measuring the height to which a fluid rises in a capillary tube above the exterior free surface. It consists of a long divided metallic stem, turning round an axis as nearly as may be parallel to its length, on a fixed tripod stand: and, attached to the stem, a spirit-level. Upon the stem slides a metallic piece bearing a telescope of which the length is approximately enough perpendicular to the axis. The telescope tube is as nearly as may be perpendicular to the length of the stem. By levelling screws in two feet of the

Catheto-meter.

Catheto-meter.

tripod the bubble of the spirit-level is brought to one position of its glass when the stem is turned all round its axis. This secures that the axis is vertical. In using the instrument the telescope is directed in succession to the two objects whose difference of level is to be found, and in each case moved (generally by a delicate screw) up or down the stem, until a horizontal wire in the focus of its eye-piece coincides with the image of the object. The difference of readings on the vertical stem (each taken generally by aid of a vernier sliding-piece) corresponding to the two positions of the telescope gives the required difference of level.

Balance.

430. The common *Gravity Balance* is an instrument for testing the equality of the gravity of the masses placed in the two pans. We may note here a few of the precautions adopted in the best balances to guard against the various defects to which the instrument is liable; and the chief points to be attended to in its construction to secure delicacy, and rapidity of weighing.

The balance-beam should be very stiff, and as light as possible consistently with the requisite stiffness. For this purpose it is generally formed either of tubes, or of a sort of lattice-framework. To avoid friction, the axle consists of a knife-edge, as it is called; that is, a wedge of hard steel, which, when the balance is in use, rests on horizontal plates of polished agate. A similar contrivance is applied in very delicate balances at the points of the beam from which the scale-pans are suspended. When not in use, and just before use, the beam with its knife-edge is lifted by a lever arrangement from the agate plates. While thus secured it is loaded with weights as nearly as possible equal (this can be attained by previous trial with a coarser instrument), and the accurate determination is then readily effected. The last fraction of the required weight is determined by a rider, a very small weight, generally formed of wire, which can be worked (by a lever) from the outside of the glass case in which the balance is enclosed, and which may be placed in different positions upon one arm of the beam. This arm is graduated to tenths, etc., and thus shows at once the value of the rider in any case as depending on its moment or leverage, § 232.

431. Qualities of a balance: Balance.

1. *Stability.*—For stability of the beam alone without pans and weights, its centre of gravity must be below its bearing knife-edge. For stability with the heaviest weights the line joining the points at the ends of the beam from which the pans are hung must be below the knife-edge bearing the whole.

2. *Sensibility.*—The beam should be sensibly deflected from a horizontal position by the smallest difference between the weights in the scale-pans. The definite measure of the sensibility is the angle through which the beam is deflected by a stated difference between the loads in the pans.

3. *Quickness.*—This means rapidity of oscillation, and consequently speed in the performance of a weighing. It depends mainly upon the depth of the centre of gravity of the whole below the knife-edge and the length of the beam.

In our Chapter on Statics we shall give the investigation. The sensibility and quickness will there be calculated for any given form and dimensions of the instrument.

A fine balance should turn with about a 500,000th of the greatest load which can safely be placed in either pan. In fact few measurements of any kind are correct to more than *six* significant figures.

The process of *Double Weighing,* which consists in counterpoising a mass by shot, or sand, or pieces of fine wire, and then substituting weights for it in the same pan till equilibrium is attained, is more laborious, but more accurate, than single weighing; as it eliminates all errors arising from unequal length of the arms, etc.

Correction is required for the weights of air displaced by the two bodies weighed against one another when their difference is too large to be negligible.

432. In the *Torsion-balance,* invented and used with great effect by Coulomb, a force is measured by the torsion of a glass fibre, or of a metallic wire. The fibre or wire is fixed at its upper end, or at both ends, according to circumstances. In general it carries a very light horizontal rod or needle, to the extremities of which are attached the body on Torsion-balance.

Torsion-balance.

which is exerted the force to be measured, and a counterpoise. The upper extremity of the torsion fibre is fixed to an index passing through the centre of a divided disc, so that the angle through which that extremity moves is directly measured. If, at the same time, the angle through which the needle has turned be measured, or, more simply, if the index be always turned till the needle assumes a definite position determined by marks or sights attached to the case of the instrument—we have the amount of torsion of the fibre, and it becomes a simple statical problem to determine from the latter the force to be measured; its direction, and point of application, and the dimensions of the apparatus, being known. The force of torsion as depending on the angle of torsion was found by Coulomb to follow the law of simple proportion up to the limits of perfect elasticity—as might have been expected from Hooke's Law (see *Properties of Matter*), and it only remains that we determine the amount for a particular angle in absolute measure. This determination is in general simple enough in theory; but in practice requires considerable care and nicety. The torsion-balance, however, being chiefly used for comparative, not absolute, measure, this determination is often unnecessary. More will be said about it when we come to its applications.

433. The ordinary spiral spring-balances used for roughly comparing either small or large weights or forces, are, properly speaking, only a modified form of torsion-balance*, as they act almost entirely by the torsion of the wire, and not by longitudinal extension or by flexure. Spring-balances we believe to be capable, if carefully constructed, of rivalling the ordinary balance in accuracy, while, for some applications, they far surpass it in sensibility and convenience. They measure directly *force*, not *mass;* and therefore if used for determining masses in different parts of the earth, a correction must be applied for the varying force of gravity. The correction for temperature must not be overlooked. These corrections may be avoided by the method of double weighing.

* Binet, *Journal de l'École Polytechnique*, x. 1815: and J. Thomson, *Cambridge and Dublin Math. Journal* (1848).

Pendulum.

434. Perhaps the most delicate of all instruments for the measurement of force is the *Pendulum.* It is proved in kinetics (see Div. II.) that for any pendulum, whether oscillating about a mean vertical position under the action of gravity, or in a horizontal plane, under the action of magnetic force, or force of torsion, the square of the number of *small* oscillations in a given time is proportional to the magnitude of the force under which these oscillations take place.

For the estimation of the relative amounts of gravity at different places, this is by far the most perfect instrument. The method of coincidences by which this process has been rendered so excessively delicate will be described later.

Bifilar Balance.

435. The *Bifilar Suspension,* an arrangement for measuring small horizontal forces, or couples in horizontal planes, in terms of the weight of the suspended body, is due originally to Sir William Snow Harris, who used it in one of his electrometers, as a substitute for the simple torsion-balance of Coulomb. It was used also by Gauss in his bifilar magnetometer for measuring the horizontal component of the terrestrial magnetic force*. In this instrument the bifilar suspension is adjusted to keep a bar-magnet in a position approximately perpendicular to the magnetic meridian. The small natural augmentations and diminutions of the horizontal component are shown by small azimuthal motions of the bar. On account of some obvious mechanical and dynamical difficulties this instrument was not found very convenient for absolute determinations, but from the time of its first practical introduction by Gauss and Weber it has been in use in all Magnetic Observatories for measuring the natural variations of the horizontal magnetic component. It is now made with a much smaller magnet than the great bar weighing twenty-five pounds originally given with it by Gauss; but the bars in actual use at the present day are still enormously too large† for their duty. The weight of the

Bifilar Magnetometer.

* Gauss, *Resultate aus den Beobachtungen des magnetischen Vereins im Jahre* 1837. Translated in Taylor's *Scientific Memoirs*, Vol. II., Article VI.

† The suspended magnets used for determining the direction and the intensity of the horizontal magnetic force in the Dublin Magnetic Observatory,

Bifilar Magnetometer.

bar with attached mirror ought not to exceed eight grammes, so that two single silk fibres may suffice for the bearing threads. The only substantial alteration, besides the diminution of its magnitude, which has been made in the instrument since Gauss and Weber's time is the addition of photographic apparatus and clockwork for automatic record of its motions. For absolute determinations of the horizontal component force, Gauss's method of deflecting a freely suspended magnet by a magnetic bar brought into proper positions in its neighbourhood, and again making an independent set of observations to determine the period of oscillation of the same deflecting bar when suspended by a fine fibre and set to vibrate through a small horizontal angle on each side of the magnetic meridian, is the method which has been uniformly in use both in magnetic observatories and in travellers' observations with small portable apparatus since it was first invented by Gauss*.

Absolute measurement of Terrestrial Magnetic Force.

Bifilar Balance.

In the bifilar balance the two threads may be of unequal lengths, the line joining their upper fixed ends need not be horizontal, and their other ends may be attached to any two points of the suspended body: but for most purposes, and particularly for regular instruments such as electrometers and magnetometers with bifilar suspension, it is convenient to have, as nearly as may be, the two threads of equal length, their fixed ends at the same level, and their other ends attached to the suspended body symmetrically with reference to its centre of gravity (as illustrated in the last set of drawings of § 345[x]). Supposing the instrument-maker to have fulfilled these conditions of symmetry as nearly as he can with reference to the four points of attachment of the threads, we have still to adjust properly the lengths of the threads. For this purpose remark that a small difference in the lengths will throw the suspended body into an unsymmetrical

as described by Dr Lloyd in his *Treatise on Magnetism* (London, 1874), are each of them 15 inches long, $\frac{7}{8}$ of an inch broad, and $\frac{1}{4}$ of an inch in thickness, and must therefore weigh about a pound each. The corresponding magnets used at the Kew Observatory are much smaller. They are each 5·4 inches long, 0·8 inch broad, and 0·1 inch thick, and therefore the weight of each is about 0·012 pound, or nearly 55 grammes.

* *Intensitas Vis Magneticae Terrestris ad Mensuram Absolutam revocata*, Commentationes Societatis Gottingensis, 1832.

position, in which, particularly if its centre of gravity be very low (as it is in Sir W. Thomson's Quadrant Electrometer), much more of its weight will be borne by one thread than by the other. This will diminish very much the amount of the horizontal couple required to produce a stated azimuthal deflection in the regular use of the instrument, in other words will increase its sensibility above its proper amount, that is to say, the amount which it would have if the conditions of symmetry were fully realized. Hence the proper adjustment for equalizing the lengths of the threads in a symmetrical bifilar balance, or for giving them their right difference in an unsymmetrical arrangement, in order to make the instrument as accurate as it can be, is to alter the length of one or both of the threads, until we attain to the condition of *minimum sensibility*, that is to say minimum angle of deflection under the influence of a given amount of couple.

Bifilar Balance.

The great merit of the bifilar balance over the simple torsion-balance of Coulomb for such applications as that to the horizontal magnetometer in the continuous work of an observatory, is the comparative smallness of the influence it experiences from changes of temperature. The torsional rigidity of iron, copper, and brass wires is diminished about $\frac{1}{2}$ per cent. with 10° elevation of temperature, while the linear expansions of the same metals are each less than $\frac{1}{50}$ per cent. with the same elevation of temperature. Hence in the unifilar torsion-balance, if iron, copper, or brass (the only metals for which the change of torsional rigidity with change of temperature has hitherto been measured) is used for the material of the bearing fibre, the sensibility is augmented $\frac{1}{2}$ per cent. by 10° elevation of temperature.

On the other hand, in the bifilar balance, if torsional rigidity does not contribute any sensible proportion to the whole directive couple (and this condition may be realized as nearly as we please by making the bearing wires long enough and making the distance between them great enough to give the requisite amount of directive couple), the sensibility of the balance is affected only by the linear expansions of the substances concerned. If the equal distances between the two pairs of points

Bifilar Balance.

of attachment, in the normal form of bifilar balance (or that in which the two threads are vertical when the suspended body is uninfluenced by horizontal force or couple), remained constant, the sensibility would be augmented with elevation of temperature in simple proportion to the linear expansions of the bearing wires; and this small influence might, if it were worth while to make the requisite mechanical arrangements, be perfectly compensated by choosing materials for the frames or bars bearing the attachments of the wires so that the proportionate augmentation of the distance between them should be just half the elongation of either wire, because the sensibility, as shown by the mathematical formula below, is simply proportional to the length of the wires and inversely proportional to the square of the distance between them. But, even without any such compensation, the temperature-error due to linear expansions of the materials of the bifilar balance is so small that in the most accurate regular use of the instrument in magnetic observatories it may be almost neglected; and at most it is less than $\frac{1}{25}$ of the error of the unifilar torsion-balance, at all events if, as is probably the case, the changes of rigidity with changes of temperature in other metals are of similar amounts to those for the three metals on which experiments have been made. In reality the chief temperature-error of the bifilar magnetometer depends on the change of the magnetic moment of the suspended magnet with change of temperature. It seems that the magnetism of a steel magnet diminishes with rise of temperature and augments with fall of temperature, but experimental information is much wanted on this subject.

The amount of the effect is very different in different bars, and it must be experimentally determined for each bar serving in a bifilar magnetometer. The amount of the change of magnetic moment in the bar which had been most used in the Dublin Magnetic Observatory was found to be ·000029 per degree Fahrenheit or at the rate of ·000052 per degree Centigrade, being about the same amount as that of the change of torsional rigidity with temperature of the three metals referred to above.

Let a be the half length of the bar between the points of attachment of the wires, θ the angle through which the bar has been turned (in a horizontal plane) from its position of equilibrium, l the length of one of the wires, ι its inclination to the vertical.

Bifilar Balance.

Then $l\cos\iota$ is the difference of levels between the ends of each wire, and evidently, by the geometry of the case,

$$\tfrac{1}{2}l\sin\iota = a\sin\tfrac{1}{2}\theta.$$

Now if Q be the couple tending to turn the bar, and W its weight, the principle of mechanical effect gives

$$Qd\theta = -Wd(l\cos\iota)$$
$$= Wl\sin\iota d\iota.$$

But, by the geometrical condition above,

$$l^2\sin\iota\cos\iota d\iota = a^2\sin\theta d\theta.$$

Hence
$$\frac{Q}{a^2\sin\theta} = \frac{W}{l\cos\iota},$$

or
$$Q = \frac{Wa^2}{l}\frac{\sin\theta}{\sqrt{1-\frac{4a^2}{l^2}\sin^2\frac{\theta}{2}}},$$

which gives the couple in terms of the deflection θ.

If the torsion of the wires be taken into account, it is sensibly equal to θ (since the greatest inclination to the vertical is small), and therefore the couple resulting from it will be $E\theta$. This must be added to the value of Q just found in order to get the whole deflecting couple.

436. Ergometers are instruments for measuring energy. *White's friction brake* measures the amount of work actually performed in any time by an engine or other "prime mover," by allowing it during the time of trial to waste all its work on friction. *Morin's ergometer* measures work without wasting any of it, in the course of its transmission from the prime mover to machines in which it is usefully employed. It consists of a simple arrangement of springs, measuring at every instant the *couple* with which the prime mover turns the shaft that transmits its work, and an integrating machine from which the work done by this couple during any time can be read off.

Ergometers.

Let L be the couple at any instant, and ϕ the whole angle through which the shaft has turned from the moment at which the reckoning commences. The integrating machine shows at any moment the value of $\int Ld\phi$, which (§ 240) is the whole work done.

Ergometers.

437. White's friction brake consists of a lever clamped to the shaft, but not allowed to turn with it. The moment of the force required to prevent the lever from going round with the shaft, multiplied by the whole angle through which the shaft turns, measures the whole work done against the friction of the clamp. The same result is much more easily obtained by wrapping a rope or chain several times round the shaft, or round a cylinder or drum carried round by the shaft, and applying measured forces to its two ends in proper directions to keep it nearly steady while the shaft turns round without it. The difference of the moments of these two forces round the axis, multiplied by the angle through which the shaft turns, measures the whole work spent on friction against the rope. If we remove all other resistance to the shaft, and apply the proper amount of force at each end of the dynamimetric rope or chain (which is very easily done in practice), the prime mover is kept running at the proper speed for the test, and having its whole work thus wasted for the time and measured.

APPENDIX B′.

CONTINUOUS CALCULATING MACHINES.

I. Tide-predicting Machine.

Tide-predicting Machine.

The object is to predict the tides for any port for which the tidal constituents have been found from the harmonic analysis from tide-gauge observations; not merely to predict the times and heights of high water, but the depths of water at any and every instant, showing them by a continuous curve, for a year, or for any number of years in advance.

This object requires the summation of the simple harmonic functions representing the several constituents* to be taken into account, which is performed by the machine in the following manner:—For each tidal constituent to be taken into account the machine has a shaft with an overhanging crank, which carries a pulley pivoted on a parallel axis adjustable to a greater or less distance from the shaft's axis, according to the greater or less range of the particular tidal constituent for the different ports for which the machine is to be used. The several shafts, with their axes all parallel, are geared together so that their periods are to a sufficient degree of approximation proportional to the periods of the tidal constituents. The crank on each shaft can be turned round on the shaft and clamped in any position: thus it is set to the proper position for the epoch of the particular tide which it is to produce. The axes of the several shafts are horizontal, and their vertical planes are at successive distances one from another, each equal to the diameter of one of the pulleys (the diameters of these being equal). The shafts are in two rows, an upper and a lower, and the grooves of the pulleys are all in one plane perpendicular to their axes.

Suppose, now, the axes of the pulleys to be set each at zero distance from the axis of its shaft, and let a fine wire or chain,

* See Report for 1876 of the Committee of the British Association appointed for the purpose of promoting the Extension, Improvement, and Harmonic Analysis of Tidal Observations.

Tide-predicting Machine.

with one end hanging down and carrying a weight, pass alternately over and under the pulleys in order, and vertically upwards or downwards (according as the number of pulleys is even or odd) from the last pulley to a fixed point. The weight is to be properly guided for vertical motion by a geometrical slide. Turn the machine now, and the wire will remain undisturbed with all its free parts vertical and the hanging weight unmoved. But now set the axis of any one of the pulleys to a distance $\frac{1}{2} T$ from its shaft's axis and turn the machine. If the distance of this pulley from the two on each side of it in the other row is a considerable multiple of $\frac{1}{2} T$, the hanging weight will now (if the machine is turned uniformly) move up and down with a simple harmonic motion of amplitude (or semi-range) equal to T in the period of its shaft. If, next, a second pulley is displaced to a distance $\frac{1}{2} T'$, a third to a distance $\frac{1}{2} T''$, and so on, the hanging weight will now perform a complex harmonic motion equal to the sum of the several harmonic motions, *each* in its proper period, which would be produced separately by the displacements T, T', T''. Thus, if the machine was made on a large scale, with T, T',... equal respectively to the actual semi-ranges of the several constituent tides, and if it was turned round slowly (by clockwork, for example), each shaft going once round in the actual period of the tide which it represents, the hanging weight would rise and fall exactly with the water-level as affected by the whole tidal action. This, of course, could be of no use, and is only suggested by way of illustration. The actual machine is made of such magnitude, that it can be set to give a motion to the hanging weight equal to the actual motion of the water-level reduced to any convenient scale: and provided the whole range does not exceed about 30 centimetres, the geometrical error due to the deviation from perfect parallelism in the successive free parts of the wire is not so great as to be practically objectionable. The proper order for the shafts is the order of magnitude of the constituent tides which they produce, the greatest next the hanging weight, and the least next the fixed end of the wire: this so that the greatest constituent may have only one pulley to move, the second in magnitude only two pulleys, and so on.

One machine of this kind has already been constructed for the British Association, and another (with a greater number of shafts to include a greater number of tidal constituents) is being constructed for the Indian Government. The British Association machine, which is kept available for general use, under charge of the Science and Art Department in South Kensington, has ten shafts, which taken in order, from the hanging weight, give respectively the following tidal constituents*:

Tide-predicting Machine.

1. The mean lunar semi-diurnal.
2. The mean solar semi-diurnal.
3. The larger elliptic semi-diurnal.
4. The luni-solar diurnal declinational.
5. The lunar diurnal declinational.
6. The luni-solar semi-diurnal declinational.
7. The smaller elliptic semi-diurnal.
8. The solar diurnal declinational.
9. The lunar quarter-diurnal, or first shallow-water tide of mean lunar semi-diurnal.
10. The luni-solar quarter-diurnal, shallow-water tide.

The hanging weight consists of an ink-bottle with a glass tubular pen, which marks the tide level in a continuous curve on a long band of paper, moved horizontally across the line of motion of the pen, by a vertical cylinder geared to the revolving shafts of the machine. One of the five sliding points of the geometrical slide is the point of the pen sliding on the paper stretched on the cylinder, and the couple formed by the normal pressure on this point, and on another of the five, which is about four centimetres above its level and one and a half centimetres from the paper, balances the couple due to gravity of the ink-bottle and the vertical component of the pull of the bearing wire, which is in a line about a millimetre or two farther from the paper than that in which the centre of gravity moves. Thus is ensured, notwithstanding small inequalities on the paper, a pressure of the pen on the paper very approximately constant and as small as is desired.

Hour marks are made on the curve by a small horizontal movement of the ink-bottle's lateral guides, made once an hour; a somewhat greater movement, giving a deeper notch, serves to mark the noon of every day.

The machine may be turned so rapidly as to run off a year's tides for any port in about four hours.

Each crank should carry an adjustable counterpoise, to be

* See Report for 1876 of the British Association's Tidal Committee.

Tide-predicting Machine.

adjusted so that when the crank is not vertical the pulls of the approximately vertical portions of wire acting on it through the pulley which it carries shall, as exactly as may be, balance on the axis of the shaft, and the motion of the shaft should be resisted by a slight weight hanging on a thread wrapped once round it and attached at its other end to a fixed point. This part of the design, planned to secure against "lost time" or "back lash" in the gearings, and to preserve uniformity of pressure between teeth and teeth, teeth and screws, and ends of axles and "end-plates," was not carried out in the British Association machine.

II. Machine for the Solution of Simultaneous Linear Equations*.

Equation-Solver.

Let $B_1, B_2, \ldots B_n$ be n bodies each supported on a fixed axis (in practice each is to be supported on knife-edges like the beam of a balance).

Let $P_{11}, P_{21}, P_{31}, \ldots P_{n1}$ be n pulleys each pivoted on B_1;
$P_{12}, P_{22}, P_{32}, \ldots P_{n2}$ " " B_2;
$P_{13}, P_{23}, P_{33}, \ldots P_{n3}$ " " B_3;
....

" $C_1, C_2, C_3, \ldots C_n$, be n cords passing over the pulleys;
" $D_1, P_{11}, P_{12}, P_{13}, \ldots P_{1n}, E_1$, be the course of C_1;
" $D_2, P_{21}, P_{22}, P_{23}, \ldots P_{2n}, E_2$, " " C_2;
....

" $D_1, E_1, D_2, E_2, \ldots D_n, E_n$, be fixed points;
" $l_1, l_2, l_3, \ldots l_n$ be the lengths of the cords between D_1, E_1, and $D_2, E_2, \ldots$ and D_n, E_n, along the courses stated above, when $B_1, B_2, \ldots B_n$, are in particular positions which will be called their zero positions;
" $l_1 + e_1, l_2 + e_2, \ldots l_n + e_n$ be their lengths between the same fixed points, when $B_1, B_2, \ldots B_n$ are turned through angles x_1, $x_2, \ldots x_n$ from their zero positions;

(11), (12), (13), ... (1n),
(21), (22), (23), ... (2n),
(31), (32), (33), ... (3n),
....

* Sir W. Thomson, *Proceedings of the Royal Society*, Vol. XXVIII., 1878.

Equation-Solver.

quantities such that

$$\left.\begin{array}{l}(11)\,x_1+(12)\,x_2+\ldots+(1n)\,x_n=e_1\\(21)\,x_1+(22)\,x_2+\ldots+(2n)\,x_n=e_2\\(31)\,x_1+(32)\,x_2+\ldots+(3n)\,x_n=e_3\\\ldots\ldots\ldots\ldots\ldots\ldots\ldots\ldots\ldots\ldots\\(n1)\,x_1+(n2)\,x_2+\ldots+(nn)\,x_n=e_n\end{array}\right\}\ldots\ldots\ldots\ldots\ldots\text{(I)}.$$

We shall suppose $x_1, x_2, \ldots x_n$ to be each so small that (11), (12),...(21), etc., do not vary sensibly from the values which they have when $x_1, x_2, \ldots x_n$, are each infinitely small. In practice it will be convenient to so place the axes of $B_1, B_2, \ldots B_n$, and the mountings of the pulleys on $B_1, B_2, \ldots B_n$, and the fixed points D_1, E_1, D_2, etc., that when $x_1, x_2, \ldots x_n$ are infinitely small, the straight parts of each cord and the lines of infinitesimal motion of the centres of the pulleys round which it passes shall be all parallel. Then $\frac{1}{2}(11), \frac{1}{2}(21), \ldots \frac{1}{2}(n1)$ will be simply equal to the distances of the centres of the pulleys $P_{11}, P_{21}, \ldots P_{n1}$, from the axis of B_1; $\frac{1}{2}(12), \frac{1}{2}(22) \ldots \frac{1}{2}(n2)$ the distances of $P_{12}, P_{22}, \ldots P_{n2}$ from the axis of B_2; and so on.

In practice the mountings of the pulleys are to be adjustable by proper geometrical slides, to allow any prescribed positive or negative value to be given to each of the quantities (11), (12), ...(21), etc.

Suppose this to be done, and each of the bodies $B_1, B_2, \ldots B_n$ to be placed in its zero position and held there. Attach now the cords firmly to the fixed points $D_1, D_2, \ldots D_n$ respectively; and, passing them round their proper pulleys, bring them to the other fixed points $E_1, E_2, \ldots E_n$, and pass them through infinitely small smooth rings fixed at these points. Now hold the bodies $B_1, B_2, \ldots$ each fixed, and (in practice by weights hung on their ends, outside $E_1, E_2, \ldots E_n$) pull the cords through $E_1, E_2, \ldots E_n$ with any given tensions* $T_1, T_2, \ldots T_n$. Let $G_1, G_2, \ldots G_n$ be moments round the fixed axes of $B_1, B_2, \ldots B_n$ of the forces required to hold the bodies fixed when acted on by the cords thus

* The idea of force here first introduced is not essential, indeed is not technically admissible to the purely kinematic and algebraic part of the subject proposed. But it is not merely an ideal kinematic construction of the algebraic problem that is intended; and the design of a kinematic machine, for success in practice, essentially involves dynamical considerations. In the present case some of the most important of the purely algebraic questions concerned are very interestingly illustrated by these dynamical considerations.

Equation-Solver.

stretched. The principle of "virtual velocities," just as it came from Lagrange (or the principle of "work"), gives immediately, in virtue of (I),

$$\left.\begin{array}{l} G_1 = (11)\,T_1 + (21)\,T_2 + \ldots + (n1)\,T_n \\ G_2 = (12)\,T_1 + (22)\,T_2 + \ldots + (n2)\,T_n \\ \ldots\ldots\ldots\ldots\ldots\ldots\ldots\ldots\ldots\ldots\ldots\ldots \\ G_n = (1n)\,T_1 + (2n)\,T_2 + \ldots + (nn)\,T_n \end{array}\right\} \ldots\ldots\ldots\ldots\ldots (II).$$

Apply and keep applied to each of the bodies, B_1, B_2, ... B_n (in practice by the weights of the pulleys, and by counter-pulling springs), such forces as shall have for their moments the values G_1, G_2 ... G_n, calculated from equations (II) with whatever values seem desirable for the tensions T_1, T_2, ... T_3. (In practice, the straight parts of the cords are to be approximately vertical, and the bodies B_1, B_2, are to be each balanced on its axis when the pulleys belonging to it are removed, and it is advisable to make the tensions each equal to half the weight of one of the pulleys with its adjustable frame.) The machine is now ready for use. To use it, pull the cords simultaneously or successively till lengths equal to e_1, e_2, ... e_n are passed through the rings E_1, E_2, ... E_n, respectively.

The *pulls* required to do this may be positive or negative; in practice, they will be infinitesimal downward or upward pressures applied by hand to the stretching weights which remain permanently hanging on the cords.

Observe the angles through which the bodies B_1, B_2, ... B_n are turned by this given movement of the cords. These angles are the required values of the unknown x_1, x_2, ... x_n, satisfying the simultaneous equations (I).

The actual construction of a practically useful machine for calculating as many as eight or ten or more of unknowns from the same number of linear equations does not promise to be either difficult or over-elaborate. A fair approximation having been found by a first application of the machine, a very moderate amount of straightforward arithmetical work (aided very advantageously by Crelle's multiplication tables) suffices to calculate the residual errors, and allow the machines (with the setting of the pulleys unchanged) to be re-applied to calculate the corrections (which may be treated decimally, for convenience): thus, 100 times the amount of the correction on each of the original unknowns may be made the new unknowns, if the magnitudes thus

Equation-Solver.

falling to be dealt with are convenient for the machine. There is, of course, no limit to the accuracy thus obtainable by successive approximations. The exceeding easiness of each application of the machine promises well for its real usefulness, whether for cases in which a single application suffices, or for others in which the requisite accuracy is reached after two, three, or more, of successive approximations.

The accompanying drawings represent a machine for finding six* unknowns from six equations. Fig. 1 represents in elevation and plan one of the six bodies B_1, B_2, etc. Fig. 2 shows in elevation and plan one of the thirty-six pulleys P, with its cradle on geometrical slide (§ 198). Fig. 3 shows in front-elevation the general disposition of the instrument.

Fig. 1. One of the six moveable bodies, B.

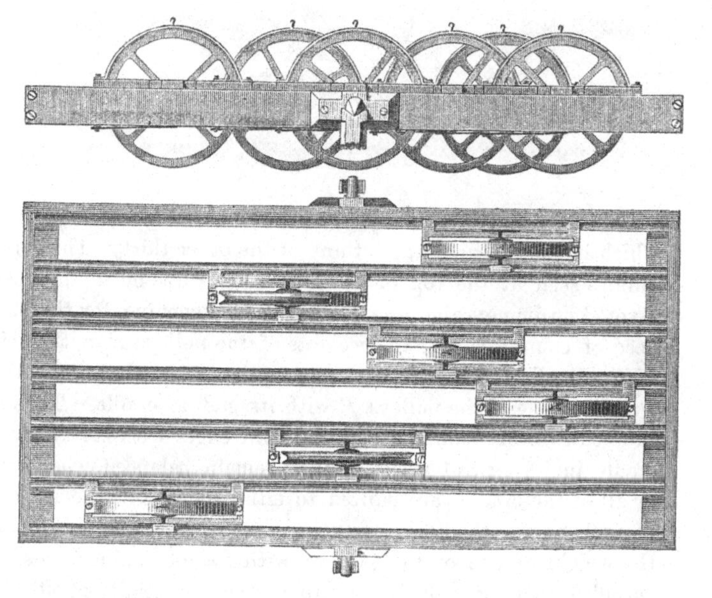

Elevation.

Plan.

* This number has been chosen for the first practical machine to be constructed, because a chief application of the machine may be to the calculation of the corrections on approximate values already found of the six elements of the orbit of a comet or asteroid.

Equation-Solver.

FIG. 2. One of the thirty-six pulleys, P, with its sliding cradle.
Full Size.

In Fig. 3 only one of the six cords, and the six pulleys over which it passes, is shown, not any of the other thirty. The three pulleys seen at the top of the sketch are three out of eighteen pivoted on immoveable bearings above the machine, for the purpose of counterpoising the weights of the pulleys P, with their sliding cradles. Each of the counterpoises is equal to twice the weight of one of the pulleys P with its sliding cradle. Thus if the bodies B are balanced on their knife-edges with each sliding cradle in its central position, they remain balanced when one or all of the cradles are shifted to either side; and the tension of each of the thirty-six essential cords is exactly equal to half the weight of one of the pulleys with its adjustable frame, as specified above (the deviations from exact verticality of all the free portions of the thirty-six essential cords and the eighteen counterpoising cords being neglected).

FIG. 3. General disposition of machine.

Equation-Solver.

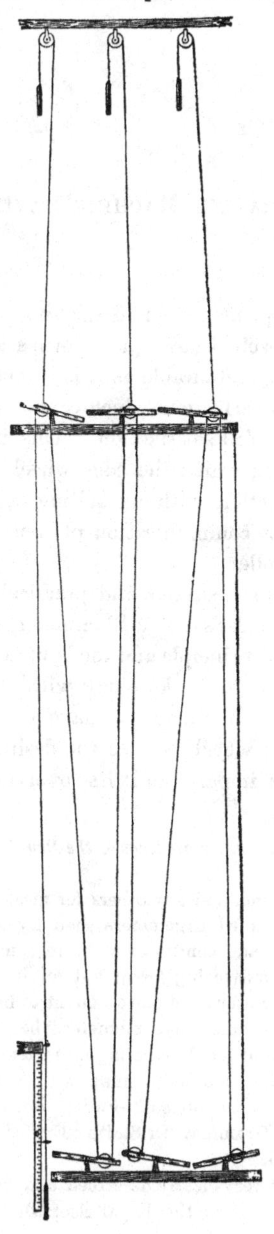

III. An Integrating Machine having a New Kinematic Principle*.

Disk-, Globe-, and Cylinder-Integrating Machine.

The kinematic principle for integrating ydx, which is used in the instruments well known as Morin's Dynamometer† and Sang's Planimeter‡, admirable as it is in many respects, involves one element of imperfection which cannot but prevent our contemplating it with full satisfaction. This imperfection consists in the sliding action which the edge wheel or roller is required to take in conjunction with its rolling action, which alone is desirable for exact communication of motion from the disk or cone to the edge roller.

The very ingenious, simple, and practically useful instrument well known as Amsler's Polar Planimeter, although different in its main features of principle and mode of action from the instruments just referred to, ranks along with them in involving the like imperfection of requiring to have a sidewise sliding action of its edge rolling wheel, besides the desirable rolling action on the surface which imparts to it its revolving motion—a surface

* Professor James Thomson, *Proceedings of the Royal Society*, Vol. xxiv., 1876, p. 262.

† Instruments of this kind, and any others for measuring mechanical work, may better in future be called Ergometers than Dynamometers. The name "dynamometer" has been and continues to be in common use for signifying a spring instrument for measuring *force;* but an instrument for measuring *work*, being distinct in its nature and object, ought to have a different and more suitable designation. The name "dynamometer," besides, appears to be badly formed from the Greek; and for designating an instrument for *measurement of force*, I would suggest that the name may with advantage be changed to *dynamimeter*. In respect to the mode of forming words in such cases, reference may be made to Curtius's Grammar, Dr Smith's English edition, § 354, p. 220.—J. T., 26th February, 1876.

‡ Sang's Planimeter is very clearly described and figured in a paper by its inventor, in the Transactions of the Royal Scottish Society of Arts, Vol. iv. January 12, 1852.

which in this case is not a disk or cone, but is the surface of the paper, or any other plane face, on which the map or other plane diagram to be evaluated in area is drawn.

Disk-, Globe-, and Cylinder-Integrating Machine.

Professor J. Clerk Maxwell, having seen Sang's Planimeter in the Great Exhibition of 1851, and having become convinced that the combination of slipping and rolling was a drawback on the perfection of the instrument, began to search for some arrangement by which the motion should be that of perfect rolling in every action of the instrument, corresponding to that of combined slipping and rolling in previous instruments. He succeeded in devising a new form of planimeter or integrating machine with a quite new and very beautiful principle of kinematic action depending on the mutual rolling of two equal spheres, each on the other. He described this in a paper submitted to the Royal Scottish Society of Arts in January 1855, which is published in Vol. IV. of the Transactions of that Society. In that paper he also offered a suggestion, which appears to be both interesting and important, proposing the attainment of the desired conditions of action by the mutual rolling of a cone and cylinder with their axes at right angles.

The idea of using pure rolling instead of combined rolling and slipping was communicated to me by Prof. Maxwell, when I had the pleasure of learning from himself some particulars as to the nature of his contrivance. Afterwards (some time between the years 1861 and 1864), while endeavouring to contrive means for the attainment in meteorological observatories of certain integrations in respect to the motions of the wind, and also in endeavouring to devise a planimeter more satisfactory in principle than either Sang's or Amsler's planimeter (even though, on grounds of practical simplicity and convenience, unlikely to turn out preferable to Amsler's in ordinary cases of taking areas from maps or other diagrams, but something that I hoped might possibly be attainable which, while having the merit of working by pure rolling contact, might be simpler than the instrument of Prof. Maxwell and preferable to it in mechanism), I succeeded in devising for the desired object a new kinematic method, which has ever since appeared to me likely sometime to prove valuable when occasion for its employment might be found. Now, within the last few days, this principle, on being suggested to my brother as perhaps capable of being usefully employed towards the development of tide-calculating machines

Disk-, Globe-, and Cylinder-Integrator.

which he had been devising, has been found by him to be capable of being introduced and combined in several ways to produce important results. On his advice, therefore, I now offer to the Royal Society a brief description of the new principle as devised by me.

The new principle consists primarily in the transmission of motion from a disk or cone to a cylinder by the intervention of a loose ball, which presses by its gravity on the disk and cylinder, or on the cone and cylinder, as the case may be, the pressure being sufficient to give the necessary frictional coherence at each point of rolling contact; and the axis of the disk or cone and that of the cylinder being both held fixed in position by bearings in stationary framework, and the arrangement of these axes being such that when the disk or the cone and the cylinder are kept steady, or, in other words, without rotation on their axes, the ball can roll along them in contact with both, so that the point of rolling contact between the ball and the cylinder shall traverse a straight line on the cylindric surface parallel necessarily to the axis of the cylinder—and so that, in the case of a disk being used, the point of rolling contact of the ball with the disk shall traverse a straight line passing through the centre of the disk—or that, in case of a cone being used, the line of rolling contact of the ball on the cone shall traverse a straight line on the conical surface, directed necessarily towards the vertex of the cone. It will thus readily be seen that, whether the cylinder and the disk or cone be at rest or revolving on their axes, the two lines of rolling contact of the ball, one on the cylindric surface and the other on the disk or cone, when both considered as lines traced out in space fixed relatively to the framing of the whole instrument, will be two parallel straight lines, and that the line of motion of the ball's centre will be straight and parallel to them. For facilitating explanations, the motion of the centre of the ball along its path parallel to the axis of the cylinder may be called the ball's longitudinal motion.

Now for the integration of ydx: the distance of the point of contact of the ball with the disk or cone from the centre of the disk or vertex of the cone in the ball's longitudinal motion is to represent y, while the angular space turned by the disk or cone from any initial position represents x; and then the angular space turned by the cylinder will, when multiplied by a suitable constant numerical coefficient, express the integral in terms of any required unit for its evaluation.

Disk-, Globe-, and Cylinder-Integrator.

The longitudinal motion may be imparted to the ball by having the framing of the whole instrument so placed that the lines of longitudinal motion of the two points of contact and of the ball's centre, which are three straight lines mutually parallel, shall be inclined to the horizontal sufficiently to make the ball tend decidedly to descend along the line of its longitudinal motion, and then regulating its motion by an abutting controller, which may have at its point of contact, where it presses on the ball, a plane face perpendicular to the line of the ball's motion. Otherwise the longitudinal motion may, for some cases, preferably be imparted to the ball by having the direction of that motion horizontal, and having two controlling flat faces acting in close contact without tightness at opposite extremities of the ball's diameter, which at any moment is in the line of the ball's motion or is parallel to the axis of the cylinder.

It is worthy of notice that, in the case of the disk-, ball-, and cylinder-integrator, no theoretical nor important practical fault in the action of the instrument would be involved in any deficiency of perfect exactitude in the practical accomplishment of the desired condition that the line of motion of the ball's point of contact with the disk should pass through the centre of the disk. The reason of this will be obvious enough on a little consideration.

The plane of the disk may suitably be placed inclined to the horizontal at some such angle as 45°; and the accompanying sketch, together with the model, which will be submitted to the Society by my brother, will aid towards the clear understanding of the explanations which have been given.

My brother has pointed out to me that an additional operation, important for some purposes, may be effected by arranging that the machine shall give a continuous record of the growth of the integral by introducing additional mechanisms suitable for continually describing a curve such that for each point of it the abscissa shall represent the value of x, and the ordinate shall represent the integral attained from $x=0$ forward to that value of x. This, he has pointed out, may be effected in practice by having a cylinder axised on the axis of the disk, a roll of paper covering this cylinder's surface, and a straight bar situated parallel to this cylinder's axis and resting with enough of pres-

Disk-, Globe-, and Cylinder-Integrator.

sure on the surface of the primary registering or *the indicating* cylinder (the one, namely, which is actuated by its contact with the ball) to make it have sufficient frictional coherence with that

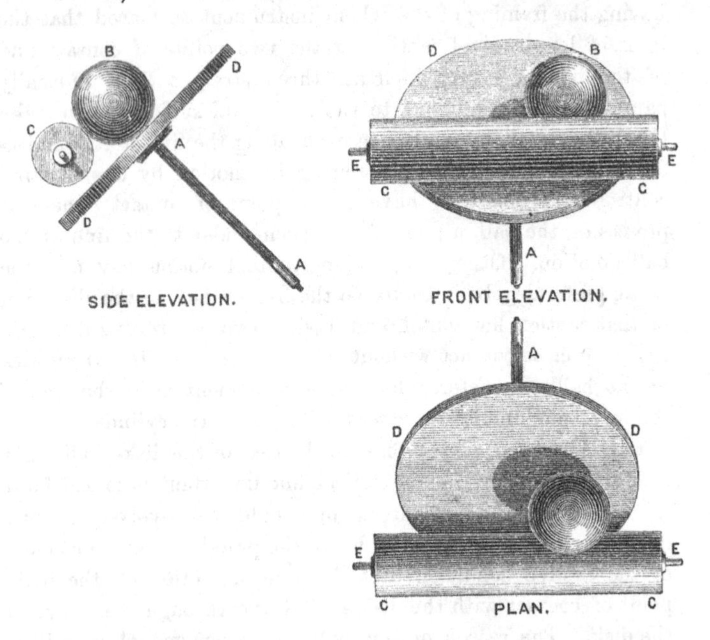

surface, and by having this bar made to carry a pencil or other tracing point which will mark the desired curve on the secondary registering or *the recording* cylinder. As, from the nature of the apparatus, the axis of the disk and of the secondary registering or recording cylinder ought to be steeply inclined to the horizontal, and as, therefore, this bar, carrying the pencil, would have the line of its length and of its motion alike steeply inclined with that axis, it seems that, to carry out this idea, it may be advisable to have a thread attached to the bar and extending off in the line of the bar to a pulley, passing over the pulley, and having suspended at its other end a weight which will be just sufficient to counteract the tendency of the rod, in virtue of gravity, to glide down along the line of its own slope, so as to leave it perfectly free to be moved up or down by the frictional coherence between itself and the moving surface of the indicating cylinder worked directly by the ball.

IV. AN INSTRUMENT FOR CALCULATING $\left(\int \phi(x)\psi(x)\,dx\right)$, THE INTEGRAL OF THE PRODUCT OF TWO GIVEN FUNCTIONS*.

Machine to calculate Integral of Product of two Functions.

In consequence of the recent meeting of the British Association at Bristol, I resumed an attempt to find an instrument which should supersede the heavy arithmetical labour of calculating the integrals required to analyze a function into its simple harmonic constituents according to the method of Fourier. During many years previously it had appeared to me that the object ought to be accomplished by some simple mechanical means; but it was not until recently that I succeeded in devising an instrument approaching sufficiently to simplicity to promise practically useful results. Having arrived at this stage, I described my proposed machine a few days ago to my brother Professor James Thomson, and he described to me in return a kind of mechanical integrator which had occurred to him many years ago, but of which he had never published any description. I instantly saw that it gave me a much simpler means of attaining my special object than anything I had been able to think of previously. An account of his integrator is communicated to the Royal Society along with the present paper.

To calculate $\int \phi(x)\,\psi(x)\,dx$, the rotating disk is to be displaced from a zero or initial position through an angle equal to

$$\int_0^x \phi(x)\,dx,$$

while the rolling globe is moved so as always to be at a distance from its zero position equal to $\psi(x)$. This being done, the cylinder obviously turns through an angle equal to $\int_0^x \phi(x)\,\psi(x)\,dx$, and thus solves the problem.

One way of giving the required motions to the rotating disk and rolling globe is as follows:—

* Sir W. Thomson, *Proceedings of the Royal Society*, Vol. XXIV., 1876, p. 266.

Machine to calculate Integral of Product of two Functions.

On two pieces of paper draw the curves

$$y=\int_0^x \phi(x)\,dx, \text{ and } y=\psi(x).$$

Attach these pieces of paper to the circumference of two circular cylinders, or to different parts of the circumference of one cylinder, with the axis of x in each in the direction perpendicular to the axis of the cylinder. Let the two cylinders (if there are two) be geared together so as that their circumferences shall move with equal velocities. Attached to the framework let there be, close to the circumference of each cylinder, a slide or guide-rod to guide a moveable point, moved by the hand of an operator, so as always to touch the curve on the surface of the cylinder, while the two cylinders are moved round.

Two operators will be required, as one operator could not move the two points so as to fulfil this condition—at all events unless the motion were very slow. One of these points, by proper mechanism, gives an angular motion to the rotating disk equal to its own linear motion, the other gives a linear motion equal to its own to the centre of the rolling globe.

The machine thus described is immediately applicable to calculate the values H_1, H_2, H_3, etc. of the harmonic constituents of a function $\psi(x)$ in the splendid generalization of Fourier's simple harmonic analysis, which he initiated himself in his solutions for the conduction of heat in the sphere and the cylinder, and which was worked out so ably and beautifully by Poisson*, and by Sturm and Liouville in their memorable papers on this subject published in the first volume of Liouville's *Journal des Mathématiques.* Thus if

$$\psi(x)=H_1\phi_1(x)+H_2\phi_2(x)+H_3\phi_3(x)+\text{etc.}$$

be the expression for an arbitrary function ψx, in terms of the generalized harmonic functions $\phi_1(x)$, $\phi_2(x)$, $\phi_3(x)$, etc., these functions being such that

$$\int_0^l \phi_1(x)\,\phi_2(x)\,dx=0,\ \int_0^l \phi_1(x)\,\phi_3(x)\,dx=0,\ \int_0^l \phi_2(x)\,\phi_3(x)=0, \text{ etc.},$$

* His general demonstration of the reality of the roots of transcendental equations essential to this analysis (an exceedingly important step in advance from Fourier's position), which he first gave in the *Bulletin de la Société Philomathique* for 1828, is reproduced in his *Théorie Mathématique de la Chaleur*, § 90.

we have

Machine to calculate Integral of Product of two Functions.

$$H_1 = \frac{\int_0^l \phi_1(x)\psi(x)\,dx}{\int_0^l \{\phi_1(x)\}^2\,dx},$$

$$H_2 = \frac{\int_0^l \phi_2(x)\psi(x)\,dx}{\int_0^l \{\phi_2(x)\}^2\,dx},$$

etc.

In the physical applications of this theory the integrals which constitute the denominators of the formulas for H_1, H_2, etc. are always to be evaluated in finite terms by an extension of Fourier's formula for the $\int_0^X xu_i^2\,dx$ of his problem of the cylinder* made by Sturm in equation (10), § iv. of his *Mémoire sur une Classe d'Équations à différences partielles* in Liouville's *Journal*, Vol. I. (1836). The integrals in the numerators are calculated with great ease by aid of the machine worked in the manner described above.

The great practical use of this machine will be to perform the simple harmonic Fourier-analysis for tidal, meteorological, and perhaps even astronomical, observations. It is the case in which

$$\phi(x) = {\sin \atop \cos}(nx);$$

and the integration is performed through a range equal to $\frac{2i\pi}{n}$ (i any integer) that gives this application. In this case the addition of a simple crank mechanism, to give a simple harmonic angular motion to the rotating disk in the proper period $\frac{2\pi}{n}$, when the cylinder bearing the curve $y = \psi(x)$ moves uniformly, supersedes the necessity for a cylinder with the curve $y = \phi(x)$ traced on it, and an operator keeping a point always on this curve in the manner described above. Thus one operator will be enough to carry on the process; and I believe that in the application of it to the tidal harmonic analysis he will be able in an

* Fourier's *Théorie Analytique de la Chaleur*, § 319, p. 391 (Paris, 1822).

Machine to calculate Integral of Product of two Functions.

hour or two to find by aid of the machine any one of the simple harmonic elements of a year's tides recorded in curves in the usual manner by an ordinary tide-gauge—a result which hitherto has required not less than twenty hours of calculation by skilled arithmeticians. I believe this instrument will be of great value also in determining the diurnal, semi-diurnal, ter-diurnal, and quarter-diurnal constituents of the daily variations of temperature, barometric pressure, east and west components of the velocity of the wind, north and south components of the same; also of the three components of the terrestrial magnetic force; also of the electric potential of the air at the point where the stream of water breaks into drops in atmospheric electrometers, and of other subjects of ordinary meteorological or magnetic observations; also to estimate precisely the variation of terrestrial magnetism in the eleven years sun-spot period, and of sun-spots themselves in this period; also to disprove (or prove, as the case may be) supposed relations between sun-spots and planetary positions and conjunctions; also to investigate lunar influence on the height of the barometer, and on the components of the terrestrial magnetic force, and to find if lunar influence is sensible on any other meteorological phenomena—and if so, to determine precisely its character and amount.

From the description given above it will be seen that the mechanism required for the instrument is exceedingly simple and easy. Its accuracy will depend essentially on the accuracy of the circular cylinder, of the globe, and of the plane of the rotating disk used in it. For each of the three surfaces a much less elaborate application of the method of scraping than that by which Sir Joseph Whitworth has given a true plane with such marvellous accuracy will no doubt suffice for the practical requirements of the instrument now proposed.

V. MECHANICAL INTEGRATION OF LINEAR DIFFERENTIAL EQUATIONS OF THE SECOND ORDER WITH VARIABLE COEFFICIENTS*.

Mechanical Integration of Linear Differential Equations of Second Order.

Every linear differential equation of the second order may, as is known, be reduced to the form

$$\frac{d}{dx}\left(\frac{1}{P}\frac{du}{dx}\right) = u \text{ .... (1)},$$

where P is any given function of x.

On account of the great importance of this equation in mathematical physics (vibrations of a non-uniform stretched cord, of a hanging chain, of water in a canal of non-uniform breadth and depth, of air in a pipe of non-uniform sectional area, conduction of heat along a bar of non-uniform section or non-uniform conductivity, Laplace's differential equation of the tides, etc. etc.), I have long endeavoured to obtain a means of facilitating its practical solution.

Methods of calculation such as those used by Laplace himself are exceedingly valuable, but are very laborious, too laborious unless a serious object is to be attained by calculating out results with minute accuracy. A ready means of obtaining approximate results which shall show the general character of the solutions, such as those so well worked out by Sturm†, has always seemed to me a desideratum. Therefore I have made many attempts to plan a mechanical integrator which should give solutions by successive approximations. This is clearly done now, when we have the instrument for calculating $\int \phi(x)\,\psi(x)\,dx$, founded on my brother's disk-, globe-, and cylinder-integrator, and described in a previous communication to the Royal Society; for it is easily proved‡ that if

* Sir W. Thomson, *Proceedings of the Royal Society*, Vol. XXIV., 1876, p. 269.

† *Mémoire sur les équations différentielles linéaires du second ordre*, Liouville's *Journal*, Vol. I. 1836.

‡ Cambridge Senate-House Examination, Thursday afternoon, January 22nd, 1874.

Mechanical Integration of Linear Differential Equations of Second Order.

$$\left.\begin{aligned} u_2 &= \int_0^x P\left(C - \int_0^x u_1\, dx\right) dx, \\ u_3 &= \int_0^x P\left(C - \int_0^x u_2\, dx\right) dx, \\ &\text{etc.,} \end{aligned}\right\} \quad \text{.............} \quad (2)$$

where u_1 is any function of x, to begin with, as for example $u_1 = x$; then u_2, u_3, etc. are successive approximations converging to that one of the solutions of (1) which vanishes when $x = 0$.

Now let my brother's integrator be applied to find $C - \int_0^x u_1\, dx$, and let its result feed, as it were, continuously a second machine, which shall find the integral of the product of its result into Pdx. The second machine will give out continuously the value of u_2. Use again the same process with u_2 instead of u_1, and then u_3, and so on.

After thus altering, as it were, u_1 into u_2 by passing it through the machine, then u_2 into u_3 by a second passage through the machine, and so on, the thing will, as it were, become refined into a solution which will be more and more nearly rigorously correct the oftener we pass it through the machine. If u_{i+1} does not sensibly differ from u_i, then each is sensibly a solution.

So far I had gone and was satisfied, feeling I had done what I wished to do for many years. But then came a pleasing surprise. Compel agreement between the function fed into the double machine and that given out by it. This is to be done by establishing a connexion which shall cause the motion of the centre of the globe of the first integrator of the double machine to be the same as that of the surface of the second integrator's cylinder. The motion of each will thus be necessarily a solution of (1). Thus I was led to a conclusion which was quite unexpected; and it seems to me very remarkable that the general differential equation of the second order with variable coefficients may be rigorously, continuously, and in a single process solved by a machine.

Take up the whole matter *ab initio*: here it is. Take two of my brother's disk-, globe-, and cylinder-integrators, and connect the fork which guides the motion of the globe of each of the integrators, by proper mechanical means, with the circumference of the other integrator's cylinder. Then move one integrator's disk through an angle $= x$, and simultaneously move the other

Mechanical Integration of Linear Differential Equations of Second Order.

integrator's disk through an angle always $= \int_0^x P dx$, a given function of x. The circumference of the second integrator's cylinder and the centre of the first integrator's globe move each of them through a space which satisfies the differential equation (1).

To prove this, let at any time g_1, g_2 be the displacements of the centres of the two globes from the axial lines of the disks; and let dx, Pdx be infinitesimal angles turned through by the two disks. The infinitesimal motions produced in the circumferences of two cylinders will be

$$g_1 dx \text{ and } g_2 P dx.$$

But the connexions pull the second and first globes through spaces respectively equal to those moved through by the circumferences of the first and second cylinders. Hence

$$g_1 dx = dg_2, \text{ and } g_2 P dx = dg_1;$$

and eliminating g_2,

$$\frac{d}{dx}\left(\frac{1}{P}\frac{dg_1}{dx}\right) = g_1,$$

which shows that g_1 put for u satisfies the differential equation (1).

The machine gives the complete integral of the equation with its two arbitrary constants. For, for any particular value of x, give arbitrary values G_1, G_2. [That is to say mechanically; disconnect the forks from the cylinders, shift the forks till the globes' centres are at distances G_1, G_2 from the axial lines, then connect, and move the machine.]

We have for this value of x,

$$g_1 = G_1, \text{ and } \frac{dg_1}{dx} = G_2 P;$$

that is, we secure arbitrary values for g_1 and $\frac{dg_1}{dx}$ by the arbitrariness of the two initial positions G_1, G_2 of the globes.

VI. Mechanical Integration of the general Linear Differential Equation of any Order with Variable Coefficients*.

Mechanical Integration of General Linear Differential Equation of Any Order.

Take any number i of my brother's disk-, globe-, and cylinder-integrators, and make an integrating chain of them thus:—Connect the cylinder of the first so as to give a motion equal to its own† to the fork of the second. Similarly connect the cylinder of the second with the fork of the third, and so on. Let g_1, g_2, g_3, up to g_i, be the positions‡ of the globes at any time. Let infinitesimal motions $P_1 dx$, $P_2 dx$, $P_3 dx$, ... be given simultaneously to all the disks (dx denoting an infinitesimal motion of some part of the mechanism whose displacement it is convenient to take as independent variable). The motions ($d\kappa_1$, $d\kappa_2$, ... $d\kappa_i$) of the cylinders thus produced are

$$d\kappa_1 = g_1 P_1 dx,\quad d\kappa_2 = g_2 P_2 dx, \ldots d\kappa_i = g_i P_i dx \;\ldots\ldots(1).$$

But, by the connexions between the cylinders and forks which move the globes, $d\kappa_1 = dg_2$, $d\kappa_2 = dg_3$, ... $d\kappa_{i-1} = dg_i$; and therefore

$$\left.\begin{aligned} &dg_2 = g_1 P_1 dx,\ dg_3 = g_2 P_2 dx,\ \ldots\ dg_i = g_{i-1} P_{i-1} dx \\ \text{and}\quad &d\kappa_1 = g_1 P_1 dx,\ d\kappa_2 = g_2 P_2 dx,\ \ldots\ d\kappa_i = g_i P_i\, dx. \end{aligned}\right\}\ldots(2).$$

Hence

$$g_1 = \frac{1}{P_1}\frac{d}{dx}\frac{1}{P_2}\frac{d}{dx}\cdots\frac{1}{P_{i-1}}\frac{d}{dx}\frac{1}{P_i}\frac{d\kappa_i}{dx} \;\ldots\ldots\ldots\ldots(3).$$

Suppose, now, for the moment that we couple the last cylinder with the first fork, so that their motions shall be equal—that is to say, $\kappa_i = g_1$. Then, putting u to denote the common value of these variables, we have

$$u = \frac{1}{P_1}\frac{d}{dx}\frac{1}{P_2}\frac{d}{dx}\cdots\frac{1}{P_{i-1}}\frac{d}{dx}\frac{1}{P_i}\frac{du}{dx} \;\ldots\ldots\ldots\ldots(4).$$

* Sir W. Thomson, *Proceedings of the Royal Society*, Vol. xxiv., 1876, p. 271.

† For brevity, the motion of the circumference of the cylinder is called the cylinder's motion.

‡ For brevity, the term "position" of any one of the globes is used to denote its distance, positive or negative, from the axial line of the rotating disk on which it presses.

Mechanical Integration of General Linear Differential Equation of Any Order.

Thus an endless chain or cycle of integrators with disks moved as specified above gives to each fork a motion fulfilling a differential equation, which for the case of the fork of the ith integrator is equation (4). The differential equations of the displacements of the second fork, third fork, ... $(i-1)$th fork may of course be written out by inspection from equation (4).

This seems to me an exceedingly interesting result; but though $P_1, P_2, P_3, \ldots P_i$ may be any given functions whatever of x, the differential equations so solved by the simple cycle of integrators cannot, except for the case of $i=2$, be regarded as the general linear equation of the order i, because, so far as I know, it has not been proved for any value of i greater than 2 that the general equation, which in its usual form is as follows,

$$Q_1 \frac{d^i u}{dx^i} + Q_2 \frac{d^{i-1} u}{dx^{i-1}} + \ldots Q_i \frac{du}{dx} - u = 0 \;\ldots\ldots\ldots\ldots (5),$$

can be reduced to the form (4). The general equation of the form (5), where $Q_1, Q_2, \ldots Q_i$ are any given forms of x, may be integrated mechanically by a chain of connected integrators thus:—

First take an open chain of i simple integrators as described above, and simplify the movement by taking

$$P_1 = P_2 = P_3 = \ldots = P_2 = 1,$$

so that the speeds of all the disks are equal, and dx denotes an infinitesimal angular motion of each. Then by (2) we have

$$g_i = \frac{d\kappa_i}{dx},\; g_{i-1} = \frac{d^2\kappa_i}{dx^2},\; \ldots,\; g_2 = \frac{d^{i-1}\kappa_i}{dx^{i-1}},\; g_1 = \frac{d^i\kappa_i}{dx^i} \;\ldots\ldots (6).$$

Now establish connexions between the i forks and the ith cylinder, so that

$$Q_1 g_1 + Q_2 g_2 + \ldots + Q_{i-1} g_{i-1} + Q_i g_i = \kappa_i \;\ldots\ldots\ldots (7).$$

Putting in this for g_1, g_2, etc. their values by (6), we find an equation the same as (5), except that κ_i appears instead of u. Hence the mechanism, when moved so as to fulfil the condition (7), performs by the motion of its last cylinder an integration of the equation (5). This mechanical solution is complete; for we may give arbitrarily any initial values to $\kappa_i, g_i, g_{i-1}, \ldots g_3, g_2$; that is to say, to

$$u,\; \frac{du}{dx},\; \frac{d^2u}{dx^2},\; \ldots \frac{d^{i-1}u}{dx^{i-1}}.$$

Mechanical Integration of General Linear Differential Equation of Any Order.

Until it is desired actually to construct a machine for thus integrating differential equations of the third or any higher order, it is not necessary to go into details as to plans for the mechanical fulfilment of condition (7); it is enough to know that it can be fulfilled by pure mechanism working continuously in connexion with the rotating disks of the train of integrators.

Addendum.

Mechanical Integration of any Differential Equation of Any Order.

The integrator may be applied to integrate any differential equation of any order. Let there be i simple integrators; let x_1, g_1, κ_1 be the displacements of disk, globe, and cylinder of the first, and so for the others. We have

$$g_1 = \frac{d\kappa_1}{dx_1}, \quad g_2 = \frac{d\kappa_2}{dx_2}, \text{ etc.}$$

Now by proper mechanism establish such relations between

$$x_1, g_1, \kappa_1, x_2, g_2, \text{ etc.}$$

that

$$f^{(1)}(x_1, g_1, \kappa_1, x_2, \ldots) = 0,$$
$$f^{(2)}(x_1, g_1, \kappa_1, x_2, \ldots) = 0,$$
$$\ldots\ldots\ldots\ldots\ldots\ldots\ldots\ldots$$
$$f^{(2i-1)}(x_1, g_1, \kappa_1, x_2, \ldots) = 0$$

($2i-1$ relations).

This will leave just one degree of freedom; and thus we have $2i-1$ simultaneous equations solved. As one particular case of relations take

$$x_1 = x_2 = \ldots (i-1 \text{ relations}),$$

and $$g_2 = \kappa_1, \quad g_3 = \kappa_2, \quad \text{etc. } (i-1 \text{ relations});$$

so that

$$g_1 = \frac{d^i\kappa_i}{dx^i}, \quad g_2 = \frac{d^{i-1}\kappa_i}{dx^{i-1}}, \text{ etc.}$$

Thus one relation is still available. Let it be

$$f(x, g_1, g_2, \ldots g_i, \kappa_i) = 0.$$

Thus the machine solves the differential equation

$$f\left(x, \frac{d^iu}{dx^i}, \frac{d^{i-1}u}{dx^{i-1}}, \ldots \frac{du}{dx}, u\right) = 0 \text{ (putting } u \text{ for } \kappa_i).$$

Or again, take $2i$ double integrators. Let the disks of all be connected so as to move with the same speed, and let t be the

Mechanical Integration of any Differential Equation of Any Order.

displacement of any one of them from any particular position. Let

$$x, y, x', y', x'', y'', \ldots x^{(i-1)}, y^{(i-1)}$$

be the displacements of the second cylinders of the several double integrators. Then (the second globe-frame of each being connected to its first cylinder) the displacements of the first globe-frames will be

$$\frac{d^2x}{dt^2}, \frac{d^2y}{dt^2}, \frac{d^2x'}{dt^2}, \frac{d^2y'}{dt^2}, \text{ etc.}$$

Let now X, Y, X', Y', etc. be each a given function of

$$x, y, x', y', x'', \text{ etc.}$$

By proper mechanism make the first globe of the first double integrator-frame move so that its displacement shall be equal to X, and so on. The machine then solves the equations

$$\frac{d^2x}{dt^2} = X, \frac{d^2y}{dt^2} = Y, \frac{d^2x}{dt^2} = X', \text{ etc.}$$

For example, let

$$\begin{aligned} X = &(x'-x)f\{(x'-x)^2 + (y'-y)^2\} \\ &+ (x''-x)f\{(x''-x)^2 + (y''-y)^2\} \\ &+ \ldots\ldots\ldots\ldots \\ Y = &(y'-y)f\{(x'-x)^2 + (y'-y)^2\} \\ &+ (y''-y)f\{(x''-x)^2 + (y''-y)^2\} \\ &+ \ldots\ldots\ldots\ldots \\ X' = &\text{etc.}, \quad Y' = \text{etc.}, \end{aligned}$$

where f denotes any function.

Construct in (frictionless) steel the surface whose equation is

$$z = \xi f(\xi^2 + \eta^2)$$

(and repetitions of it, for practical convenience, though *one* theoretically suffices). By aid of it (used as if it were a cam, but for two independent variables) arrange that one moving auxiliary piece (an x-auxiliary I shall call it), capable of moving to and fro in a straight line, shall have displacement always equal to

$$(x'-x)f\{(x'-x)^2 + (y'-y)^2\},$$

that another (a y-auxiliary) shall have displacement always equal to

$$(y'-y)f\{(x'-x)^2 + (y'-y)^2\},$$

Mechanical Integration of any Differential Equation of Any Order.

that another (an x-auxiliary) shall have displacement equal to

$$(x''-x)f\{(x''-x)^2+(y''-y)^2\},$$

and so on.

Then connect the first globe-frame of the first double integrator, so that its displacement shall be equal to the sum of the displacements of the x-auxiliaries; that is to say, to

$$\begin{aligned}&(x'-x)f\{(x'-x)^2+(y'-y)^2\}\\&+(x''-x)f\{(x''-x)^2+(y''-y)^2\}\\&+\text{etc.}\end{aligned}$$

This may be done by a cord passing over pulleys attached to the x-auxiliaries, with one end of it fixed and the other attached to the globe-frame (as in my tide-predicting machine, or in Wheatstone's alphabetic telegraph-sending instrument).

Then, to begin with, adjust the second globe-frames and the second cylinders to have their displacements equal to the initial velocity-components and initial co-ordinates of i particles free to move in one plane. Turn the machine, and the positions of the particles at time t are shown by the second cylinders of the several double integrators, supposing them to be free particles attracting or repelling one another with forces varying according to any function of the distance.

The same may clearly be done for particles moving in three dimensions of space, since the components of force on each may be mechanically constructed by aid of a cam-surface whose equation is

$$z=\xi f(\eta)$$

and taking η for the distance between any two particles, and

$$\xi=x'-x$$

or $$=y'-y$$

or $$=x''-x, \text{ etc.}$$

Thus we have a complete mechanical integration of the problem of finding the free motions of any number of mutually influencing particles, not restricted by any of the approximate suppositions which the analytical treatment of the lunar and planetary theories requires.

VII. HARMONIC ANALYZER*.

Harmonic Analyzer.

This is a realization of an instrument designed rudimentarily in the author's communication to the Royal Society ("Proceedings," February 3rd, 1876), entitled "On an Instrument for Calculating ($\int \phi (x) \psi (x) dx$), the Integral of the Product of two given Functions."

It consists of five disk-, globe-, and cylinder-integrators of the kind described in Professor James Thomson's paper "On an Integrating Machine having a new Kinematic Principle," of the same date, and represented in the woodcuts of Appendix B′, III.

The five disks are all in one plane, and their centres in one line. The axes of the cylinders are all in a line parallel to it. The diameters of the five cylinders are all equal, so are those of the globes; hence the centres of the globes are in a line parallel to the line of the centres of the disks, and to the line of the axes of the cylinders.

One long wooden rod, properly supported and guided, and worked by a rack and pinion, carries five forks to move the five globes and a pointer to trace the curve on the paper cylinder. The shaft of the paper cylinder carries at its two ends cranks at right angles to one another; and a toothed wheel which turns a parallel shaft, and a third shaft in line with the first, by means of three other toothed wheels. This third shaft carries at its two ends two cranks at right angles to one another.

Another toothed wheel on the shaft of the paper drum turns another parallel shaft, which, by a slightly oblique toothed wheel working on a crown wheel with slightly oblique teeth, turns one of the five disks uniformly (supposing to avoid circumlocution the paper drum to be turning uniformly). The cylinder of the integrator, of which this one is the disk, gives the continuously growing value of $\int y dx$.

Each of the four cranks gives a simple harmonic angular motion to one of the other four disks by means of a slide and crosshead, carrying a rack which works a sector attached to the disk. Hence, the cylinders moved by the disks, driven by the

* Sir W. Thomson, *Proceedings of the Royal Society*, Vol. XXVII., 1878, p. 371.

Harmonic Analyzer.

first mentioned pair of cranks, give the continuously growing values of

$$\int y \cos \frac{2\pi x}{c}\, dx, \text{ and } \int y \sin \frac{2\pi x}{c}\, dx;$$

where c denotes the circumference of the paper drum: and the two remaining cylinders give

$$\int y \cos \frac{2\pi \omega x}{c}\, dx, \text{ and } \int y \sin \frac{2\pi \omega x}{c}\, dx;$$

where ω denotes the angular velocity of the shaft carrying the second pair of shafts, that of the first being unity.

The machine, with the toothed wheels actually mounted on it when shown to the Royal Society, gave $\omega = 2$, and was therefore adopted for the meteorological application. By removal of two of the wheels and substitution of two others, which were laid on the table of the Royal Society, the value of ω becomes $\frac{39 \times 109}{40 \times 110}$* (according to factors found by Mr E. Roberts, and supplied by him to the author, for the ratio of the mean lunar to the mean solar periods relatively to the earth's rotation). Thus, the same machine can serve for analyzing out simultaneously the mean lunar and mean solar semi-diurnal tides from a tide-gauge curve. But the dimensions of the actual machine do not allow range enough of motion for the majority of tide-gauge curves, and they are perfectly sufficient and suitable for meteorological work. The machine, with the train giving $\omega = 2$, is therefore handed over to the Meteorological Office to be brought immediately into practical work by Mr Scott (as soon as a brass cylinder of proper diameter to suit the $24h$ length of his curves is substituted for the wooden model cylinder in the machine as shown to the Royal Society): and the construction of a new machine for the tidal analysis, to have eleven disk-, globe-, and cylinder-integrators in line, and four crank shafts having their axes in line with the paper drum, according to the preceding description, in proper periods to analyse a tide curve by one process for mean level, and for the two components of each of the five chief tidal constituents—that is to say,

Tidal Harmonic Analyzer.

* The actual numbers of the teeth in the two pairs of wheels constituting the train are 78 : 80 and 109 : 110.

Tidal Harmonic Analyzer.

(1) The mean solar semi-diurnal;
(2) „ „ lunar „
(3) „ „ lunar quarter-diurnal, shallow-water tide;
(4) „ „ lunar declinational diurnal;
(5) „ „ luni-solar declinational diurnal;

is to be immediately commenced. It is hoped that it may be completed without need to apply for any addition to the grant already made by the Royal Society for harmonic analyzers.

Counterpoises are applied to the crank shafts to fulfil the condition that gravity on cranks, and sliding pieces, and sectors, is in equilibrium. Error from "back lash" or "lost time" is thus prevented simply by frictional resistance against the rotation of the uniformly rotating disk and of the tertiary shafts, and by the weights of the sectors attached to the oscillating disks.

Addition, April, 1879. The machine promised in the preceding paper has now been completed with one important modification:—Two of the eleven constituent integrators, instead of being devoted, as proposed in No. 3 of the preceding schedule, to evaluate the lunar quarter-diurnal shallow-water tide, are arranged to evaluate the solar declinational diurnal tide, this being a constituent of great practical importance in all other seas than the North Atlantic, and of very great scientific interest. For the evaluation of quarter-diurnal tides, whether lunar or solar, and of semi-diurnal tides of periods the halves of those of the diurnal tides, that is to say of all tidal constituents whose periods are the halves of those of the five main constituents for which the machine is primarily designed, an extra paper-cylinder, of half the diameter of the one used in the primary application of the machine, is constructed. By putting in this secondary cylinder and repassing the tidal curve through the machine the secondary tidal constituents (corresponding to the first "overtones" or secondary harmonic constituents of musical sounds) are to be evaluated. Similarly tertiary, quaternary, etc. tides (corresponding to the second and higher overtones in musical sounds) may be evaluated by passing the curve over cylinders of one-third and of smaller sub-multiples of the diameter of the primary cylinder. These secondary and tertiary tidal constituents are only perceptible at places where the rise and fall is influenced by a large area of sea, or a considerable length of

Secondary, tertiary, quaternary, etc. tides, due to influence of shallow water,—analogous to musical overtones.

Tidal Harmonic Analyzer.

channel through which the whole amount of the rise and fall is notable in proportion to the mean depth. They are very perceptible at almost all commercial ports, except in the Mediterranean, and to them are due such curious and practically important tidal characteristics as the double high waters at Southampton and in the Solent and on the south coast of England from the Isle of Wight to Portland, and the protracted duration of high water at Havre.

END OF PART I.

CAMBRIDGE: PRINTED BY C. J. CLAY, M.A., AT THE UNIVERSITY PRESS.

Printed in the United States
By Bookmasters